# FUNDAMENTALS OF THE THEORY OF PLASTICITY

L. M. Kachanov

*Professor of Mathematics and Mechanics*
*Leningrad University, USSR*

DOVER PUBLICATIONS, INC.
Mineola, New York

*Bibliographical Note*

This Dover edition, first published in 2004, is an unabridged republication of the *Foundations of the Theory of Plasticity*, originally published by North-Holland Publishing Company, Amsterdam, in 1971. The list of errata has been specially prepared by the author for this edition.

*Library of Congress Cataloging-in-Publication Data*

Kachanov, L.M. (Lazar' Markovich)
[Osnovy teorii plastichnosti. English]
Fundamentals of theory of plasticity / L.M. Kachanov.
p. cm.
Originally published: Foundations of the theory of plasticity. Amsterdam : North-Holland Pub. Co., 1971, in series: North-Holland series in applied mathemetics and mechanics, v. 12.
Includes bibliographical references and index.
ISBN 0-486-43583-0 (pbk.)
1. Plasticity. I. Title.

QA931.K2313 2004
531'.385—dc22

2004041438

Manufactured in the United States of America
Dover Publications, Inc., 31 East 2nd Street, Mineola, N.Y. 11501

# Contents

# Preface to the Second Edition

This book is based on the lectures on the theory of plasticity given by the author in the Mechanics-Mathematics Faculty of the University of Leningrad.

The author has not attempted an exhaustive treatment of the whole of plasticity theory. The book is mainly concerned with plastic deformation of metals at normal temperatures, as related to questions of strength of machines and structures. The prime concern of the book is to give a simple account of the basic equations of plasticity theory and of the most well-developed methods for solving them; also to discuss problems characterizing the special nature of the plastic state and the diversity of the mechanical problems involved.

The second edition has been substantially revised, reflecting important developments in the progress of the theory during recent years. Thus, more attention has been given to yield surfaces and to the associated yield law, and the chapters devoted to plane stress and the axisymmetric problem have been extended. The section on extremum principles and energetic methods of solution has been essentially rewritten. A new chapter has been included on the theory of shakedown, which is of great significance in connection with the role of varying loads in the origin of rupture. In the last ten years marked progress has been made with the model of a rigid-plastic body in dynamic problems; and so some additions have been introduced in Chapter XI.

In order to avoid overloading the course with unnecessary details, some of the more difficult problems have been omitted from the second edition, and the chapters on the stability of equilibrium and composite plastic media have been shortened.

It is assumed that the reader is familiar with the basic ideas of strength of materials and of the theory of elasticity. In order to facilitate study of the book, the more difficult passages that may be omitted on a first reading have been indicated by small type or by asterisks.

The list of references is not exhaustive. To assist orientation in the vast literature on the theory of plasticity, books and review articles in which supplementary references can be found are given in special lists.

The author acknowledges the readers who have pointed out defects in the first edition, and is grateful to A.A. Vakulenko, A.I. Kuznetsov, V.I. Rozenblyum, and G.S. Shapiro for their valuable comments in connection with the manuscript. A.I. Kuznetsov also read the proofs; the author expresses sincere thanks to him for a number of important corrections.

L.M. Kachanov

# Basic Notation

| | |
|---|---|
| $\sigma_x, \sigma_y, \sigma_z, \tau_{xy}, \tau_{yz}, \tau_{xz}; \sigma_{ij}$ | – components of the stress tensor. |
| $s_x, s_y, s_z, \tau_{xy}, \tau_{yz}, \tau_{xz}; \epsilon_{ij}$ | – components of the stress deviatoric. |
| $\epsilon_x, \epsilon_y, \epsilon_z, \frac{1}{2}\gamma_{xy}, \frac{1}{2}\gamma_{yz}, \frac{1}{2}\gamma_{xz}; \epsilon_{ij}$ | – components of the strain tensor. |
| $e_x, e_y, e_z, \frac{1}{2}\gamma_{xy}, \frac{1}{2}\gamma_{yz}, \frac{1}{2}\gamma_{xz}; e_{ij}$ | – components of the strain deviatoric. |
| $\xi_x, \xi_y, \xi_z, \frac{1}{2}\eta_{xy}, \frac{1}{2}\eta_{yz}, \frac{1}{2}\eta_{xz}; \xi_{ij}$ | – components of the rate of strain tensor. |
| $u_x, u_y, u_z; u_i$ | – displacements components. |
| $v_x, v_y, v_z; v_i$ | –velocity components. |
| $\delta_{ij}$ | – Kronecker symbol. |
| $T, \Gamma, H$ | – the intensities respectively of the tangential stress, the shearing strain and the shear strain rate. |
| $\sigma_s, \tau_s$ | – the yield limits respectively for tension and pure shear. |

# Errata

Page 21 (eq. 2.15): $\mu_\epsilon = \sqrt{3} \cot (\omega_\epsilon + \frac{1}{3}\pi), \epsilon_1 \geqslant \epsilon_2 \geqslant \epsilon_3$

Page 38 (paragraph 2, line 5): ". . . *to an increasing parameter* $t > 0$"

Page 68 (paragraph 3, line 1): ". . . of the coordinates alone, and $t > 0$ is some increasing parameter . . ."

Page 147 (eq. 31.3): $\sigma_z = \frac{1}{2}(\sigma_x + \sigma_y)$.

Page 260 (paragraph 4, line 2): $\xi_z = \xi_3 = 0$,

Page 290 (paragraph 4, line 1): ". . . the limit moment is $M_*^0 = \frac{1}{4}\sigma_s h^2$."

Page 333 (paragraph 8, lines 1, 2): ". . . of the inequality can be replaced by $\int X_{ni} v_i' dS_F$."

Page 383 (paragraph 2, line 3): ". . . it is necessary that, at $\sigma_2 < 0$"

Page 460 (paragraph 1, line 3): ". . . and plastic elements (fig. 260a)."

# Introduction

## 1. Theory of Plasticity

It is well known that solid bodies are elastic only if the applied loads are small. Under the influence of more or less substantial forces, bodies experience inelastic, plastic deformations. Plastic properties are extremely varied, and depend both on the materials under investigation and the ambient conditions (temperature, duration of the process and so on). Thus, the plastic deformation of durable metals (steel, various hard alloys, etc.), in conditions of normal temperature, is virtually independent of time; while metals operating under conditions of high temperature (components of boilers, steam and gas turbines) experience plastic deformation which increases with time (creep), i.e. roughly speaking they flow rather like a viscous fluid.

When speaking of plasticity theory, one usually has in mind the theory of time-independent plastic deformations (athermal plasticity). Just such plastic deformations are considered in the present book; only in the last chapter are the effects of viscosity discussed. Plastic flow in which time has an influence is studied in the theory of creep, the theory of viscoplasticity, and rheology.

Plasticity theory aims at the mathematical study of stresses and displacements in plastically deformable (in the above sense) bodies.

The theory of plasticity is an aspect of the mechanics of deformable bodies and is closely related to the theory of elasticity, in which the stresses and strains in perfectly elastic bodies are studied; many of the basic concepts of elasticity theory are also to be found in the theory of plasticity.

The methods of plasticity theory are those commonly used in analyzing the mechanics of deformable media. The first problem is to establish the basic laws of plastic deformation on the basis of experimental data (and, if possible, certain considerations borrowed from theoretical physics). With the help of these laws, which have a phenomenological character, a system of equations is obtained. The solution of these equations, so as to obtain a picture of the plastic deformation of a body in various circumstances, is then the other major problem of plasticity theory.

Let us dwell on some particular features of the theory of plasticity. In the first place, a large part of the theory (in contrast with elasticity theory) is taken up with the question of establishing laws of plastic deformation in a complex stress state. This is a difficult question, and it should be noted that laws which agree satisfactorily with experimental data (with certain limitations) have been established mainly for metals, although they probably retain their validity for many other materials. Another characteristic of plasticity theory is the non-linearity of the principal laws, and hence of the basic equations of the theory. The solution of these equations presents great mathematical difficulties; the classical methods of mathematical physics are useless here. The development of methods of investigation which overcome these difficulties for specific problems is therefore extremely important. Under these conditions the use of new computing techniques is also very significant.

## 2. Application of the Theory of Plasticity

Plasticity theory has important applications in engineering and physics.

The resolution of many questions relating to the strength of various kinds of machines and structures is based on the inferences of plasticity theory. The theory opens up prospects of a more complete use of the strength resources of bodies and leads to a progressive method of calculating details of machines and structures as regards their load-bearing capacity. This method is characterized by its simplicity, and also it often permits an approach to the problem of the direct design of optimum structures (the theory of optimal design, see [73]).

The economic value of the use of processes involving plastic deformation of metals in hot and cold conditions is well known (rolling, drawing, forging, stamping, cutting of metals, etc.); the analysis of the forces necessary to accomplish these processes and of the corresponding deformation distributions constitutes another very important area for the application of plasticity theory.

The study of the strength-properties of materials is based on the results of plasticity theory, since as a rule plastic deformation precedes fracture.

Analysis of the behaviour of structures under impacting, impulsive loads requires the development of the dynamical theory of plasticity.

In recent years plasticity theory has been successfully employed in research into the mechanism of pressure in rocks, a matter of considerable interest for the mining industry.

Finally, it should be noted that there is a series of works projecting the use of the methods of plasticity theory in geophysical and geological problems.

## 3. A Brief Historical Account

The earliest contributions to a mathematical theory of plasticity were made in the seventies of the last century and are associated with the names of Saint Venant, for his investigation of the equations of plane strain [156,157], and M. Levy, who, following the ideas of Saint Venant established the equations for the three-dimensional case [129]; he also introduced the method of linearization for the plane problem [130].

Subsequently the development of the theory of plasticity proceeded slowly. Some progress was made at the beginning of this century, when the works of Haar and von Karman ([162], 1909) and R. von Mises ([136], 1913) were published. The former was attempt to obtain equations for plasticity theory from a certain variational principle. In von Mises' work a new yield criterion was clearly formulated [1] (the condition of the constancy of the intensity of the tangential stresses).

During the last twenty years the theory of plasticity has been intensively developed, at first mainly in Germany. In the works of H. Hencky [94,96], L. Prandtl [144], R. von Mises [136] and other authors important results were obtained, both in the basic equations of plasticity theory and in the methods of solving the plane problem. At the same time the first systematic researches into the laws of plastic deformation in a complex stress state were undertaken, and also the first successful applications of plasticity theory to

[1] We note that a similar condition was given earlier, though in a less precise form, and not in connection with the construction of the mathematical theory of plasticity.

engineering problems. In recent years plasticity theory has come to attract the attention of a wide circle of scientists and engineers; intensive theoretical and experimental research is proceeding in many countries, including the Soviet Union. The theory of plasticity, together with gas dynamics, is becoming the most energetically developed branch of continuum mechanics.

# 1

# Basic Concepts in the Mechanics of Continuous Media

In this chapter we outline the basic formulae of the theory of stress and strain; in the process we identify the most important information for the development of the theory of plasticity.

## §1. Stress

### 1.1. *Stress*

At a given point in a continuous medium the state of stress is characterized by a symmetric stress tensor

$$T_\sigma = \begin{Vmatrix} \sigma_x & \tau_{xy} & \tau_{xz} \\ \tau_{xy} & \sigma_y & \tau_{yz} \\ \tau_{xz} & \tau_{yz} & \sigma_z \end{Vmatrix}, \tag{1.1}$$

where $\sigma_x$, $\sigma_y$, $\sigma_z$ are the normal components, and $\tau_{xy}$, $\tau_{yz}$, $\tau_{xz}$ the tangential components of stress in a rectangular coordinate system with axes $x$, $y$, $z$.

The stress vector **p** on an arbitrarily oriented surface with unit normal **n** (fig. 1) is given by Cauchy's formulae:

$$
\begin{aligned}
p_x &= \sigma_x n_x + \tau_{xy} n_y + \tau_{xz} n_z , \\
p_y &= \tau_{xy} n_x + \sigma_y n_y + \tau_{yz} n_z , \\
p_z &= \tau_{xz} n_x + \tau_{yz} n_y + \sigma_z n_z ,
\end{aligned}
\tag{1.2}
$$

where $n_x$, $n_y$, $n_z$ are the components of the unit normal **n** and are equal respectively to the direction cosines cos $(n, x)$, cos $(n, y)$, cos $(n\ z)$.

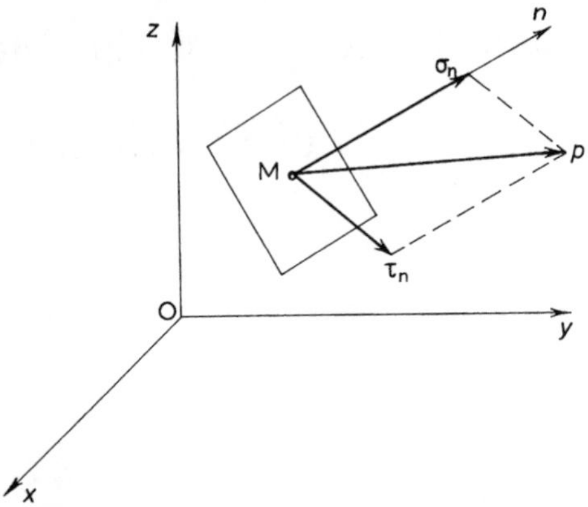


Fig. 1.

The projection of the vector **p** in the direction of the normal gives the normal stress $\sigma_n$, acting on the surface in question:

$$\sigma_n = \sigma_x n_x^2 + \sigma_y n_y^2 + \sigma_z n_z^2 + 2\tau_{xy} n_x n_y + 2\tau_{yz} n_y n_z + 2\tau_{xz} n_x n_z . \tag{1.3}$$

The magnitude of the tangential stress $\tau_n$ equals

$$\tau_n = \sqrt{p_x^2 + p_x^2 + p_z^2 - \sigma_n^2} . \tag{1.4}$$

At each point of the medium there exist three mutually perpendicular surface elements on which the tangential stresses are zero. The directions of the normals to these surfaces constitute the principal directions of the stress tensor and do not depend on the choice of the coordinate system $x, y, z$. This means that any stress state at the given point may be induced by stretching the neighbourhood of the point in three mutually perpendicular directions. The corresponding stresses are called *principal normal stresses*; we shall denote them by $\sigma_1, \sigma_2, \sigma_3$ and number the principal axes so that

$$\sigma_1 \geqslant \sigma_2 \geqslant \sigma_3 . \tag{1.5}$$

The stress tensor, referred to principal axes, has the form

$$T_\sigma = \begin{Vmatrix} \sigma_1 & 0 & 0 \\ 0 & \sigma_2 & 0 \\ 0 & 0 & \sigma_3 \end{Vmatrix} .$$

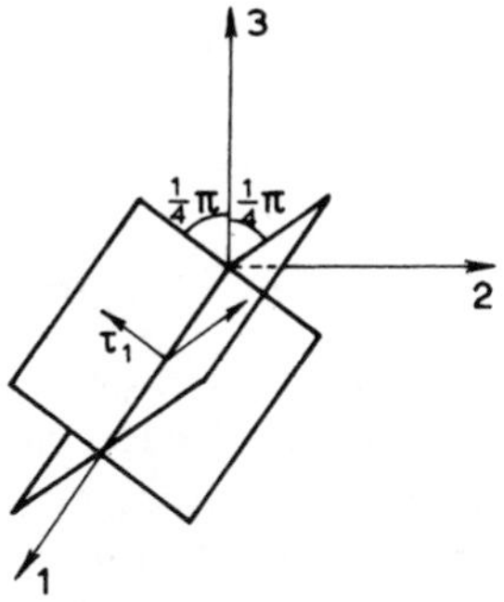


Fig. 2.

It is not difficult to show from formulae (1.2)–(1.4) that in cross-sections, which bisect the angles between the principal planes and pass respectively through the principal axes 1, 2, 3 (fig. 2), the tangential stresses have magnitudes

$$\tfrac{1}{2}|\sigma_2-\sigma_3| , \qquad \tfrac{1}{2}|\sigma_3-\sigma_1| , \qquad \tfrac{1}{2}|\sigma_1-\sigma_2| .$$

The tangential stresses on these cross-sections have extremum values and are called *principal tangential stresses*. We define these by the formulae

$$\tau_1 = \tfrac{1}{2}(\sigma_2-\sigma_3) , \qquad \tau_2 = \tfrac{1}{2}(\sigma_3-\sigma_1) , \qquad \tau_3 = \tfrac{1}{2}(\sigma_1-\sigma_2) . \quad (1.6)$$

With change in the orientation of a surface, the intensity of the tangential stress $\tau_n$ acting on it also changes. The maximum value of $\tau_n$ at a given point is called the *maximum tangential stress* $\tau_{max}$. If condition (1.5) is satisfied, then

$$\tau_{max} = -\tau_2 .$$

It is not difficult to determine from equation (1.3) that the normal stresses on the planes on which the principal tangential stresses (1.6) act, respectively equal

$$\tfrac{1}{2}(\sigma_2+\sigma_3)\,, \qquad \tfrac{1}{2}(\sigma_3+\sigma_1)\,, \qquad \tfrac{1}{2}(\sigma_1+\sigma_2)\,. \tag{1.7}$$

Once the principal stresses $\sigma_1$, $\sigma_2$, $\sigma_3$ and the principal directions 1, 2, 3 have been assigned, the stress tensor can be determined. This designation is notable for its mechanical clarity. The principal stresses $\sigma_i$ ($i = 1, 2, 3$) are the roots of the cubic equation

$$\begin{vmatrix} \sigma_x-\lambda & \tau_{xy} & \tau_{xz} \\ \tau_{xy} & \sigma_y-\lambda & \tau_{yz} \\ \tau_{xz} & \tau_{yz} & \sigma_z-\lambda \end{vmatrix} = 0$$

or

$$-\lambda^3 + I_1(T_\sigma)\,\lambda^2 + I_2(T_\sigma)\,\lambda + I_3(T_\sigma) = 0\,. \tag{1.8}$$

It is evident that the normal stresses $\sigma_n$ on a given surface do not depend on the choice of the coordinate system and are changed only by a rotation of the surface. The principal stresses $\sigma_1$, $\sigma_2$, $\sigma_3$ are extremum values of the normal stress $\sigma_n$ and hence are independent of the choice of the coordinate system. Equation (1.8) may be obtained as the condition for extremum values of $\sigma_n$. Consequently the coefficients of the cubic equation (1.8) do not change on transformation from one orthogonal coordinate system to another, i.e. they are invariants. These coefficients

$$\begin{aligned} I_1(T_\sigma) &= \sigma_1 + \sigma_2 + \sigma_3 \equiv 3\sigma\,, \\ I_2(T_\sigma) &= -(\sigma_1\sigma_2 + \sigma_2\sigma_3 + \sigma_3\sigma_1)\,, \\ I_3(T_\sigma) &= \sigma_1\sigma_2\sigma_3\,, \end{aligned} \tag{1.9}$$

written for brevity in principal axes, are called respectively the linear, quadratic and cubic invariants of the tensor; it is convenient to work with them because they are completely rational functions of the stress components and moreover are symmetrical (i.e. are not changed by a permutation of the independent variables).

The quantity

$$\sigma = \tfrac{1}{3}(\sigma_x + \sigma_y + \sigma_z)$$

is called the *mean* (or hydrostatic) *pressure* at a point. The significance of the other invariants will be explained below.

1.2. *The stress deviatoric*

Since, as a rule, materials possess different mechanical properties as regards shear and uniform, cubical compression, it is advantageous to write the stress tensor in the form [1])

$$T_\sigma = \sigma T_1 + D_\sigma , \tag{1.10}$$

where $\sigma T_1 = \begin{Vmatrix} \sigma & 0 & 0 \\ 0 & \sigma & 0 \\ 0 & 0 & \sigma \end{Vmatrix}$ is a spherical tensor corresponding to the mean pressure at the point, and

$$D_\sigma = \begin{Vmatrix} \sigma_x - \sigma & \tau_{xy} & \tau_{xz} \\ \tau_{xy} & \sigma_y - \sigma & \tau_{yz} \\ \tau_{xz} & \tau_{yz} & \sigma_z - \sigma \end{Vmatrix} \tag{1.11}$$

is a tensor characterising the tangential stress at the given point, termed the *stress deviatoric*.

The normal components of the latter (i.e. $\sigma_x - \sigma$, $\sigma_y - \sigma$, $\sigma_z - \sigma$) will sometimes be designated by $s_x$, $s_y$, $s_z$. The principal directions of the stress deviatoric $D_\sigma$ and of the stress tensor $T_\sigma$ coincide, but the principal values $s_i$ differ from $\sigma_i$ by the mean pressure, and they are determined, obviously, by the cubic equation

$$-\lambda^3 + I_2(D_\sigma)\,\lambda + I_3(D_\sigma) = 0 , \tag{1.12}$$

whose roots are all real.

The invariants of the deviatoric are easily obtained from (1.9), if we replace $\sigma_1, \sigma_2, \sigma_3$ by $s_1, s_2, s_3$ respectively:

$$\begin{aligned} &I_1(D_\sigma) = 0 , \\ &I_2(D_\sigma) = \tfrac{1}{6}\left[(\sigma_1 - \sigma_2)^2 + (\sigma_2 - \sigma_3)^2 + (\sigma_3 - \sigma_1)^2\right] , \\ &I_3(D_\sigma) = s_1 s_2 s_3 . \end{aligned} \tag{1.13}$$

It is evident that the stress deviatoric is characterized by only five independent quantities.

[1]) $T_1$ is so-called unit tensor

$$T_1 = \begin{Vmatrix} 1 & 0 & 0 \\ 0 & 1 & 0 \\ 0 & 0 & 1 \end{Vmatrix} ,$$

for which any direction is principal and which has diagonal elements equal to unity in an arbitrary orthogonal system of coordinates $x$, $y$, $z$.

The non-negative quantity

$$T = +\sqrt{I_2(D_\sigma)} =$$

$$= \frac{1}{\sqrt{6}}\sqrt{(\sigma_x-\sigma_y)^2 + (\sigma_y-\sigma_z)^2 + (\sigma_z-\sigma_x)^2 + 6\,(\tau_{xy}{}^2 + \tau_{yz}{}^2 + \tau_{xz}{}^2)} \quad (1.14)$$

is called the *tangential stress intensity* [1]).

The tangential stress intensity is zero only when the state of stress is a state of hydrostatic pressure.

For pure shear

$$\sigma_1 = \tau\,, \qquad \sigma_2 = 0\,, \qquad \sigma_3 = -\tau\,,$$

where $\tau$ is the shear stress. It follows that

$$T = \tau\,.$$

In the case of simple tension (compression) in the direction of the $x$-axis

$$\sigma_x = \sigma_1\,; \qquad \sigma_y = x_z = \tau_{xy} = \tau_{yz} = \tau_{xz} = 0\,;$$

then

$$T = |\sigma_1|/\sqrt{3}\,. \quad (1.15)$$

Since the cubic equation (1.12) has real roots, its solution can be expressed in trigonometrical form. Using well-known algebraic formulae we can express the principal components of the deviatoric in terms of an invariant [43, 44]:

$$s_1 = \frac{2}{\sqrt{3}}\,T\cos\left(\omega_\sigma - \tfrac{1}{3}\pi\right),$$

$$s_2 = \frac{2}{\sqrt{3}}\,T\cos\left(\omega_\sigma + \tfrac{1}{3}\pi\right), \quad (1.16)$$

$$s_3 = \frac{2}{\sqrt{3}}\,T\cos\omega_\sigma\,.$$

The angle $\omega_\sigma$ is specified by the equation

$$-\cos 3\omega_\sigma = \frac{3\sqrt{3}\,I_3(D_\sigma)}{2T^3}\,. \quad (1.17)$$

[1]) Sometimes one considers the *reduced stress* (or stress intensity) equal to $\sqrt{3}\,T$; in the case of simple tension (compression) the reduced stress equals $|\sigma_1|$.

From (1.6) and (1.16) it is not difficult to find the principal tangential stresses:

$$\begin{aligned}\tau_1 &= -T \sin(\omega_\sigma - \tfrac{1}{3}\pi)\,,\\ \tau_2 &= -T \sin(\omega_\sigma + \tfrac{1}{3}\pi)\,,\\ \tau_3 &= T \sin \omega_\sigma\,.\end{aligned} \tag{1.18}$$

The angle $\omega_\sigma$ lies in the range

$$0 \leqslant \omega_\sigma \leqslant \tfrac{1}{3}\pi\,. \tag{1.19}$$

In fact, since $\sigma_1 \geqslant \sigma_2 \geqslant \sigma_3$, then $\tau_1 \geqslant 0$, $\tau_2 \leqslant 0$, $\tau_3 \geqslant 0$, i.e. $\sin(\omega_\sigma - \frac{1}{3}\pi) \leqslant 0$, $\sin(\omega_\sigma + \frac{1}{3}\pi) \geqslant 0$, $\sin \omega_\sigma \geqslant 0$, whence (1.19) follows.

We have already shown that $\tau_{max} = -\tau_2$, and so from (1.19) we obtain the inequality

$$1 \leqslant \frac{T}{\tau_{max}} \leqslant \frac{2}{\sqrt{3}}\,, \tag{1.20}$$

established by A.A. Il'yushin by another method. Thus the tangential stress intensity $T$ and the maximum shear stress $\tau_{max}$ are not significantly different. Indeed

$$T \approx 1.08\, \tau_{max} \tag{1.21}$$

with a maximum error of about 7%.

As shown by V.V. Novozhilov [139], the tangential stress intensity $T$ is proportional to the mean square of the tangential stresses evaluated over the surface of a small sphere surrounding the point of the body under consideration.

### 1.3. *Tensor notation*

The analysis and presentation of general questions in plasticity theory are greatly simplified and clarified by the use of tensor notation. This notation is becoming increasingly widespread in contemporary scientific literature on plasticity theory, and therefore it will be used in various parts of this book.

The cartesian coordinates $x, y, z$ will be denoted by $x_1, x_2, x_3$ and written as $x_i$, where the index $i$ takes the values 1, 2, 3. Of course in place of $i$ it is possible to take another letter (for example, $j$, $j = 1, 2, 3$; usually latin letters are used). By $n_i$ (or, say, $n_j$) we denote the components of unit vector normal to a surface; it is obvious that $n_i$ equal the direction cosines of the normal.

It is now possible to specify the components of the stress tensor by $\sigma_{ij}$, $i, j = 1, 2, 3$. Because of the reciprocity law of the tangential stresses, $\sigma_{ij} = \sigma_{ji}$. The relation between the tensor notation and the "technical" notation used earlier is evidently: $\sigma_{11} = \sigma_x$, $\sigma_{12} = \tau_{xy}$, etc. We shall, moreover, refer to the stress tensor as the tensor $\sigma_{ij}$.

Cauchy's formulae (1.2) can now be rewritten in the form

$$p_j = \sum_{i=1}^{3} \sigma_{ij} n_i , \qquad j = 1, 2, 3 .$$

Widespread use is made of the summation rule, introduced by A. Einstein. The summation sign is omitted, with the convention that every repeated latin index in a monomial indicates summation over the numbers 1, 2, 3. Then the preceeding formula is written in the form

$$p_j = \sigma_{ij} n_i . \tag{1.22}$$

A repeated index $i$ is called a *dummy* index (or summation index); it may be substituted by any other (usually latin) letter. In each monomial the same dummy index must not be repeated more than twice. The index $j$ is sometimes called *free*.

It is easy to see that the normal stress $\sigma_n$ (1.3) equals

$$\sigma_n = \sigma_{ij} n_i n_j ; \tag{1.23}$$

here there are two dummy indices $i, j$ and it follows that two summations are to be carried out. There are no free indices.

The mean pressure equals

$$\sigma = \tfrac{1}{3}\sigma_{ii} = \tfrac{1}{3}(\sigma_{11} + \sigma_{22} + \sigma_{33}) .$$

The *Kronecker symbol* (delta symbol) is defined by the relations

$$\delta_{ij} = \begin{cases} 1 \text{ when } i = j , \\ 0 \text{ when } i \neq j . \end{cases}$$

A tensor with these components in a coordinate system $x_i$ is called the *unit tensor* (see $T_1$).

The stress deviatoric has the components

$$s_{ij} = \sigma_{ij} - \sigma\delta_{ij} . \tag{1.24}$$

The linear invariant of the deviatoric is zero, i.e. $s_{ii} = 0$. It is not difficult to see that the tangential stress intensity in the new notation equals

$$T = (\tfrac{1}{2} s_{ij} s_{ij})^{\frac{1}{2}} . \tag{1.25}$$

The mean pressure can also be expressed in the form

$$\sigma = \tfrac{1}{3}\sigma_{ij}\delta_{ij}\,.$$

1.4. *Geometrical interpretation*

Let us now return to an examination of the quantities $\sigma$, $T$, $\omega_\sigma$, through which the principal stresses are evaluated. It is possible to give the quantities $\sigma$, $T$, $\omega_\sigma$ a simple geometrical interpretation. For this purpose we introduce the space of the principal stresses $\sigma_1$, $\sigma_2$, $\sigma_3$. Then the stress at a given point in this *stress space* can be represented by a vector $\overline{\mathrm{OP}}$, with components $\sigma_1$, $\sigma_2$, $\sigma_3$ respectively (fig. 3). The surface

$$\sigma_1 + \sigma_2 + \sigma_3 = 0 \tag{1.26}$$

passes through the origin of coordinates and is equally inclined to the axes. Since the sum of the squares of the cosines of the angles between the normal $n$ and the axes equals unity, then $\cos(n, \sigma_i) = 1/\sqrt{3}$. It follows that the unit vector normal to the surface in question is given by

$$\mathbf{n} = \frac{1}{\sqrt{3}}(\mathbf{i}_1 + \mathbf{i}_2 + \mathbf{i}_3)\,,$$

where $\mathbf{i}_1$, $\mathbf{i}_2$, $\mathbf{i}_3$ are unit vectors along the axes $\sigma_1$, $\sigma_2$, $\sigma_3$. The straight line

$$\sigma_1 = \sigma_2 = \sigma_3$$

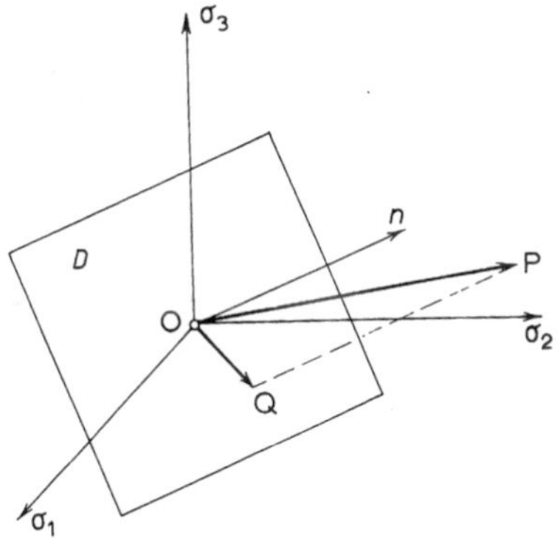

passes through the origin and is perpendicular to the surface under consideration. Points on this line, called the *hydrostatic axis*, correspond to the hydrostatic stress state.

We introduce the vector $\overline{\mathrm{OP}}$ in the form

$$\overline{\mathrm{OP}} = \sigma_1 \mathbf{i}_1 + \sigma_2 \mathbf{i}_2 + \sigma_3 \mathbf{i}_3 .$$

The projection of $\overline{\mathrm{OP}}$ on the normal is proportional to the mean pressure:

$$(\overline{\mathrm{OP}}, \mathbf{n}) = \sqrt{3}\,\sigma .$$

We now introduce the vector $\overline{\mathrm{OQ}}$:

$$\overline{\mathrm{OQ}} = s_1 \mathbf{i}_1 + s_2 \mathbf{i}_2 + s_3 \mathbf{i}_3 ,$$

which characterizes the deviatoric $D_\sigma$. It is easily seen that $\overline{\mathrm{OP}} = \overline{\mathrm{OQ}} + \sqrt{3}\,\sigma\mathbf{n}$.

We note that

$$(\overline{\mathrm{OQ}}, \mathbf{n}) = 0 ,$$

i.e. the vector $\overline{\mathrm{OQ}}$ lies in the plane (1.26); we shall call the latter the *deviatoric plane*.

The length of the vector $\overline{\mathrm{OQ}}$ is proportional to the shear stress intensity

$$|\overline{\mathrm{OQ}}| = \sqrt{2}\,T . \tag{1.27}$$

The angle $\omega_\sigma$ defines the location of the vector $\overline{\mathrm{OQ}}$ in the deviatoric plane.

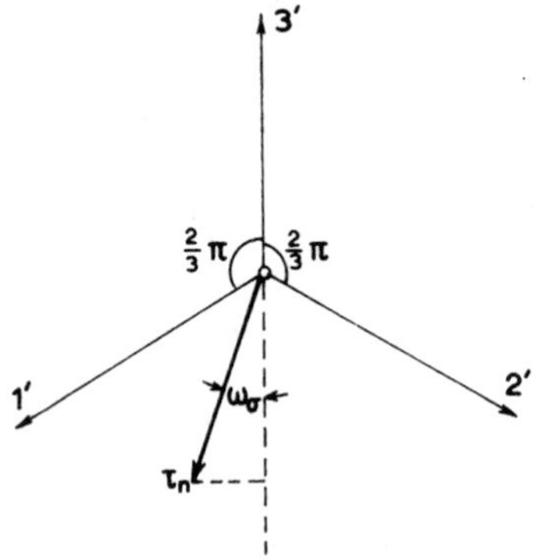


Fig. 4.

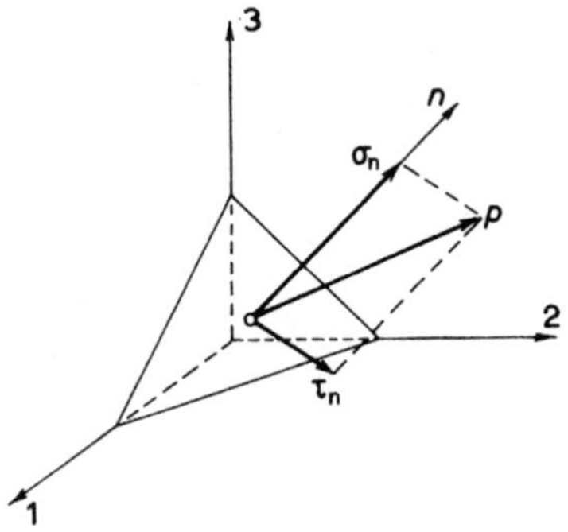


Fig. 5.

Indeed, suppose that the axes 1′, 2′ and 3′ are the projections of the axes $\sigma_1$, $\sigma_2$, $\sigma_3$ on the plane $D$ (fig. 4). We shall calculate the projection of the vector $\overline{OQ}$ on the axis 3′. Since

$$\cos(\sigma_3, 3') = \sqrt{\tfrac{2}{3}}, \qquad \cos(\sigma_1, 3') = \cos(\sigma_2, 3') = -\frac{1}{\sqrt{6}},$$

then

$$\text{proj.}\, \overline{OQ} = \sum_{i=1}^{3} s_i \cos(\sigma_i, 3') = -\sqrt{2}\, T \cos \omega_\sigma .$$

Thus the angle between the vector $\overline{OQ}$ and the negative 3′ axis equals $\omega_\sigma$. The vector $\overline{OQ}$ cannot be inclined to the negative 3′ axis by more than 60°.

Consider the plane which at a given point of the medium is equally inclined to the principal axes. This plane will be called the *octahedral* plane (since it is the face of a right octahedron, fig. 5). The projections of the stress vector **p** (cf. fig. 3) acting on the octahedral plane are, by Cauchy's formulae (1.2), equal respectively to $\sigma_1/\sqrt{3}, \sigma_2/\sqrt{3}, \sigma_3/\sqrt{3}$. Hence the normal stress on this surface is

$$\sigma_n = \sigma ,$$

i.e. equals the mean pressure; the shear stress is proportional to $T$:

$$\tau_n = \sqrt{\tfrac{2}{3}}\, T .$$

1.5. *Mohr's circle*

A graphical representation of the stresses in different cross-sections passing through a given point is provided by Mohr's diagram. Suppose that at this point the directions of the coordinate axes coincide with the principal directions; then because of formulae (1.3) and (1.2) we have

$$\sigma_n = \sigma_1 n_1^2 + \sigma_2 n_2^2 + \sigma_3 n_3^2 ,$$
$$\sigma_n^2 + \tau_n^2 = \sigma_1^2 n_1^2 + \sigma_2^2 n_2^2 + \sigma_3^2 n_3^2 ,$$

while

$$1 = n_1^2 + n_2^2 + n_3^2 .$$

From this system of equations we find the squares of the direction cosines:

$$n_1^2 = \frac{\tau_n^2 + (\sigma_n - \sigma_2)(\sigma_n - \sigma_3)}{(\sigma_1 - \sigma_2)(\sigma_1 - \sigma_3)} ,$$
$$n_2^2 = \frac{\tau_n^2 + (\sigma_n - \sigma_3)(\sigma_n - \sigma_1)}{(\sigma_2 - \sigma_3)(\sigma_2 - \sigma_1)} , \qquad (1.28)$$
$$n_3^2 = \frac{\tau_n^2 + (\sigma_n - \sigma_1)(\sigma_n - \sigma_2)}{(\sigma_3 - \sigma_1)(\sigma_3 - \sigma_2)} .$$

Since $\sigma_1 \geqslant \sigma_2 \geqslant \sigma_3$, and the left-hand sides of these equalities are non-negative, it follows that

$$\tau_n^2 + (\sigma_n - \sigma_2)(\sigma_n - \sigma_3) \geqslant 0 ,$$
$$\tau_n^2 + (\sigma_n - \sigma_3)(\sigma_n - \sigma_1) \leqslant 0 ,$$
$$\tau_n^2 + (\sigma_n - \sigma_1)(\sigma_n - \sigma_2) \geqslant 0 ,$$

i.e. the stresses $\sigma_n$, $\tau_n$ lie inside the region bounded by the semi-circles and shown cross-hatched in fig. 6; the points on any circle have corresponding surface elements which contain the respective principal axis. The direction cosines of the surface element, with given $\sigma_n$, $\tau_n$, are calculated from formulae (1.28). It is evident that the radii of the circles are not changed when a uniform additional pressure is imposed on the body and the whole figure merely shifts along the horizontal axis $\sigma_n$.

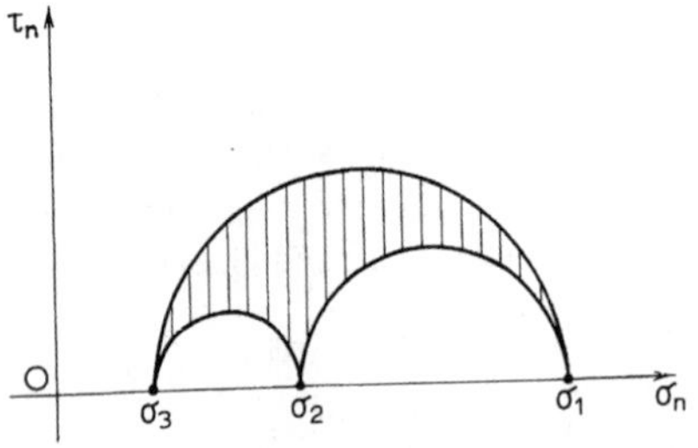


Fig. 6.

It is possible to express the inter-relationship between the principal components of the stress tensor by means of the coefficient

$$\mu_\sigma = 2\,\frac{\sigma_2 - \sigma_3}{\sigma_1 - \sigma_3} - 1 , \qquad (1.29)$$

introduced by Lode and Nadai. This characterizes the position of the point $\sigma_2$ in Mohr's diagram, and loses its meaning only in the case of hydrostatic pressure.

For the same values of $\mu_\sigma$ Mohr's diagrams are similar. It is evident that at a fixed value of $\mu_\sigma$ the character of the stress is precisely defined up to a multiplicative factor and an additive hydrostatic pressure. In this sense it is possible to regard $\mu_\sigma$ as representing the shape of the stress tensor (or deviatoric), and as being characteristic of the "nature of the stress state". The multiplicative factor, which characterizes the "scale", is proportional to the intensity $T$, as is evident from (1.16).

The parameter $\mu_\sigma$ varies in the range −1 to + 1; thus, for pure tension ($\sigma_1 > 0, \sigma_2 = \sigma_3 = 0$)

$$\mu_\sigma = -1 \ ,$$

for pure compression ($\sigma_1 = \sigma_2 = 0, \sigma_3 = -\sigma_1$)

$$\mu_\sigma = +1 \ ,$$

for pure shear ($\sigma_1 > 0, \sigma_2 = 0, \sigma_3 = -\sigma_1$)

$$\mu_\sigma = 0 \ .$$

The parameter $\mu_\sigma$ is a function of the invariants $I_2(D_\sigma), I_3(D_\sigma)$ and bears a simple relationship to the angle $\omega_\sigma$. From (1.29) and (1.16) it follows that

$$\mu_\sigma = \sqrt{3} \cot (\omega_\sigma + \tfrac{1}{3}\pi) \ . \tag{1.30}$$

The angle $\omega_\sigma$ is sometimes called the angle of the nature of the stress state. Notice that for tension $\omega_\sigma = \frac{1}{3}\pi$, for shearing $\omega_\sigma = \frac{1}{6}\pi$, and for compression $\omega_\sigma = 0$.

## §2. Strain

### 2.1. *The strain tensor*

Suppose that on deformation the points of a medium undergo a displacement **u**, whose components we shall designate by $u_x, u_y, u_z$. The deformation of the medium is characterized by the symmetric *strain tensor*

$$T_\epsilon = \left\| \begin{matrix} \epsilon_x & \frac{1}{2}\gamma_{xy} & \frac{1}{2}\gamma_{xz} \\ \frac{1}{2}\gamma_{xy} & \epsilon_y & \frac{1}{2}\gamma_{yz} \\ \frac{1}{2}\gamma_{xz} & \frac{1}{2}\gamma_{yz} & \epsilon_z \end{matrix} \right\| ,$$

with components

$$\epsilon_x = \frac{\partial u_x}{\partial x} + \frac{1}{2}\left[\left(\frac{\partial u_x}{\partial x}\right)^2 + \left(\frac{\partial u_y}{\partial x}\right)^2 + \left(\frac{\partial u_z}{\partial x}\right)^2\right], \dots,$$
$$\gamma_{xy} = \frac{\partial u_x}{\partial y} + \frac{\partial u_y}{\partial x} + \left[\frac{\partial u_x}{\partial x}\frac{\partial u_x}{\partial y} + \frac{\partial u_y}{\partial x}\frac{\partial u_y}{\partial y} + \frac{\partial u_z}{\partial x}\frac{\partial u_z}{\partial y}\right], \dots \tag{2.1}$$

The strain tensor, as with any symmetric tensor, can be reduced to diagonal form:

$$T_\epsilon = \begin{Vmatrix} \epsilon_1 & 0 & 0 \\ 0 & \epsilon_2 & 0 \\ 0 & 0 & \epsilon_3 \end{Vmatrix},$$

where $\epsilon_1$, $\epsilon_2$, $\epsilon_3$ are called *principal elongations.* This means that any strain may be realised by simple extensions in three mutually perpendicular directions (principal directions).

The differences

$$\gamma_1 = \epsilon_2 - \epsilon_3, \qquad \gamma_2 = \epsilon_3 - \epsilon_1, \qquad \gamma_3 = \epsilon_1 - \epsilon_2 \tag{2.2}$$

are called *principal shears*. The greatest in value of the shears at a given point will be called the *maximum shear* $\gamma_{max}$.

2.2. *Small deformation*

In the case of small deformation the components $\epsilon_x, \epsilon_y, \dots, \gamma_{xz}$ are small compared with unity; if, moreover, the angles of rotation are sufficiently small (an analysis of this question is given in the course on elasticity theory by V.V. Novozhilov [27]), then it is possible to neglect the products $(\partial u_x/\partial x)^2, (\partial u_x/\partial x)(\partial u_x/\partial y), \dots$, in formula (2.1). Consequently

$$\epsilon_x = \frac{\partial u_x}{\partial x}, \qquad \epsilon_y = \frac{\partial u_y}{\partial y}, \qquad \epsilon_z = \frac{\partial u_z}{\partial z};$$
$$\gamma_{xy} = \frac{\partial u_x}{\partial y} + \frac{\partial u_y}{\partial x}, \quad \gamma_{yz} = \frac{\partial u_y}{\partial z} + \frac{\partial u_z}{\partial y}, \quad \gamma_{xz} = \frac{\partial u_x}{\partial z} + \frac{\partial u_z}{\partial x}. \tag{2.3}$$

Here $\epsilon_x$, $\epsilon_y$, $\epsilon_z$ represent the relative elongations respectively in the directions of the axes $x$, $y$, $z$, and $\gamma_{xy}$, $\gamma_{yz}$, $\gamma_{xz}$ the relative shears ($\gamma_{xy}$ is the change in angle between the axes $x$, $y$, and so on); the relative change in volume equals

$$\epsilon = \epsilon_x + \epsilon_y + \epsilon_z\,. \tag{2.4}$$

These simple formulae are inadequate if we wish to describe substantial changes in the shape of massive bodies; for then the strain components are comparable in size with unity and it is necessary to proceed from the general relations (2.1). We also emphasize that even with small elongations and shears, the linear relations (2.3) often prove to be inadequate in problems of deformation and stability of flexible bodies (rods, plates, shells), because elements of the bodies undergo considerable displacements and rotations. Subsequently, when talking about small deformation, we shall mean deformation such that (2.3) are applicable.

We shall often use tensor notation for the strain components:

$$\epsilon_{ij} = \frac{1}{2}\left(\frac{\partial u_i}{\partial x_j} + \frac{\partial u_j}{\partial x_i}\right) , \tag{2.5}$$

where $x_i$ are cartesian coordinates, and $u_i$ are the components of the vector displacement. It is easily seen that

$$\epsilon = \epsilon_{ij}\delta_{ij} .$$

2.3. *Invariants*

The invariants of the strain tensor are formed just as for the stress tensor, and in principal axes have the form

$$\begin{aligned} I_1(T_\epsilon) &= \epsilon_1 + \epsilon_2 + \epsilon_3 , \\ I_2(T_\epsilon) &= -(\epsilon_1\epsilon_2 + \epsilon_2\epsilon_3 + \epsilon_3\epsilon_1) , \\ I_3(T_\epsilon) &= \epsilon_1\epsilon_2\epsilon_3 . \end{aligned} \tag{2.6}$$

It is convenient to represent the strain tensor as the sum

$$T_\epsilon = \tfrac{1}{3}\epsilon T_1 + D_\epsilon , \tag{2.7}$$

where $\frac{1}{3}\epsilon T_1$ is a spherical tensor corresponding to volumetric dilatation, and the *strain deviatoric*

$$D_\epsilon = \begin{Vmatrix} \epsilon_x - \frac{1}{3}\epsilon & \frac{1}{2}\gamma_{xy} & \frac{1}{2}\gamma_{xz} \\ \frac{1}{2}\gamma_{xy} & \epsilon_y - \frac{1}{3}\epsilon & \frac{1}{2}\gamma_{yz} \\ \frac{1}{2}\gamma_{xz} & \frac{1}{2}\gamma_{yz} & \epsilon_z - \frac{1}{3}\epsilon \end{Vmatrix}$$

characterizes the change in shape of elements of the medium caused by shear. The invariants of the strain deviatoric are

$$\begin{aligned} I_1(D_\epsilon) &= 0 , \\ I_2(D_\epsilon) &= \tfrac{1}{6}[(\epsilon_1-\epsilon_2)^2 + (\epsilon_2-\epsilon_3)^2 + (\epsilon_3-\epsilon_1)^2] , \\ I_3(D_\epsilon) &= (\epsilon_1-\tfrac{1}{3}\epsilon)(\epsilon_2-\tfrac{1}{3}\epsilon)(\epsilon_3-\tfrac{1}{3}\epsilon) . \end{aligned} \tag{2.8}$$

In plasticity theory an important role is played by the quadratic invariant $I_2(D_\epsilon)$, which may be considered as a global characteristic of the distortion in shape of an element of the medium. The non-negative quantity

$$\Gamma = +2\sqrt{I_2(D_\epsilon)} = \\ = \sqrt{\tfrac{2}{3}}\sqrt{(\epsilon_x-\epsilon_y)^2 + (\epsilon_y-\epsilon_z)^2 + (\epsilon_z-\epsilon_x)^2 + \tfrac{3}{2}(\gamma_{xy}^2 + \gamma_{yz}^2 + \gamma_{xz}^2)} \tag{2.9}$$

is called the *shear strain intensity* [1]).

In the case of pure shear

$$\epsilon_x = \epsilon_y = \epsilon_z = \gamma_{yz} = \gamma_{xz} = 0\,, \qquad \gamma_{xy} = \gamma\,.$$

Introducing these values into (2.1) we find:

$$\Gamma = |\gamma|\,.$$

The numerical factor before the root in (2.9) is chosen so that in pure shear the intensity $\Gamma$ equals the magnitude of the shear $\gamma$.

The relation (2.7) can also be written in the form

$$\epsilon_{ij} = \tfrac{1}{3}\epsilon\delta_{ij} + e_{ij}\,, \tag{2.10}$$

where $e_{ij}$ are the strain deviatoric components. In this notation the first equality in (2.8) appears as $e_{ii} = 0$, and the shear strain intensity equals

$$\Gamma = (2e_{ij}e_{ij})^{\frac{1}{2}}\,. \tag{2.11}$$

2.4. *Geometrical interpretation*

A geometrical interpretation, analogous to that discussed above for the stress tensor, can be developed for any symmetrical tensor, and in particular for the strain tensor.

As in our earlier analysis we obtain

$$\begin{aligned} e_1 &= \frac{1}{\sqrt{3}}\Gamma\cos(\omega_\epsilon - \tfrac{1}{3}\pi)\,, \\ e_2 &= \frac{1}{\sqrt{3}}\Gamma\cos(\omega_\epsilon + \tfrac{1}{3}\pi)\,, \\ e_3 &= -\frac{1}{\sqrt{3}}\Gamma\cos\omega_\epsilon\,, \end{aligned} \tag{2.12}$$

[1]) We sometimes consider the reduced strain (or strain intensity) which equals $\Gamma/\sqrt{3}$. In the case of simple tension (compression) of a rod of incompressible material, the reduced strain equals $|\epsilon_1|$.

where

$$-\cos 3\omega_\epsilon = \frac{12\sqrt{3}\, I_3(D_\epsilon)}{\Gamma^3} . \tag{2.13}$$

So, just as before,

$$0 \leqslant \omega_\epsilon \leqslant \tfrac{1}{3}\pi$$

and we find the approximate relation

$$\Gamma \approx 1.08 \cdot \gamma_{\max} . \tag{2.14}$$

Mohr's diagram retains its significance; it is necessary to measure along the abscissa the relative elongation $\epsilon$ in a given direction and along the ordinate half modulus of the shear $\gamma_n$ on the plane perpendicular to $n$.

Corresponding to the parameter $\mu_\sigma$ we introduce the parameter

$$\mu_\epsilon = 2\frac{\epsilon_2 - \epsilon_3}{\epsilon_1 - \epsilon_3} - 1 ,$$

associated with the angle of the nature of the shear $\omega_\epsilon$ by the relation

$$\mu_\epsilon = \sqrt{3} \cot\left(\omega_\epsilon + \tfrac{1}{3}\pi\right) . \tag{2.15}$$

2.5. *The strain compatibility conditions*

The strain components must satisfy the six homologous relations of Saint Venant:

$$\frac{\partial^2 \epsilon_x}{\partial y^2} + \frac{\partial^2 \epsilon_y}{\partial x^2} = \frac{\partial^2 \gamma_{xy}}{\partial x \partial y} ; \ldots ,$$

$$2\frac{\partial^2 \epsilon_x}{\partial y \partial z} = \frac{\partial}{\partial x}\left(-\frac{\partial \gamma_{yz}}{\partial x} + \frac{\partial \gamma_{xz}}{\partial y} + \frac{\partial \gamma_{xy}}{\partial z}\right) ; \ldots \tag{2.16}$$

The rest of the relations are obtained by cyclic substitution of the indices.

2.6. *Strain components in cylindrical and spherical coordinates*

Later we shall need expressions for the strain components in cylindrical and spherical coordinates; we quote them without derivation [20, 48].

*Cylindrical coordinates* $r$, $\varphi$, $z$. Suppose that the coordinates of the displacement vector $u_r$, $u_\varphi$, $u_z$ are independent of $\varphi$; then the relative extensions and shear strains have the form

$$\epsilon_r = \frac{\partial u_r}{\partial r}, \qquad \epsilon_\varphi = \frac{u_r}{r}, \qquad \epsilon_z = \frac{\partial u_z}{\partial z},$$
$$\gamma_{r\varphi} = \frac{\partial u_\varphi}{\partial r} - \frac{u_\varphi}{r}, \qquad \gamma_{\varphi z} = \frac{\partial u_\varphi}{\partial z}, \qquad \gamma_{rz} = \frac{\partial u_r}{\partial z} + \frac{\partial u_z}{\partial r}. \tag{2.17}$$

*Spherical coordinates* $r$, $\varphi$, $\chi$. In the case of interest, namely that of spherical symmetry, the components of the displacement vector $u_\varphi = u_\chi = 0$, and

$$\epsilon_r = \frac{\partial u_r}{\partial r}, \qquad \epsilon_\varphi = \epsilon_\chi = \frac{u_r}{r}; \qquad \gamma_{r\varphi} = \gamma_{\varphi\chi} = \gamma_{r\chi} = 0. \tag{2.18}$$

## §3. Rate of strain

### 3.1. *The strain-rate tensor*

Let the particles of the medium move with velocity **v**, whose components are

$$v_x = v_x(x, y, z, t), \quad v_y = v_y(x, y, z, t), \quad v_z = v_z(x, y, z, t).$$

In an infinitesimal time d$t$ the medium experiences an infinitesimal strain, determined by the translations $u_x$d$t$, $u_y$d$t$, $u_z$d$t$. The components of this strain, calculated from (2.3), have a common factor d$t$, and on dividing this out we get the components of the symmetrical *strain-rate tensor*

$$T_\xi = \begin{Vmatrix} \xi_x & \frac{1}{2}\eta_{xy} & \frac{1}{2}\eta_{xz} \\ \frac{1}{2}\eta_{xy} & \xi_y & \frac{1}{2}\eta_{yz} \\ \frac{1}{2}\eta_{xz} & \frac{1}{2}\eta_{yz} & \xi_z \end{Vmatrix},$$

where

$$\xi_x = \frac{\partial v_x}{\partial x}, \qquad \xi_y = \frac{\partial v_y}{\partial y}, \qquad \xi_z = \frac{\partial v_z}{\partial z};$$
$$\eta_{xy} = \frac{\partial v_x}{\partial y} + \frac{\partial v_y}{\partial x}, \qquad \eta_{yz} = \frac{\partial v_y}{\partial z} + \frac{\partial v_z}{\partial y}, \qquad \eta_{xz} = \frac{\partial v_x}{\partial z} + \frac{\partial v_z}{\partial x}. \tag{3.1}$$

The quantities $\xi_x$, $\xi_y$, $\xi_z$ determine the *rates of relative elongations* of an elementary volume in the directions of the coordinate axes; $\eta_{xy}$, $\eta_{yz}$, $\eta_{xz}$ determine the *angular rates of change* of initially right angles. The *rate of relative volume dilatation is*

$$\xi = \xi_x + \xi_y + \xi_z = \text{div}\,\mathbf{v} \tag{3.2}$$

Besides the rate of pure shear characterized by the tensor $T_\xi$, a volume element experiences rigid body displacement, determined by the translatory velocity **v**, and a rotation with angular velocity

$$\omega = \tfrac{1}{2}\,\mathrm{curl}\,\mathbf{v}\,.$$

The *acceleration of a particle* of the medium is given by the total (material) derivative of the velocity

$$w_x = \frac{\partial v_x}{\partial t} + v_x\frac{\partial v_x}{\partial x} + v_y\frac{\partial v_x}{\partial y} + v_z\frac{\partial v_x}{\partial z}; \ldots \tag{3.3}$$

Here the first term on the right characterizes the local change, and the remainder represents the translatory part, which takes into account the change produced by following the particle motion through space.

In tensor notation the components of the rate of strain are

$$\xi_{ij} = \frac{1}{2}\left(\frac{\partial v_i}{\partial x_j} + \frac{\partial v_j}{\partial x_i}\right),$$

where $v_i$ are the components of the velocity vector.

### 3.2. *Invariants of the strain-rate tensor*

The invariants of the tensor $T_\xi$ and of the deviatoric $D_\xi$ may be obtained from formulae (2.6) and (2.8) on replacing $\epsilon_x, \ldots, \gamma_{xz}$ by $\xi_x, \ldots, \eta_{xz}$. We shall write out only the expression for the *shear strain-rate intensity:*

$$H = +\,2\sqrt{I_2(D_\xi)} = \\ = \sqrt{\tfrac{2}{3}}\sqrt{(\xi_x - \xi_y)^2 + (\xi_y - \xi_z)^2 + (\xi_z - \xi_x)^2 + \tfrac{3}{2}(\eta_{xy}^2 + \eta_{yz}^2 + \eta_{xz}^2)}\,. \tag{3.4}$$

Mohr's diagram and the coefficient $\mu_\xi$ can also be applied to the strain-rates. The quantity $\omega_\xi$ and corresponding formulae for the principal values of the deviatoric $D_\xi$ can be introduced as before.

### 3.3. *Strain and rate of strain*

Since velocities are the total derivatives of the displacement with time,

$$v_i = \frac{\mathrm{d}u_i}{\mathrm{d}t},$$

then

$$\xi_{ij} = \frac{1}{2}\left(\frac{\partial}{\partial x_j}\frac{\mathrm{d}u_i}{\mathrm{d}t} + \frac{\partial}{\partial x_i}\frac{\mathrm{d}u_j}{\mathrm{d}t}\right). \tag{3.5}$$

It is evident that

$$\xi_{ij} \neq \frac{\mathrm{d}}{\mathrm{d}t} \epsilon_{ij} .$$

In the case of small deformation simple relations exist between the strain components and the strain-rate components, namely,

$$v_i = \frac{\partial}{\partial t} u_i ,$$

and

$$\xi_{ij} = \frac{\partial}{\partial t} \epsilon_{ij} . \tag{3.6}$$

The acceleration is given by the formula

$$w_i = \frac{\partial^2 u_i}{\partial t^2} . \tag{3.7}$$

The translatory part in the expression for the total derivative is omitted on the grounds that, with small strains, it is usually possible to assume that the coordinate derivatives of displacement and velocity can be neglected.

One should finally record that $\xi_i \neq (\partial/\partial t)\, \epsilon_i$, since the principal axes of the strain tensor and the strain-rate tensor do not, in general, coincide.

3.4. *Increments in the strain components*

The mechanical properties of metals in conditions of relatively slow, plastic deformation at not too high a temperature, are practically independent of the rate of deformation, as will be explained below. In this case the interest lies in fact, not in the rate of strain, but in the infinitesimal increments $\xi_{ij}\mathrm{d}t$ (conventionally we shall denote them by $\mathrm{d}\epsilon_{ij}$, bearing in mind that generally speaking these quantities are not differentials of the components of strain). These are determined in accordance with (3.5) by the formula

$$\mathrm{d}\epsilon_{ij} = \frac{1}{2} \left( \frac{\partial}{\partial x_j} \mathrm{d}u_i + \frac{\partial}{\partial x_i} \mathrm{d}u_j \right) . \tag{3.8}$$

They generate a tensor $T_{\mathrm{d}\epsilon}$ and have a simple physical meaning. The relations (3.8) are useful for describing large strains, which may be obtained by integrating the infinitesimal changes (3.8).

In formulae (3.8) the increments in the strain components are evaluated with respect to the instantaneous state; the system of coordinates $x_i$ is assumed to be fixed in the volume element.

Consider, for example, a uniform tension along the axis of a cylinder, the axis coinciding with $x_1$; then

$$\mathrm{d}\epsilon_1 = \mathrm{d}l/l\,,$$

where $l$ is the instantaneous length of the cylinder and $\mathrm{d}l$ is the infinitesimal change in it. Integration leads to the so-called *natural elongation*

$$\int_{l_0}^{l} \frac{\mathrm{d}l}{l} = \ln \frac{l}{l_0}\,,$$

where $l_0$ is the initial length.

If the principal axes do not rotate under deformation, the integrals $\int \mathrm{d}\epsilon_i$ have a simple physical meaning, being equal to the corresponding natural elongations $\ln(l_i/l_{i0})$. It is evident that with this quite simple law strains are additive: the sum of successive natural elongations equals the resultant natural elongation.

In the general case the integrals $\int \mathrm{d}\epsilon_{ij}$ can not be evaluated and do not have a specific physical meaning; these integrals can be found if the strain path is known, i.e. if the components $\mathrm{d}\epsilon_{ij}$ are known as functions of some parameter (for example, the load). This limits the range of application of natural elongations as measures of strain to the case of fixed principal directions.

The invariants of the tensor $T_{\mathrm{d}\epsilon}$ (deviatoric $D_{\mathrm{d}\epsilon}$), which are obtained from the corresponding invariants of the tensor $T_\epsilon$ on passing to the components $\mathrm{d}\epsilon_{ij}$, will be denoted by

$$\mathrm{d}\epsilon\,, \qquad \overline{\mathrm{d}\Gamma}\,, \qquad \mu_{\mathrm{d}\epsilon}\,, \qquad \omega_{\mathrm{d}\epsilon}\,.$$

We emphasize once more that the quantities $\mathrm{d}\epsilon_{ij}$ must not be regarded as differentials of the strain components $\epsilon_{ij}$. The latter is true only for small strains, when (2.3) are valid; in this case simple superposition of strains is permissible and the integrals $\int \mathrm{d}\epsilon_{ij}$ are the strain components.

3.5. *Compatibility conditions for the rates of strain*

The strain-rate components, like the strain components (§2), can not be specified arbitrarily. They must satisfy six compatibility conditions, fully analogous to those of Saint-Venant (2.16):

$$\frac{\partial^2 \xi_x}{\partial y^2} + \frac{\partial^2 \xi_y}{\partial x^2} = \frac{\partial^2 \eta_{xy}}{\partial x \partial y}\,;\ \ldots\,,$$

$$2\frac{\partial^2 \xi_x}{\partial y \partial z} = \frac{\partial}{\partial x}\left(-\frac{\partial \eta_{yz}}{\partial x} + \frac{\partial \eta_{xz}}{\partial y} + \frac{\partial \eta_{xy}}{\partial z}\right);\ \ldots \tag{3.9}$$

### 3.6. *The case of incompressible media*

For incompressible media $\xi = 0$, i.e.

$$\partial v_i / \partial x_i = 0 . \tag{3.10}$$

With this condition the components $\xi_{ij}$ are components of the deviatoric of the strain-rate, and the *intensity of the shear strain-rate* is

$$H = (2\xi_{ij}\xi_{ij})^{\frac{1}{2}} . \tag{3.11}$$

## §4. Differential equations of motion. Boundary and initial conditions

### 4.1. *The differential equations of motion*

We shall denote the density of the medium by $\rho$, the components of body force per unit mass by $F_x$, $F_y$, $F_z$, and the components of acceleration of a particle of the medium by $w_x$, $w_y$, $w_z$. The motion of an element of the medium is determined by the forces applied to it; once these forces are specified, we obtain the differential equations of motion of a continuous medium, first derived by Cauchy:

$$\begin{aligned}
&\frac{\partial \sigma_x}{\partial x} + \frac{\partial \tau_{xy}}{\partial y} + \frac{\partial \tau_{xz}}{\partial z} + \rho\,(F_x - w_x) = 0 , \\
&\frac{\partial \tau_{xy}}{\partial x} + \frac{\partial \sigma_y}{\partial y} + \frac{\partial \tau_{yz}}{\partial z} + \rho\,(F_y - w_y) = 0 , \\
&\frac{\partial \tau_{xz}}{\partial x} + \frac{\partial \tau_{yz}}{\partial y} + \frac{\partial \sigma_z}{\partial z} + \rho\,(F_z - w_z) = 0 .
\end{aligned} \tag{4.1}$$

We emphasize that these equations describe the motion of the elements about the medium, regarded as a solid particle.

In tensor notation these equations may be written in the form

$$\frac{\partial \sigma_{ij}}{\partial x_i} + \rho\,(F_j - w_j) = 0 . \tag{4.2}$$

Later we shall require the differential equations of equilibrium in cylindrical and spherical coordinates; we quote these equations without derivation (see [21, 48]).

### 4.2. *Equations of equilibrium in cylindrical coordinates*

In cylindrical coordinates $r$, $\varphi$, $z$ the equilibrium equations have the form

$$
\begin{aligned}
&\frac{\partial \sigma_r}{\partial r} + \frac{1}{r}\frac{\partial \tau_{r\varphi}}{\partial \varphi} + \frac{\partial \tau_{rz}}{\partial z} + \frac{\sigma_r - \sigma_\varphi}{r} + \rho F_r = 0\,,\\
&\frac{\partial \tau_{r\varphi}}{\partial r} + \frac{1}{r}\frac{\partial \sigma_\varphi}{\partial \varphi} + \frac{\partial \tau_{\varphi z}}{\partial z} + \frac{2\tau_{r\varphi}}{r} + \rho F_\varphi = 0\,,\\
&\frac{\partial \tau_{rz}}{\partial r} + \frac{1}{r}\frac{\partial \tau_{\varphi z}}{\partial \varphi} + \frac{\partial \sigma_z}{\partial z} + \frac{\tau_{rz}}{r} + \rho F_z = 0\,.
\end{aligned}
\tag{4.3}
$$

4.3. *Equations of equilibrium in spherical coordinates*

In spherical coordinates $r$ (radius), $\varphi$ (longitude), $\chi$ (latitude) the equations of equilibrium have in the case of spherical symmetry the form

$$
\frac{\mathrm{d}\sigma_r}{\mathrm{d}r} + \frac{2}{r}(\sigma_r - \sigma_\varphi) + \rho F_r = 0\,,
\tag{4.4}
$$

while

$$
\sigma_\varphi = \sigma_\chi\,, \qquad \tau_{r\varphi} = \tau_{\varphi\chi} = \tau_{r\chi} = 0\,.
$$

4.4. *Boundary conditions*

Besides the equations quoted above, we require some *boundary conditions,* which can be of various types.

On the boundary S of a body loads $p_x$, $p_y$, $p_z$ may be given. In this case equations (1.2) must be satisfied on S; these are the conditions for the equilibrium of an elementary tetrahedron adjacent to the boundary and under the influence of internal and external forces.

Alternatively the displacements (or velocities) of points on the body's boundary may be prescribed.

Finally the boundary conditions may be mixed, with the loads given on one part and the displacements (or speeds) given on another part on the boundary.

4.5. *Initial conditions*

If the process of deformation is non-stationary and is described by equations containing derivatives with respect to time or with respect to a load parameter, it is necessary to prescribe the initial state of the body.

## §5. The mechanical constitutive equations of a body

5.1. *The mechanical constitutive equations*

The quantities considered above (force, stress, displacement, rotation, strain, strain-rate, etc.) are necessary for a description of the dynamic and kinematic states of an elementary particle of the body and may be called the mechanical variables. As we have seen, they are connected by the three equations of motion (4.1) only. For the construction of a complete phenomenological theory of the motion of a continuum it is necessary to have in addition the relationship between the dynamic and kinematic states of a particle. A collection of such relations may be termed "mechanical constitutive equations"; it is necessary to distinguish them from the equations of motion (4.1), which derive from d'Alembert's principle and which describe the mechanics of translation and rotation of particles of the medium.

The mechanical properties of real bodies are extremely complicated. It is not, however, necessary to attempt to formulate constitutive equations describing *all* details of the mechanical behaviour of a body under the influence of loads. On the contrary, it is expedient to choose the simplest mechanical model, one that reflects only the most essential properties. Then it becomes possible to develop satisfactorily a general mathematical theory. Such simple models form the basis for subsequent refinements, and this explains the great importance that models of ideal elastic bodies and ideal fluids have occupied in mechanics and its applications.

### 5.2. *Elastic body; ideal and viscous fluid*

Continuum mechanics has long been concerned with the motions of ideal and viscous fluids and also the deformations of ideal elastic bodies. For the latter, we have as a constitutive relation the *generalized Hooke's law*:

$$\epsilon = 3k\sigma \,, \tag{5.1}$$

$$D_\sigma = 2GD_\epsilon \,, \tag{5.2}$$

where $k$, $G$ are constants of the material [1]).

In this form, the law emphasizes the distinction between the resistance of an elastic body to changes in volume and to changes in shape (shearing). The constants $k$, $G$ can be regarded as independent.

For an *ideal fluid* we have the characteristic equation

$$f(\sigma, \mu) = 0 \tag{5.3}$$

and the condition of zero viscosity

$$D_\sigma = 0 \,.$$

[1]) The coefficient of volumetric compression, $k = (1-2\nu)/E$, where $E$ is Young's modulus, $\nu$ is Poisson's coefficient; the shear modulus $G = E/2(1+\nu)$.

For a *viscous fluid*, in addition to the characteristic equation (5.3), there is the generalized Newton's law

$$D_\sigma = 2\mu' D_\xi \,, \tag{5.4}$$

where $\mu'$= const. is the coefficient of viscosity.

5.3. *Concluding remarks*

The above examples characterize the simplest mechanical properties of real bodies. In particular the solid bodies described here have only the property of ideal elasticity. Unfortunately solid bodies can be treated as elastic only within narrow limits, and it is necessary to consider the important question of plastic deformation. To do this it is first of all necessary to establish equations for the plastic state. In principle one can pose the question of deriving such equations on the basis of solid-state physics. But the process of plastic deformation is extremely complicated, being connected above all with various defects of the crystal lattice. This fact, together with the complexity of the structure of modern metallic alloys, makes clear the difficulty of the problem. There remains a second method, namely the use of equations for plasticity based on experimental data. Such methods were introduced for the model of an ideal elastic body and for ideal and viscous fluids considered above. It was only much later that these equations were evolved from physical considerations.

Finally, we remark that thermodynamic analysis contributes greatly to the foundations of the equations of the plastic state, in particular the thermodynamics of irreversible processes, which has been successfully developed in recent years [42, 57, 69, 89].

## PROBLEMS

1. Show that the principal directions of a tensor and its deviatoric coincide.
2. Show that from the relation

$$D_\epsilon = \psi D_\sigma \,,$$

where $\psi$ is a scalar, it follows that the deviatorics $D_\epsilon$, $D_\sigma$ have identical principal directions, and that $\mu_\epsilon = \mu_\sigma$.

3. Show by direct transformation from the coordinate system $x, y, z$ to another co-

ordinate system $\xi$, $\eta$, $\zeta$ that the mean pressure and the shear stress intensity are invariants:

$$\sigma_x + \sigma_y + \sigma_z = \sigma_\xi + \sigma_\eta + \sigma_\zeta ,$$

$$(\sigma_x - \sigma_y)^2 + \ldots + 6(\tau_{xy}^2 + \ldots) = (\sigma_\xi - \sigma_\eta)^2 + \ldots + 6(\tau_{\xi\eta}^2 + \ldots) .$$

4. Find the radial displacement in the case of strain with spherical symmetry for an incompressible medium. Calculate the natural strains.

5. Find the radial displacement in the case of small axi-symmetrical strains of an incompressible medium; the displacement $u_z$ in the direction of the $z$-axis is assumed to be zero.

How do the results change if $u_z$ is constant?

# 2

# Equations of the Plastic State

## §6. The mechanical properties of solids

6.1. *Changes in density and shape of a solid*

It is common practice to draw a distinction between solids and fluids, although from a physical point of view this categorization is to some extent conditional. Solid bodies and bodies sufficiently liquid to form drops are differentiated by the effect of external forces on them, namely by unequal resistance to change of shape. Water offers little resistance to change of shape, whereas to change the shape of a piece of steel the application of great force is required. Experiments by Bridgeman and others have shown that volumetric compression of solids (not porous) and liquids is an elastic deformation, with the dependence of the relative volumetric change on the pressure being very nearly linear [6, 25]. Thus, *change of density of a body is an elastic deformation, determined by the mean pressure.* It is usually possible to neglect as insignificant changes in density caused by plastic deformation.

The change of shape of a body is caused by shear strains. For isotropic materials shear strains are almost independent of pressure, provided the pressure is not very high. Bridgeman's experiments show that the increase of the shear modulus at a pressure of $10^5$ atm., compared with its value at zero pressure, is + 2.2% for coiled steel, + 1.8% for nickel and so on. The influence of pressure may prove to be important for questions of the movement of rock at great depths in the earth.

It should be noted that for anisotropic materials the shear strains do depend on pressure; pressure also plays an important role in questions of limiting equilibrium of freely-flowing media.

### 6.2. *Elastic and plastic deformations*

A picture of the resistance of a solid to change of shape is given by experiments involving the extension of cylindrical specimens by the action of a gradually increasing force $P$. The upper part of fig. 7 illustrates tension diagrams for soft steel and copper at room temperature.

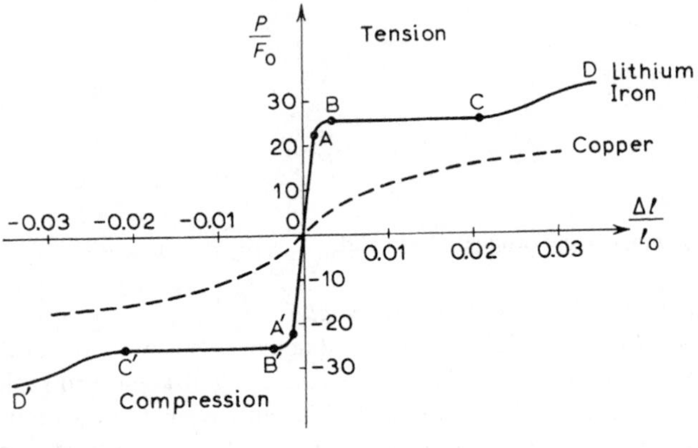


Fig. 7.

The stress $P/F_0$, where $F_0$ is the original cross-sectional area of the rod, is measured along the vertical axis, and the relative elongation $\Delta l/l_0$, where $l_0$ is the initial length of the specimen, is taken along the horizontal axis. Point A corresponds to the so-called *limit of proportionality* and lies a little below the elastic limit B, beyond which residual strains appear and the elongations rapidly increase; the characteristic *yield plateau* BC is displayed, after which the stress again increases. Section CD corresponds to the hardening state of the material. The diagram for the compression of such materials is on the whole similar to the tension diagram, although the values of the stresses at points A′, B′, C′, D′ are usually somewhat larger than those at the corresponding points A, B, C, D. Transition to the yield plateau is sometimes associated with a sharp peak. The stress which characterizes the plateau BC will be called the *yield limit* [1]).

For certain materials (e.g. annealed copper, aluminium, high alloy steels, etc.) the tension curve does not possess a yield plateau and sometimes it does not have a linear section.

If the load is reduced, then in general the unloading curve ABC (fig. 8) is nearly a straight line, whose slope is the same as that of the line of the elastic section; the magnitude of the residual strain is given by the distance OC.

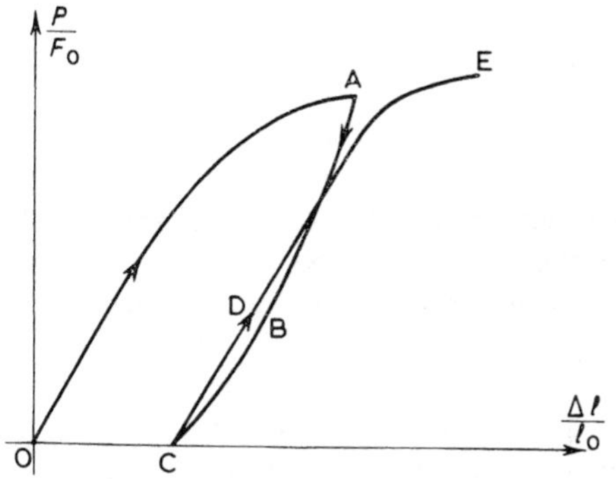


Fig. 8.

Experiments with pure shear (torsion of a tube) result in strain curves completely analogous to the tension curves.

In plasticity theory strain curves are usually schematic. In fig. 9 we show such a schematic diagram of the dependence between shear strain $\gamma$ and tangential stress $\gamma$ in experiments with pure shear. Initially, for $\tau < \tau_s$, the material follows Hooke's law

$$\tau = G\gamma \,. \tag{6.1}$$

Then a phase AB of yield sets in, which is characterized by the growth of shear strain at constant tangential stress

$$\tau = \text{const.} = \tau_s \,. \tag{6.2}$$

[1]) Notice that this definition does not conform with the concept, widely used in engineering, of a conditional yield limit as the stress corresponding to a residual strain of 0.2%.

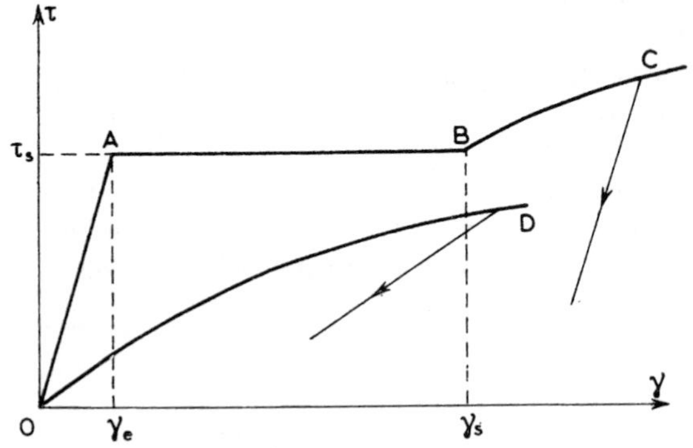


Fig. 9.

This state continues until $\gamma$ reaches the value $\gamma_s$, which we shall call the shear yield limit. After this the material passes to a strain-hardening phase BC, where the dependence between $\tau$ and $\gamma$ may be written in the form

$$\tau = g(\gamma)\,\gamma\,. \tag{6.3}$$

The function $g(\gamma)$ is sometimes called the *modulus of plasticity;* experiments give $0 \leqslant g(\gamma) \leqslant G$. In the absence of a yield plateau, the strain-hardening phase BC directly joins the linear elastic section OA.

### 6.3. *Strain-hardening*

For metals the unloading curve ABC (fig. 8) is in general a straight line; if the specimen is again loaded, then the stress curve CDE will differ little from the line ABC. In this way, owing to the original stretching, the metal seems to acquire elastic properties and to increase its elastic limit; to a significant extent it in fact loses the capacity for plastic deformation. This phenomenon is called *strain hardening*.

In the course of time it is observed that the hardening is partially relieved. This phenomenon, called *relaxation* of the material, becomes more noticeable with increase in temperature. At high temperatures the acquired hardening disappears (*annealing* of the material).

### 6.4. *Strain anisotropy*

Hardening usually has a directional character. Therefore as a result of plastic deformation, material acquires so-called *strain anisotropy*. One of the results of strain anisotropy is the *Bauschinger effect*, namely that an initial plastic deformation of one sign reduces the resistance of the material with respect to a subsequent plastic deformation of the opposite sign. Thus plastic tension of a rod leads to a noticeable drop in the yield plateau in a subsequent compression of that rod.

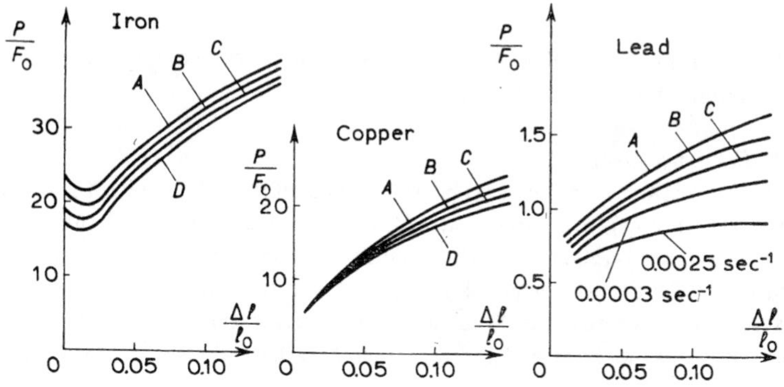


Fig. 10.

6.5. *Influence of the strain rate*

If tests take place at normal time intervals at room temperature, the mechanical properties of steel and other refractory metals are almost independent of the strain rate. Fig. 10 illustrates the experimental results of Ziebel and Pomp for the following rates of relative compression: $A \sim 1.25$ sec$^{-1}$ $B \sim 0.2$ sec$^{-1}$, $C \sim 0.025$ sec$^{-1}$, $D$ near zero. Nevertheless the rate of testing is very important in experiments with very malleable materials (lead, tin, etc.), in prolonged tests at high temperatures with steel, copper and other metals, and at large strain rates. The influence of the rate of testing is strongly dependent on temperature; namely it diminishes with reduction in temperature and at sufficiently low temperatures virtually disappears. The same dependence on rate also emerges in the increase of resistance to deformation with increasing strain-rate.

This is indicative of the fact that in "normal" conditions, the plastic deformation of "hard" metals is practically unrelated to the thermal motion of atoms (athermal plasticity).

6.6. *Creep*

At sufficiently high temperatures, plastic deformation is observed with quite small stress after a sufficient length of time. This phenomenon is called *creep*, and manifests itself in some cases in the growth of deformation in the course of time at a fixed load, and in other cases in a continuous reduction of stress at constant strain (*relaxation*). At high temperatures creep determines the reliability and life of machines. In connection with this we may note the rapid development of a *theory of creep*.

## §7. Experimental studies of plastic deformations in complex stress states. Simple and complex loading

### 7.1. *The experiments*

In the last ten years much work has been devoted to the study of flow and hardening in a complex stress state. Most researchers have carried out experiments using thin-walled tubes (fig. 11); by a combination of tension, torsion and internal pressure it is possible to produce in the wall of the tube an arbitrary, plane (more accurately, "almost plane") stress state. Thus, under the influence of an axial force $P$ and a torque $M$ we have the stress ($P + M$-tests)

$$\sigma_\varphi \approx 0\,, \qquad \sigma_z = \frac{P}{2\pi a h}\,, \qquad \tau_{\varphi z} = \frac{M}{2\pi a^2 h}\,,$$

where $a$ is the mean radius of the tube and $h$ is its thickness. Under the action of an axial force $P$ and an internal pressure $p$ ($P + p$-tests)

$$\sigma_\varphi \approx p\frac{a}{h}\,, \qquad \sigma_z = \frac{P}{2\pi a h}\,, \qquad \tau_{\varphi z} \approx 0\,.$$

The stress $\sigma_r$, of order $p$, is negligible compared with the stresses $\sigma_\varphi$, $\sigma_z$, since $a/h \gg 1$.

By measuring the strains in the tube (the change in diameter, length of the

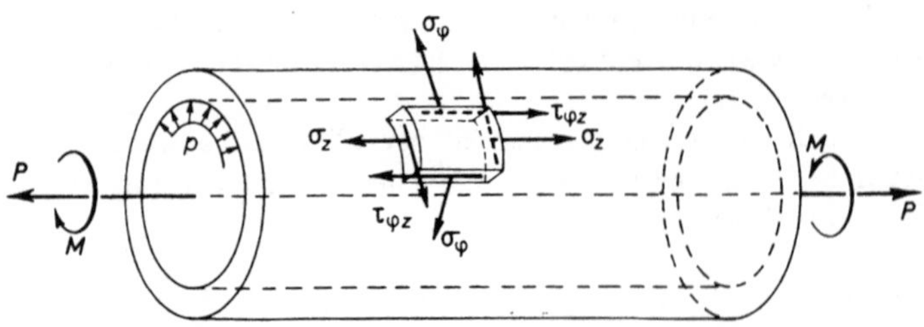


Fig. 11.

tube and its angle of twist) and comparing them with a known stress state, we can estimate the laws of plastic strain.

In recent years attempts have been made to load the tube with some external pressure $q$ in addition to the internal pressure $p$. From this the behaviour of the material in a triaxial stress state is successfully obtained. Addition of the outer pressure substantially complicates the experiments.

Investigation has also been carried out on the tension and torsion of a solid cylinder, subject to a pressure on the lateral surface. Such experiments are not difficult to carry out, but yield less, since the stress distribution in a solid cylinder is not uniform and cannot be directly calculated from the measured loads.

### 7.2. *Simple and complex loading*

Simple loading is characterized by the fact that the components of stress increase during each experiment in proportion to a single parameter (in experiments with thin-walled tubes it is evident that the external loads also increase in proportion to the same parameter). Consequently the form of the stress tensor and its principal directions remain constant in time.

In complex loading the directions of the principal axes and the interrelations between the principal stresses can change.

We shall take as an example the $P+M$-tests. In $P, M$ coordinates the loading process is represented by a certain curve OC (fig. 12). Simple loading corresponds to a straight line, say $OO_1$. Any other loading path corresponds to complex loading.

Experiments with simple loading are easier to achieve, as the arrangement of the testing device is simpler (in this case one power source is sufficient, e.g. one hydraulic press).

We give an example of complex loading: a thin-walled tube is first twisted,

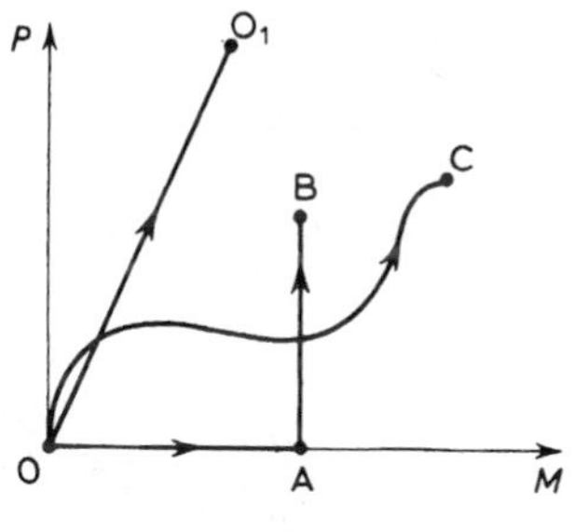


Fig. 12.

then subjected to tension at constant moment $M$; in fig. 12 this case is represented by the line OAB. Loading of this type is sometimes called *step-shaped.*

*Remarks.* As has been shown, the influence of hydrostatic pressure on the process of plastic deformation is negligible. Because of this, the criterion of simple loading can be stated in a somewhat weakened form: *with simple loading the components of the stress deviatoric change in proportion to an increasing parameter t*

$$s_{ij} = s_{ij}^0 t ,$$

where $s_{ij}^0$ is a constant deviatoric.

Then the principal axes of the stress deviatoric and the coefficient of Lode and Nadai $\mu_\sigma$ (the shape of the stress tensor) do not vary; the mean pressure $\sigma$ can change in an arbitrary way.

## §8. Yield criteria. Yield surface and yield curve

The strain curves given above were associated with a uniaxial stress state. It is important to know the behaviour of the material *for a complex stress state.* In particular it is necessary to have an idea of what conditions characterize the change of the material from an elastic state to a yield state (plateau AB, fig. 9). In the yield state $\sigma_1 = \text{const.} = \sigma_s$ for simple tension and $\tau = \text{const.} = \tau_s$ for pure shear.

Here the question arises of a possible form of the condition which characterizes the transition to the elastic limit with a complex stress state. This condition, satisfied in the yield state, is called the *yield criterion* (or plasticity condition). For an isotropic medium this condition must be a symmetric function of the principal stresses

$$f(\sigma_1, \sigma_2, \sigma_3) = \text{const.} = K ,$$

where $K$ is a constant of the material and is connected with the yield limit. Because of the basic symmetry of the functions, a stress component can be replaced by its invariants, and the last condition can also be written in the form

$$f[\sigma, I_2(T_\sigma), I_3(T_\sigma)] = K .$$

It has been observed above that in the majority of problems the influence of mean pressure on the process of change of shape is negligible; then the yield condition takes the form

$$f[I_2(D_\sigma), I_3(D_\sigma)] = K\,, \tag{8.1}$$

i.e. in fact it depends only on the *difference* of the principal stresses. Note that the yield criterion is often written in the shorter form

$$f[I_2(D_\sigma), I_3(D_\sigma)] = 0\,,$$

in which the presence of the "yield limit" parameter $K$ is implicit.

If we take advantage of the geometrical interpretation of the stress state developed above, then equation (8.1) will be the equation of a cylinder, the axis of which is the straight line $\sigma_1 = \sigma_2 = \sigma_3$, perpendicular to the deviatoric plane, since the mean pressure does not enter into (8.1). It is sufficient to examine the trace of this cylinder on the deviatoric plane. This will be a curve C, symmetrical relative to the axes 1′, 2′, 3′, and called the *yield curve* (fig. 13).

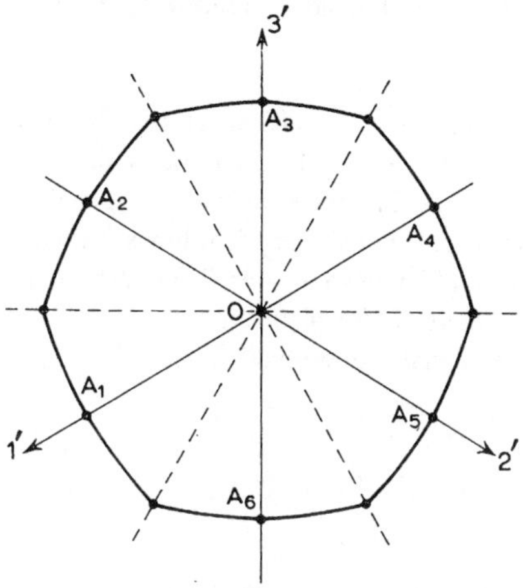


Fig. 13.

The yield curve C possesses the following properties:

(1) The curve C does not pass through the origin of coordinates O, since the yield condition is attained with considerable shear stresses.

(2) Let us assume that the properties of the material are identical under compression and tension. Then the curve C must be symmetrical with respect to straight lines perpendicular to the axes 1′, 2′, 3′, since a yield state also occurs on changing the sign of the stress.

(3) The yield curve must be *convex*, i.e. it must lie on one side of a tangent (or a reference line if C consists of straight sections).

This restriction follows from the condition of non-negative work of plastic deformation (Drucker's postulate, see §18).

Since the principal directions for an isotropic body are equivalent and the yield limits on compression and tension are equal, the yield curve must pass through six points $A_1, A_2, \ldots, A_6$ on the axes $1', 2', 3'$ equidistant from the origin (fig. 13).

In consequence of the above the yield curve consists of twelve equal arcs. Thus, it is sufficient in experimental studies of the yield criteria to trace the behaviour of the material on one of these arcs.

Some generalizations concerning plasticity conditions will be examined in §16.

## §9. Condition of constant maximum tangential stress (Tresca-Saint Venant criterion)

From his experiments on the efflux of metals through an orifice, the French engineer Tresca inferred the hypothesis that in the yield state the maximum tangential stress has the same value at all points of the medium for a given material. This value equals $\frac{1}{2}\sigma_s$, as follows from an examination of the case of simple tension. Somewhat later Saint Venant gave a mathematical formulation of this condition for plane strain.

In the three-dimensional case we have

$$
\begin{aligned}
2|\tau_1| &= |\sigma_2-\sigma_3| \leqslant \sigma_s\,,\\
2|\tau_2| &= |\sigma_3-\sigma_1| \leqslant \sigma_s\,,\\
2|\tau_3| &= |\sigma_1-\sigma_2| \leqslant \sigma_s\,,
\end{aligned}
\tag{9.1}
$$

where here (and in the next section) the conditions $\sigma_1 \geqslant \sigma_2 \geqslant \sigma_3$ can remain unfulfilled (otherwise we should always have $2\tau_{max} = \sigma_1 - \sigma_3$).

In the *elastic state* all the conditions (9.1) are satisfied with the inequality signs.

In the *yield state* the equality signs must hold in one or two of these conditions. Since $\sigma_s > 0$, we cannot have all three principal tangential stresses simultaneously equal to the constant $\sigma_s$ (because it is impossible for the sum of an odd number of terms of equal modulus to be equal to zero, and we have $\tau_1 + \tau_2 + \tau_3 = 0$).

From (9.1) we obtain the following relation between the yield limit $\sigma_s$ under tension and the yield limit $\tau_s$ under pure shear (recall that in this case $\sigma_1 = \tau, \sigma_2 = 0, \sigma_3 = -\tau$, i.e. $\tau_{max} = \tau$):

$$\sigma_s = 2\tau_s . \tag{9.2}$$

The conditions (9.1) define a right hexahedral prism with axis $\sigma_1 = \sigma_2 = \sigma_3$ perpendicular to the deviatoric plane. (It is easy to see, for example, that the equation $\sigma_2 - \sigma_3 = \pm \sigma_s$ represents a pair of parallel planes passing through the axis $\sigma_1$ and the line $\sigma_1 = \sigma_2 = \sigma_3$.) The trace of the prism on the deviatoric plane is a right hexagon (fig. 14). The impossibility of satisfying simultaneously all three equality signs in (9.1) is geometrically obvious.

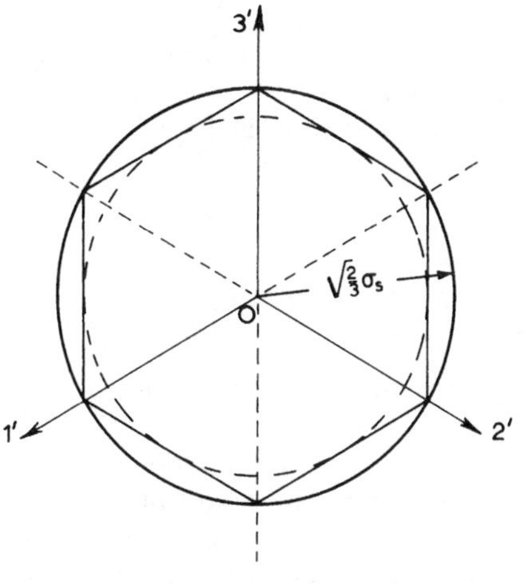


Fig. 14.

The planes concerned intercept segments of length $\sigma_s$ on the axes $\sigma_1, \sigma_2, \sigma_3$; since $\cos(\sigma_3, 3') = \sqrt{\frac{2}{3}}$, it is easy to see that the radius of the circle circumscribing the hexagon is $\sqrt{\frac{2}{3}}\,\sigma_s$.

We note one further fact: the maximum tangential stress equals half the difference of the greatest and least principal stresses; *the intermediate principal stress has no influence on the yield state.*

On the whole the Tresca-Saint Venant criterion satisfactorily characterizes the yield state of a material and agrees with the observations of Lueders. More thorough experimental research reveals a small systematic deviation from the Tresca-Saint Venant criterion in the behaviour of malleable metals in yield state. In particular, experiments suggest some influence of the intermediate principal stress on the yield state.

## §10. Condition of constant intensity of tangential stresses (von Mises criterion)

The use of the Tresca-Saint Venant criterion, expressed by inequalities, is associated with certain mathematical difficulties in three-dimensional problems. This situation gave von Mises [1]) the idea of replacing the hexahedral prism with the circumscribed circular cylinder:

$$(\sigma_1-\sigma_2)^2+(\sigma_2-\sigma_3)^2+(\sigma_3-\sigma_1)^2=2\sigma_s^2\,. \tag{10.1}$$

The intersection of this cylinder with the deviatoric plane is the circle circumscribing the hexagon (fig. 14).

Von Mises' criterion may be written in the form

$$T=\frac{\sigma_s}{\sqrt{3}}. \tag{10.2}$$

In the case of pure shear $T=\tau$ and (10.2) gives

$$\tau_s=\frac{\sigma_s}{\sqrt{3}}=0.577\sigma_s\,. \tag{10.3}$$

Von Mises regarded Saint Venant's criterion as exact and (10.1) as approximate; but numerous experiments showed that, in the yield state for polycrystalline materials, the von Mises' criterion was satisfied generally rather better than the condition of constant maximum tangential stress. In particular, the relation (10.3) turns out to be better than (9.2), agreeing with experiments carried out with malleable metals. Thus von Mises' criterion has acquired an independent significance. At the same time it should be noted that in certain cases the Tresca-Saint Venant criterion is in accord with experimental results. Therefore the von Mises and Tresca-Saint Venant criteria may be regarded as being equally valid formulations of the yield condition.

Note that the left-hand side of equation (10.1) corresponds precisely, up to a constant multiplicative factor, with the energy of elastic change of shape. Thus the yield state is *attained* at some constant energy of elastic change of shape.

[1]) It was later found that Huber had proposed a criterion similar to (10.1) as early as 1904.

Earlier (§1) it was noted that the quantities $T$ and $\tau_{max}$ are close in value. From this it follows that the yield conditions of Tresca-Saint Venant and von Mises are not significantly different. The difference can be further reduced if we take the circle which lies midway between the circumscribed and inscribed circles (fig. 14). This corresponds to the approximate formula $T \approx 1.08\,\tau_{max}$, considered in §1.

## §11. Conditions of hardening. Loading surface

11.1. *Loading and unloading*

Plastic deformation leads to the hardening of a metal and the increase of its elastic limit (in the direction of the deformation). In simple tension (fig. 15a) the elastic limit is $\sigma_{1M}$ when the state M is attained; the range of values $0$–$\sigma_{1M}$ can be called elastic. If stress changes are within these limits, only elastic deformation will occur. With further loading at the point M plastic deformation will take place. Thus the stress is like a current elastic limit, which depends on previous plastic deformation and which distinguishes between loading (accompanied by further plastic deformation) and unloading (accompanied by pure elasticity).

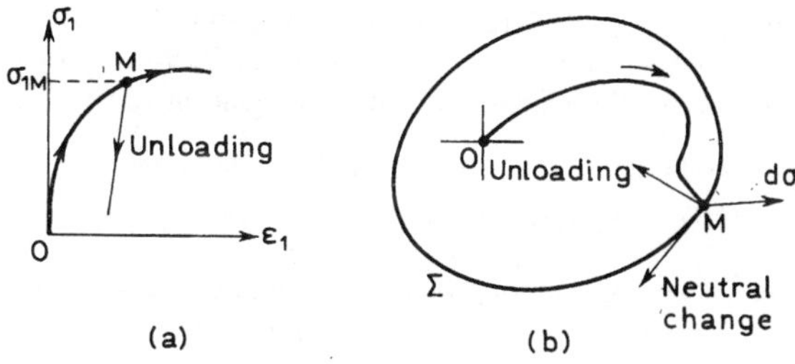


Fig. 15.

In a complex stress state it is considerably more difficult to delineate these concepts; for example, the same value of the intensities $T$ and $\Gamma$ may now correspond to a range of stress and strain states.

In connection with this point the following question arises. Let a body be in a plastic state, characterized at the instant under consideration by stresses $\sigma_{ij}$. If we give the latter infinitesimal increments $d\sigma_{ij}$ (*additional loading*), will this additional load cause further plastic deformation?

The complexity of the physical processes that occur in plastic deformation and the inadequacy of experimental results do not allow a complete answer to this question. Nevertheless it is possible to indicate the required criterion under rather wide loading conditions.

11.2. *Loading surface*

Turning to the complex stress state, we introduce the concept of a *loading surface* $\Sigma$ (sometimes called a flow surface). This is a surface in the space of the stresses $\sigma_{ij}$ which, for a given state of the medium, separates the regions of elastic and plastic deformation (fig. 15b). The origin of coordinates 0 corresponds to zero stresses. The additional loading $d\sigma_{ij}$ produces either elastic strain (unloading if the vector $d\sigma_{ij}$ is directed inwards from $\Sigma$) or plastic strain (loading if the vector $d\sigma_{ij}$ is directed outwards from $\Sigma$). An increment $d\sigma_{ij}$ which lies in the tangent plane to the loading surface (*neutral changes* [1])) leads to elastic strain only (*continuity condition*, cf. § 17).

The loading surface is not fixed (as in the case of ideal plasticity), but expands and is displaced as hardening develops. The shape and position of the loading surface $\Sigma$ depend, in general, not only on the current stress state, but also on the whole preceding history of deformation. The loading surface is convex (cf. § 18).

We shall restrict ourselves here to the simplest variety of loading surface. The construction of more general loading surfaces, taking into account the development of strain anisotropy, will be considered in § 17.

Let the loading surface $\Sigma$ experience uniform (*isotropic*) expansion with plastic deformation of the material; then its equation can be written in the form

$$f[I_2(D_\sigma), I_3(D_\sigma)] = F(q)\,, \tag{11.1}$$

where $F$ is an increasing function of some parameter $q$ characterizing the preceding plastic deformation. The yield condition (8.1) follows from (11.1) on setting $F(q) = \text{const.} = K$.

11.3. *Unloading*

With unloading the deformation of an element of the medium continues because of the accumulation of elastic potential energy. This, of course, can only be seen from experimental results.

[1]) An example of neutral loading is the following: a rod elongated by a stress $\sigma_z$ is additionally loaded with a small torsion. Then $\sigma_z \neq 0$, $d\sigma_z = 0$, $\tau = 0$, $d\tau \neq 0$, and obviously $dT \sim \sigma_z d\sigma_z + \tau d\tau = 0$, cf. § 12.

On the basis of the latter, one can assume (since the gradient of the branch AC, fig. 8, is approximately the same as that of the elastic section) that the components of elastic strain do not depend on the plastic deformation. This allows the hypothesis that the components of the total strain $\epsilon_{ij}$ (on condition that it is small) are compounded of an elastic part $\epsilon^{\mathrm{e}}_{ij}$ and a plastic part $\epsilon^{\mathrm{p}}_{ij}$:

$$\epsilon_{ij} = \epsilon^{\mathrm{e}}_{ij} + \epsilon^{\mathrm{p}}_{ij} \,. \tag{11.2}$$

The elastic strain components are related to the stress components by the generalized Hooke's law; the values of the elastic constants can be regarded as invariant. With unloading, changes occur only in the elastic strain components, i.e.

$$\epsilon^{\mathrm{e}}_{ij} = \frac{1}{2G}\left(\sigma_{ij} - \frac{3\nu}{1+\nu}\,\sigma\delta_{ij}\right). \tag{11.3}$$

The components of total strain in unloading are determined [1]) by equation (11.2); here the components $\epsilon^{\mathrm{p}}_{ij}$ do not change, and are equal to the respective plastic strains attained at the initial instant of unloading. The components $\epsilon^{\mathrm{e}}_{ij}$ are found from equation (11.3), where the $\sigma_{ij}$ are the stresses at the *end of unloading*.

## §12. Conditions of isotropic hardening

### 12.1. *A simple version of the isotropic hardening condition*

A simpler formulation of the isotropic hardening condition (11.1) contains only the quadratic invariant of the stress deviatoric. In this case equation (11.1) can be written in the form

$$T = f(q) \,. \tag{12.1}$$

The loading surface is now the surface of a circular cylinder whose axis coincides with the hydrostatic axis (§1). With plastic deformation the radius of the cylinder increases. Depending on the choice of the hardening parameter $q$, various conditions of hardening are obtained.

We note that, just as for the yield condition, it is possible in the hardening conditions to cross to a neighbouring value – to the maximum tangential stress $\tau_{\max}$ (then $I_3(D_\sigma)$ is also affected).

[1]) If the deformations are not small equation (11.2) has to be written in incremental form, cf. (13.2).

12.2. *The "single curve" hypothesis*

If the value attained by the shear strain intensity $\Gamma$ is taken as a measure of the hardening, then we obtain a relation of the form

$$T = g(\Gamma)\,\Gamma\,, \tag{12.2}$$

where $g(\Gamma)$ is some positive function, characteristic for a given material. If we construct the curve (12.2) in the coordinates $T$, $\Gamma$, then for different stress states we obtain the same ("single") curve. Since the shape of the curve does not depend on the stress state it is possible to determine $g(\Gamma)$, for example, from experiments with simple-tension or pure shear (§6).

Equation (12.2) can be formally regarded as a general condition encompassing different phases of strain. Thus, putting

$$g(\Gamma) = \tau_s/\Gamma\,,$$

we obtain von Mises' yield criterion $T = \tau_s$; while putting

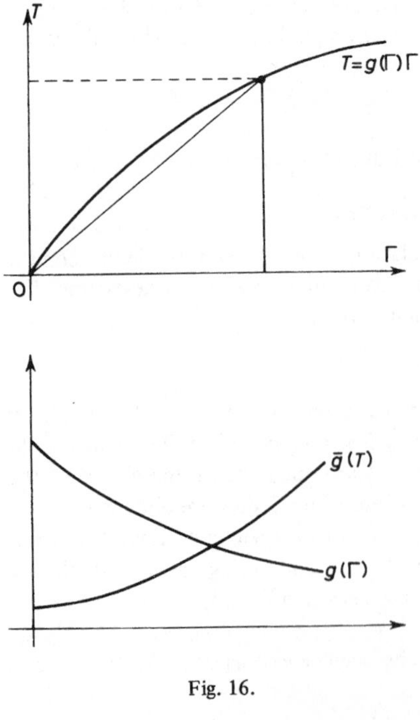


Fig. 16.

$$g(\Gamma) = G\,,$$

we obtain the case of Hooke's elastic medium, where $T = G\Gamma$.

The function $g(\Gamma)$ is sometimes called the *modulus of plasticity* (cf. §6); for real materials $\mathrm{d}T/\mathrm{d}\Gamma \geqslant 0$, where the equality sign is applicable only in the yield state. In the hardening state the strain curve becomes concave downwards (fig. 16). Because of this the slope of the tangent is smaller than the slope of the secant, i.e.

$$\frac{\mathrm{d}T}{\mathrm{d}\Gamma} - \frac{T}{\Gamma} = g'(\Gamma)\,\Gamma < 0\,.$$

Thus, $g'(\Gamma) < 0$ and $g(\Gamma)$ is a decreasing function of $\Gamma$, with $0 < g(\Gamma) \leqslant G$; an inverse function exists:

$$\Gamma = \bar{g}(T)\,T\,, \tag{12.3}$$

whereupon it is easy to see that

$$\bar{g}(T) \geqslant 1/G\,, \qquad \bar{g}'(T) > 0\,, \qquad g(\Gamma)\,\bar{g}(T) = 1\,.$$

The condition of hardening (12.2) is satisfied with sufficient accuracy in in the *simple loading* of an isotropic material.

It should be emphasized that the relation (12.2) is also frequently used when the principal axes of stress are rotated and the similarity of the stress state is disturbed. The reason lies in the fact that the experiments confirm the hardening condition (12.2) for somewhat more general loadings.

On increase of the intensity of shear strain $\Gamma$ hardening develops and increases the intensity of the tangential stresses $T$. Consequently, on loading $\mathrm{d}T > 0$, on unloading $\mathrm{d}T \leqslant 0$, while when $\mathrm{d}T = 0$ neutral changes occur.

### 12.3. *Energetic condition of hardening*

One may take as a measure of hardening $q$ the work of plastic deformation

$$A_\mathrm{p} = \int \sigma_{ij}\mathrm{d}\epsilon_{ij}^\mathrm{p}\,. \tag{12.4}$$

The condition of hardening (12.1) is then

$$T = f(A_\mathrm{p})\,. \tag{12.5}$$

The function $f$ can be determined, for example, from the tension curve; then $T = \sigma_1/\sqrt{3}$, and the work $A_\mathrm{p}$ is as a function of the relative elongation $\epsilon_1$. The condition of hardening (12.5) may be also written in the form

$$A_\mathrm{p} = \Phi(T)\,,$$

where the function $\Phi(T)$ is characteristic for a given material, and independent of the nature of the stress state.

Since the work of plastic deformation is positive, it follows that $\Phi(T) > 0$. With developing plastic deformation the work $A_p$ increases, and the loading surface expands, i.e. the intensity $T$ increases. Consequently, $\Phi'(T) > 0$. Under loading $dA_p = \Phi'(T)\, dT > 0$ and

$$dT > 0\,. \tag{12.6}$$

When $dT < 0$ a body unloads according to the elastic law. When $dT = 0$ the increment of work of plastic deformation becomes zero. The neutral changes $dT = 0$ lead to elastic deformation.

The energetic condition of hardening is more general than the preceding condition (12.2), and conforms with experiments for a wider class of loading. It is necessary to remember however that condition (12.5) does not take into account the development of strain anisotropy and may be used only for relatively uncomplicated loading paths (without abrupt zigzags and in the absence of significant changes of direction of the loading path). It is also necessary to bear in mind that significant displacements with respect to the loading surface are accompanied by some plastic deformations.

12.4. *Odquist's condition*

As a measure of the hardening $q$ it is possible to take the parameter

$$q = \int \mathrm{d}\Gamma_p = \int \sqrt{2\mathrm{d}\epsilon_{ij}^p \mathrm{d}\epsilon_{ij}^p}\,, \tag{12.7}$$

characterizing the accumulated plastic deformation.

## §13. The theory of plastic flow

13.1. *General relations*

The process of plastic deformation is irreversible, the greater part of the work of deformation being transformed into heat. The stress in the final state depends on the path of deformation. In connection with this, the equations describing plastic deformation cannot in principle be finite relations connecting the components of stress and strain (as are the Hooke's law relations), but must be differential (and, moreover, non-integrable) relations.

The equations of the theory of plastic flow establish a connection between infinitesimal increments of strain and stress, the stresses themselves and certain parameters of the plastic state.

Let us consider the starting points of this theory:

(1) *The body is isotropic.*

(2) *The relative volumetric change is small and is an elastic deformation proportional to the mean pressure:*

$$\epsilon = 3k\sigma ,$$

or

$$d\epsilon = 3k d\sigma . \tag{13.1}$$

(3) *The total increments in the strain components* $d\epsilon_{ij}$ *are compounded of the increments in the components of elastic strain* $d\epsilon_{ij}^{e}$ *and the components of plastic strain* $d\epsilon_{ij}^{p}$:

$$d\epsilon_{ij} = d\epsilon_{ij}^{e} + d\epsilon_{ij}^{p} . \tag{13.2}$$

The increment in elastic strain components is connected with the increment in elastic stress according to Hooke's law:

$$d\epsilon_{ij}^{e} = \frac{1}{2G}\left(d\sigma_{ij} - \frac{3\nu}{1+\nu}\,\delta_{ij}d\sigma\right) . \tag{13.3}$$

(4) *The stress deviatoric* $D_\sigma$ *and the deviatoric of the plastic strain increments* $D_{d\epsilon}^{p}$ *are proportional,* i.e.

$$D_{d\epsilon}^{p} = d\lambda\cdot D_\sigma , \tag{13.4}$$

where $d\lambda$ is some infinitesimal scalar multiplier. This formula generalizes experimental results on complex loading, in which the directions of the principal axes and the relations between the principal stresses have changed. According to experiments the increments to the plastic strain components ("plastic strain-rates") are proportional to the stresses at a given moment of time. In other words, the stress distribution determines the instantaneous increments to the plastic strain components.

From (13.4) we have the relations

$$d\epsilon_{ij}^{p} = d\lambda\cdot s_{ij} \tag{13.5}$$

(since $d\epsilon^{p} = 0$). Calculating now the increment in the work of plastic deformation, we find

$$dA_p = \sigma_{ij}d\epsilon_{ij}^{p} = d\lambda\cdot\sigma_{ij}s_{ij} = 2d\lambda\cdot T^2 . \tag{13.6}$$

Thus the multiplier $d\lambda$ is related to the magnitude of the increment in the work of plastic deformation; since $dA_p \geqslant 0$ so also $d\lambda \geqslant 0$. From (13.2) we obtain the total increments in the strain components:

$$d\epsilon_{ij} = d\epsilon_{ij}^{e} + d\lambda \cdot s_{ij}\,, \tag{13.7}$$

where the increments in the elastic strain components must be taken in accordance with Hooke's law (13.3).

It is easy to see, moreover, that the increment to the work of deformation is

$$dA = dA_e + dA_p\,, \tag{13.8}$$

where $dA_p$ is given by (13.6), and the increment in the work of elastic deformation is given by $dA_e = d\Pi$, with elastic potential

$$\Pi = \frac{3}{2} k\sigma^2 + \frac{1}{2G} T^2\,. \tag{13.9}$$

When $d\lambda = 0$ equations (13.7) reduce to Hooke's law in differential form. In the general case equations (13.7) are incomplete since they contain an undetermined multiplier, whose evaluation requires an additional relation.

### 13.2. *The yield state, equations of Prandtl-Reuss*

We take as our additional relation von Mises' yield condition

$$T = \tau_s\,.$$

Then

$$d\lambda = \frac{dA_p}{2\tau_s^2}\,, \tag{13.10}$$

i.e. the quantity $d\lambda$ is proportional to the increment in the work of plastic deformation; since the latter is given by the expression $\sigma_{ij} d\epsilon_{ij}^{p}$ we do not have a unique dependence of the increments in the strain components on the stress components and their increments in the yield state under consideration [1]).

If von Mises' criterion is satisfied, $dT = 0$ and plastic deformation occurs. If $dT < 0$, the medium leaves the yield state and unloading begins, proceeding in accordance with Hooke's law. Equations (13.7), subject to the von Mises' yield criterion, were proposed by Reuss [39] in 1930; for plane problems these equations were introduced by Prandtl in 1924.

### 13.3. *The Saint Venant-von Mises theory of plasticity*

If in the Prandtl-Reuss equations the elastic strain components are ignored

[1]) This property may be regarded as the definition of a perfectly plastic body; the yield condition is then a consequence, see [50].

(which is permissible during the development of plastic deformation), we obtain the equations of the *Saint Venant-von Mises theory of plasticity*

$$\mathrm{d}\epsilon_{ij} = \mathrm{d}\lambda \cdot s_{ij} ,$$

usually written, on division by $\mathrm{d}t$, in the form

$$\xi_{ij} = \lambda' s_{ij} , \tag{13.11}$$

where the quantity

$$\lambda' = \frac{1}{2\tau_s^2} \frac{\mathrm{d}A_\mathrm{p}}{\mathrm{d}t} = \frac{1}{2\tau_s^2} \sigma_{ij}\xi_{ij} = \frac{1}{2\tau_s^2} s_{ij}\xi_{ij}$$

is proportional to the rate of work of plastic deformation, i.e. it characterizes the dissipation. Eliminating the stress components from the latter equation with the aid of (13.11), we easily find

$$\lambda' = H/2\tau_s .$$

Consequently equation (13.11) can also be written as

$$\xi_{ij}/H = s_{ij}/2\tau_s . \tag{13.12}$$

For the case of plane strain with the yield condition $\tau_{\max} = \text{const.}$, equations (13.11) were given by Saint-Venant [156] in 1871. For the general case these equations were established by M. Levy [129] and von Mises [136].

It is evident that the strain rates $\xi_{ij}$ are not uniquely determined for given stress; but if the strain rates $\xi_{ij}$ are given, the components of the stress deviatoric $s_{ij}$ are uniquely determined. It is easy to see that the components $s_{ij}$, which are determined by formulae (13.12), identically satisfy the von Mises' yield criterion. We observe also that in the yield state (i.e. when the von Mises' yield criterion is satisfied) the indeterminacy of the strain-rate components, which is connected with the indeterminacy of quantity $\lambda'$, is necessary for the fulfilment of the strain compatibility conditions.

The Saint Venant-von Mises equations have wide applications in the mathematical theory of plasticity and its various applications.

### 13.4. *The hardening state*

We take as the additional relation the condition of isotropic hardening (12.5), according to which

$$\mathrm{d}A_\mathrm{p} = \Phi'(T)\, \mathrm{d}T .$$

Introducing this value into (13.6) and defining

$$\frac{\Phi'(T)}{2T^2} = F(T) ,$$

we obtain

$$\mathrm{d}\lambda = F(T)\,\mathrm{d}T . \tag{13.13}$$

Thus

$$\mathrm{d}\epsilon_{ij} = \mathrm{d}\epsilon_{ij}^{\mathrm{e}} + F(T)\,\mathrm{d}T\cdot s_{ij} . \tag{13.14}$$

These relations are true when

$$\mathrm{d}T \geqslant 0 .$$

If $\mathrm{d}T = 0$, we have neutral changes of the stress state; in this case the increments in the strain components must be related with increments in the stress components by Hooke's law, since neutral charges proceed elastically (§12). Equations (13.4) are found to be in agreement with these conclusions.

We see that *in the case of hardening the above relations determine uniquely the dependence of the increments in the strain components on the stresses and their increments.*

In the hardening state there is no condition connecting the stress components (as in the case of perfect plasticity) and the quantity $\mathrm{d}\lambda$ is wholly determinate.

Moreover, in transition from loading to neutral changes and to unloading, increments in the strain components vary continuously. This does not happen for the equations of plastic deformation theory (see §14).

13.5. *Concluding remarks*

The Prandtl-Reuss equations (13.7) in the case of perfect plasticity, and equations (13.14) in the case of hardening, relate the stress components to infinitesimal increments in the components of stress and strain, i.e. they are not finite relations (in contrast with the equations of deformation theory). Relations (13.7) and (13.14), are in general not integrable, i.e. in other words they do not reduce to finite relations between the components of stress and strain. This mathematical fact reflects the dependence of the results on the history of the deformation. If, for example, in the stress space we cross from some initial point O (fig. 17), characterized by zero stresses, to a point $O_1$ (with stresses $\sigma_{ij}^{(1)}$) by two paths I and II, then according to the equations of plastic flow theory the strain components at the point O will be different.

Equations (13.7) and (13.14) do not contain the time; if, however, they are divided by $\mathrm{d}t$, it is possible formally to change from increments $\mathrm{d}\epsilon_{ij}$ to

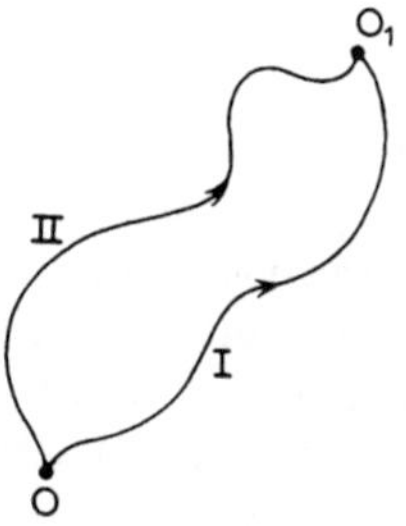


Fig. 17.

strain rates $\xi_{ij}$. Then the equations will superficially resemble the equations of viscous fluid flow. To some extent this analogy justifies the expression "theory of plastic flow". It should be emphasized that the variable $t$ can here mean either the time or a monotonically-increasing loading parameter, or even some other monotonically-increasing quantity (for example a characteristic dimension of the plastic zone). The transformation to "strain-rates" is often convenient, since it permits the use of the descriptive terminology of hydrodynamics. The equations of plastic flow theory are in principle different from the equations of viscous flow. In contrast with the latter, it is always possible to discard d$t$ and return to formulae (13.7) and (13.14), which do not contain the time.

In the sequel we shall usually speak about *flow theory* (in place of the theory of plastic flow). This expression is obviously not entirely suitable but it is brief and is used extensively here and abroad.

In the case of hardening it is possible to calculate the strain for given loading paths, i.e. for given $\sigma_{ij} = \sigma_{ij}(t)$, where $t$ is some parameter (for example, time); in principle it is also possible to find the stresses if the path of deformation is given, i.e. $\epsilon_{ij} = \epsilon_{ij}(t)$.

The Saint Venant-von Mises' equations of plasticity theory have a significantly greater simplicity of structure and represent finite relations between the stress components and the strain-rates. It should be emphasized also that in these equations time is not an essential feature and may be eliminated (by cancelling out) or replaced by some monotonically changing parameter.

## §14. Deformation theory of plasticity

### 14.1. *General relations*

Consider the slow tension of a rod (fig. 18a). Loading occurs along the segment OAB and unloading corresponds to the line BC. The area OABC represents the work lost in deformation. Experiments show that a large part of this work is transformed into heat, but, in the absence of heat transfer, there is very little change in temperature (about 2°C for a strain $\epsilon_1 = 4\%$) of the test specimen. Therefore, with monotonic increase in the external load, the curve OAB remains unchanged irrespective of whether the work of deformation has gone into heat or into elastic potential energy of the rod. On the other hand, in unloading, when deformation of the medium occurs owing to the accumulation in it of elastic energy, dissipation of energy becomes important, and the larger it is the more the unloading line BC diverges from the loading line OAB. Thus, the equation $\sigma_1 = f(\epsilon_1)$ for the loading branch can represent both plastic and non-linear elastic deformation of a rod. From these observations we can attempt to construct equations for plastic deformation in the form of finite relations between stress and strain. Such equations would be substantially simpler than the equations of plastic flow theory. Pursuing this idea, we shall consider the equations of plastic deformation as some generalization of Hooke's law. We shall base ourselves on the following propositions:

(1) *The body is isotropic.*

(2) *The relative volumetric change is an elastic deformation, proportional to the mean pressure:*

$$\epsilon = 3k\sigma \, . \tag{14.1}$$

This assumption, as noted earlier, conforms well with experiment.

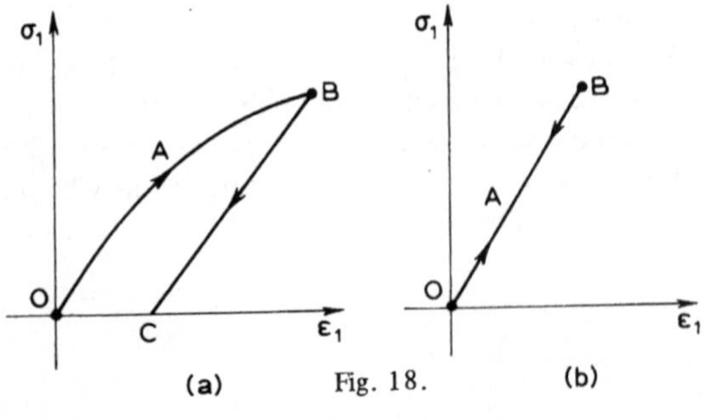


Fig. 18.

(3) *The stress and strain deviatorics are proportional*:

$$D_\epsilon = \psi D_\sigma , \tag{14.2}$$

Thus the elements of the strain deviatoric equal the corresponding elements of the stress deviatoric multiplied by a scalar $\psi$; the latter is some as yet undetermined function of the invariants of the stress and strain tensors. It is evident that the stress and strain deviatorics are coaxial (i.e. they have the same principal directions), and their principal values are respectively proportional, viz.

$$e_i = \psi s_i \qquad (i = 1, 2, 3) . \tag{14.3}$$

Hence it follows immediately that the principal shears are proportional to the principal tangential stresses, or in other words that the Mohr diagrams for the stress and strain states are similar, i.e.

$$\mu_\epsilon = \mu_\sigma .$$

Putting $\psi$ = const. = $1/2G$, we arrive at Hooke's law (5.2). Thus equation (14.2) represents a natural and simple generalization of this law.

If we put $\psi = 1/2G + \phi$, we obtain the strain components in the form of a sum of elastic strain components $\epsilon_{ij}^{e}$ and plastic strain components

$$\epsilon_{ij}^{p} = e_{ij}^{p} = \varphi s_{ij} . \tag{14.4}$$

The third proposition above must be treated as a known idealization of experimental data. We shall write (14.2) in components

$$e_{ij} = \psi s_{ij} . \tag{14.5}$$

Eliminating the volumetric dilatation from this with the aid of (14.5), we easily derive Hencky's relations [96]:

$$\epsilon_{ij} = k\sigma\delta_{ij} + \psi s_{ij} . \tag{14.6}$$

The relations (14.5) can easily be solved with respect to the stresses:

$$\sigma_{ij} = \frac{\epsilon}{3k}\,\delta_{ij} + \frac{1}{\psi}\,e_{ij} . \tag{14.7}$$

Calculating the intensity of shear strain with the aid of (14.5), we obtain the important relation

$$\Gamma = 2\psi T . \tag{14.8}$$

We now calculate, with the aid of (14.6), the increment to the work of deformation:

$$\mathrm{d}A = \sigma_{ij}\mathrm{d}\epsilon_{ij} = \mathrm{d}\,(U + \psi T^2) + T^2\mathrm{d}\psi , \tag{14.9}$$

where $U$ denotes the elastic energy of volumetric compression:

$$U = \tfrac{3}{2}k\sigma^2 = \epsilon^2/6k \, . \tag{14.10}$$

Eliminating the function $\psi$ from (14.9), we find

$$\mathrm{d}A = \sigma \mathrm{d}\epsilon + T\mathrm{d}\Gamma \, . \tag{14.11}$$

Here the first term is the increment in the *elastic energy of volumetric compression*, and the second is the increment in the *work of shape deformation.*

The above equations are incomplete since they contain the unknown function $\psi$; to determine the latter a supplementary relation of the form

$$\psi = \psi(T, \Gamma, \mu_\sigma) \tag{14.12}$$

is required. The invariants $\sigma$, $\epsilon$ are not present here, since it is possible to disregard the influence of the mean pressure on the process of change of shape; notice that, generally speaking, the relation (14.12) can contain even more complicated variables, for example the work of plastic deformation $A_\mathrm{p}$.

### 14.2. *Linear elastic state (Hooke's law)*

Let

$$\psi = \text{const.} = 1/2G \, .$$

In this case we arrive at Hooke's law

$$\epsilon_{ij} = \frac{1}{2G}\left(\sigma_{ij} - \frac{3\nu}{1+\nu}\,\sigma\delta_{ij}\right) . \tag{14.13}$$

Solving these relations for the stress components, we obtain another form of the law:

$$\sigma_{ij} = \lambda\epsilon\delta_{ij} + 2\mu\epsilon_{ij} \, , \tag{14.14}$$

where $\lambda$ and $\mu = G$ are the Lamé elastic constants.

The tangential stress intensity is here proportional to the intensity of shear strain

$$T = G\Gamma \, . \tag{14.15}$$

The increment in the work of deformation is the total differential of the elastic potential

$$\Pi = \Pi(\epsilon_{ij}) = \epsilon^2/6k + \tfrac{1}{2}G\,\Gamma^2 \, . \tag{14.16}$$

Comparing the formula for $\mathrm{d}\Pi$ with formula (14.9), we conclude that there is a relation

$$\sigma_{ij} = \partial \Pi / \partial \epsilon_{ij} , \tag{14.17}$$

expressing in essence Hooke's law in the form (14.14).

14.3. *Yield state*

Let us take as our supplementary relation the von Mises' yield criterion

$$T = \tau_s .$$

In accordance with (14.8), we have in this case

$$\psi = \Gamma / 2\tau_s , \tag{14.18}$$

i.e. in the yield state the function $\psi$ is a measure of the intensity of shear. Here too there is a *potential of the work of deformation*

$$\Pi = \epsilon^2 / 6k + \tau_s \Gamma , \tag{14.19}$$

equal to the sum of the energy of elastic volumetric compression and the work of change of shape $\tau_s \Gamma$.

The quantity $\psi = \Gamma/2\tau_s$ may be introduced into Hencky's relation (14.6); but we do not achieve a unique determination of strain components from the stress components, natural enough if we recall that on the yield plateau (fig. 9) there is no single-valued relation between stress and strain.

Further, from (14.7) we find

$$\sigma_{ij} = \frac{\epsilon}{3k} \delta_{ij} + \frac{2\tau_s}{\Gamma} e_{ij} . \tag{14.20}$$

Notice that the stresses described by these formulae are *single-valued functions of the strain components and identically satisfy the von Mises' yield criterion.*

In the case of an incompressible medium ($k = 0$), the stresses are determined by the strains, apart from the hydrostatic pressure:

$$\sigma_{ij} - \sigma \delta_{ij} = \frac{2\tau_s}{\Gamma} \epsilon_{ij} . \tag{14.21}$$

Formulae (14.17), valid also for the yield state, lead here to the relations (14.20).

14.4. *Hardening state*

We take as supplementary relation the hardening condition (12.2)

$$\Gamma = \bar{g}(T)\, T .$$

By virtue of (14.8) we obtain

$$\psi = \tfrac{1}{2}\,\bar{g}(T)\,. \tag{14.22}$$

In agreement with (14.6) we have

$$\epsilon_{ij} = k\sigma\delta_{ij} + \tfrac{1}{2}\bar{g}(T)\, s_{ij}\,. \tag{14.23}$$

Hence it is easy to find the inverse relations

$$\sigma_{ij} = \frac{\epsilon}{3k}\,\delta_{ij} + 2g(\Gamma)\, e_{ij}\,. \tag{14.24}$$

The relations obtained for $\mathrm{d}T/\mathrm{d}\Gamma \neq 0$ determine the *mutually single-valued dependence between the stress and strain components.*

In the hardening state the increment to the work of deformation (14.9) is, owing to (14.22), the total differential of some function $\Pi = \Pi(\epsilon_{ij})$ – the potential of the work of deformation. It is easy to see that

$$\Pi = \epsilon^2/6k + \int g(\Gamma)\,\Gamma\,\mathrm{d}\Gamma\,. \tag{14.25}$$

The second term characterizes the work of the change of shape of an element of the body.

The formulae (14.17) obviously apply to the hardening state under consideration.

Notice that when $g(\Gamma) = G$ we obtain Hooke's elastic medium, and when $g(\Gamma) = \tau_s/\Gamma$ we arrive at the yield state.

In § 12 we also considered another hardening condition (12.5). It is easy to see that we would have arrived at the same result by using the second hardening condition in our model. Indeed the work of plastic deformation equals

$$A_\mathrm{p} = \int T\,\mathrm{d}\Gamma - T^2/2G\,.$$

By virtue of (12.3) $\Gamma$ is a single-valued function of the intensity $T$, and hence $A_\mathrm{p}$ depends only on the intensity of the tangential stresses; this agrees with the condition (12.5).

### 14.5. *Discussion of the equations*

The above equations of plastic deformation theory were formulated by Hencky [96] in 1924 for the yield state; some time later they were generalized to the case of hardening.

The equations of this theory are non-linear, but owing to their relative simplicity they have found wide application, in spite of certain basic deficiencies.

The equations of deformation theory fully describe plastic deformation under simple loading (§ 12), when the components of the stress deviatoric increase in proportion to a single parameter; they are also applicable in cases when there is some deviation from simple loading.

Under conditions of complex loading in the hardening state, it is possible to have deformations in which the value of $T$ (or $\Gamma$) is conserved, while the components of the stress (or strain) tensor change. In so far as the single curve (12.2) required the assumption that the hardening be the same "for all directions", in so far $dT = 0$ is necessary to assume that all changes are elastic.

In connection with this, it is possible to raise various objections to plastic deformation theory.

Let us consider, for example, two loading paths up to some state $T_\sigma^{(2)}$, characterized by a value of the intensity $T_0$; one path consists of loading to the state $T_\sigma^{(1)}$ with the same intensity $T_0$, and a subsequent transition to $T_\sigma^{(2)}$ at constant intensity $T_0$; then at the end of the path we obtain plastic strains corresponding to $T_\sigma^{(1)}$. The other path is initially the same as the first, but it does not quite reach the state $T_\sigma^{(1)}$; it deviates and goes to the state $T_\sigma^{(2)}$ at intensity $T$, which gradually increases and approaches $T_0$. In as far as this path may be as close as desired to the first, it is natural to expect that the plastic strains in state $T_\sigma^{(2)}$ will be as previously obtained. However according to deformation theory equations, we obtain different values of the plastic strain corresponding to $T_\sigma^{(2)}$, since loading is proceeding all the time.

The question arises: what do the equations of plastic deformation theory represent?

Later it will be shown that *these equations are the equations for a non-linear elastic body*. It is natural that the use of these equations to describe plastic deformations under complicated zig-zag loading paths can lead to unsatisfactory results.

It is reasonable to assume that the equations of deformation theory are satisfactory for plastic deformations which develop in some definite direction. We shall return to this question later (§ 15).

We now demonstrate that the equations of deformation theory are the equations of a non-linear elastic body.

We return to the example of extension of a rod (fig. 18) by a slowly varying force. Let a fixed value of the stress $\sigma_1$ correspond to some strain $\epsilon_1$, invariant with time, and conversely a fixed value $\epsilon_1$ correspond to the time-independent stress $\sigma_1$. Every such state of the rod will be an *equilibrium state*. A process of deformation consisting of a sequence of equilibrium states is called an *equilibrium deformation process.*

We represent a process of slow unloading as taking place along the curve BAO (fig. 18a), and passing, in reverse order, through the same states as are realized under the loading OAB. If, having arrived at the initial point O, we are unable to perceive any changes, then the process is called *reversible.* Such a process can be achieved with the help of an ideal elastic body, for example with Hooke's elastic medium (fig. 18b); in the case when the stresses are not proportional to the strains, we shall speak of a *non-linear elastic body.*

As an example of a non-reversible process one may use the elastic-plastic deformation OABC (fig. 18a); for any, even infinitesimal, decrease in stress, the strain does not increase along the curve BAO, but follows the unloading line BC. We emphasize that both reversible and irreversible processes are in our case equilibrium processes.

In an isothermal process of elastic deformation there is a unique relation between stress and strain, $\sigma_1 = f(\epsilon_1)$.

We consider now an *irreversible* equilibrium process of deformation; here the stress is no longer a function only of the instantaneous values of the strain; furthermore the process of deformation depends on the direction of motion along the deformation curve, i.e. on whether loading or unloading is taking place.

We are considering an *equilibrium* irreversible process of deformation; thus the required relation must not contain time (and this means the strain-rates also); it is sufficient to indicate the behaviour of the material in loading and unloading.

*Thus, it is possible to represent the equilibrium irreversible process of the extension of a rod at each segment of loading and unloading by the constitutive equation of some ideal non-linear elastic body.*

Thermodynamic considerations, which were developed earlier, also relate to the deformation of a body in a complex stress state; here also one may pose the question of representing an equilibrium plastic deformation by state equations of a non-linear elastic body. In connection with these points, it is necessary to ascertain what are the possible forms of state equations for a non-linear elastic body. To provide an appropriate thermodynamic analysis it is necessary to determine the properties of the medium under consideration.

We assume that the state of the body is fully defined by six independent state parameters (generalized state coordinates), for which it is possible to take either the strain components $\epsilon_{ij}$, or the stress components $\sigma_{ij}$.

We shall decide on the set of parameters $\epsilon_{ij}$ [1]) and shall retain the previous initial positions 1, 2, 3 (§ 14.1), from which the relations (14.6)–(14.9) result.

In the sequel, however, we shall not base ourselves on experimentally determined yield and hardening conditions, but shall take advantage of the condition of reversibility of the deformation process in studying our ideal elastic body. This will suffice for our thermodynamic analysis.

Consider the elementary parallelepiped $dx_1dx_2dx_3$, conceived as being cut out from the medium. Stresses $\sigma_{ij}$ act on its faces. The increment in the internal energy of the element $dU dx_1dx_2dx_3$ is made up of the increment in the work of deformation $dA\, dx_1dx_2dx_3$, and the increment in the quantity of heat $dQ dx_1dx_2dx_3$ absorbed by the element, i.e.

$$dU = dA + dQ\,,$$

where $dA$ is determined by (14.9), and the stress components, in general, depend on the state parameter $\epsilon_{ij}$.

According to the first law of thermodynamics the internal energy $U$ is completely de-

[1]) For simplicity we shall consider an isothermal process; this imposes an insignificant constraint, and one may carry out an analogous thermodynamic analysis for changing temperature [117].

termined by the instantaneous state of the system [1]; consequently $\mathrm{d}U$ must be a total differential.

The existence of internal energy $U = U(\epsilon_{ij})$, being a consequence of the conservation of energy, is the case for every process. The second law of thermodynamics allows one to distinguish the *reversible processes*; namely only for reversible processes is the ratio $\mathrm{d}Q/\Theta$, where $\Theta$ is the absolute temperature, a total differential of a function of state, the entropy $J$:

$$\mathrm{d}Q/\Theta = \mathrm{d}J\,.$$

Since for an isothermal process $\Theta$ = const., so $\mathrm{d}Q$ is a total differential. But then $\mathrm{d}A$ must also be a total differential. The formula for $\mathrm{d}A$ was obtained earlier (14.9); it now follows that a non-linear-elastic body can only have the state

$$1)\ \psi = \text{const.}\,, \qquad 2)\ T = \text{const.}\,, \qquad 3)\ \psi = \tfrac{1}{2}f(T)\,,$$

where $f(T)$ is some function.

These states correspond respectively to the linear elastic state (Hooke's law), the yield state and the hardening state considered above on the basis of experimental data. The thermodynamic analysis not only eliminates these supplementary hypotheses and leads to both yield and hardening conditions, but, what is more important, elucidates the nature of the equations of plastic deformation theory and the possibility of using the equations of a non-linear-elastic body [2] in plasticity theory. Finally the concepts developed here make apparent the existence of a potential of the work of deformation.

## §15. Connection between flow theory and deformation theory

### 15.1. *The case of simple loading*

For simple loading (§7) the components of the stress deviatoric change in proportion to an increasing parameter $t$, i.e.

$$s_i = t s_i^0 \qquad (i = 1, 2, 3)\,,$$

where $s_i^0$ are some fixed principal values of the deviatoric $D_\sigma$. In this case the plasticity theories we have been considering coincide for *conditions of small deformation* [3] [12].

Indeed the principal axes of the stress deviatoric are fixed and the ratio of

[1]) That is, independent of the path which takes the body from one state to another; otherwise a *perpetuum mobile* of the first kind would be possible, i.e. the creation of energy from nothing.

[2]) We emphasize that in the model of a non-linear-elastic body even "the yield state" is a particular elastic state. A model of a non-linear-elastic body is a spring with a non-linear characteristic. It is possible to develop a certain analogy between the "yield state" ($T$ = const.) and the potential field due to gravity force (gravity being constant).

[3]) L.I. Sedov showed [155] that simple loading with finite deformations of bodies cannot as a rule be achieved.

its principal values do not change ($\mu_\sigma$ = const.); according to the equations of flow theory (13.5) we have

$$\mathrm{d}\epsilon_i^{\mathrm{p}} = \mathrm{d}\lambda \cdot t s_i^0 . \tag{15.1}$$

Let *hardening be absent*, then from von Mises' yield criterion it follows immediately that $t$ = const., i.e. the stresses $s_i = t s_i^0$ are constant. The quantity $\mathrm{d}\lambda$ is proportional to the increment in the work of plastic deformation $\mathrm{d}A_{\mathrm{p}}$, namely $\mathrm{d}\lambda = \mathrm{d}A_{\mathrm{p}}/2\tau_{\mathrm{s}}^2$. Integrating the increments in the components of plastic strain $\mathrm{d}\epsilon_i^{\mathrm{p}}$ we obtain the components of plastic strain $\epsilon_i^{\mathrm{p}}$; integration of the work elements $\mathrm{d}A_{\mathrm{p}}$ leads to the plastic work $A_{\mathrm{p}}$. The latter is a scalar function; we designate it by $2\tau_{\mathrm{s}}^2\varphi$. Then the relation (15.1) takes the form

$$\epsilon_i^{\mathrm{p}} = \varphi s_i ,$$

but these are equations of deformation theory (if the components that pertain to the elastic part of the deformation and obey Hooke's law are subtracted). This was noted by Hohenemser and Prager [39] in 1932.

If *hardening occurs*, then according to (13.13) $\mathrm{d}\lambda = F(T_0 t)\, T_0\, \mathrm{d}t$, where $T_0$ is the intensity of the tangential stresses for the state $s_i^0$. Introducing a new variable $T_0 t = \tau$ and integrating the increments in the components of plastic strain $\mathrm{d}\epsilon_i^{\mathrm{p}}$ in the relations (15.1), we obtain on the left the components $\epsilon_i^{\mathrm{p}}$ themselves. On the right-hand side (after dividing by the factor $s_i^0/T_0$) integration leads to some function $\tau$. Returning to the original variable we obtain

$$\epsilon_i^{\mathrm{p}} = F_*(T_0 t)\, s_i ,$$

where $F_*$ is some function, i.e. the equations of deformation theory in the case of hardening.

Conversely, if equivalence of the two theories is required, we equate the increments in the plastic strain components (13.5) to the increments in the plastic strain components evaluated according to the equations of deformation theory (14.4). We then obtain

$$\mathrm{d}\lambda\, s_{ij} = s_{ij}\mathrm{d}\varphi + \varphi\, \mathrm{d}s_{ij} .$$

From this it follows that

$$\frac{\mathrm{d}\lambda - \mathrm{d}\varphi}{\varphi} = \frac{\mathrm{d}s_{ij}}{s_{ij}} .$$

On the left-hand sides of these equations we have an infinitesimal increment to some scalar quantity. Integrating we find that the stresses $s_{ij}$ have the form

$$s_{ij} = s_{ij}^0 \Psi ,$$

corresponding to simple loading ($\Psi$ is a scalar).

Thus, *the two theories coincide only in the case of simple loading*,

With complex loading, deformation theory and flow theory lead to different results. Anticipating somewhat, we can say that these results are very close in one important practical deformation situation. In strain space the deformation path is represented by some line (fig. 19); beginning at some instant, let the deformation path approach a straight line (as shown). We then say that the deformation develops in a definite direction. *If this occurs then the stress states derived from the two theories converge*. In this case the influence of the complex history of the deformation quickly fades, and a constant stress state is established which is determined by the fixed strain rates characteristic of the rectilinear segment (cf. below).

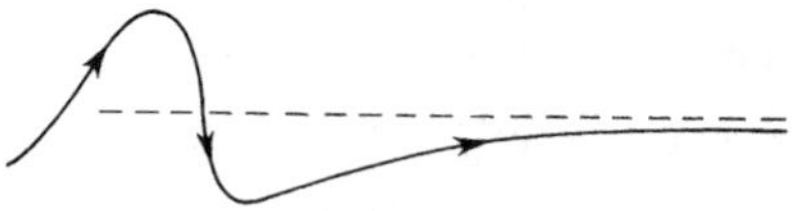

Fig. 19.

It is interesting to observe that if we begin with more general representations of the loading surface, having singularities (cf. §17), then there is a certain class of loading paths for which the equations of flow theory reduce to the equations of deformation theory (cf. work of B. Budiansky [88], V.D. Klyushnikov [125], Yu.N. Rabotnov [33]).

15.2. *Example. Combined torsion and extension of a thin-walled tube*

As an example which illustrates the properties of the plasticity equations introduced above, we consider the symmetric deformation of a circular thin-walled tube under the action of a twisting moment and an axial tension. This case corresponds to the so-called $P+M$-tests (§7).

As we have seen earlier, we can here take stress components $\sigma_z$ and $\tau_{\varphi z}$ (in cylindrical polar coordinates $r$, $\varphi$, $z$) which are different from zero. The stress components $\sigma_r$, $\sigma_\varphi$, $\tau_{r\varphi}$ and $\tau_{rz}$ are neglected, since they are small compared with $\sigma_z$ and $\tau_{\varphi z}$. The strain components $\gamma_{r\varphi}$, $\gamma_{rz}$ are small in comparison with $\gamma_{\varphi z}$. To simplify the calculations we shall suppose the material to be incompressible; this does not affect significantly the general picture of the deformation. Then from the incompressibility condition and the equations (14.23) of deformation theory it follows that $\epsilon_r = \epsilon_\varphi = -\frac{1}{2}\epsilon_z$. If we proceed from the equations of flow theory (13.7) we obtain in a similar fashion $d\epsilon_r = d\epsilon_\varphi = -\frac{1}{2}\, d\epsilon_z$. Let us restrict ourselves to the case of ideal plasticity only, and introduce the dimensionless quantities

$$q = \frac{\sigma_z}{\sigma_s}, \qquad \tau = \frac{\tau_{\varphi z}}{\tau_s}, \qquad \zeta = \frac{\epsilon_z}{\epsilon_s}, \qquad \gamma = \frac{\gamma_{\varphi z}}{\gamma_s},$$

where $E\epsilon_s = \sigma_s$, $G\gamma_s = \tau_s$.

*Solution from deformation theory.* Proceeding from (14.20) we easily obtain

$$q = \frac{\zeta}{\sqrt{\zeta^2 + \gamma^2}}, \tag{15.2}$$

where the yield condition is

$$q^2 + \tau^2 = 1 . \tag{15.3}$$

Consequently, putting

$$\begin{aligned} q &= \sin v \qquad \text{for} \qquad 0 \leqslant v \leqslant \tfrac{1}{2}\pi , \\ \tan \tfrac{1}{2} v &= w \qquad \text{for} \qquad 0 \leqslant w \leqslant 1 , \end{aligned} \tag{15.4}$$

we find

$$w = -\frac{\gamma}{\zeta} + \sqrt{\left(\frac{\gamma}{\zeta}\right)^2 + 1} , \tag{15.5}$$

where the quantities $\gamma$, $\zeta$ are supposed positive (with complex loading a change of sign leads to the Bauschinger effect, which is ignored in these theories).

*Solution from flow theory.* From equations (13.7) we obtain

$$d\zeta = dq + d\Lambda \cdot q , \qquad d\gamma = d\tau + d\Lambda \cdot \tau , \tag{15.6}$$

where, by virtue of (13.6) and the yield condition (15.3),

$$d\Lambda = \tfrac{2}{3} E\, d\lambda = q\, d\zeta + \tau\, d\gamma . \tag{15.7}$$

This should be taken together with the yield condition (15.3); we then obtain, with the aid of (15.6) and (15.7), the differential equation

$$\frac{dq}{d\zeta} = \left(\sqrt{1-q^2} - q\frac{d\gamma}{d\zeta}\right)\sqrt{1-q^2} .$$

It must be emphasized that in order to determine $q$ from this equation it is necessary to prescribe the deformation path $\gamma = \gamma(\zeta)$; this requirement does not arise if we proceed from deformation theory (cf. (15.2)).

The prescribed deformation path $\gamma = \gamma(\zeta)$ is assumed to be smooth and to satisfy the loading condition $\mathrm{d}\Lambda > 0$. On substituting from (15.4) we transform the last equation into a Riccati equation:

$$\frac{\mathrm{d}w}{\mathrm{d}\zeta} = -\tfrac{1}{2}w^2 - w\gamma'(\zeta) + \tfrac{1}{2}\,. \tag{15.8}$$

*Special cases.* For certain special functions $\gamma(\zeta)$ it is easy to construct *particular solutions* of this equation, of interest for the analysis of the equations of plastic flow theory and for the devising of experiments. We consider a few particularly simple solutions.

*Linear deformation path*. Let

$$\gamma(\zeta) = A + B\zeta\,;$$

if $\gamma = \gamma_0$ when $\zeta = \zeta_0$ and $\gamma = \gamma_1$ when $\zeta = \zeta_1$, then

$$A = \frac{\gamma_0\zeta_1 - \gamma_1\zeta_0}{\zeta_1 - \zeta_0}\,, \qquad B = \frac{\gamma_1 - \gamma_0}{\zeta_1 - \zeta_0}\,.$$

The differential equation takes the form

$$\frac{\mathrm{d}w}{\mathrm{d}\zeta} = -\tfrac{1}{2}(w^2 + 2Bw - 1)\,.$$

This equation is separable in its variables; its solution, satisfying the condition

$$w = w_0 \qquad \text{when} \qquad \zeta = \zeta_0\,,$$

has the form

$$w = \frac{Zw_2 \mp w_1}{Z \mp 1}\,. \tag{15.9}$$

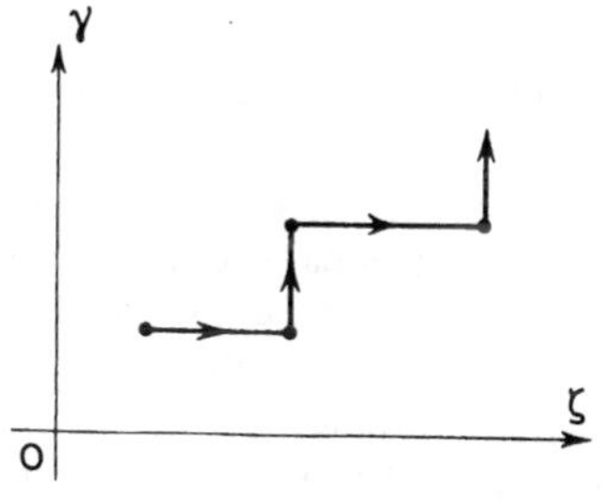


Fig. 20.

Here we have introduced the notation

$$Z = \left|\frac{w_0 - w_1}{w_0 - w_2}\right| \exp\left[-\sqrt{B^2+1}(\zeta - \zeta_0)\right] ,$$

$$\left.\begin{matrix} w_1 \\ w_2 \end{matrix}\right\} = -B \pm \sqrt{B^2-1} \begin{matrix} >0 \\ <0 \end{matrix} .$$

When $w > w_1$, we have to take the $-$ sign in (15.9), and the $+$ sign when $w < w_1$.

*Step-like path*. In experiments, step-like loading is often adopted (fig. 20); on each step either $\zeta$ or $\gamma$ is constant. The respective solutions are easily derived from (15.6) and (15.7). Thus, if $\zeta$ = const., then $d\zeta = 0$ and

$$d\gamma = d\tau/(1-\tau^2) .$$

Integrating and satisfying the condition $\tau = \tau_1$ when $\gamma = \gamma_1$, we find

$$\tau = \frac{G(\gamma)-1}{G(\gamma)+1} , \qquad q = \sqrt{1-\tau^2} , \tag{15.10}$$

where we have put

$$G(\gamma) = \frac{1+\tau_1}{1-\tau_1} \exp\left[2(\gamma-\gamma_1)\right] . \tag{15.11}$$

Now let $\gamma$ = const. Then by a similar process we obtain

$$q = \frac{Z(\zeta)-1}{Z(\zeta)+1} , \qquad \tau = \sqrt{1-q^2} ,$$

where $q = q_1$ for $\zeta = \zeta_1$, and

$$Z(\zeta) = \frac{1+q_1}{1-q_1} \exp\left[2(\zeta-\zeta_1)\right] .$$

15.3. *Convergence of results for development of deformation in a specified direction*

Fig. 21 illustrates in the $\zeta$, $\gamma$ plane various deformation paths corresponding to some of the cases that have been examined. The numbers indicate the values of $q$ according to flow theory, while the numbers in brackets give values of $q$ from deformation theory; the circle of unit radius bounds the shaded region of elastic deformation.

These data enable us to estimate, to some extent, the differences in the stress states (values of $q$) according to flow theory for different transition

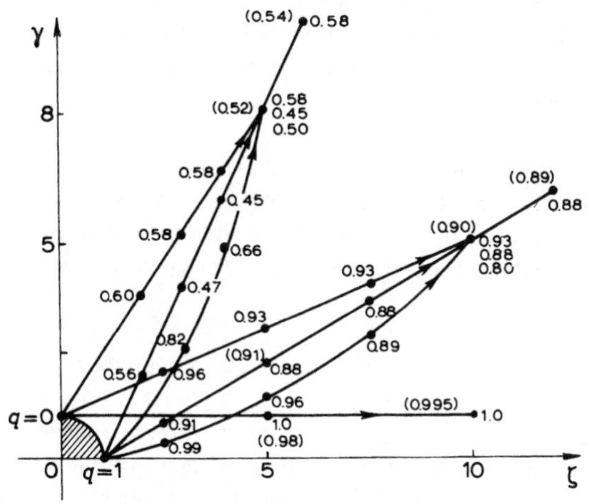

Fig. 21.

paths to the same deformation state. It is readily seen, moreover, that the stresses, as calculated from flow theory and deformation theory respectively, tend to converge with increase of strain in a prescribed direction.

If the deformation path tends to some linear path (fig. 19), it can be shown rigorously that the results are asymptotically convergent.

Beginning at some instant, let the deformation path tend to the straight line $\gamma = A + B\zeta$ as $\zeta$ increases; then $\gamma'(\zeta) \to B$ where $B > 0$.

According to (15.5), the deformation theory solution $w$ tends to the value

$$\overline{w} = -B + \sqrt{B^2 + 1}$$

with increasing $\zeta$. The flow theory solution $w$ is given by equation (15.8). If we introduce the new unknown

$$u = w - \overline{w} ,$$

then the differential equation (15.8) becomes

$$\frac{du}{d\zeta} = -[\overline{w} + \gamma'(\zeta)]\, u - \tfrac{1}{2}u^2 - [B - \gamma'(\zeta)]\, \overline{w} . \tag{15.12}$$

It is easy to see that the coefficients in this equation satisfy the conditions of Ascoli's theorem, and therefore the solution exists and tends to zero as $\zeta \to \infty$. Consequently

$$w \to \overline{w} .$$

A similar result follows form an analysis of the deformation of a tube under the action of an internal pressure and an axial force.

It can be shown [121] that even in the general case of a three-dimensional stress state, the stresses calculated from the two theories converge if the strain develops in a prescribed direction (which essentially means convergence to simple loading).

15.4. *Simple loading and proportional increase of loads*

The results derived above indicate the great importance of the class of simple (or nearly simple) loadings. In this case the relatively uncomplicated equations of deformation theory are applicable. The conditions for simple loading,

$$\sigma_{ij} = t\sigma'_{ij}\,, \qquad (15.13)$$

where the $\sigma'_{ij}$ are functions of the coordinates alone, and $t$ is some parameter, must be satisfied for every element of the body. It is usual to apply surface forces $p_i$ to the body (for simplicity body forces are neglected), with the stresses $\sigma_{ij}$ acting on elements of the body being a priori unknown. When the surface loads change arbitrarily it is naturally impossible to speak of simple loading of the body elements. But suppose that the loads increase in proportion to some parameter

$$p_i = tp'_i\,, \qquad (15.14)$$

where $p'_i$ are given functions of the coordinates of points on the surface alone.

Following A.A. Il'yshin, we can easily show that the loading will be simple if, for small strains,

1. the material is incompressible ($\epsilon = 0$),
2. the intensities of stress and strain are related by a power law of the form

$$T = A\Gamma^{\alpha}\,, \qquad (15.15)$$

where $A, \alpha > 0$ are constants.

Suppose that when $t = 1$ stresses $\sigma'_{ij}$ and strains $\epsilon'_{ij}$ are present in the body. It follows that these quantities satisfy the differential equations of equilibrium (4.2) with $F_j = 0$, $w_j = 0$, the boundary conditions (1.22), the Saint Venant compatibility conditions (2.16) and the equations of deformation theory (14.24) subject to the power law (15.15).

If the parameter is assigned some value $t$, then it is easily seen that the stresses $\sigma_{ij} = t\sigma'_{ij}$ and the strains $\epsilon_{ij} = t^{1/\alpha}\epsilon'_{ij}$ are the required solution of the problem. Actually, the differential equations of equilibrium, the boundary conditions and the Saint Venant compatibility conditions are homogeneous (with respect to stresses, external forces and strains respectively), and therefore are satisfied by the new values of the stresses (15.13), external forces (15.14) and strains. The equations of deformation theory (14.24) are also satisfied, since their right-hand sides are homogeneous functions of the strain components, of degree $\alpha$.

Returning now to the original restrictions, we note first that compressibility can usually be neglected for sufficiently developed (though small) plastic deformations. The power-law approximation (15.15) is a considerably more drastic limitation. There is a number of cases (for example, perfect plasticity) when this law gives a bad approximation. The law (15.15) leads to more or less acceptable results for developed plastic deformations and appreciable hardening of the material.

The foregoing analysis permits the assumption that simple loading is approximately achieved with developed plastic deformation and appreciable hardening under proportional increase of loads. In other cases simple (or nearly simple) loading does not apply (cf. also [155]).

15.5. *Experimental data*

Tests agree with plastic flow theory significantly better than with deformation theory. Equation (13.5) implies a similarity condition for the forms of the plastic strain-increment tensor and the stress tensor:

$$\mu^{\mathrm{p}}_{\mathrm{d}\epsilon} = \mu_\sigma \,.$$

In tests, small but systematic deviations from the similarity condition are observed. Fig. 22 shows the results of a number of tests; $\mu_\sigma$ is measured along

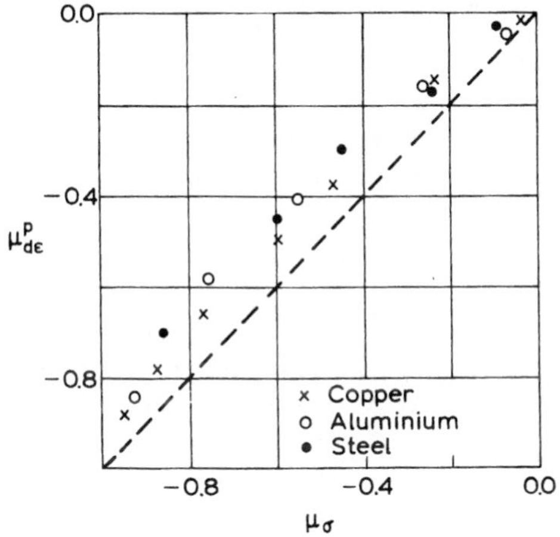


Fig. 22.

the horizontal axis, and $\mu^{\mathrm{p}}_{\mathrm{d}\epsilon}$ along the vertical; the dotted line corresponds to the similarity condition. The deviations imply small violations of the linear tensor equation (13.4). It is possible to achieve agreement with the experimental data shown in fig. 22 by introducing a non-linear tensor relationship, but this substantially complicates the original equations.

Experimental data also indicate that the directions of the principal axes of the stress tensor and the strain-increment tensor coincide.

It should be noted, however, that with complex loading, especially with intermediate unloading, the anisotropy which the material acquires in the process of plastic deformation has a considerable influence.

A description of the effects of strain anisotropy is difficult and requires a more complicated theory.

## §16. Generalizations in the case of perfect plasticity. Associated flow law

We consider a number of generalizations of the theory in the case of perfect plasticity. As before, the increments in total strain are compounded of the increments in elastic and plastic strain:

$$\mathrm{d}\epsilon_{ij} = \mathrm{d}\epsilon^{\mathrm{e}}_{ij} + \mathrm{d}\epsilon^{\mathrm{p}}_{ij} . \tag{16.1}$$

The increments in the components of elastic strain $\mathrm{d}\epsilon^{\mathrm{e}}_{ij}$ are related to the increments in the elastic stress components by Hooke's law (13.3). Furthermore, plastic volume changes are absent, i.e.

$$\mathrm{d}\epsilon^{\mathrm{p}}_{ii} = 0 . \tag{16.2}$$

### 16.1. *Yield function and plastic potential*

For a perfectly plastic medium there is, in the space of stresses $\sigma_{ij}$, *a yield surface*

$$f(\sigma_{ij}) = K \qquad (K > 0) , \tag{16.3}$$

bounding the region of plastic deformation, in which $f > K$. Plastic flow is associated with stresses corresponding to points on the yield surface.

Thus, according to the von Mises yield criterion,

$$f(\sigma_{ij}) = T^2 ; \qquad K = \tau_{\mathrm{s}}^2 . \tag{16.4}$$

The material is in an elastic state if $T < \tau_{\mathrm{s}}$, and in a plastic state if $T = \tau_{\mathrm{s}}$. In the space of principal stresses $\sigma_1, \sigma_2, \sigma_3$ this equation defines the surface of a circular cylinder with axis $\sigma_1 = \sigma_2 = \sigma_3$ (the hydrostatic cylinder, §1).

The yield surface (16.3) is convex (cf. below, § 18), i.e. it lies on one side of a tangent plane (or of a reference plane in the case of plane segments).

In addition to the yield function, we sometimes introduce the plastic potential $\Phi(\sigma_{ij})$, which enables the equations of plastic flow to be written in the form

$$\mathrm{d}\epsilon_{ij}^{\mathrm{p}} = \mathrm{d}\lambda \cdot \frac{\partial \Phi}{\partial \sigma_{ij}}, \tag{16.5}$$

where $\mathrm{d}\lambda \geqslant 0$ is some undetermined, infinitesimal scalar multiplier. The incompressibility condition (16.2) then gives

$$\frac{\partial \Phi}{\partial \sigma_{ij}} = 0. \tag{16.6}$$

The relations (16.5) can be best understood if we consider a representation of the tensors $\mathrm{d}\epsilon_{ij}^{\mathrm{p}}$, $\sigma_{ij}$ by vectors in the nine-dimensional space of the stresses $\sigma_{ij}$. Of course this representation is not complete and can be used only in a certain sense. In analyzing the equations of the plastic state, we usually use only the simplest operations on tensors, and we can establish a correspondence between these operations and operations on the representational vectors. The vector formulation is convenient, facilitates interpretation of experimental data, and is widely employed in analysis of the equations of the plastic state.

The simplest operations on tensors are the following:

1. *Multiplication of a tensor by a scalar* $\varphi$; this means multiplication of each component of the tensor by the scalar $\varphi$; i.e. if the tensor $T_a$ has components $a_{ij}$, then the tensor $\varphi T_a$ has components $\varphi a_{ij}$.

2. *Addition of tensors*; we add the components having the same indices, i.e. the tensor $T_a + T_b$ has components $a_{ij} + b_{ij}$.

3. *Contraction of two tensors* with respect to both indices; this is the sum $a_{ij} b_{ij}$. An example of a contraction is the increment in the work of plastic deformation $\mathrm{d}A_{\mathrm{p}} = \sigma_{ij} \mathrm{d}\epsilon_{ij}^{\mathrm{p}}$.

We introduce vectors **A**, **B**, having components $a_{ij}$, $b_{ij}$ in 9-dimensional space. Then the first operation corresponds to multiplication of the vector by a scalar, i.e. the vector $\varphi\mathbf{A}$. The second corresponds to addition of vectors $\mathbf{A} + \mathbf{B}$. Finally, the contraction of tensors corresponds to the scalar product of vectors:

$$(\mathbf{A}, \mathbf{B}) = a_{ij} b_{ij}.$$

*Remark*. Because of their symmetry the tensors of stress $T_\sigma$ and strain $T_\epsilon$ have only six different components. Nevertheless it is convenient to represent these tensors by vectors in 9-dimensional space, since then the scalar product of the vectors $\sigma_{ij}$ and $\epsilon_{ij}$ will immediately give the contraction. This is connected with the fact that the components of the strain tensor are not the shears themselves, but half the shears. We can also consider a six-dimensional space, taking as components of the stress vector the six components of the stress tensor multiplied by some numbers. The latter are chosen so that the scalar product of vectors correspond to the contraction of tensors. It is more convenient, however, to have the same components for the vectors as for the tensors – in 9-dimensional space.

If the principal directions of the tensor $T_a$ are fixed, we can represent the tensor $T_a$ by a vector **A** with components $a_1$, $a_2$, $a_3$ in three-dimensional space $a_i$. This interpretation has been used previously, in § 1.

Thus, the stress state in 9-dimensional space can be represented by a vector $\sigma_{ij}$. The plastic strain-increments $d\epsilon_{ij}^{p}$ can also be represented by a vector in the same space, if $d\epsilon_{ij}^{p}$ is multiplied by a constant of the appropriate dimensionality. The equation $\Phi(\sigma_{ij})$ = const. defines the surface (hypersurface) of the plastic potential. Since the direction cosines of the normal to this surface are proportional to $\partial\Phi/\partial\sigma_{ij}$, the relations (16.5) imply that the plastic flow vector $d\epsilon_{ij}^{p}$ is directed along the normal to the surface of plastic potential.

16.2. *Associated flow law*

Of great importance is the simplest case when the yield function and plastic potential coincide:

$$f = \Phi\ .$$

Apart from its simplicity, this case enables the derivation of a uniqueness theorem and extremum principles, which give the theory a certain completeness.

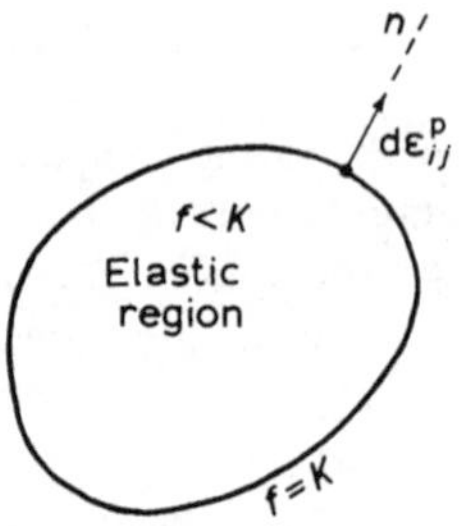


Fig. 23.

Thus,

$$d\epsilon_{ij}^{p} = d\lambda \frac{\partial f}{\partial \sigma_{ij}}, \tag{16.7}$$

and plastic flow develops along the normal to the yield surface (fig. 23).

The increments in the work of plastic deformation are

$$dA_p = \sigma_{ij} d\epsilon_{ij}^{p} = d\lambda \cdot \sigma_{ij} \frac{\partial f}{\partial \sigma_{ij}}. \tag{16.8}$$

If $f$ is a homogeneous function of the stresses of degree $m$, then

$$dA_p = d\lambda \cdot mf = d\lambda \cdot mK,$$

i.e. the multiplier $d\lambda$ is proportional to the increment in the work of plastic deformation:

$$d\lambda = \frac{dA_p}{mK}. \tag{16.9}$$

In flow theory (§13) this leads to formula (13.10), which was obtained earlier

Note that (16.7) covers both the equations of plastic flow theory (13.7) and the equations of the Saint Venant-von Mises plasticity theory (13.11). It is easy to show that in this case

$$\frac{\partial f}{\partial \sigma_{ij}} = \frac{\partial T^2}{\partial \sigma_{ij}} = s_{ij},$$

and equations (16.7) take the form $d\epsilon_{ij}^{p} = d\lambda s_{ij}$. The incompressibility condition (16.2) is obviously satisfied.

The relation (16.7) is called the *associated law of plastic flow*, because it is connected (associated) with the yield criterion. The associated flow law enables various generalizations of the plasticity equations by considering yield surfaces of more complex form.

If we consider the flow in the space of principal stresses (convenient for fixed principal directions, for example, in axisymmetric problems), then in place of (16.7) we can use the relations

$$d\epsilon_i^{p} = d\lambda \frac{\partial f}{\partial \sigma_i} \qquad (i = 1, 2, 3). \tag{16.10}$$

16.3. *Piecewise-smooth yield surfaces*

The associated flow law in the form (16.7) requires that the yield surface be smooth, i.e. that it have a continuously turning tangent plane, with a defined normal to the surface. However, we cannot really exclude from our considerations singular yield surfaces (having. edges and vertices). Thus, the Tresca-Saint Venant criterion defined the surface of a hexahedral prism (§9), the normal along its edges being undefined. As we shall see later, the use of the Tresca-Saint Venant yield criterion in place of the von Mises criterion frequently leads to substantial mathematical simplification (for example, in problems of plane stress). For this reason it is necessary to supplement the relations (16.7), valid at points of continuity of the yield surface, in such a way that plastic flow corresponding to edges and vertices of the yield surface can also be determined. For simplicity we shall restrict ourselves to consideration of an edge. This is the most common case in practice; cases of vertices can be analyzed in a similar way.

Following Prager [29] and Koiter [124] we assume that the flow on the edge is a linear combination of the flow on the right and left of the edge (fig. 24):

$$d\epsilon_{ij}^{p} = d\lambda_1 \frac{\partial f_1}{\partial \sigma_{ij}} + d\lambda_2 \frac{\partial f_2}{\partial \sigma_{ij}}. \tag{16.1}$$

Here $f_1 = \text{const.}, f_2 = \text{const.}$ are the equations of the yield surfaces on the two sides of the edge. The undetermined multipliers $d\lambda_1, d\lambda_2$ are non-negative, as a result of which the flow develops along a direction lying within the angle formed by the normals to the two adjacent faces.

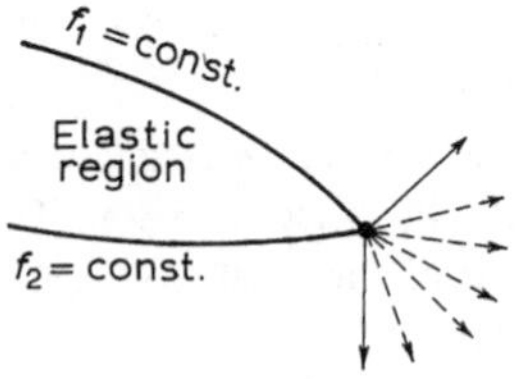


Fig. 24.

At a smooth point of the surface the plastic flow is fixed in direction, while its magnitude is undetermined; the multiplier $d\lambda$ is expressed in terms of the increment in the work of plastic deformation, and is found in the process of solving particular problems.

On an edge of the yield surface, both the direction and the magnitude are undetermined. The multipliers $d\lambda_1, d\lambda_2$ are also found in the course of solving individual problems. Along the edge "two yield criteria" are satisfied ($f_1$ = const. and $f_2$ = const.); this compels us to introduce two arbitrary multipliers in order to avoid inconsistency with the compatibility conditions for strain.

If the flow is considered in the space of principal stresses, then in place of (16.11) we shall have

$$d\epsilon_i^p = d\lambda_1 \frac{\partial f_1}{\partial \sigma_i} + d\lambda_2 \frac{\partial f_2}{\partial \sigma_i}. \tag{16.12}$$

16.4. *Example: Flow on the edge of the Tresca-Saint Venant prism*

By way of illustration we consider the flow corresponding to stress states for points on the edge of a hexahedral Tresca-Saint Venant prism. Suppose for definiteness that the edge is generated by the intersection of the planes (cf. §9)

$$f_1 \equiv \sigma_1 - \sigma_2 = \sigma_s\,; \qquad f_2 \equiv \sigma_1 - \sigma_3 = \sigma_s\,.$$

The intercepts of these planes on the deviatoric plane are shown by the solid lines in fig. 25. For the first plane the flow can be written in the form (we use (16.10), change to strain rates $\xi_i^p$ and omit the index p; $d\lambda_1 = \lambda_1' dt$, $d\lambda_2 = \lambda_2' dt$):

$$\frac{\xi}{1} = \frac{\xi_2}{1} = \frac{\xi_3}{0} = \lambda_1'\,.$$

The flow for the second plane is given by the relations

$$\frac{\xi_1}{1} = \frac{\xi_2}{0} = \frac{\xi_3}{-1} = \lambda_2'\,.$$

For points on the edge, the law (16.12) gives

$$\xi_1 = \lambda_1' + \lambda_2'\,; \qquad \xi_2 = -\lambda_1'\,; \qquad \xi_3 = -\lambda_2'\,.$$

The rate of work of plastic deformation (dissipation) for points on the edge is equal to

$$\dot{A}_p = \sigma_i \xi_i = \sigma_s(\lambda_1' + \lambda_2')\,.$$

Here $\lambda_1' + \lambda_2' = \xi_1$ is the largest principal rate of extension.

A similar analysis for the flow on the other edges easily shows that in all cases the dissipation is given by the simple formula

$$\dot{A}_p = \sigma_s \xi_{max}\,, \tag{16.13}$$

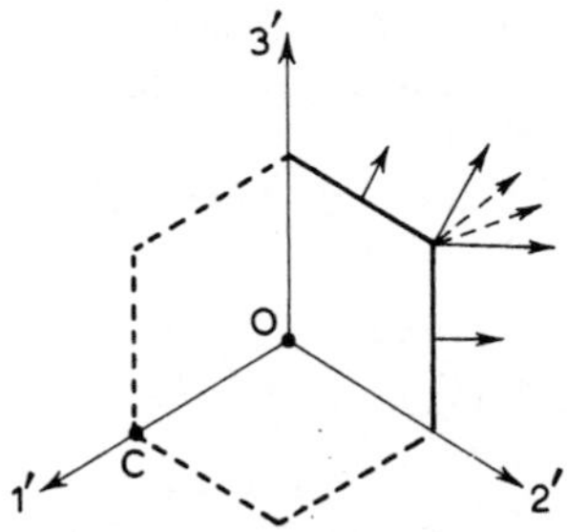


Fig. 25.

where $\xi_{max}$ is the absolute value of the numerically largest principal strain rate.

The physical interpretation of flow on an edge determined in this way gives rise to certain difficulties, which have already appeared in connection with, for example, simple tension at the corner point C (fig. 25) of the Tresca-Saint Venant hexagon. Here the flow is given by the relations

$$\xi_1 = \lambda_1' + \lambda_2' ; \qquad \xi_2 = -\lambda_1' ; \qquad \xi_3 = -\lambda_2' ,$$

where the first principal direction is directed along the axis of the rod. Thus, the transverse strains are arbitrary, satisfying only the incompressibility condition. This picture is not consistent with the usual concept of tension in an isotropic rod (according to these relations a circular rod can become elliptical). However, such paradoxical results appear in extreme cases only and relate mainly to the velocity field. Limit loads obtained on the basis of the present model are good approximations.

The Prager-Koiter model should be regarded as an idealized approximation, which is often useful. It is scarcely expedient to try to find a physical meaning in individual paradoxical results.

### 16.5. *Anisotropic medium*

On the basis of the representation given above it is possible to derive equations for the plastic flow of an anisotropic body. It is only necessary to introduce an anisotropic yield function. In the simplest version this function is given by the quadratic form

$$f = c_{ijkl}\sigma_{ij}\sigma_{kl} ,$$

where $c_{ijkl}$ are coefficients of anisotropy. Other versions of a similar type are discussed in the works of von Mises [52, 70], Hill [54] and other authors. A theory has also been developed based on a generalization of the Tresca-Saint Venant yield criterion [11].

## §17. Generalizations. Case of a hardening medium

### 17.1. *Loading surface*

The equations of flow theory, derived in §13, give a satisfactory description of the mechanical behaviour of a hardening medium only in cases of not too complex loading. But hardening which accompanies plastic deformation has a directional character, and in a triaxial (or biaxial) stress state it is a complex and, in general, not sufficiently understood effect. Its description requires a substantial extension of the theory.

At the basis of the theory lies the concept of a loading surface $\Sigma$ (fig. 15b), in the space of the stresses $\sigma_{ij}$, which separates a region of elastic deformation from a region of plastic deformation for a given state of the medium. An infinitesimal increment in the stress $d\sigma_{ij}$ leads either to elastic deformation (unloading if $d\sigma_{ij}$ is directed inside $\Sigma$) or to continuation of plastic deformation (loading if $d\sigma_{ij}$ is directed outside $\Sigma$). Increments $d\sigma_{ij}$ which lie in the tangent plane to the loading surface (*neutral changes*) can only lead to elastic deformation (i.e. if the representative point moves on the surface $\Sigma$ plastic deformation cannot occur). This condition (*continuity condition*) is necessary for a continuous transition from plastic to elastic deformation with continuous change in the direction of the vector $d\sigma_{ij}$.

The loading surface expands and is displaced with the development of hardening, and the elastic limit changes (differently in different directions). In perfect plasticity the yield surface is the limiting position of the loading surfaces if these all contract on the initial surface.

In general the shape and position of the loading surface depend not only on the current stress but also on the whole preceding history of the deformation.

### 17.2. *Example. Construction of a loading surface from experimental data*

As a simple example we consider the construction of a loading surface from experimental data. It is not possible to perform experiments with arbitrary, triaxial stress, and therefore we examine loading curves in certain sections of the loading surface. Usually we restrict ourselves to loading curves in plane stress, where one of the principal stresses is zero. Here we shall consider, in particular, experiments on thin-walled tubes under the action of an internal pressure $p$ and an axial force $P$ ($p + P$-tests, §7). The tube is in a state of plane stress, with the principal stresses equal to $\sigma_\varphi$ and $\sigma_z$. Denote by $\sigma_s$ the yield limit in uniaxial tension, equal, say, to the stress which corresponds to a residual strain of 0.2%, and consider the plane of the stresses $\sigma_z$, $\sigma_\varphi$. We construct an *initial* yield curve which bounds the elastic region for the original (non-hardening) material. To do this, experiments are performed for a series of specimens with different values of the ratio $P/p$ (fixed for each test); the corresponding loading paths, which are obviously rays, are shown by dotted lines in fig. 26a. On each ray an experimental point is located for which the residual strain is 0.2%. This number is, of course, conditional and in principle can take other values. However, an examination of various secondary effects (after-effect, creep) which complicate the picture suggests that it would be unwise to choose excessively small criteria. Joining the experimental points by a smooth curve, we obtain the initial loading curve $\Sigma_0$ (fig. 26a). If the material has experienced plastic deformation previously, the loading surface is constructed in the following way. Suppose, for example, that we have to construct a loading curve for steel which has previously experienced plastic deformation under loading along the ray OA (fig. 26b). We take the tubes and subject them to loading along OA, then we remove the load and test each tube at a certain fixed value of $P/p$ (i.e. along the ray) until a residual strain of 0.2% is attained on the ray chosen for the given tube. Marking the corresponding points on the $\sigma_z$, $\sigma_\varphi$ plane and joining them by a smooth curve, we obtain the required loading curve (fig. 26b). According to experiments, the loading curve is displaced in the direction of the prior plastic deformation.

Experiments carried out in plane stress conditions with different loading

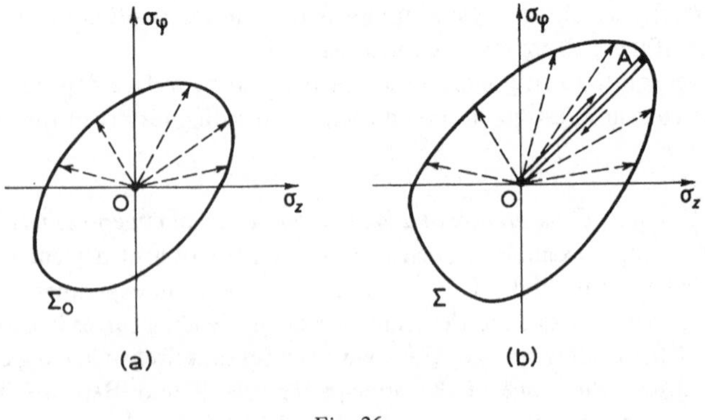


Fig. 26.

programmes enable us to build up a picture of the dependence of the loading surface on the process of plastic deformation.

### 17.3. *Some forms of the loading surface*

We consider some simple forms of the loading surface $\Sigma$. We assume that in its original state the material has yield limits which are identical in magnitude for compression and tension.

*Remark.* It is necessary to emphasize the difficulties of constructing the loading surface experimentally. Finding the yield limit from the limit of proportionality or in the presence of a small amount (0.01% or 0.02%) of residual strain $\Delta$ can lead to invalid conclusions, because of the influence of secondary effects. For this reason one sometimes begins from a rough tolerance-limit (for example, $\Delta = 0.2\%$). This is acceptable if the yield limit is sufficiently high and the hardening relatively small; the relative errors will then be low. When the hardening is large or the elastic limit low, the range 0.2% can lead to gross distortion. Some authors avoid this difficulty by adducing the values of the yield limit in accordance with different tolerance-limits (from 0.02% to 0.2%, or even 0.5%). This approach contains useful information, but contradicts the theoretical model of a unique yield surface for a given state.

The material experiences hardening in the direction of previous deformation, while in the reverse direction unhardening takes place (Bauschinger effect). For an unhardening material the yield limit is comparatively low (sometimes it can scarcely be discerned at all), and then its experimental determination depends essentially on the tolerance.

With existing experimental data there is a sharp decrease (approximately 10-fold) in the accuracy of the yield limit as the distance from the direction OA (fig.26b) of the previous plastic deformation increases. In effect, only a small, frontward part of the loading surface has its actual value, depending on the choice of the tolerance limit and characterized by a more or less sharp transition to the plastic zone. The rear part of the loading surface depends essentially on the magnitude of the tolerance. In particular, this is a conventional method for determining the Bauschinger effect.

In this connection the question arises of the relationship between the theoretical model and the experimental data, and also of the method of experimentation.

It is possible that for some materials the construction of the rear portion of the yield surface can be achieved with optimally-diminished tolerance and an appropriate estimate of the associated secondary effects. It is also possible to recommend the following general method of establishing the "true" yield limit from the test data. Suppose yield limits have been obtained experimentally with adjacent, sufficiently wide tolerance-limits (for example, $\Delta_1 = 0.1\%$ and $\Delta_2 = 0.2\%$); then we find the "true" yield limit by extrapolating to "zero" tolerance. When additional points are available the accuracy of the extrapolation can be improved.

If for some reason the construction of a yield surface cannot be reliably effected, it is probably inevitable that the concept of a yield surface has to be abandoned. In particular problems, of course, a limited use of the yield surface (front portion) is still possible.

*Isotropic hardening.* In the preceding section the equation of a fixed yield surface had the form $f(\sigma_{ij}) = K$. If we assume that, with plastic deformation, hardening develops equally in all directions and is independent of the hydrostatic pressure $\sigma$, the equation of the loading surface can be written in the form

$$f(s_{ij}) = \varphi(q) , \tag{17.1}$$

where the scalar $q > 0$ is some measure of isotropic hardening, and $\varphi$ is an increasing function.

One common measure of hardening $q$ is the work of plastic deformation $A_\text{p}$, i.e.

$$q = \int \sigma_{ij} \mathrm{d}\epsilon_{ij}^\text{p} .$$

Another, less frequent, measure is characteristic of the accumulated plastic strain (Odquist's parameter):

$$q = \int \sqrt{2\mathrm{d}\epsilon_{ij}^\text{p} \mathrm{d}\epsilon_{ij}^\text{p}} .$$

Note that equation (17.1) can contain, in general, several measures of hardening $q_1, q_2, q_3, \ldots$ . If the medium is isotropic the function must depend only on the invariants of the stress deviatoric. In particular, if we take into account only the quadratic invariant – the intensity of tangential stresses $T$ – (which is quite sufficient in the first approximation), then equation (17.1) takes the form

$$s_{ij}s_{ij} = \varphi(q) . \tag{17.2}$$

A similar condition has been considered earlier (§12).

According to (17.1) the loading surface expands uniformly ("isotropically") and remains self-similar with increasing plastic deformation (fig. 27). It should be understood that the Bauschinger effect is not involved here, since the yield limits in the direct ($\mathrm{OM}^+$) and inverse ($\mathrm{OM}^-$) loading directions are equal in magnitude.

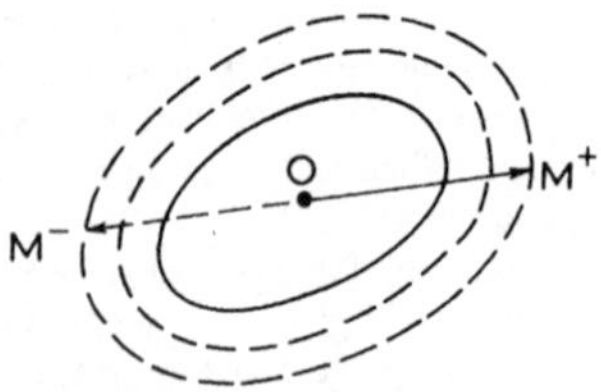


Fig. 27.

The concept of an isotropic loading surface is useful in describing simple experiments on plastic deformation where the development is in some preferred direction, and in particular for experiments on simple loading.

*Translational hardening.* Let the loading surface $\Sigma$ experience rigid displacement in the direction of deformation. In fig. 28 the solid line shows the initial position of the loading surface, and the dotted line its position after some plastic deformation. If now the material is again loaded, then the elastic limit increases in the direction of the preceding deformation ($OM^+$); if hardening has taken place, the elastic limit in the inverse direction ($OM^-$) diminishes. This model describes the Bauschinger effect, at least qualitatively. In the case under consideration the equation of the loading surface has the form

$$f(s_{ij}-a_{ij}) = K \,, \tag{17.3}$$

where $a_{ij}$ are the coordinates of the centre of the loading surface, which change with plastic deformation and generate a deviatoric. For the increments $da_{ij}$ differential relations are in general required [84, 141, 178]. The simplest version, which is well known, [115], is

$$a_{ij} = c\epsilon_{ij}^{p} \,, \tag{17.4}$$

where $c$ is a positive constant, characteristic for a given material, and $\epsilon_{ij}^{p}$ are components of plastic strain. Thus the components of a rigid displacement of the loading surface are here proportional to the components of plastic strain.

*Translation and expansion.* A combination of the two preceding cases leads to the more general scheme

$$f(s_{ij}-a_{ij}) = \varphi(q) \,. \tag{17.5}$$

The loading surface experiences translation and uniform expansion in all directions, i.e. it retains its shape.

Equation (17.5) satisfactorily described the hardening of a material in quite a wide range of variation of loading path.

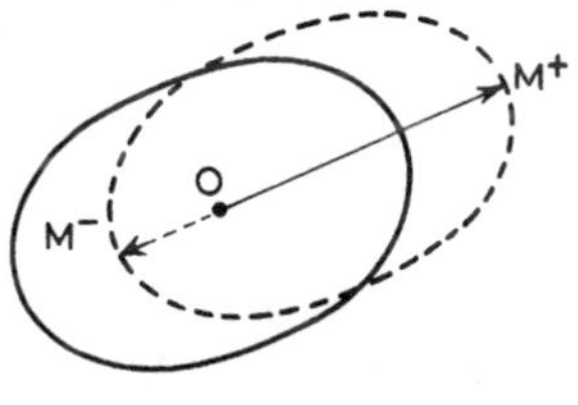


Fig. 28.

If the medium is isotropic in its initial state and its behaviour depends only on the quadratic invariant – the intensity of tangential stresses $T$ – then equation (17.5) takes the form

$$(s_{ij}-a_{ij})\,(s_{ij}-a_{ij}) = \varphi(q)\,. \tag{17.6}$$

According to experimental data the right-hand side (the "radius" of the surface) undergoes comparatively little change.

It is not difficult to examine more general equations of the loading surface, including initial anisotropy of the medium and containing not one but several hardening parameters [84, 69].

17.4. *Associated flow law*

To derive the equations which relate the increments in the strain components with the stress components and their increments, we utilize some propositions which have already been encountered earlier.

First of all we assume that the increment in total strain is made up of the increments in elastic and plastic components:

$$\mathrm{d}\epsilon_{ij} = \mathrm{d}\epsilon^{\mathrm{e}}_{ij} + \mathrm{d}\epsilon^{\mathrm{p}}_{ij}\,. \tag{17.7}$$

The increments in elastic strain components can be found from Hooke's law. Plastic volume changes are neglected, i.e.

$$\mathrm{d}\epsilon^{\mathrm{p}}_{ii} = 0\,.$$

The next assumption is the *continuity condition*, which we have touched on at the beginning of this section. Let the current state be described by some position of the loading surface $\Sigma$ (fig. 15). An infinitesimal stress-increment $\mathrm{d}\sigma_{ij}$ either is accompanied by elastic strain alone, or it induces plastic strains $\mathrm{d}\epsilon^{\mathrm{p}}_{ij}$ as well. We have already seen that if the plastic components are to vanish continuously in transition to elastic deformation, it is necessary to introduce neutral changes $\mathrm{d}\sigma_{ij}$ which lie in the tangent plane to the loading surface $\Sigma$ and which lead only to changes in the elastic strains $\mathrm{d}\epsilon^{\mathrm{e}}_{ij}$. Hence it follows that the increments in the plastic strain components $\mathrm{d}\epsilon^{\mathrm{p}}_{ij}$ must be proportional to the quantity

$$\mathrm{d}'f = \frac{\partial f}{\partial \sigma_{ij}}\mathrm{d}\sigma_{ij}\,,$$

where the prime signifies that the increment $\mathrm{d}'f$ is evaluated only with respect to increments in the stress components [1]). Then if the stress-increment $\mathrm{d}\sigma_{ij}$

[1]) That is, with constant plastic strains.

lies in the tangent plane to the loading surface, we have $d'f = 0$ and $d\epsilon_{ij}^{p} = 0$. Note that with condition (17.1),

$$\frac{\partial f}{\partial \sigma_{ij}} = \frac{\partial f}{\partial s_{ij}}, \qquad \text{and} \qquad d'f = \frac{\partial f}{\partial s_{ij}} ds_{ij} .$$

A further assumption is the associated flow law, according to which the direction of the increment vector $d\epsilon_{ij}^{p}$ is along the normal to the loading surface $\Sigma$ in the case of hardening as well. As a consequence, the components $d\epsilon_{ij}^{p}$ must be proportional to the direction cosines of the normal to $\Sigma$, differing only by a general scalar multiplier from the partial derivatives $\partial f/\partial \sigma_{ij}$. Thus

$$\begin{aligned} d\epsilon_{ij}^{p} &= g \frac{\partial f}{\partial \sigma_{ij}} d'f \qquad &&\text{for} \qquad d'f \geqslant 0 , \\ d\epsilon_{ij}^{p} &= 0 &&\text{for} \qquad d'f < 0 \ (\text{unloading}) . \end{aligned} \tag{17.8}$$

The proportionality factor $g$ is called the *hardening function*: it characterizes the level of hardening attained and, in general, depends on the history of the deformation. If the current stress state $\sigma_{ij}$ corresponds to points inside $\Sigma$, i.e. is elastic, then $d\epsilon_{ij}^{p} = 0$.

Since $d\sigma_{ij} d\epsilon_{ij}^{p} > 0$ for a hardening body (cf. Drucker's postulate, §18), it follows from (17.8) that for loading $g(d'f)^2 > 0$, i.e. $g > 0$.

Specification of a smooth loading surface under the associated flow law completely determines the plastic strain-increments; the hardening function $g$ is found from the equation of the loading surface (i.e. from the hardening condition) if we take into account that $df = d\varphi$ and use equations (17.8).

As an example, let the loading surface be given by equation (17.2), where $q$ is Odquist's parameter. Then $df = \varphi' dq = \varphi' \sqrt{2 d\epsilon_{ij}^{p} d\epsilon_{ij}^{p}}$. Forming $dq$ with the aid of (17.8), we obtain

$$g = (4\varphi' T)^{-1} .$$

If for $q$ we take the work of plastic deformation, then $dq = \sigma_{ij} d\epsilon_{ij}^{p}$. Evaluating $dq$ from (17.8), we easily find

$$g = (4\varphi' T^2)^{-1} .$$

A special case of these equations is the equations of flow theory with hardening (13.14), which we have considered in §13. These equations follow from (17.7), (17.8) with

$$f(\sigma_{ij}) = T^2 ; \qquad g = F(T)/2T . \tag{17.9}$$

Equations (17.8) were formulated by Prager.

17.5. *Singular loading surfaces*

It has been assumed above that the loading surface $\Sigma$ is regular, i.e. that it has a continuously turning normal. But loading surfaces with edges and conical points have often to be considered. Here it is convenient to distinguish the two cases.

Sometimes the loading surface has an edge, whose position on the surface is fixed. For example, we have already seen that the solution of a problem is often simplified if we change from the intensity of tangential stresses $T$ to the approximately equal quantity $\tau_{max}$ (§ 1). This corresponds to a transformation from the von Mises circle in the deviatoric plane to the Tresca-Saint Venant hexagon (fig. 25). Then the flow on the edge is determined as the linear combination of the flows on the two sides of the edge.

In the second case the singularity is of a more essential character. It is in fact possible to assume that in the neighbourhood of a point M of the loading surface where a stress-increment $d\sigma_{ij}$ occurs, a singularity (conical point) develops (fig. 29). There are several theories relating to the development of these singularities on loading surfaces [67, 69, 70], but the equations of plasticity theory become extremely complicated in this case. Nevertheless the analysis of such situations is clearly of interest. Experimental data are somewhat contradictory, and do not at present allow a definitive statement to be made about the existence of singular points.

17.6. *Concluding remarks*

The development of plasticity theory for hardening media is of great practical significance, since many modern structural metals experience perceptible

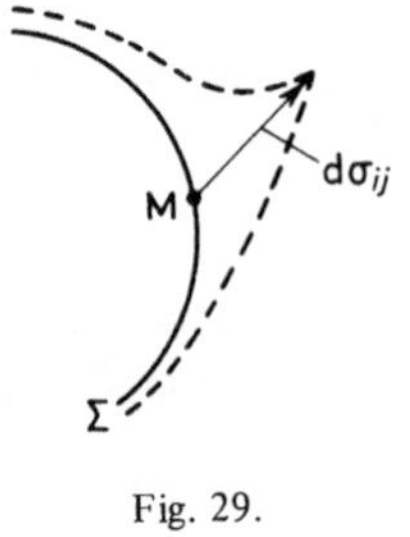


Fig. 29.

hardening. As we have already noted, the theories of hardening media developed above do not fully describe the behaviour of metals in conditions of complex loading. At the same time these equations are themselves extremely complicated, and their solution in concrete problems involves considerable mathematical difficulty. It is therefore customary in applications to proceed either from the Prandtl-Reuss equations (13.14) with an isotropic hardening condition, or from the equations of deformation theory (14.23) with the "single-curve" law (the intensity of the tangential stresses is a function of the intensity of the shear strain, § 12). The isotropic hardening law is useful for comparatively simple loading paths. The single curve model is applicable in an even narrower range of situations. It follows that solutions of boundary value problems on the basis of either theory are in fact restricted to "sufficiently simple" loading. It does not seem possible to formulate this condition more precisely. A comparison of available solutions found from the two theories usually indicates little difference between them.

A question of great importance is that of plastic deformation under cyclic loading. Experimental data on this question can be found in the book by V.V. Moskvitin [24]; this book also presents a technique for utilizing the equations of deformation theory to describe these processes. Expressions describing repeated loading of an elastic-plastic medium have been derived by R.A. Arutyunyan and A.A. Vakulenko [84] on the basis of the physically more correct flow theory. These authors begin with the equation of the loading surface (17.6), where the "coordinates of the centre" $a_{ij}$ are determined by certain differential relations. The theoretical results agree satisfactorily with experimental data.

For more complex loading paths the above theories for a hardening medium often turn out to be inadequate.

In the circumstances it is natural to try to go beyond the preceding formal representations of loading surface and associated flow law, and to consider different approaches to developing a theory of hardening media. Various treatments of this problem have been investigated, but they are outside the scope of this book. We shall limit ourselves to a few references to the literature.

First of all we note a series of works [42, 57, 69, 89] in which the methods of thermodynamics of irreversible processes are used. This approach has been widely developed in recent years, with the aim of ascertaining the structure of the equations of plastic deformation.

Another method consists in imposing sufficiently general restrictions on the structure of the tensor equations. We have already (§ 16) noted conditions in which tensors can be represented by vectors. A.A. Il'yushin [13] derived a class of tensor relations having a corresponding invariant vector formulation. This makes the analysis convenient, but not fully general.

Finally, we should draw attention to the so-called "physical theories of plasticity", in which the properties of media are inferred on the basis of an analysis of the deformation of individual crystals. A similar theory ("slip theory") has been proposed for a complex stress state by Batdorf and Budiansky [85]. The metal consists of randomly distributed crystals, in each of which the plastic slip takes place along certain planes. Statistical averaging of the slip leads to "stress-strain" relations having a complex structure. Several other versions of slip theory have been developed in the works of A.K. Malmeister [22] and other authors. Note that the equations of slip theory admit a purely phenomenological treatment [125].

Not only are the Batdorf-Budiansky equations complex, but also there is a series of important properties which they do not describe (for example, the Bauschinger effect). Further complication of the theory is needed to allow for this effect.

## §18. Drucker's postulate. Convexity of the loading surface. Justification of the associated flow law

### 18.1. *Hardening condition and Drucker's postulate*

In the preceding sections we have repeatedly considered a hardening medium, by which we have meant a medium whose elastic limit increases in the process of deformation. But no rigorous definition of hardening has been given. On the other hand a few simple examples are enough to convince us of the need for a more precise definition of a hardening material.

In fig. 30 the $\sigma$, $\epsilon$ curves symbolize the relation between stress and strain. In the case (a) the material is actually hardening. Here an additional load-

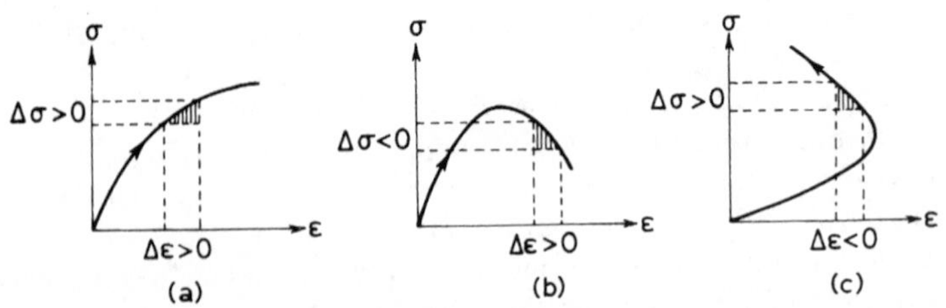


Fig. 30.

ing $\Delta\sigma > 0$ gives rise to an additional strain $\Delta\epsilon > 0$, with the product $\Delta\sigma \cdot \Delta\epsilon > 0$. The additional stress $\Delta\sigma$ does positive work – represented by the shaded triangle in the diagram. Material of this kind is called *stable*.

In case (b) the deformation curve has a descending branch, where the strain increases with decreasing stress. On this segment the additional stress does negative work, i.e. $\Delta\sigma \cdot \Delta\epsilon < 0$. This material is called *unstable*.

In case (c) the strain decreases with increasing stress, so that $\Delta\sigma \cdot \Delta\epsilon < 0$ again. In the mechanical scheme this case contradicts the law of conservation of energy, since it allows "free" extraction of useful work (for example, an extended rod raises slightly the load $P$ under an additional load $\Delta P$).

Only the first of these cases is relevant to the deformation of real materials. Drucker's postulate generalizes these ideas and provides a suitable definition of hardening.

Consider an element of a hardening medium in which there is an original stress $\sigma_{ij}^0$; to this element we apply additional stresses (in general, of arbitrary magnitude), and then remove them. We assume that changes take place sufficiently slowly for the process to be regarded as isothermal. Then it is postulated that

1. *In the process of loading* additional stresses produce positive work.

2. *For a complete cycle* of additional loading and unloading, *additional stresses* do positive work if plastic deformation takes place. *For a hardening material* the work will be equal to zero only for purely elastic changes.

We emphasize once again that what we have in mind here is not the work of the total stresses, but only the work of the additional stresses on the additional strains. Returning to case (b) in fig. 30, we observe that the work of the stress $\sigma$ is positive (i.e. $\sigma\Delta\epsilon > 0$), even though $\Delta\sigma \cdot \Delta\epsilon < 0$. According to Drucker's postulate the prolongation of plastic deformation of a hardening medium requires additional forces.

Drucker's postulate leads to an important inequality.

Let $\Sigma$ be the current position of the loading surface (fig. 31). Consider some loading path $\mathrm{A} \to \mathrm{B} \to \mathrm{C}$. The initial stress $\sigma_{ij}^0$ corresponds to an initial point A lying inside or on the surface $\Sigma$. The point B (stress $\sigma_{ij}$) is on the surface $\Sigma$. An infinitesimal stress-increment $d\sigma_{ij}$ proceeds from the point B, and causes a corresponding elastic strain $d\epsilon_{ij}^e$ and plastic strain $d\epsilon_{ij}^p$. Denote by $\Sigma'$ a new, neighbouring position of the loading surface. We now return to the point A by some path $\mathrm{C} \to \mathrm{A}$. Drucker's postulate states that the work of the additional stresses is positive over the whole cycle, i.e.

$$\oint (\sigma_{ij} - \sigma_{ij}^0)\, d\epsilon_{ij} > 0 \,.$$

For a closed path ABCA, the work of the additional stresses on the elastic

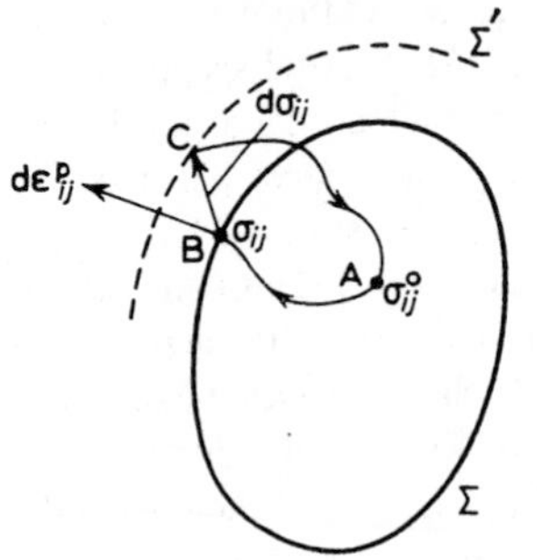


Fig. 31.

strains $d\epsilon^e_{ij}$ is zero, and consequently

$$\oint(\sigma_{ij}-\sigma^0_{ij})\, d\epsilon^p_{ij} > 0\ .$$

Since plastic deformation takes place only on an infinitesimal segment B → C, the last inequality takes the form

$$(\sigma_{ij}-\sigma^0_{ij})\, d\epsilon^p_{ij} > 0\ . \tag{18.1}$$

This inequality is sometimes called the *local maximum principle*. For a hardening medium the expression can only be equal to zero in the absence of plastic deformation.

Consider now a different cycle, in which the initial stress is $\sigma_{ij}$, corresponding to the point B on the loading surface Σ. Then by Drucker's hypothesis:

*For the loading process* B → C

$$d\sigma_{ij} d\epsilon_{ij} > 0\ . \tag{18.2}$$

*For the cycle of loading and unloading* B → C → B

$$d\sigma_{ij} d\epsilon^p_{ij} > 0 \tag{18.3}$$

(since the work on elastic strains is zero in a closed path).

18.2. *Convexity of the loading surface and necessity of an associated flow law*

From the inequality (18.1) we have that the scalar product of the vector of additional stresses $\sigma_{ij}-\sigma^0_{ij}$ (vector $\overline{AB}$ in fig. 32) and the vector of the plastic strain-increments $d\epsilon^p_{ij}$ is positive. Consequently these vectors always generate an acute angle. From this we infer the convexity of the loading surface and also the associated flow law (i.e. the orthogonality of the vector $d\epsilon_{ij}^{p}$ to the surface $\Sigma$).

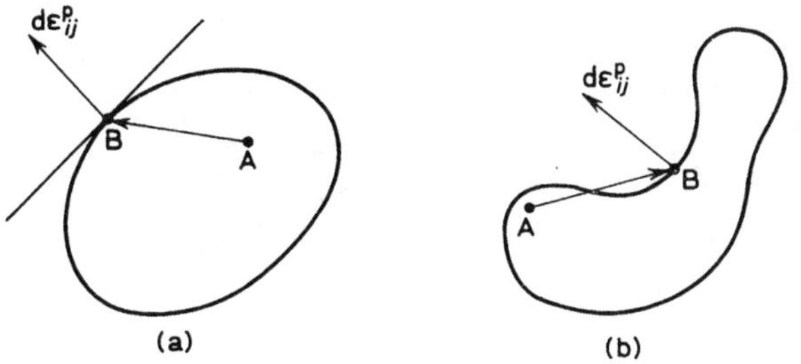


Fig. 32.

Let the loading surface $\Sigma$ be convex (i.e. $\Sigma$ lies on one side of a tangent plane (fig. 32a), or of a reference plane, as in the case of the Tresca-Saint Venant hexagonal prism). Condition (18.1) will only be satisfied if the vector $d\epsilon_{ij}^{p}$ is normal to $\Sigma$; otherwise there will always be a vector $\sigma_{ij}-\sigma_{ij}^{0}$ making an obtuse angle with $d\epsilon_{ij}^{p}$. Note that $d\epsilon_{ij}^{p}$ depends on the form of the loading surface $\Sigma$, but not on the choice of the point A inside $\Sigma$.

If the surface $\Sigma$ is not convex (fig. 32b), then whatever the inclination of the vector $d\epsilon_{ij}^{p}$ to the surface $\Sigma$ it is always possible to choose the point A such that the condition (18.1) is violated.

Equation (18.1) also imposes definite restrictions on the plastic flow along edges of the loading surface and at its conical points. It was assumed earlier (§17) that the flow on an edge is a linear combination of the flows on the left and right of the edge, i.e. the vector $d\epsilon_{ij}^{p}$ is perpendicular to the edge and lies inside the angle generated by the normals to $\Sigma$ on the two sides of the edge (fig. 33a). This picture now follows from condition (18.1). In the case of a conical point the flow vector $d\epsilon_{ij}^{p}$ must, for the same reason, lie inside the cone generated by the normals to the loading surface in the neighbourhood of the apex (fig. 33b).

18.3. *Case of perfect plasticity*

We have already observed in the preceding section that the case of perfect plasticity, when the loading (yield) surface $\Sigma$ is fixed, is a limiting case of hardening if all consecutive loading surfaces contract to their initial position. If we return to the $\sigma$, $\epsilon$ deformation curve, which symbolizes the relation be-

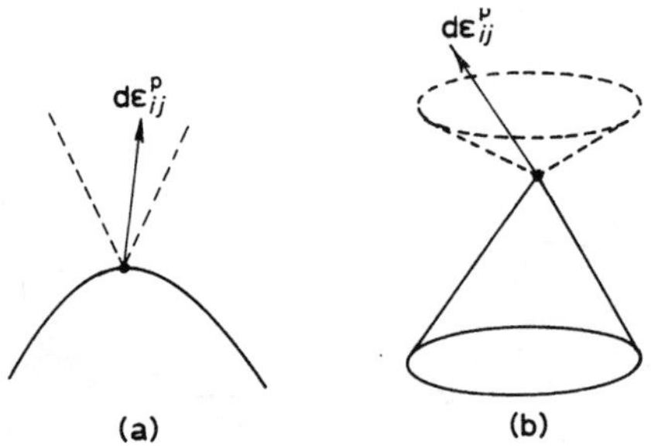


Fig. 33.

tween stress and strain, we see that in a perfectly plastic body $\Delta\sigma = 0$, and consequently $\Delta\sigma\cdot\Delta\epsilon = 0$. The stress-increment $d\sigma_{ij}$ now lies in the tangent plane to the yield surface. The requirement that the work of the additional stresses be positive must now be replaced by the requirement that it be non-negative. In this extended form Drucker's postulate continues to hold for a perfectly plastic material. The inequality (18.1) is now valid with the sign $\geqslant$, and consequently *the yield surface must be convex*. Precisely as before, the plastic flow vector $d\epsilon^p_{ij}$ is normal to the yield surface, i.e. *the associated flow law holds.*

In perfect plasticity the condition (18.3) becomes the equation

$$d\sigma_{ij}d\epsilon^p_{ij} = 0\,, \tag{18.4}$$

expressing the orthogonality of the two vectors. This condition is valid for a loading process B → C and for every cycle B → C → B.

18.4. *Concluding remarks*

Drucker's postulate is essentially a generalization of simple facts. It leads to important conclusions regarding the convexity of the loading surface and the necessity of the associated law of plastic flow. It is evident that the equations of plasticity could now be constructed in a different way from that of previous sections. It would be sufficient to begin with the concept of a loading surface and to assume Drucker's postulate and the continuity condition (§17). The equations of plastic flow (§13) would ensue directly from these propositions.

We note finally that the inequalities (18.1) and (18.2) which follow from Drucker's postulate are a convenient basis for a simple derivation of uniqueness theorems and maximum principles (ch. 8).

## § 19. The equations of thermoplasticity

The elements of many machines and installations are loaded in conditions of high and, frequently, variable temperature. Plastic deformation often occurs in such circumstances. In the presence of a temperature field the analysis of plastic behaviour of metals becomes substantially more complicated, because the yield limit depends on temperature. Henceforth we assume that the temperature is not too high, so that creep can be neglected. This condition can be relaxed if the deformation takes place in the course of a short time interval; for then deformation due to creep does not have time to develop and can be disregarded.

19.1. *Equations of plastic flow theory*

With variable temperature the relative volumetric change is given by the well-known expression

$$\epsilon = 3k\sigma + 3\alpha\theta \ , \tag{19.1}$$

where $k$ is the coefficient of volumetric compression, $\alpha$ the coefficient of linear thermal expansion, and $\theta$ the temperature.

The components of the strain deviatoric $e_{ij}$ obviously do not involve thermal expansions, and consequently the increments in these components are made up of elastic strain-increments and plastic strain-increments:

$$\mathrm{d}e_{ij} = \mathrm{d}e_{ij}^{\mathrm{e}} + \mathrm{d}e_{ij}^{\mathrm{p}} \ . \tag{19.2}$$

The components of the stress and elastic strain deviatorics are related by Hooke's law, i.e.

$$\mathrm{d}e_{ij}^{\mathrm{e}} = \frac{1}{2G}\,\mathrm{d}s_{ij} \ , \tag{19.3}$$

where $G$ is the shear modulus.

Consider now the plastic components $\mathrm{d}e_{ij}^{\mathrm{p}}$. As in the isothermal case, the theory is based on the concept of a *loading surface* $\Sigma$ in stress space, which bounds the region of elastic deformation. In the non-isothermal case the loading surface depends also on the temperature, i.e. it is defined by a relation of the form

$$f(s_{ij}, \theta, q, \ldots) = 0 \ . \tag{19.4}$$

We limit ourselves to consideration of the simple case of isotropic hardening, corresponding to condition (17.2). Then the equation of the loading surface can be written in the form

$$f \equiv s_{ij}s_{ij} - \varphi(q, \theta) = 0 . \tag{19.5}$$

With developing plastic deformation a representative point is located on the loading surface (19.5), so that

$$\mathrm{d}f = \frac{\partial f}{\partial s_{ij}} \mathrm{d}s_{ij} + \frac{\partial f}{\partial \theta} \mathrm{d}\theta + \frac{\partial f}{\partial q} \mathrm{d}q = 0 . \tag{19.6}$$

We examine the loading and unloading criteria. The first two terms in (19.6) are denoted by $\mathrm{d}'f$.

*Unloading*. In this case the representative point moves inwards with respect to the loading surface, i.e. $\mathrm{d}f < 0$; the plastic strains remain unchanged; consequently $\mathrm{d}q = 0$ and

$$\mathrm{d}'f = \frac{\partial f}{\partial s_{ij}} \mathrm{d}s_{ij} + \frac{\partial f}{\partial \theta} \mathrm{d}\theta < 0 .$$

*Neutral changes*. If the representative point is displaced along the loading surface $\Sigma$, then $\mathrm{d}f = 0$. Here no plastic strains occur, i.e. $\mathrm{d}q = 0$, and thus we have neutral changes. In this case

$$\mathrm{d}'f = \frac{\partial f}{\partial s_{ij}} \mathrm{d}s_{ij} + \frac{\partial f}{\partial \theta} \mathrm{d}\theta = 0 .$$

*Loading*. If plastic deformation takes place, the representative point continues to lie on the moving surface $\Sigma$, i.e. $\mathrm{d}f = 0$.

Plastic loading is described by the condition

$$\mathrm{d}'f = \frac{\partial f}{\partial s_{ij}} \mathrm{d}s_{ij} + \frac{\partial f}{\partial \theta} \mathrm{d}\theta > 0 .$$

We now proceed to formulate the dependence relation for the increments in plastic strain components. As in the isothermal case (§17), these increments must be proportional to the quantity $\mathrm{d}'f$, which characterizes the transition from loading to unloading. In the non-isothermal case also we assume the validity of the associated flow law. As a consequence the displacement vector $\mathrm{d}e^{\mathrm{p}}_{ij}$ must be directed along the normal to the loading surface $\Sigma$ in stress space. That is, the quantities $\mathrm{d}e^{\mathrm{p}}_{ij}$ must be proportional to the direction cosines of the normal to $\Sigma$, in other words, proportional to $\partial f/\partial s_{ij}$. Thus

$$\begin{aligned} \mathrm{d}e^{\mathrm{p}}_{ij} &= 0 && \text{for} \quad \mathrm{d}'f < 0 \text{ (unloading)} , \\ \mathrm{d}e^{\mathrm{p}}_{ij} &= g \frac{\partial f}{\partial s_{ij}} \mathrm{d}'f && \text{for} \quad \mathrm{d}'f \geqslant 0 , \end{aligned} \tag{19.7}$$

where $g > 0$ is the hardening function, which characterizes the level of hardening attained and which depends on the history of the deformation and heating. The function $g$ is related to the equation of the loading surface (§17).

The construction of equations of thermoplasticity which are valid for a sufficiently wide range of variations in stress and temperature is a very difficult task. Various aspects of this problem have been examined in the works [5, 69, 87, 90].

19.2. *Case of perfect plasticity*

If there is no hardening, the yield surface is defined by an equation of the form

$$f(s_{ij}, \theta) = 0 .$$

In particular, for the von Mises yield criterion

$$f = \tfrac{1}{2} s_{ij} s_{ij} - k^2(\theta) = 0 , \tag{19.8}$$

where the yield limit $k$ is a function of the temperature $\theta$. From the associated flow law (16.7) and the condition (19.8) we find

$$\begin{aligned} &\mathrm{d}e^{\mathrm{p}}_{ij} = 0 && \text{for unloading} \\ &\mathrm{d}e^{\mathrm{p}}_{ij} = \mathrm{d}\lambda \cdot s_{ij} && \text{for} \quad f = 0 \quad \text{and} \quad \mathrm{d}f = 0 , \end{aligned} \tag{19.9}$$

where the multiplier $\mathrm{d}\lambda$ is proportional to the work of plastic deformation.

19.3. *Deformation theory equations*

As in flow theory, the relative volume changes are here determined in accordance with (19.1), and the components of the strain deviatoric are made up of elastic and plastic strain components:

$$e_{ij} = e^{\mathrm{e}}_{ij} + e^{\mathrm{p}}_{ij} . \tag{19.10}$$

The components of the elastic strain deviatoric are given by Hooke's law (19.3), while the components of the plastic strain deviatoric are given by relations analogous to (14.5):

$$e^{\mathrm{p}}_{ij} = \varphi s_{ij} , \tag{19.11}$$

where for the case of hardening $\varphi = \varphi(T, \theta)$. The deficiencies of deformation theory are even more apparent in the non-isothermal case. Finite temperature changes give rise to single-valued plastic strains. Nevertheless, deformation theory is applied quite widely in calculating thermal stresses at the elastic limit. But it should be emphasized that this procedure is subject to strong restrictions: the loading must be close to simple, and the temperature must change monotonically.

We have been considering the case of hardening. If we have a situation of perfect plasticity, the function $\varphi$ remains indeterminate, but a yield condition is applicable (such as the von Mises yield condition (19.8)).

19.4. *Concluding remarks*

In thermal problems it it usually not possible to neglect elastic deformations. Nevertheless there are some cases in which the development of plastic flow can be described by a rigid-plastic model.

Singular loading (yield) surfaces can be taken into account, just as in the isothermal case. For example, we can consider the Tresca-Saint Venant hexagonal prism, when the flow on an edge is represented just as before (§§16, 17).

The reader will find the solutions to various problems and further generalizations in the literature on thermoplasticity [5, 74, 87, 90].

## PROBLEMS

1. Obtain the von Mises plasticity condition for the case of axisymmetric deformation of a thin-walled tube (§ 7).

2. Obtain the Tresca-Saint Venant plasticity condition ($\tau_{max}$ = const.) for a thin-walled, closed spherical shell under the action of an internal pressure.

3. A thin-walled, closed, spherical shell, made of a hardening material, experiences an internal pressure. Find how the change in the shell's diameter depends on the pressure.

4. The equations of deformation theory (§ 14) and of flow theory (§ 13) are equivalent in the case of uniaxial tension. What is then the relation between the functions $g(T)$ and $F(T)$?

5. A thin, plane sheet is stretched uniformly in all directions in its own plane. Obtain the von Mises and Tresca-Saint Venant yield criteria.

6. A thin plane sheet lying in the $x$, $y$-plane, experiences uniform tension $q$ in the $x$-direction and a uniform compression $p$ in the $y$-direction. Obtain the von Mises and Tresca-Saint Venant yield criteria. What are the directions of the surfaces on which the maximum tangential stress acts?

7. In the plane of the principal stresses $\sigma_1$, $\sigma_2$ the yield curve is defined by the conditions $|\sigma_1| = \sigma_s$, $|\sigma_2| = \sigma_s$. Write down the flow equations in the various regimes according to the associated law.

8. Find the hardening function $g$ in the case of a loading surface of the form $(s_{ij} - c\epsilon_{ij}^p)(s_{ij} - c\epsilon_{ij}^p) = K$.

# 3

# Equations of Elastic-Plastic Equilibrium. The Simplest Problems

## §20. System of equations for plastic equilibrium

In regions of elastic deformation Hooke's law is valid, and the stress and strain fields are described by the system of equations of elasticity theory.

In regions of plastic deformation the equations of deformation theory or of plastic flow theory (or, possibly, more complicated relations) apply. In these cases the systems of equations which characterize the stress and strain fields are substantially more complex. We shall briefly examine these systems.

### 20.1. *Deformation theory*

In this case we obtain, for loading, equations which are superficially rather similar to the equations of elasticity theory. Here we can also derive equations of plastic equilibrium which contain only displacements or only stresses.

The differential equations of equilibrium in displacements are generalizations of the well-known Lamé equations in elasticity theory, and can be expressed in the following form. We utilize formulae (14.17)

$$\sigma_{ij} = \frac{\partial \Pi}{\partial \epsilon_{ij}}, \tag{20.1}$$

where $\Pi$, the potential of the work of deformation, is a function of the components of strain. Substituting (20.1) into the equilibrium equations (4.2) we arrive at a system of three equations:

$$\frac{\partial}{\partial x_i}\left(\frac{\partial \Pi}{\partial \epsilon_{ij}}\right) + \rho F_j = 0 . \tag{20.2}$$

Eliminating the strain components with the aid of formula (2.3), we obtain a system of three non-linear partial differential equations of second order with respect to the unknown functions $u_i$. For the yield state and the hardening state the systems will be different, because the potentials of the work of deformation are different. If we take for $\Pi$ the expression (14.16) for the elastic potential, the equations (20.2) reduce to the Lamé differential equations.

In order to derive the system of equations in stress components it is necessary to supplement the differential equations of equilibrium with certain relations analogous to the Beltrami-von Mises identities of elasticity theory. To do this we need to introduce into the Saint Venant compatibility conditions (2.16) the strain components from Hencky's equation (14.6). In the yield state it is necessary to add the von Mises plasticity equation, which is required for definition of the function $\psi$. In the hardening state the function $\psi$ is determined directly from the stress $2\psi = \bar{g}(T)$, and hence an additional relation is unnecessary.

Because of their complexity the respective systems of equations have not been written out in detail; it is more convenient to formulate the equations directly in specific problems.

It should be understood that for the differential equations of plastic equilibrium the classical methods of solution of elasticity theory are not available. On the other hand numerical methods of solution are very appropriate. Various methods of successive approximation have also been used with success. As a rule, of course, the application of these methods depends on the use of electronic calculating machines.

It is worth mentioning here that modifications of a different method, developed by Southwell [60], have also been successful. In addition it is possible to utilize variational formulations of the corresponding boundary-value problems to construct approximate solutions (cf. §§67, 68).

The solutions of the non-linear equations of deformation theory in the case of hardening can be obtained by different versions of a method of successive approximation. The solution of problems in plasticity theory then reduces to the solution of a sequence of linear problems, each of which can be interpreted as some problem in elasticity theory ("the method of elastic solutions" [12, 86]).

Let us consider, briefly, some of these approaches.

*Method of additional loads.* Equations (14.24) are written in the form

$$\sigma_{ij} = \left(\frac{\epsilon}{2k}\,\delta_{ij} + 2Ge_{ij}\right) + \underline{2\,[g(\Gamma) - G]\,e_{ij}}\,.$$

The underlined terms in this expression represent divergences from Hooke's law. We substitute these relations into the differential equations of equilibrium and boundary conditions (1.2). We then transfer to the right-hand sides of the equations the components arising from the underlined terms, and regard them as known. The equations so obtained can now be integrated as equations of elasticity theory in displacement, but with additional volume and surface forces. In the zeroth approximation we neglect these additional loads and solve the problem of elasticity theory. The values $u_i^{(0)}$ thus obtained are then substituted in the right-hand sides, and the first approximation is the solution of the elastic problem with given loads. The process is then repeated.

*Method of additional strains.* We write (14.23) in the form

$$\epsilon_{ij} = \left(k\sigma\delta_{ij} + \frac{1}{2G}\,s_{ij}\right) + \underline{\frac{1}{2}\left[g(T) - \frac{1}{G}\right] s_{ij}}$$

and solve the problem in stresses. The differential equations of equilibrium and the boundary conditions (1.2) remain unchanged, but the continuity equations contain additional terms which can be interpreted as additional stresses, and can be determined by successive approximation.

*Method of variable coefficients of elasticity.* The system of equations can be written in the form of the equations of elasticity theory with variable "coefficients of elasticity". A successive approximation method can then be applied.

The convergence of these approximation methods has only been partially investigated.

### 20.2. *Theory of plastic flow*

Here the equations are substantially more complicated. Since the equations of plastic flow contain both the stress components and infinitesimal increments in these components, combined in non-integrable form, it is in general not possible to solve these equations for the stresses. Consequently it is impossible to formulate a system of equations in displacements, analogous to (20.2).

A system of equations involving stresses alone can be formulated; however, in addition to derivatives of the stress components with respect to the coordinates, it will also contain derivatives of the infinitesimal increments in the stress components.

In particular problems various methods of numerical integration are usually used, following step by step the development of the plastic state with successive small increments in the load parameter. Examples of such calculations can be found in Hill's book [54]. At each stage it is necessary to solve a certain problem for an elastic isotropic body with variable coefficients of elasticity (complicated, of course, by possible regions of unloading).

The problem is somewhat simplified if it is possible to neglect increments in the components of elastic strain by comparison with increments in the components of plastic strain.

## §21. Continuity conditions at the boundary of elastic and plastic regions

As long as the intensity of the tangential stresses $T$ nowhere reaches the yield limit $\tau_s$, the body remains wholly in an elastic state. With increase of the loads on the body, regions of plasticity will in general develop, and these are separated from the elastic part of the body by a surface

$$\varphi = 0 \qquad \text{or} \qquad \psi = \text{const.} = 1/2G\,, \tag{21.1}$$

on the basis of deformation theory, and by a surface

$$A_p = 0\,, \tag{21.2}$$

on the basis of the equations of flow theory [1]).

We clarify how the components of stress and strain change on passage through the surface $\Sigma$ which separates the regions $V_1$, $V_2$ of the different states of the medium.

We construct at an arbitrary point of this surface an orthogonal system of coordinates $x$, $y$, $z$ such that the $z$ axis is along the direction of the normal to $\Sigma$, while the $x$ and $y$ axes lie in the tangent plane (fig. 24). Quantities in the region $V_1$ will be designated with a single prime, while quantities in $V_2$ will have two primes.

The equilibrium equations for an element of the surface $\Sigma$ obviously lead to the conditions

$$\sigma_z' = \sigma_z''\,, \qquad \tau_{xz}' = \tau_{xz}''\,, \qquad \tau_{yz}' = \tau_{yz}''\,. \tag{21.3}$$

[1]) Note that $\varphi$ and $\psi$ (as well as $A_p$) vary continuously, i.e. that the elastic state passes continuously into the plastic state.

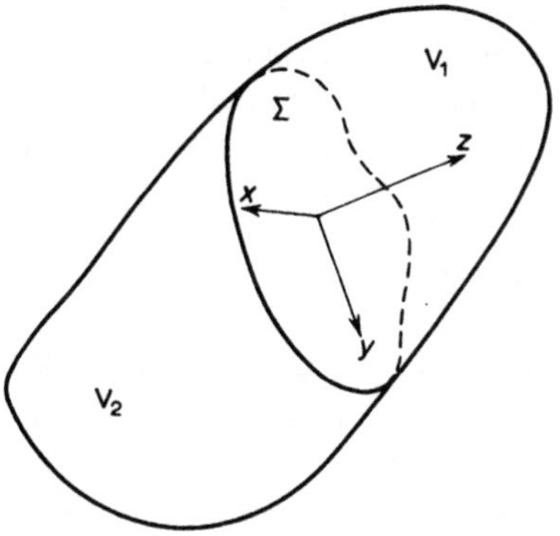


Fig, 34.

We shall assume that the displacements are continuous functions (i.e. that there are no "cracks" or "slips"). Then an arbitrary arc on the surface Σ must experience the same elongation, independently of from which side the surface Σ is approached; this requirement will be fulfilled if

$$\epsilon_x' = \epsilon_x'' , \qquad \epsilon_y' = \epsilon_y'' , \qquad \gamma_{xy}' = \gamma_{xy}'' \tag{21.4}$$

on Σ.

Consider first the equations of *deformation theory*; here $\varphi = 0$ on Σ, so that from the above relations and Hencky's equations it follows that

$$\begin{aligned} \sigma_x' - \nu(\sigma_y' + \sigma_z') &= \sigma_x'' - \nu(\sigma_y'' + \sigma_z'') , \\ \sigma_y' - \nu(\sigma_x' + \sigma_z') &= \sigma_y'' - \nu(\sigma_x'' + \sigma_z'') , \\ \tau_{xy}' &= \tau_{xy}'' . \end{aligned} \tag{21.5}$$

Utilizing the first of relations (21.3), we find from (21.5) that $\sigma_x' = \sigma_x''$, $\sigma_y' = \sigma_y''$, i.e. *on the separation surface Σ all the components of stress are continuous*. It then follows from Hencky's equations that *all strain components are continuous on Σ*.

Turn now to the equations of *plastic flow theory*. For elements lying on Σ on the side of the plastic region, the components of plastic strain are zero. Consider an arbitrary point of the body; initially this point experiences elastic strain, and as the load increases up to the yield limit the separation surface Σ approaches the point. Since the elastic state passes continuously into the yield state, the components of stress and strain on both sides of the surface Σ are related by Hooke's law. But then the arguments applied to the preceding case are again completely relevant, together with the conclusion that all the components of stress and strain are continuous on $\Sigma$.

A similar analysis establishes the continuity of stress and strain components in transition from yield state to hardening state.

## §22. Residual stresses and strains

If on loading a body experiences inhomogeneous deformation, then unloading will be accompanied, in general, by the appearance not only of residual strains, but also of residual stresses.

Suppose we have a state of maximum loading (which can be represented schematically by the point B in fig. 18a), followed by unloading. Let there be body forces $\overline{\mathbf{F}}$, surface forces $\overline{\mathbf{F}}_\mathrm{n}$, stress components $\overline{\sigma}_{ij}$ and strain components $\overline{\epsilon}_{ij}$. In unloading the body conforms to Hooke's law (§11); let the unloading be completed with the vanishing of all the external forces; whereupon the body acquires residual stresses $\sigma_{ij}^0$ and residual strains $\epsilon_{ij}^0$. Assuming the strains to be small we can represent the unloading as the application of reverse forces $-\overline{\mathbf{F}}, -\overline{\mathbf{F}}_\mathrm{n}$.

The original stresses $\overline{\sigma}_{ij}$ and strains $\overline{\epsilon}_{ij}$ can be regarded as some initial (intrinsic) stresses and strains in the body. It is well known that for *small elastic strains* (when, as a rule, the superposition principle for the application of loads is valid) the presence of the initial stresses and strains does not affect the values of the stresses and strains arising from the external forces. In other words, the *elastic* stresses and strains produced by the external forces can be determined by assuming that there are no initial stresses and strains in the body [1]).

Thus, it is possible to find the stresses $\sigma_{ij}^*$ and the strains $\epsilon_{ij}^*$, corresponding to the imaginary applied forces $-\overline{\mathbf{F}}, -\overline{\mathbf{F}}_\mathrm{n}$ without considering the original distribution of stresses $\overline{\sigma}_{ij}$ and strains $\overline{\epsilon}_{ij}$. Since superposition is possible, the residual stresses and strains are equal to the respective sums:

$$\sigma_{ij}^0 = \overline{\sigma}_{ij} + \sigma_{ij}^* ; \qquad \epsilon_{ij}^0 = \overline{\epsilon}_{ij} + \epsilon_{ij}^* . \tag{22.1}$$

The residual displacements $u_i^0$ will obviously be equal to

$$u_i^0 = \overline{u}_i + u_i^* . \tag{22.2}$$

[1]) In elasticity theory it is assumed that there are no stresses and strains in a body in the absence of external forces (hypothesis of the natural state of a body).

The results have meaning only so long as *Hooke's law is not violated* during unloading, i.e. so long as the intensity of the residual stresses $T^0$ does not exceed a certain value, depending on the properties of the material. If this condition is violated then unloading is accompanied by *secondary* plastic deformation. Analysis of unloading in this case is considerably more complicated (see [24, 84]).

We consider as a simple example a system consisting of three rods of equal length $l$ and equal cross-sectional area $F$ (fig. 35). We shall assume the vertical rod to be superfluous and let $x$ be the stress in it. From the equilibrium conditions we find that the stresses in the rods are

$$s_1 = s_3 = p - x\,; \qquad s_2 = x\,; \qquad p = P/F\,. \tag{22.3}$$

For an elastic system

$$s_2 = \tfrac{2}{3}p\,; \qquad s_1 = s_3 = \tfrac{1}{3}p\,; \qquad p \leqslant \tfrac{3}{2}\sigma_s\,. \tag{22.4}$$

When $p = \frac{3}{2}\sigma_s$ the rod 2 passes to a plastic state, and then $s_2 = \sigma_s$. The stresses in the rods in the elastic-plastic system equal

$$\bar{s}_2 = \sigma_s\,; \qquad \bar{s}_1 = \bar{s}_3 = p - \sigma_s\,. \tag{22.5}$$

This solution is correct as long as $\bar{s}_1 = \bar{s}_3 \leqslant \sigma_s$; when $\bar{s}_1 = \bar{s}_3 = \sigma_s$, the limit load is reached and the whole lattice passes into the plastic state. Thus, $p \leqslant 2\sigma_s$.

Subtracting from (22.5) the stresses (22.4) in the elastic state, we find the residual stresses

$$s_2^0 = \sigma_s - \tfrac{2}{3}p\,; \qquad s_1^0 = s_3^0 = \tfrac{2}{3}p - \sigma_s\,. \tag{22.6}$$

Because of the condition $p \leqslant 2\sigma_s$ secondary plastic deformations do not occur with unloading in this case.

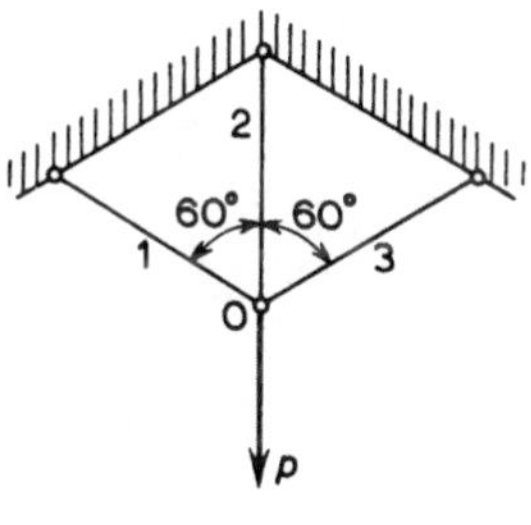


Fig. 35.

## §23. Plastic-rigid body

For sufficiently small loads a body will be in an elastic state. With increase of loads regions of plastic deformation develop in the body; for the non-hardening material we are considering in this section these will be regions of yield. The boundaries of the latter are unknown *a priori* and are determined by the conditions of continuity (§21). The mathematical difficulties which emerge in solving such mixed problems are very great; solutions are known only for the simplest cases. Because of this, further possible simplifications in the formulation of the problem acquire great significance.

First of all one should mention the frequently used assumption of *incompressibility of the material* ($k = 0$). This leads to a considerable simplification of the equations, and in many problems it is a completely acceptable approximation. However, the principal difficulty, that of having to solve a mixed elastic-plastic problem, is not removed.

In recent times there has been significant development of the model of a *plastic-rigid body*; in this model elastic deformations are completely disregarded. The equations of the plastic state are then substantially simplified; we have, for example, the Saint Venant-von Mises equations (13.12).

In other words, the modulus of elasticity is taken to be infinite ($E \to \infty$), which corresponds to changing from a deformation curve with an elastic segment (fig. 36a) to a deformation curve with only one yield plateau (fig. 36b). Dotted lines with arrows show how unloading takes place in both cases.

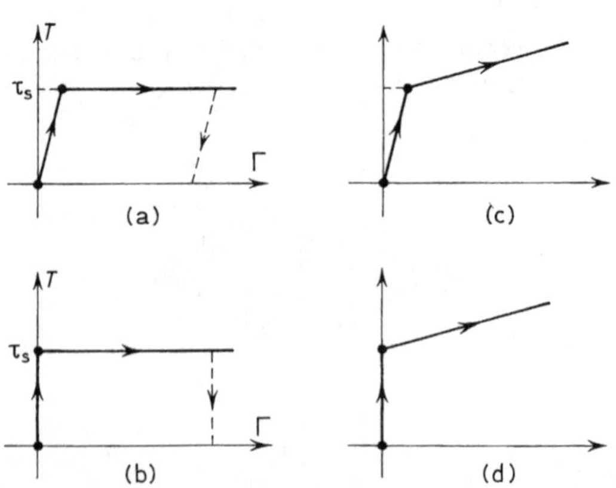


Fig. 36.

In such an arrangement the body remains completely undeformed ("rigid"), so long as the stress state in it does not anywhere come to satisfy the yield condition and the possibility of plastic flow does not arise. If it does, then some parts of the body remain rigid and it is necessary to find solutions in the plastic zones such that the velocities on their boundaries match the velocities of motion of the rigid parts.

Naturally this scheme is not always suitable. It leads to an acceptable approximate solution if the plastic region is such that nothing restrains the development of plastic deformations. A problem that may serve as an example of such a type is the extension of a bar with a sufficiently large hole (figs. 102 and 103); here the plastic deformation is localized in the weakened cross-section. Owing to this the plastic deformations may considerably exceed the elastic deformations, which justifies the use of the model of a plastic-rigid body.

If the plastic region is enclosed within the elastic region (as in the case of a space with a spherical cavity under the action of an internal pressure, fig. 41) or if plastic flow is hindered by the geometric peculiarities of the body's shape or the special character of the boundary conditions, then this plastic-rigid model of a body may lead to considerable errors.

A consistent interpretation of the plastic-rigid model of a body is associated with a series of difficulties. In the first place we note that a solution based on this model may not in general agree with the solution of the elastic-plastic problem in the limit $E \to \infty$. In a series of cases (for example, in pure bending of a rod) elastic regions disappear only with infinitely large curvature, i.e. the indicated limiting transition requires the analysis of large deformations (or the formulation of special conditions for a simultaneous increase in $E$). There are no theorems that would allow us to estimate the closeness of solutions of the elastic-plastic problem to solutions of the plastic-rigid model. Furthermore it is necessary that the stresses in the rigid part should have an acceptable character on their continuation from the plastic zone and should not reach the yield condition, i.e. that $T < \tau_s$. This condition is difficult to test, since in the rigid parts the stress distribution is undetermined. Associated with this is the absence, characteristically for the plastic-rigid model, of a unique velocity field.

Nevertheless the concept of plastic rigidity of a body has permitted the construction of a series of new solutions (for both static and dynamic problems; see, for example, §78), well confirmed by experiment, and has led to a more accurate formulation of many problems in plasticity theory.

In conclusion we note that analogous to the plastic-rigid model of a body (characterized by the yield plateau), the model of a *rigid-hardening body* is sometimes introduced. This is shown in fig. 36d, for the case of linear hardening. Here also elastic strain is completely neglected.

## §24. Elastic-plastic bending of a beam

We consider the problem of elastic-plastic bending of a beam; for simplicity we shall take a beam whose cross-section possesses two axes of symmetry (fig. 37).

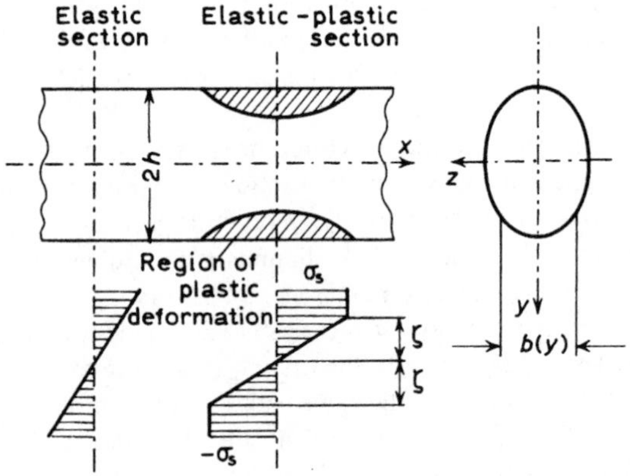


Fig. 37.

### 24.1. *Pure bending*

We shall consider pure bending of a beam of constant cross-section. Let all stress components, except $\sigma_x$, equal zero, while $\sigma_x$ is a function only of the coordinate $y$. For an elastic beam

$$\sigma_x = My/J \,,$$

where $M$ is the bending moment and $J$ is the moment of inertia of the section.

When hardening is absent, the yield condition gives that in the plastic zones

$$|\sigma_x| = \sigma_s \,.$$

It is evident that with increasing $M$ the loading on each element is simple, and consequently we can proceed from the equations of deformation theory.

It is not difficult to see that, on the hypothesis of planar sections and by the equations of this theory, the strain components will be

$$\epsilon_x = -\frac{d^2 v}{dx^2} y\,, \qquad \epsilon_y = \epsilon_z = -\tfrac{1}{2}\epsilon_x + \tfrac{1}{2}k\sigma_s\,, \qquad \gamma_{xy} = \gamma_{yz} = \gamma_{xz} = 0\,,$$

where $v = v(x)$ is the displacement of the axis of the beam (sagging). Since the strain components are linear functions of the coordinate $y$, Saint Venant's identities are satisfied. In addition the continuity conditions are identically satisfied on the surface of separation. Thus [1])

$$\sigma_x = \begin{cases} \sigma_s y/\zeta & \text{for} \quad |y| \leqslant \zeta\,, \\ \sigma_s \cdot \operatorname{sign} y & \text{for} \quad |y| \geqslant \zeta\,. \end{cases} \tag{24.1}$$

where $\zeta$ is the distance from the neutral plane of the beam to the yield zone on the given section. The bending moment $M$ is assumed positive; with negative moment a minus sign should be placed in front of $\sigma_s$.

The moment of stresses is equal to the bending moment

$$M = \frac{\sigma_s}{\zeta} J_e + \sigma_s S_p\,, \tag{24.2}$$

where $J_e$ is the moment of inertia of the elastic nucleus and $\frac{1}{2}S_p$ is the static moment of one of the plastic zones with respect to the axis $z$:

$$J_e = 2 \int_0^{\zeta} b(y)\, y^2\, dy\,, \qquad S_p = 2 \int_{\zeta}^{h} b(y)\, y\, dy\,.$$

Here $b(y)$ is the width of the cross-section, and $2h$ is the full height of the profile. Thus a cross-section of given form implies a specific relationship, $M = M(\zeta)$ or, inversely $\zeta = \zeta(M)$. By Hooke's law we have for the nucleus:

$$\sigma_x = -Ey \frac{d^2 v}{dx^2}\,.$$

On the boundary of the elastic nucleus $y = \zeta$, $\sigma_x = \sigma_s$, therefore the curvature of the axis of the beam is determined by the equation

$$\frac{d^2 v}{dx^2} = -\frac{\sigma_s}{E} \frac{1}{\zeta(M)}\,. \tag{24.3}$$

[1]) The function sign $y$ is given by the equalities: sign = + 1 for $y > 0$, sign = −1 for $y < 0$, sign 0 = 0.

The solution obtained satisfies all the equations of elastic-plastic equilibrium. With a negative moment the sign in front of $\sigma_s$ should be changed.

When the bending moment in the beam is removed residual strains and stresses develop. These are calculated according to the model discussed in §22. Suppose that an elastic-plastic distribution of stresses $\bar{\sigma}_x$ (continuous curve in fig. 38a) corresponds to a given bending moment $M$. On this graph the stress distribution $\sigma_x^*$ *in an elastic beam* with the same bending moment is indicated by a dotted line. Subtracting these curves we obtain the graph of the residual stresses $\sigma_x^0$ (fig. 38b). The residual curvature of the beam can be found by subtracting the curvature of the elastic beam from the curvature of the elastic-plastic beam.

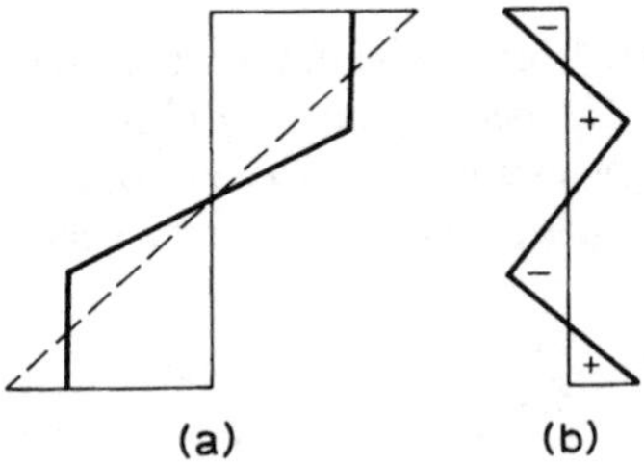


Fig. 38.

As the bending moment increases the zone of plastic deformation expands (i.e. $\zeta$ decreases); in the limit $\zeta = 0$ and the bending moment is

$$M_* = \sigma_s S_p \,. \tag{24.4}$$

This value of the bending moment is called its *limit value*; it corresponds precisely to the plastic state of the beam, when the stress curve in cross-sections has the form shown in fig. 39.

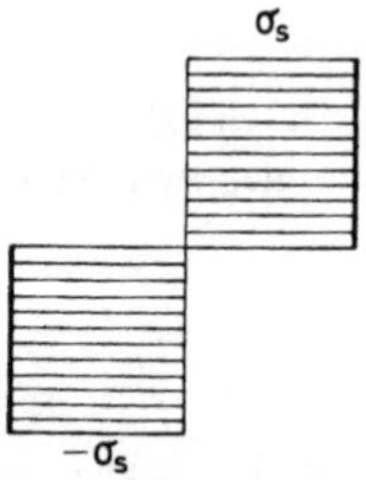


Fig. 39.

The neutral plane here is the plane of stress discontinuity, and the curvature for the limit state tends to infinity. It must be emphasized, however, that even for comparatively small plastic deformations the value of the bending moment is close to its limit, so that the concept of a limit moment has practical significance. Consider as an example a rectangular cross-section; here

$$M_* = bh^2\sigma_s\,, \qquad M/M_* = 1 - \tfrac{1}{3}(\zeta/h)^2\,. \tag{24.5}$$

Hence it is clear that even when $\zeta/h = \frac{1}{3}$, the bending moment differs from its limiting value by less than 4%.

24.2. *Transverse bending*

Bending under the action of a transverse load is more complex and is accompanied, in particular, by tangential stresses $\tau_{xy}$; but in most applications (i.e. in sufficiently long beams) the latter can be neglected. This is due to the fact that the hypotheses of the theory of thin rods have on the whole a geometric character.

The bending moment varies along the length of the beam and so does $\zeta$. The distribution of plastic zones along the length of a beam of given cross-section can be easily calculated if we introduce the bending moment as a function of $x$ into the relation $\zeta = \zeta(M)$. It is necessary to distinguish between the segment of the beam which is elastically deformed and the segment which undergoes elastic-plastic deformation (fig. 37). The differential equation of bending of an elastic beam holds on the former, while on elastic-plastic segments it is necessary to proceed from the differential equation (24.3). In statically determinate problems the right-hand side of the equation will here be a known function of $x$; in statically indeterminate problems we have to introduce superfluous unknowns. In either case the differential equation (24.3) is easy to solve. At points where the elastic and elastic-plastic segments meet, the deflection and the angle of slope of the tangent line to the elastic segment must be continuous.

The limit moment can be found from our previous formula (24.2); in contrast with pure bending, we have here the characteristic appearance of a "plastic joint", which forms in the section, experiencing the action of the maximum bending moment.

As an example consider the bending of a beam of rectangular cross-section under a concentrated force $2P$ (fig. 40). Here

$$\frac{1}{\sqrt{3}}\,\frac{\zeta}{h} = \sqrt{1 - \frac{M}{M_*}}\,, \qquad M = Pl\left(1 - \frac{x}{l}\right).$$

Thus the separation-boundary is a parabola; in the limit state the vertex of

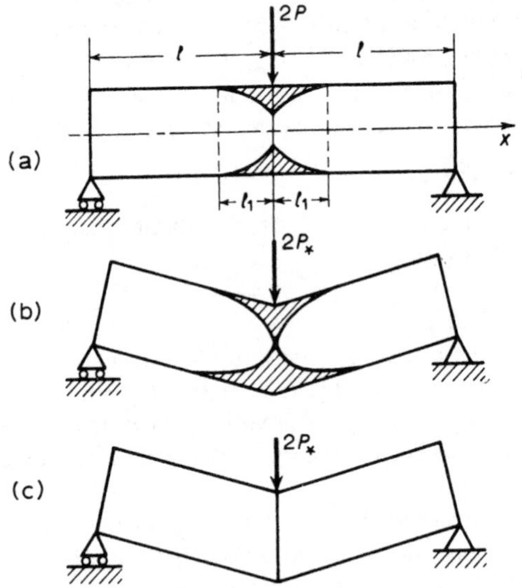


Fig. 40.

the parabola passes through the origin. The value of the limit load is found from the condition of development of a plastic joint ($\zeta = 0$)

$$P_* l = M_* .$$

At this instant the load-bearing power of the beam is depleted, and the beam is converted into a "mechanism" with a plastic joint (fig. 40b). The length of the elastic-plastic segment $2l_1$ attains its maximum value $2 \times 0.3l$.

It is natural to assume that the limit load is the breaking load for a beam, and that consequently the choice of the latter's dimensions should be governed by a safety factor based on the limit load.

The method of calculation based on limit loads has considerable advantages over the method based on limit (maximum) stresses. The latter have a local character and do not typify the strength of the whole structure when it is built of plastic metal and works under a dead load. In these conditions local overstresses are not dangerous, and their evaluation from the elastic model gives an incorrect representation of the reserve strength of the structure.

Experimental data substantially confirm both the distribution of plastic zones and the magnitudes of the bending and limit moments. The preceding results can easily be generalized to the case of a cross-section with one axis of symmetry. It is also quite easy to take into account, approximately, the tangential stresses under bending.

Calculations of beams and frames with limit loads have been extensively carried out and are to be found in a series of monographs. We mention here in particular the books by A.A. Gvozdev [7], B. Neal [26] and P. Hodge [55]; extensive references can be found in these. See also the work by G.S. Shapiro [170].

It is interesting to examine the behaviour of a beam proceeding from the model of a rigid-plastic body (§23). In this model the beam remains rigid (undeformed) as long as the bending moment does not reach the limit value $M_*$. Plastic deformation then occurs in the section under the force (fig. 40c), and the beam is "partially broken". Localization of the plastic deformation is, of course, connected with the fact that the beam is being regarded as a one-dimensional continuum, with tangential stresses neglected. A more complete picture of the limit equilibrium of a rigid-plastic beam will be considered later (in ch. 5).

### 24.3. *Bending of beams of hardening material*

The bending of beams of hardening material can be examined on the basis of assumptions similar to those introduced earlier. We shall not discuss this question, referring only to sources in the literature [25, 26].

We note that when the hardening is insignificant the bending of beams can be treated according to the foregoing formulae, provided that the stress $\sigma_s$ is taken as the mean stress on the hardening segment in the deformation interval under consideration.

## §25. Hollow sphere under pressure

### 25.1. *Formulation of the problem*

Consider the elastic-plastic equilibrium of a hollow sphere under the action of an internal pressure $p$. Because of the spherical symmetry ($r, \varphi, \chi$ are spherical coordinates) the shears $\gamma_{r\varphi}, \gamma_{\varphi\chi}, \gamma_{\chi r}$ and the tangential stresses $\tau_{r\varphi}, \tau_{\varphi\chi}, \tau_{\chi r}$ are zero, and $\epsilon_\varphi = \epsilon_\chi$, $\sigma_\varphi = \sigma_\chi$. Each element of the sphere experiences simple loading since the principal directions are invariant and the coefficient $\mu_\sigma = \pm 1$ (the upper sign corresponds to the case $\sigma_r > \sigma_\varphi$ and the lower to the case $\sigma_r < \sigma_\varphi$). Thus the solution to this problem can be directly obtained from the equations of deformation theory.

The intensity of the tangential stresses in this problem is

$$T = \frac{1}{\sqrt{3}} |\sigma_\varphi - \sigma_r| .$$

The normal stresses $\sigma_r$, $\sigma_\varphi$ satisfy the equilibrium equation

$$\frac{d\sigma_r}{dr} + 2\frac{\sigma_r - \sigma_\varphi}{r} = 0 , \tag{25.1}$$

and the components of strain

$$\epsilon_r = \frac{du}{dr} , \qquad \epsilon_\varphi = \frac{u}{r} ,$$

where $u$ is the radial displacement, satisfy the continuity condition

$$\frac{d\epsilon_\varphi}{dr} + \frac{\epsilon_\varphi - \epsilon_r}{r} = 0 . \tag{25.2}$$

The boundary conditions have the form

$$\text{when} \quad r = a \quad \sigma_r = -p , \tag{25.3}$$

$$\text{when} \quad r = b \quad \sigma_r = 0 . \tag{25.4}$$

25.2. *Initial elastic state*

If the pressure is moderate the sphere is in an elastic state.

Using the above equations together with Hooke's law we easily find

$$\sigma_r = \tilde{p} \left(1 - \frac{b^3}{r^3}\right) ,$$

$$\sigma_\varphi = \tilde{p} \left(1 + \frac{1}{2}\frac{b^3}{r^3}\right) , \tag{25.5}$$

$$u = \tilde{p} r \left(k + \frac{1}{4G}\frac{b^3}{r^3}\right) ,$$

where

$$\tilde{p} = p \frac{a^3}{b^3 - a^3} .$$

The stress distribution is shown by a dotted line in fig. 41. The intensity of the tangential stresses

$$T = \tilde{p} \frac{\sqrt{3}}{2} \frac{b^3}{r^3}$$

is a maximum when $r = a$.

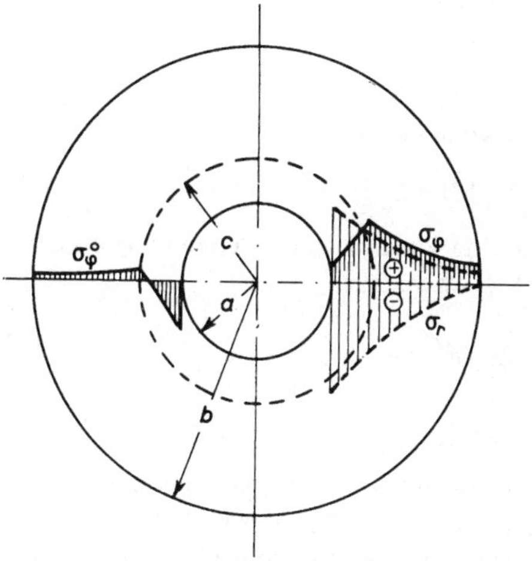


Fig. 41.

When the pressure

$$\tilde{p} < \frac{2}{3}\left(\frac{a}{b}\right)^3 \sigma_s \equiv \tilde{p}_0$$

the sphere remains in an elastic state. When $\tilde{p} = \tilde{p}_0$ the material of the sphere goes over into a plastic state on the inner surface $r = a$. On further increase the region of plastic deformation is enlarged.

25.3. *Elastic-plastic state*

The yield condition (neglecting hardening) takes the form

$$\sigma_\varphi - \sigma_r = \pm\, \sigma_s\,. \tag{25.6}$$

The sign of the difference $\sigma_\varphi - \sigma_r$ at the onset of plastic deformation is known from the solution of the elastic problem (25.5). With increasing pressure the zone of plasticity grows but the sign of $\sigma_\varphi - \sigma_r$ remains the same by virtue of the continuity of $T$. Note that this method of choosing the sign is based on knowledge of the "history" of the onset of the plastic zone, and is, of course, available in other problems; thus,

$$\sigma_\varphi - \sigma_r = +\, \sigma_s\,. \tag{25.7}$$

With the aid of this condition we can reduce the equilibrium equation to the form

$$\frac{d\sigma_r}{dr} - 2\frac{\sigma_s}{r} = 0 ,$$

from which it follows immediately that

$$\sigma_r = 2\sigma_s \ln r + C_1 ,$$

where $C_1$ is an arbitrary constant. On determining its value from the boundary condition (25.3), we obtain

$$\sigma_r = 2\sigma_s \ln\frac{r}{a} - p ,$$
$$\sigma_\varphi = \sigma_r + \sigma_s . \tag{25.8}$$

Here we have an example of a "statically determinate problem", where the stresses in the yield zone are completely determined by the equilibrium equations and the yield conditions (without involving the strains). Statically determinate problems constitute an important class of problems, characteristic for the yield state.

To determine the strains and displacements in the yield zone we make use of Hencky's relations:

$$\epsilon_r = \frac{du}{dr} = \psi(\sigma_r - \sigma) + k\sigma ,$$
$$\epsilon_\varphi = \frac{u}{r} = \psi(\sigma_\varphi - \sigma) + k\sigma . \tag{25.9}$$

Since the strain components have to satisfy the continuity condition (25.2), it follows that if we substitute $\epsilon_r$, $\epsilon_\varphi$ from (25.9) and $\sigma_r$, $\sigma_\varphi$ from (25.8) we obtain the differential equation

$$\frac{d\psi}{dr} + \frac{3}{r}\psi + \frac{6k}{r} = 0 ,$$

whose solution has the form

$$\psi = -2k + \frac{C_2}{r^3} , \tag{25.10}$$

where $C_2$ is an arbitrary constant.

To solve the mixed elastic-plastic problem it is necessary to write down the solution of the elastic problem in the region ($c \leqslant r \leqslant b$), where the boundary

$c$ has to be determined. This solution can be obtained from formulae (25.5), if we replace $-p$ and $a$ by the quantities $q$ and $c$, where $q$ is the stress $\sigma_r$ on the boundary of the regions of elasticity and yield.

To determine the unknown constants $c$, $q$, $C_2$ we have the condition of continuity of state

$$\psi = 1/2G \qquad \text{when} \qquad r = c\,,$$

the condition of continuity of radial stress

$$\sigma_r\big|_{r=c-0} = \sigma_r\big|_{r=c+0}$$

and the condition of continuity of the displacement

$$u\big|_{r=c-0} = u\big|_{r=c+0}\,.$$

From the first of these we find

$$\psi = -2k + \left(\frac{1}{2G} + 2k\right)\left(\frac{c}{r}\right)^3 .$$

The others lead to the equations

$$\begin{aligned} q &= 2\sigma_s \ln\frac{c}{a} - p\,, \\ \ln\frac{c}{a} - \frac{1}{3}\left(\frac{c}{b}\right)^3 &= \frac{p}{2\sigma_s} - \frac{1}{3}\,. \end{aligned} \tag{25.11}$$

The stress distribution in the elastic-plastic state is shown in fig. 41.

## 25.4. *Influence of compressibility*

The above solution enables us to estimate the influence of the compressibility of the material. First of all we note that the stresses in the elastic and plastic zones, as well as the radius of the extent of the latter, are independent of the coefficient of volumetric compression $k$. Moreover equation (25.5) shows that in the elastic zone the ratio of the displacement $u$ to the displacement $u'$ for an incompressible sphere ($k = 0$) is

$$u/u' = 1 + 4kG\,(r/b)^3\,.$$

When $\nu = 0.3$ the maximum value of this ratio is attained when $r = b$ and is equal to 1.615; thus in a given elastic-plastic problem the displacements depend significantly on the value of the Poisson coefficient. We can obviously assume that in other problems also neglect of changes in volume introduces insignificant errors in the determination of the *basic* components of the stress state, if the loads on the surface of the body are prescribed.

25.5. *Residual stresses and strains*

Let the pressure $p$ be removed; then residual stresses and strains arise in the sphere. To calculate these it is necessary to find the stresses $\sigma_r^*, \sigma_\varphi^*$ in an elastic sphere subject to a *tension* $p$. These stresses can be determined from formulae (25.5) if the sign in front of $p$ is reserved.

From (22.1) the residual stresses have the form

$$\left.\begin{aligned} \sigma_r^0 &= 2\sigma_s \ln\frac{r}{a} - p - \tilde{p}\left(1 - \frac{b^3}{r^3}\right), \\ \sigma_\varphi^0 &= 2\sigma_r \ln\frac{r}{a} - p + \sigma_s - \tilde{p}\left(1 + \frac{1}{2}\frac{b^3}{r^3}\right) \end{aligned}\right\} \quad \text{for} \quad a \leqslant r \leqslant c\,, \tag{25.12}$$

$$\left.\begin{aligned} \sigma_r^0 &= -(\tilde{p} + \tilde{q})\left(1 - \frac{b^3}{r^3}\right), \\ \sigma_\varphi^0 &= -(\tilde{p} + \tilde{q})\left(1 + \frac{1}{2}\frac{b^3}{r^3}\right) \end{aligned}\right\} \quad \text{for} \quad r \geqslant c\,,$$

where we have put

$$\tilde{q} \equiv q\frac{c^3}{b^3 - c^3}\,.$$

These formulae are valid so long as the intensity of the residual tangential stresses does not exceed the yield limit (in accordance with the condition $T^0 < \tau_s$). Since max $|\sigma_\varphi^0 - \sigma_r^0|$ is attained when $r = a$, it follows that these formulae are valid when

$$\tilde{p} < 2\tilde{p}_0\,. \tag{25.13}$$

The distribution of residual stresses $\sigma_\varphi^0$ is given on the left-hand side of fig. 41; in the neighbourhood of the cavity the residual stresses are compressible.

If now a pressure is reintroduced, which does not exceed the original value, then no new plastic deformations take place in the sphere. Indeed the first effect of new loading will be the emergence of additional stresses and strains, in accordance with the equations of elasticity theory and independent of the presence of intrinsic stresses. On the other hand the attainment of the elastic limit will be determined also by the sizes of the intrinsic (in this case residual) stresses, which must be added to the stresses induced by the new loading. The sphere behaves as if it were strengthened in comparison with its first loading.

This effect is called *strengthening of the structure*, or *autofrettage*. It is widely used as a technique for enhancing the strength of a structure by means of preliminary plastic deformation. The term *shakedown* is also used. The structure shakes down to a variable load as a result of the emergence of a favourable field of residual stresses. In a given problem condition (25.13) determines the *region of shakedown*. General theorems on shakedown will be presented in ch. 9.

25.6. *Limit load*

If we increase the pressure $p$, the plastic zone will expand ($c \to b$) until it reaches the outer surface of the sphere ($c = b$). The solution (25.8) will then be valid up to $r = b$; because of the boundary condition (25.4) we have

$$2\sigma_s \ln (b/a) - p = 0 .$$

This equation defines the *limit pressure* for which the sphere is entirely in a yield state:

$$p_* = 2\sigma_s \ln (b/a) .$$

It should be emphasized that the limit pressure can be found very simply; it is not necessary to consider the elastic solutions, and in our problem it is not even necessary to consider the strains.

Let us look at the displacements of the outer surface of the sphere. Initially, when the sphere is in the elastic state, the displacement is proportional to the pressure. After this it has a transition section, corresponding to the elastic-plastic state (fig. 42). When the limit value $p_*$ is reached the displace-

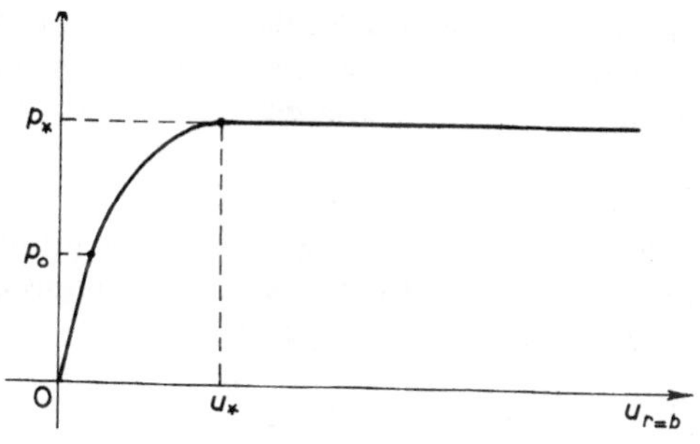


Fig. 42.

ment is equal to $u_*$ (fig. 42); subsequently the displacement is indeterminate, since it contains a function $\psi$ with the integration constant $C_2$; in a purely plastic state the latter can be found without the imposition of any conditions (it is also necessary to prescribe the displacement on $r = a$, for example).

Thus, when the limit load $p_*$ is reached the sphere loses the ability to resist increasing external forces; it "opens up", its load-bearing capacity has been exhausted. When considering the strength of a sphere under static pressure, it is natural to think in terms of the limit load $p_*$ in introducing some safety factor. Note that the model of a plastic-rigid body leads to the same value of the limit load.

25.7. *Solution of the problem in displacements*

The solution of our problem in displacements is easily found. From (14.20) we have

$$\sigma_r = \frac{1}{3k}(\epsilon_r + 2\epsilon_\varphi) \mp \frac{2}{3}\sigma_s ,$$

$$\sigma_\varphi = \frac{1}{3k}(\epsilon_r + 2\epsilon_\varphi) \pm \frac{1}{3}\sigma_s .$$

Introducing here the values of the strain components and then substituting $\sigma_r$, $\sigma_\varphi$ in the equilibrium equation (25.1), we obtain the differential equation

$$r^2 \frac{d^2u}{dr^2} + 2r\frac{du}{dr} - 2u \mp 6k\sigma_s r = 0 ;$$

its solution is

$$u = C_1' r + C_2' \frac{1}{r^2} \pm 2k\sigma_s r \ln r ,$$

where $C_1'$, $C_2'$ are arbitrary constants. Further calculations lead to the determination of the arbitrary constants from the continuity condition at $r = c$ and the boundary condition at $r = a$. These lead to the previous results.

25.8. *Plastic deformation around a spherical cavity in an unbounded body*

We superpose on our preceding solutions in elastic and plastic regions a uniformly applied tension $+p$. The plasticity condition remains unchanged and in the yield zone we have

$$\sigma_r = 2\sigma_s \ln(r/a) ,$$

$$\sigma_\varphi = \sigma_r + \sigma_s .$$

The inner surface of the sphere is free from stresses and the outer surface is subject to the applied tensile stresses $p$. Taking the limit $b \to \infty$ we obtain in the elastic part of the space

$$\sigma_r = p + (q-p)\frac{c^3}{r^3},$$

$$\sigma_\varphi = p - \tfrac{1}{2}(q-p)\frac{c^3}{r^3}.$$

The relations (25.11) take the form

$$q = p - \tfrac{2}{3}\sigma_s, \qquad c = a \exp\left(\frac{p}{2\sigma_s} - \frac{1}{3}\right).$$

Since $c \geqslant a$, the condition for emergence of a yield zone is

$$p \geqslant \tfrac{2}{3}\sigma_s.$$

The solid lines in fig. 43 indicate the stress distribution, while the dotted line shows the stresses in an ideal elastic body. Thus the coefficient of stress concentration is decreased as a result of plastic deformation. In the lower part of fig. 43 a diagram of the residual stresses $\sigma_\varphi^0$ is illustrated.

Note that for a sphere the problem is soluble in quadratures in the presence of hardening, a temperature jump and body forces [16]. The case of large deformation of a hollow sphere has also been examined (cf., for example, the book by Hill [54]).

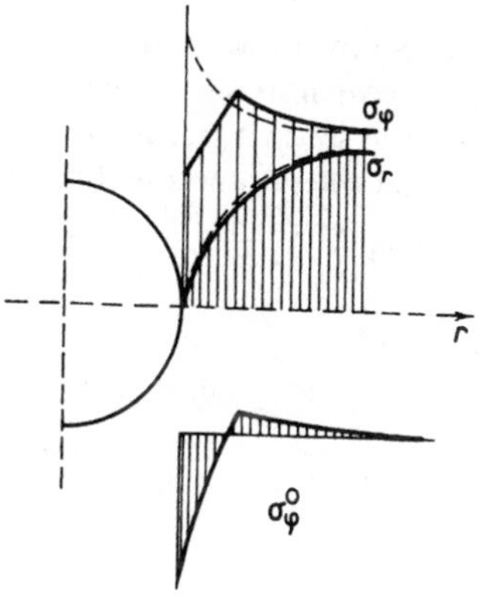


Fig. 43.

## §26. Cylindrical tube under pressure

### 26.1. *Formulation of the problem*

A cylindrical tube subject to internal pressure is an important element of many machines and structures; it is therefore natural that a large number of theoretical and experimental investigations should have been devoted to evaluating the plastic deformation of tubes. A rigorous analysis of this problem presents considerable difficulties and results are achieved by numerical methods or the method of successive approximation. However it is possible to obtain a simple approximate solution if we introduce a number of simplifications, which are justified by the results of numerical integration.

We shall examine in detail the case of a long tube; a force $p\pi a^2$ acts along the axis of the tube, where $a$ is its internal radius. Other cases are briefly discussed at the end of the section.

### 26.2. *Initial elastic state*

The stress distribution in an elastic tube is described by the well-known Lamé solution

$$\begin{aligned} \sigma_r &= -\tilde{p}\,(b^2/r^2-1)\,, \\ \sigma_\varphi &= \tilde{p}\,(b^2/r^2+1)\,, \\ \sigma_z &= \tilde{p} = \tfrac{1}{2}(\sigma_r+\sigma_\varphi)\,, \end{aligned} \tag{26.1}$$

where

$$\tilde{p} = pa^2/(b^2-a^2)\,.$$

Graphs of the stresses are shown dotted in fig. 44.

The intensity of the tangential stresses is calculated with the aid of the expressions (26.1). It is easy to see that, according to the yield condition, the plastic state is reached on the interior surface of the tube when the pressure is

$$p_0 = \tau_s\,(1-a^2/b^2)\,.$$

### 26.3. *Case of a thin-walled tube*

A purely plastic state for a thin-walled tube is characterized by stresses

$$\sigma_r \approx 0\,, \qquad \sigma_\varphi \approx \frac{pa}{b-a}\,, \qquad \sigma_z \approx \frac{pa}{2(b-a)} \approx \tfrac{1}{2}(\sigma_r+\sigma_\varphi)\,,$$

where

$$p = 2\tau_s\,(b/a-1)\,.$$

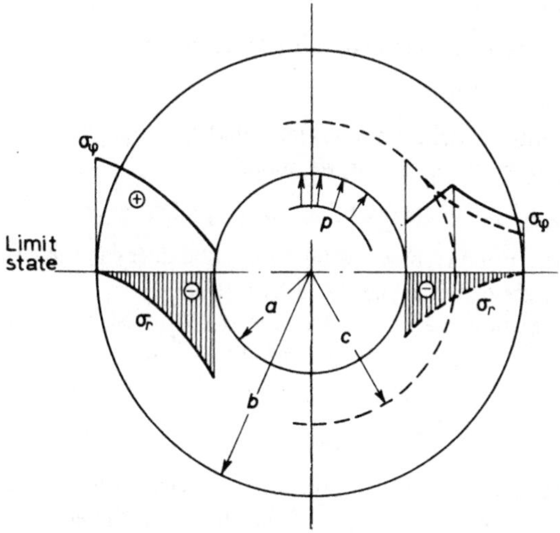


Fig. 44.

26.4. *Elastic-plastic state of a thick-walled tube*

As we have pointed out already, an exact solution of this problem involves substantial difficulties. An approximate solution can be based on the following considerations, which are confirmed by solutions obtained through numerical integration.

In the elastic state $\sigma_z$ is half the sum $\sigma_r + \sigma_\varphi$; this is also true for a purely plastic state of a thin-walled tube. It can be assumed that in other cases $2\sigma_z = \sigma_r + \sigma_\varphi$ too. The parameter $\mu_\sigma$ is then constant ($\mu_\sigma = 0$), and, consequently, the loading is simple and we can proceed directly from the equations of deformation theory. Note that the mean pressure is $\sigma = \sigma_z$.

Plastic deformation develops in the annulus $a \leqslant r \leqslant c$.

In the elastic zone $c \leqslant r \leqslant b$ the formulae (26.1) are valid if in place of $\tilde{p}$ we introduce

$$\tilde{q} = \frac{-qc^2}{b^2-c^2},$$

where $q$ is the radial stress on the separation line $r = c$.

In the plastic zone we have the differential equation of equilibrium

$$\frac{d\sigma_r}{dr}+\frac{\sigma_r-\sigma_\varphi}{r}=0\,.$$

The von Mises yield criterion in our case takes the form

$$\sigma_\varphi-\sigma_r=2\tau_s\,. \tag{26.2}$$

But now the differential equation is immediately integrable; using the boundary condition $\sigma_r=-p$ when $r=a$, we obtain

$$\sigma_r=-p+2\tau_s\ln(r/a)\,,\qquad \sigma_\varphi=\sigma_r+2\tau_s\,. \tag{26.3}$$

The stress distribution is shown in fig. 44 by solid lines (on the left in the limit state, on the right $\sigma_\varphi$ in the elastic-plastic state).

On the separation line $r=c$ the stresses $\sigma_r$, $\sigma_\varphi$ must be continuous; these requirements will be fulfilled if $c$ satisfies the equation

$$\ln\frac{c}{a}+\frac{1}{2}\left(1-\frac{c^2}{b^2}\right)=\frac{p}{2\tau_s}. \tag{26.4}$$

This gives the radius $c$ of the plastic zone; moreover, (26.3) enables the evaluation of $q$ and the determination of the stresses in the elastic region. From Hooke's law

$$\epsilon_\varphi=\frac{u}{r}=\frac{1}{E}\left[\sigma_\varphi-\nu(\sigma_r+\sigma_z)\right]$$

we obtain the displacements in the elastic region.

By Hencky's relations the components of strain in the plastic zone are

$$\epsilon_r=k\sigma-\tau_s\psi\,,\qquad \epsilon_\varphi=k\sigma+\tau_s\psi\,.$$

Substituting these values into the continuity condition

$$\frac{d\epsilon_\varphi}{dr}+\frac{\epsilon_\varphi-\epsilon_r}{r}=0$$

and calculating the mean pressure with the aid of formula (26.3), we obtain for the function $\psi$ the differential equation

$$\frac{d\psi}{dr}+\frac{2}{r}\psi+\frac{2k}{r}=0\,.$$

Hence we find that

$$\psi=-k+\frac{C}{r^2},$$

where $C$ is an arbitrary constant.

It is still necessary to satisfy the condition of continuous transition from the plastic state to the elastic, $\psi = 1/2G$ at $r = c$, and the condition of continuity of the displacement $u$ when $r = c$. Since the mean pressure $\sigma$ is continuous, it follows from the second condition that when $r = c$

$$\tau_s \psi = \frac{1}{4G}(\sigma_\varphi - \sigma_r) .$$

But on the separation line the yield condition is satisfied; hence it follows that $\psi = 1/2G$ when $r = c$, i.e. the first condition. Consequently both conditions will be fulfilled if

$$C = c^2 \left(\frac{1}{2G} + k\right) .$$

The corresponding axial elongation is equal to $\epsilon_z = k\sigma$ and for a long tube this must be constant. In our approximate solution this requirement is not satisfied in the plastic zone. For an incompressible material ($k = 0$) the solution is exact, since $\epsilon_z = 0$.

26.5. *Limit state*

The limit state is attained when $c = b$; from (26.4) we obtain the limit pressure

$$p_* = 2\tau_s \ln (b/a) .$$

This formula is widely used in calculating the strength of thick-walled cylindrical tubes and vessels. The stress distribution $\sigma_r$, $\sigma_\varphi$ in the limit state is shown on the left-hand side of fig. 44.

26.6. *Other cases*

We dwell briefly on other cases of plastic deformation of a tube.

If the tube experiences plane strain, it is necessary to begin with the condition $\epsilon_z = 0$. In this case the solution obtained earlier will be valid for an incompressible material. The corresponding elongation will be small if compressibility is taken into account; this fact enables the earlier solution for $\sigma_r$, $\sigma_\varphi$ to be used as an approximation for the case of plane strain.

If the ends of the tube are free the axial force is zero. In this case an approximate solution based on the assumption $\sigma_z = 0$ gives good results.

Note that in all cases exact calculations using plastic flow theory and deformation theory give close results.

26.7. *Concluding remarks*

As we have pointed out this problem has been studied by many workers. Inclusion of the hardening does not give rise to any significant additional difficulties.

Numerical methods of calculation for the tube, using plastic flow theory, have been considered by Hill, Lee and Tupper [54] and Thomas [194]; numerical methods in deformation theory have been used by V.V. Sokoloskii [44], Allen and Sopwith [176] and others. Large deformations of a tube have been discussed in [54]. The influence of thermal stresses has also been examined.

The book by A.A. Ilyshin and P.M. Ogibalov [15] treats various problems of elastic-plastic deformation of hollow cylinders.

## PROBLEMS

1. Consider the bending of a cantilever of rectangular transverse cross-section under a force applied at the end. Find the regions of plastic deformation, the limiting value of the force, and the deflection of the cantilever in an elastic-plastic state.

2. Find the limit load for a uniformly loaded beam of circular transverse cross-section, supported at an end.

3. Derive the differential equation for the deflection of a beam

$$D\frac{d^2v}{dx^2} = \pm |M|^{1/\mu}$$

(where $D$ is the "rigidity") on the condition that the stress $\sigma_x$ is related to the strain $\epsilon_x$ by the expression

$$\sigma_x = B|\epsilon_x|^{\mu-1}\epsilon_x \qquad (0<\mu\leqslant 1),$$

where $B$, $\mu$ are constants.

4. A solid, non-uniformly heated sphere (the temperature $\theta$ is a function of radius) experiences elastic-plastic strain. Find the stress distribution if the von Mises criterion is satisfied in the plastic zone, and $\theta = \theta_0\ [1-(r/b)^\beta]$, where $\beta > 0$, $\theta_0 > 0$ are constants.

5. Find the stress distribution in a long ($\epsilon_z = 0$) rotating tube subject to elastic-plastic strain. (Assume the condition of incompressibility, and that the von Mises yield criterion is satisfied in the plastic zone.) Determine the angular speed of rotation for which the limit state is attained.

6. Show that the problem of deformation of a thin-walled tube under the action of an internal pressure and an axial force reduces to the integration of a Riccati equation (cf. the case in § 15).

# 4

# Torsion

## §27. Torsion of prismatic rods. Basic equations

### 27.1. *Preliminaries*

We consider the *torsion of a prism of arbitrary cross-section*. Let the lower end of the rod be clamped and the $z$-axis be parallel to the axis of the rod (fig. 45); the rod twists under the action of a moment $M$. Following the hypotheses of Saint Venant in the theory of elastic torsion, we assume that the cross-sections experience rigid rotation in their planes, but are distorted in the direction of the $z$-axis:

$$u_x = -\omega z y\,, \qquad u_y = \omega z x\,, \qquad u_z = w(x, y; \omega)\,,$$

where $\omega$ is the torsion per unit length of the rod, and $w(x, y; \omega)$ is an unknown function. Then the components of strain will be

$$\begin{gathered} \epsilon_x = \epsilon_y = \epsilon_z = \gamma_{xy} = 0\,, \\ \gamma_{xz} = \frac{\partial w}{\partial x} - \omega y\,, \qquad \gamma_{yz} = \frac{\partial w}{\partial y} + \omega x\,. \end{gathered} \tag{27.1}$$

The function $w(x, y; \omega)$ characterizes the distortion (warping) of a cross-section. Beginning with the equations of flow theory (13.7), we can easily

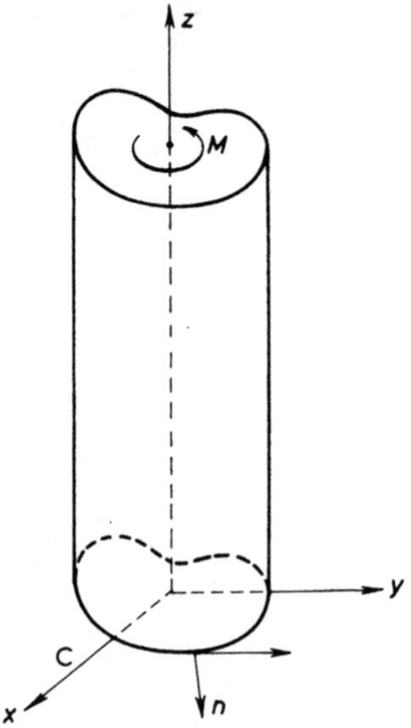


Fig. 45.

show that the normal stresses and the tangential stress $\tau_{xy}$ are equal to zero:

$$\sigma_x = \sigma_y = \sigma_z = \tau_{xy} = 0 . \tag{27.2}$$

Consequently the tangential stress vector $\boldsymbol{\tau}_z = \tau_{xz}\mathbf{i} + \tau_{yz}\mathbf{j}$ acts in sections $z = \text{const.}$ (fig. 46). The intensities $T$ and $\Gamma$ are respectively

$$T^2 = \tau_{xz}^2 + \tau_{yz}^2 , \qquad \Gamma^2 = \gamma_{xz}^2 + \gamma_{yz}^2 . \tag{27.3}$$

It is easy to see that the third invariant of the stress deviatoric is equal to zero; therefore it follows from (1.17) that $\omega_\sigma = \text{const.} = \frac{1}{6}\pi$, i.e. that the form of the stress deviatoric is retained at all times.

From formulae (1.16) we obtain:

$$\sigma_1 = T , \qquad \sigma_2 = 0 , \qquad \sigma_3 = -T ,$$

i.e. we have a state of pure shear. Examining the principal directions, we find that $\cos(i, z)$, $i = 1, 2, 3$, are constant, while the remaining direction cosines are proportional to one of the ratios $\tau_{xz}/T$, $\tau_{yz}/T$. The maximum tangential stress is equal to

$$\tau_{\max} = |\boldsymbol{\tau}_z| = T . \tag{27.4}$$

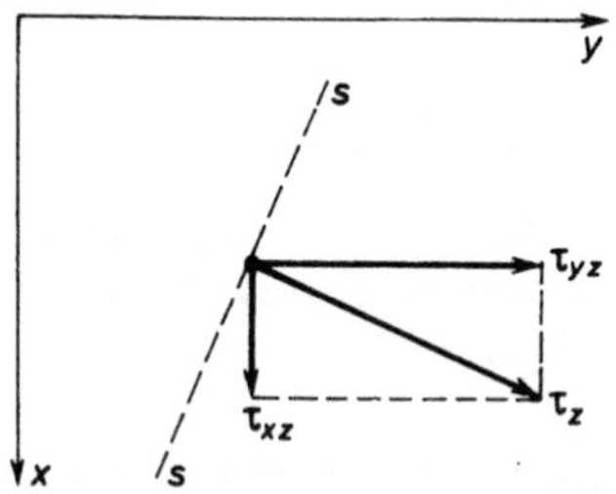


Fig. 46.

The maximum tangential stresses act along planes $z$ = const. and along the cylindrical surfaces with generators parallel to the $z$-axis and with polar curve perpendicular at each point to the vector $\tau_z$. The intercepts of these cylindrical surfaces (slip surfaces) on the plane $z = 0$ are called *slip lines*.

27.2. *Basic equations*

The components of tangential stress must satisfy the differential equation of equilibrium

$$\frac{\partial \tau_{xz}}{\partial x} + \frac{\partial \tau_{yz}}{\partial y} = 0 . \tag{27.5}$$

From (27.1) the continuity condition follows:

$$\frac{\partial \gamma_{xz}}{\partial y} - \frac{\partial \gamma_{yz}}{\partial x} = -2\omega . \tag{27.6}$$

The equilibrium equation (27.5) implies that the expression

$$\tau_{xz} \mathrm{d}y - \tau_{yz} \mathrm{d}x = \mathrm{d}F$$

is a total differential of the stress function $F(x, y)$, i.e.

$$\tau_{xz} = \frac{\partial F}{\partial y} , \qquad \tau_{yz} = -\frac{\partial F}{\partial x} . \tag{27.7}$$

Here $-\mathrm{d}F$ is the flux of the tangential stress $\boldsymbol{\tau}_z$ through an element of arc $\mathrm{d}s$. The curves on a level surface of stresses (surface $z = F(x, y)$) are called *stress curves* (*stress trajectories*). Along the stress trajectories $F$ = const. or $\mathrm{d}F = 0$, consequently, $\tau_{yz}/\tau_{xz} = \mathrm{d}y/\mathrm{d}x$, i.e. *the vector* $\boldsymbol{\tau}_z$ *is directed along the tangent to the stress trajectories.*

The lateral surface of the rod is free of stresses and therefore along a contour C

$$\tau_{xz} \cos (n, x) + \tau_{yz} \cos (n, y) = 0 .$$

Since $dy = ds \cos (n, x)$, $dx = -ds \cos (n, y)$, it is obvious that the vector $\boldsymbol{\tau}_z$ is directed along the tangent to the contour. From (27.7) we obtain:

$$\partial F/\partial s = 0 ,$$

i.e. $F$ = const. on the contour. In other words, *the contour is one of the stress trajectories*. For a simply-connected contour it is possible to write

$$F = 0 .$$

The twisting moment $M$ is balanced by the moment of the stresses:

$$M = \iint (x\tau_{yz} - y\tau_{xz})\, dxdy ,$$

where the integration extends over the whole area of the cross-section. Substituting from (27.7) and integrating by parts, we obtain

$$M = -\oint_C F\, [x \cos (n, x) + y \cos (n, y)]\, ds + 2\iint F dxdy .$$

For a simply-connected contour this formula reduces to

$$M = 2\iint F dxdy , \tag{27.8}$$

i.e. the twisting moment is numerically equal to twice the volume enclosed within the stress surface $z = F(x, y)$.

If the contour is multiply-connected (fig. 47) then the stress function can take different constant-values $F_0, F_1, \ldots, F_m$ on the contours – the outer $C_0$

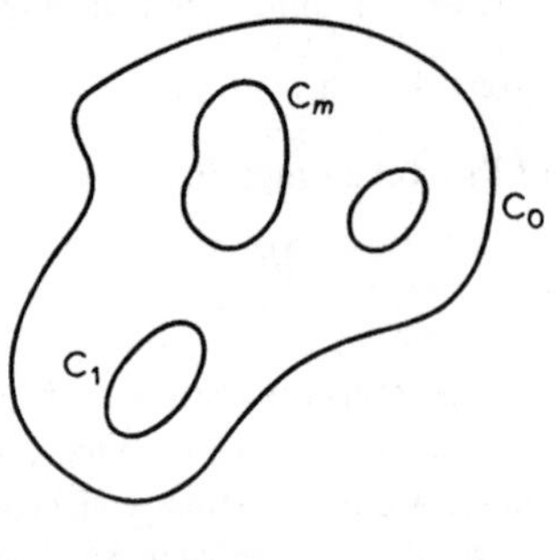


Fig. 47.

and the inner $C_1, \ldots, C_m$. One of these constants may be arbitrarily assigned, since the additive constant in the stress function does not affect the solution of the torsion problem; let $F_0 = 0$. Then we obtain

$$M = 2 \sum_{i=1}^{m} F_i \Omega_i + 2 \iint F \mathrm{d}x \mathrm{d}y , \tag{27.9}$$

where $\Omega_i$ is the area bounded by the contour $C_i$.

27.3. *Elastic torsion*

In elastic torsion we have by Hooke's law

$$\gamma_{xz} = \frac{1}{G} \frac{\partial F_\mathrm{e}}{\partial y} , \qquad \gamma_{yz} = -\frac{1}{G} \frac{\partial F_\mathrm{e}}{\partial x} ,$$

where we distinguish the "elastic" stress function by the suffix "e".

Substituting these expressions in the continuity equation, we obtain the differential equation of torsion:

$$\frac{\partial^2 F_\mathrm{e}}{\partial x^2} + \frac{\partial^2 F_\mathrm{e}}{\partial y^2} = -2G\omega . \tag{27.10}$$

Since the boundary conditions for $F_\mathrm{e}$ do not contain $\omega$, it follows from (27.10) that the stress function $F_\mathrm{e}$ has $\omega$ as a multiplier. The ratios $\tau_{xz}/T$, $\tau_{yz}/T$ are independent of $\omega$, and consequently in elastic torsion the principal directions at each point are fixed. Further, from (27.1) and Hooke's law it follows that in elastic torsion, warping is proportional to the angle of torsion $\omega$.

If a hole, having the outlines of a cross-section of the rod, is cut in a plate, the hole is then covered with a film (membrane) at tension $N$, and the film is subjected to the action of uniform pressure $q$, then a small bending of the film $v(x, y)$ satisfies the differential equation

$$\frac{\partial^2 v}{\partial x^2} + \frac{\partial^2 v}{\partial y^2} = -\frac{q}{N}$$

and the contour condition $v = 0$. Consequently, the stress function $F_\mathrm{e}$ satisfies the same equations as the bending $v$. This analogy, observed by Prandtl, enables an experimental solution of a torsion problem to be found with the help of a soap film or any other film, in cases where the solution of the Poisson equation (27.10) for a given contour is difficult.

## §28. Plastic torsion

### 28.1. *The stress state*

We consider torsion of a rod, on the assumption that the *whole section is in the yield state*. Since $\tau_{max} = T$, the von Mises and Tresca-Saint Venant yield criteria have the same form:

$$\tau_{xz}^2 + \tau_{yz}^2 = k^2 , \qquad (28.1)$$

where $k = \sigma_s/\sqrt{3}$ according to von Mises criterion, and $k = \frac{1}{2}\sigma_s$ according to the Tresca-Saint Venant criterion.

Substituting the formulae (27.7) in the yield criterion we obtain the differential equation of the "plastic" stress function:

$$\left(\frac{\partial F_p}{\partial x}\right)^2 + \left(\frac{\partial F_p}{\partial y}\right)^2 = k^2 , \qquad (28.2)$$

where on the contour

$$F_p = \text{const.}$$

It is easy to see, that the surface of plastic stresses

$$z = F_p(x, y)$$

is the *surface with constant angle of slope* (the surface of natural slope, "the roof"), constructed on the contour of a cross-section [1]). It is easy to construct such a surface for a simply-connected contour by pouring sand on a horizontal sheet of cardboard set out in the shape of a cross-section. This surface is obviously independent of the angle of torsion. In the case of a multiply-connected contour $F_p$ takes different constant values on the contours and the construction is somewhat complicated; numerous examples of plastic stress surfaces have been given in the book by A. Nadai [25].

The stress trajectories $F_p$ = const. are equidistant curves, parallel to the contour of the cross-section. The slip lines coincide with the normals to the contour.

Notice that the first order equation (28.2) has one family of characteristics: straight-lines, which coincide with the slip lines. The stress trajectories and slip lines (dotted) for a rectangular contour are shown in fig. 48. The

[1]) By virtue of (28.1) the direction cosines of the normal to the surface $z = F_p(x, y)/k$ are equal to

$$\cos(n, x) = -\frac{1}{\sqrt{2}\,k}\frac{\partial F_p}{\partial x}, \qquad \cos(n, y) = -\frac{1}{\sqrt{2}\,k}\frac{\partial F_p}{\partial y}, \qquad \cos(n, z) = \frac{1}{\sqrt{2}} .$$

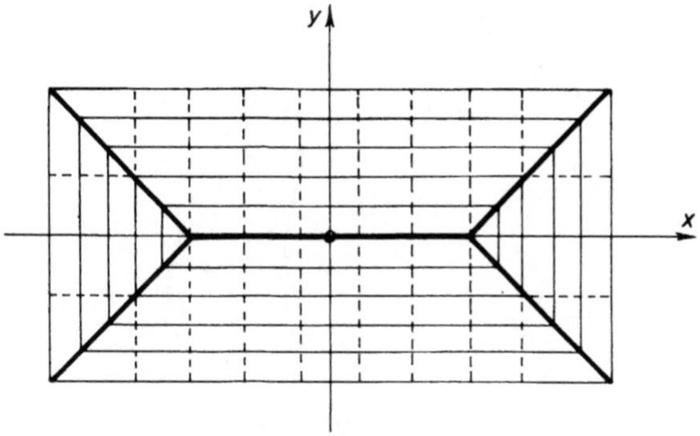


Fig. 48.

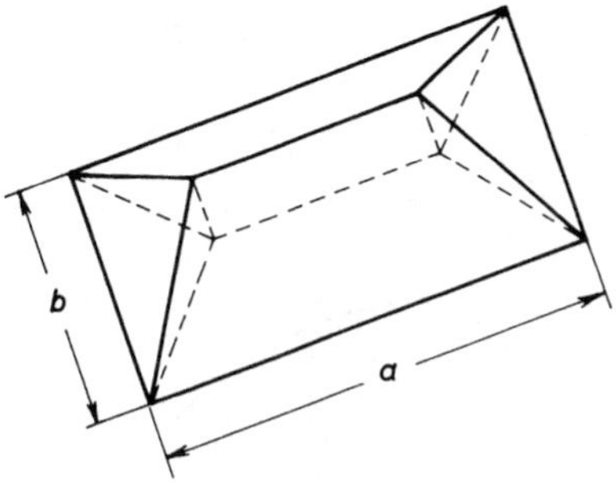


Fig. 49.

stress surface has edges (fig. 49), and the projections of these edges on the $x$, $y$-plane are called *lines of rupture*. These are indicated by the heavy lines in fig. 48.

Thus, the tangential stress vector in the plastic region is constant in magnitude, and its direction is perpendicular to the normal to the contour of the region (fig. 50). It follows that the stresses are calculated most simply from the contour of the region. For example, in torsion of a rod of rectangular profile (fig. 48), $\tau_{xz} = 0$, $\tau_{yz} = k$ in the right-triangular region and $\tau_{xz} = -k$, $\tau_{yz} = 0$ in the upper trapezoidal region.

If the contour has a re-entrant angle, the stress trajectories flow around it along the arc of a circle (fig. 51). In fig. 52 we show the "roof" for a cross-section in the shape of a corner. A portion of the surface of a circular cone

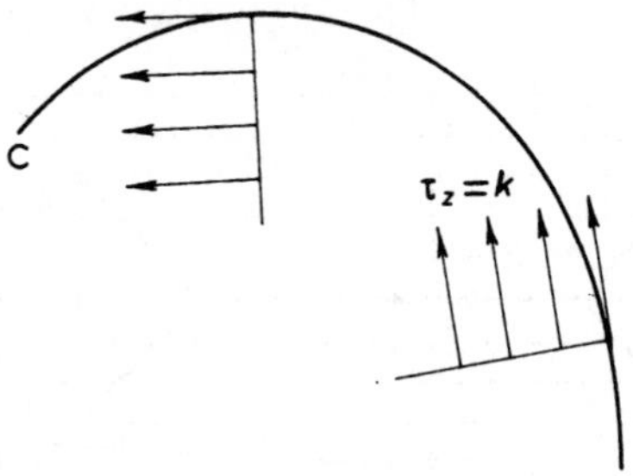


Fig. 50.

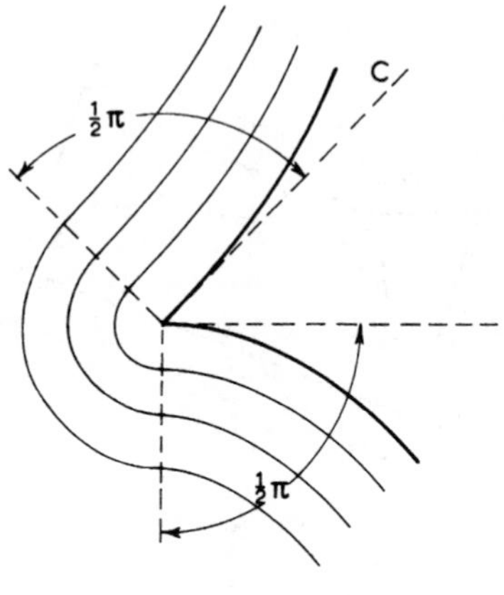


Fig. 51.

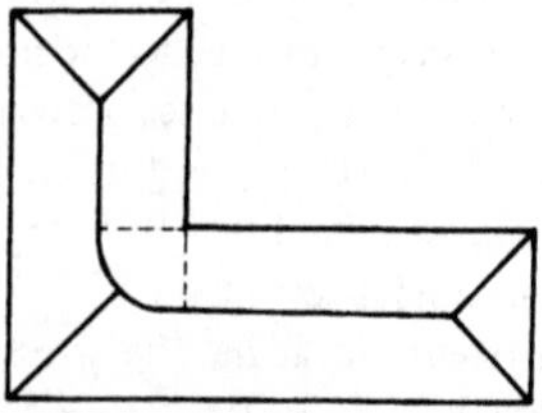

Fig. 52.

emerges from the re-entrant angle, while an edge emerges from the convex angle of the contour.

Along the lines of rupture the components $\tau_{xy}$, $\tau_{yz}$ have discontinuities, that is, the *direction* of the tangential stress $\boldsymbol{\tau}_z$ changes discontinuously.

28.2. *Limit moment*

The purely plastic state of the rod which we have been considering is called a *limit state*. There is a corresponding limit twisting moment (for a simply-connected contour)

$$M_* = 2\iint F_{\mathrm{p}} \mathrm{d}x\mathrm{d}y \,, \tag{28.3}$$

equal to twice the volume under "the roof" constructed on the given contour.

Evaluation of $M_*$ is easily effected. Thus, for a rectangle (fig. 49)

$$M_* = \tfrac{1}{6}k(3a-b)\,b^2 \,,$$

while for a circle of radius $a$ (fig. 53)

$$M_* = \tfrac{2}{3}k\pi a^3 \,, \qquad M_0 = \tfrac{1}{2}k\pi a^3 \,.$$

The limit moment characterizes the load-bearing capacity of the rod under torsion; the moment corresponding to the appearance of plastic deformations is denoted by $M_0$.

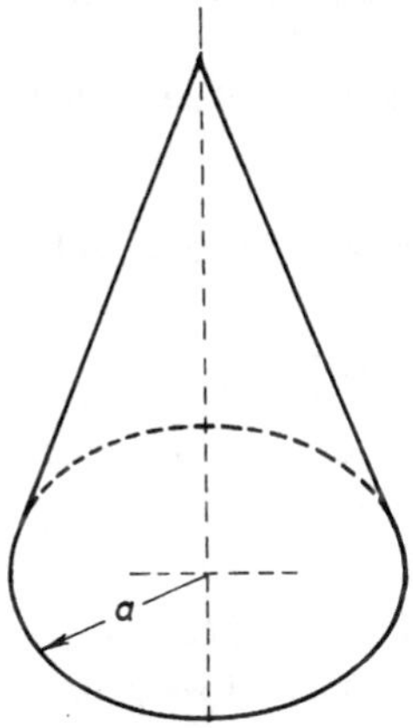


Fig. 53.

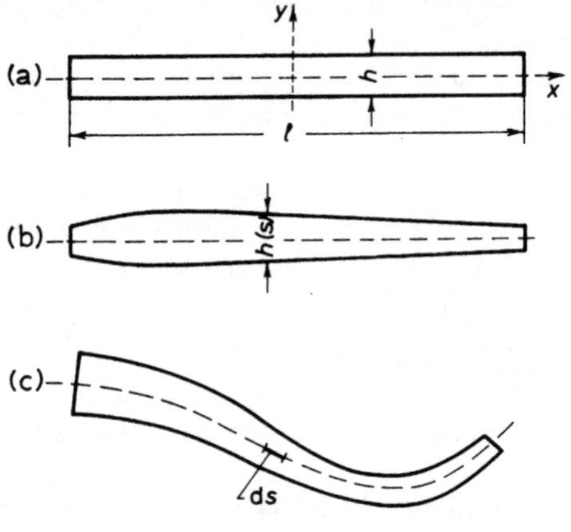


Fig. 54.

28.3. *Thin open profile*

For a very thin and long rectangle (fig. 54a) $M_* \approx \frac{1}{2}klh^2$. If the thickness changes slowly, then (fig. 54b)

$$M_* = \frac{1}{2}k \int_0^l h^2(s)\,\mathrm{d}s\,. \tag{28.4}$$

This formula is also valid for a bent profile (fig. 54b), as the form of the stress surface shows.

For a thin-walled circular tube with a cut ($c$ is the radius of the mean line),

$$M'_* = \pi k h^2 c\,.$$

It is interesting to compare this value with the limit moment $M_*^0$ for the whole tube of the same section:

$$M_*^0 = 2\pi k h c^2 = 2M'_*\,c/h\,,$$

i.e. the cut tube has a lower load-carrying capacity:

$$M_*^0 \gg M'_*\,.$$

28.4. *Determination of the axial displacement (warping)*

In the limit state the question of warping is not of much interest. We investigate here the relations which permit us to find the axial displacement in the plastic zone for elastic-plastic torsion (§29).

In the elastic zone Hooke's law gives:

$$\frac{\gamma_{xz}}{\gamma_{yz}} = \frac{\tau_{xz}}{\tau_{yz}} . \tag{28.5}$$

In the plastic zone we have to begin with the equations of flow theory (13.7)

$$d\gamma_{xz} = \frac{1}{G} d\tau_{xz} + d\lambda \cdot \tau_{xz} ,$$

$$d\gamma_{yz} = \frac{1}{G} d\tau_{yz} + d\lambda \cdot \tau_{yz} .$$

But in this zone the tangential stresses at a given point do not change; consequently their increments are equal to zero and

$$d\gamma_{xz} = \frac{\tau_{xz}}{\tau_{yz}} d\gamma_{yz} .$$

This also gives (28.5). Substituting now in (28.5) the components of stress given by (27.1), we obtain the differential equation for the axial displacement

$$\tau_{yz}\left(\frac{\partial w}{\partial x} - \omega y\right) - \tau_{xz}\left(\frac{\partial w}{\partial y} + \omega x\right) = 0 , \tag{28.6}$$

where the tangential stresses are known functions. This first-order partial differential equation is easy to solve (see for example, [31]).

## §29. Elastic-plastic torsion

### 29.1. *Nadai's analogy*

In elastic-plastic torsion which precedes the limit state, there will be elastic and plastic zones in a section of the rod.

In the elastic zones the stress function $F_e$ satisfies the differential equation of elastic torsion (27.10). In the plastic zones the stress function $F_p$ is calculated from the differential equation of the "roof" (28.2).

On the boundary of the elastic and plastic zones the stresses are continuous, i.e.

$$\frac{\partial F_e}{\partial x} = \frac{\partial F_p}{\partial x} , \qquad \frac{\partial F_e}{\partial y} = \frac{\partial F_p}{\partial y} .$$

It follows that, on a curve of separation,

$$F_{\mathrm{p}} = F_{\mathrm{e}} + \text{const.} \tag{29.1}$$

If $F_{\mathrm{p}} = F_{\mathrm{e}}$ at some point, then this condition holds along the whole boundary.

Analytic solutions of elastic-plastic torsion problems present great mathematical difficulties. A visual representation of elastic-plastic torsion is given by *Nadai's analogy.*

A rigid roof (for example, of glass) with a constant angle of slope is constructed over the given contour. A membrane is stretched over the base of the roof and is loaded with a uniformly distributed pressure. For small pressures the membrane does not touch the roof – this corresponds to elastic torsion (fig. 55a). As the pressure increases an instant is reached when the membrane begins to adjoin the roof at a number of points, which corresponds to the development of plastic deformations. As the pressure continues to increase the membrane becomes more and more in contact with the roof (fig. 55b, c). Here the projections of the contact zones correspond to regions of plastic deformation, while the remainder will be an elastic nucleus. It is clear that the relevant differential equations, the condition $F_{\mathrm{e}} = F_{\mathrm{p}}$ and the contour condition are all satisfied. The twisting moment will be equal to twice the volume under the membrane.

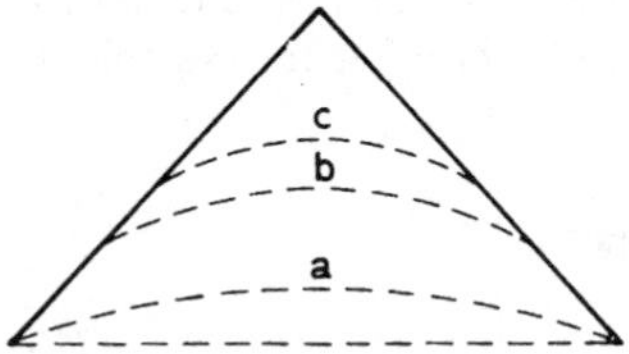


Fig. 55.

For simply-connected profiles the elastic zones degenerate into lines of rupture in the limit (infinite angle of torsion).

Nadai's analogy can be used for an experimental solution to the problem of elastic-plastic torsion (see [25]).

Fig. 56 shows the development of plastic regions for a rectangular section. The shading indicates the direction of the slip lines; the latter are easily displayed if the plastically twisted rod is cut across the section subjected to etching. Fig. 57 shows photographs of etchings on ground edges for rods of rectangular cross-section with various angles of torsion; with increase of the angle

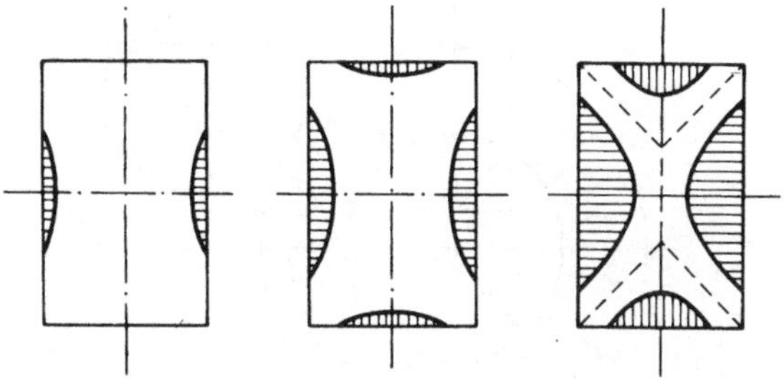

Fig. 56.

Fig. 57.

of torsion, dark zones, corresponding to plastically deformed slip layers, spread increasingly over the cross-section.

29.2. *Rod of circular section*

The tangential stress (fig. 58) is

$$\tau_z = \begin{cases} rk/c & \text{for} \quad r \leqslant c\,, \\ k & \text{for} \quad r \geqslant c\,. \end{cases}$$

The twisting moment is

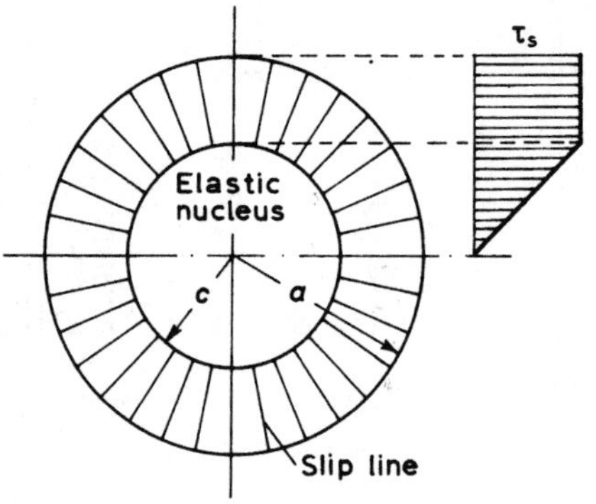


Fig. 58.

$$M = 2\pi \int_0^a \tau_z r^2 \mathrm{d}r = M_* \left(1 - \frac{1}{4}\frac{c^3}{a^3}\right) ,$$

where the limit moment $M_* = \frac{2}{3}\pi a^3 k$. The angle of torsion is found by considering the deformations of the elastic nucleus:

$$\omega = \frac{\tau_z}{Gr} = \frac{k}{Gc} .$$

The slip lines coincide with the radial directions (fig. 58). The limiting state ($c = 0$) is attained at infinite angle of torsion, when the elastic nucleus degenerates to a point of rupture. It should be emphasized, however, that with the development of torsion the moment $M$ rapidly approaches the limit moment $M_*$ (thus, when $c/a = \frac{1}{2}$, $M = \frac{31}{32}M_*$); the load-carrying capacity of the rod is practically exhausted at comparatively small angles of torsion.

29.3. *Inverse method of solution of the elastic-plastic problem*

It was emphasized earlier that if the directions of the normals to the contour are known, then the stresses in the plastic region are easily determined, because in this case we know the magnitude of the tangential stress $\tau_z$ at each point. This enables the development of inverse methods of solving elastic-plastic problems. We shall consider here a simple approach, proposed by V.V. Sokoloskii. Let us suppose that we know the elastic nucleus, bounded by the contour L, and the solution of the differential equation of elastic torsion (27.10) which satisfies the plasticity condition on the contour of the nucleus

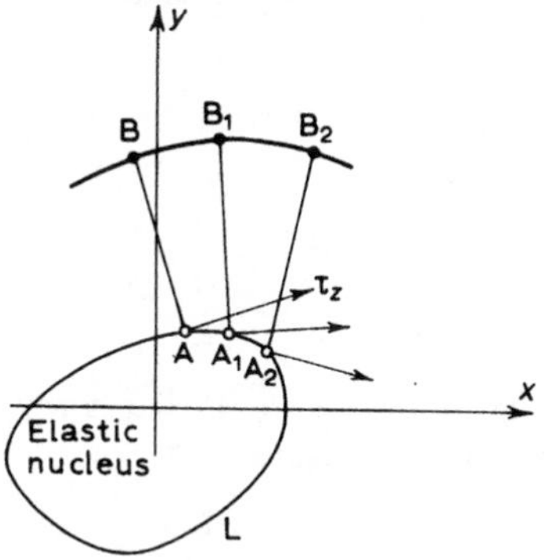


Fig. 59.

L. Along L we calculate the directions of the tangential stress vector, and construct normals AB, $A_1B_1$, . . . to it (fig. 59). The orthogonal trajectory $BB_1B_2$ . . . to the family of normals, if it is closed, gives the outline of the contour C of the rod.

29.4. *Example: elastic-plastic torsion of an oval rod*

Using this method V.V. Sokolovskii found the solution to the problem of elastic-plastic torsion of a rod having oval cross-section (fig. 60).

Let the contour L be the ellipse

$$\frac{x^2}{a^2}+\frac{y^2}{b^2}=1\ .$$

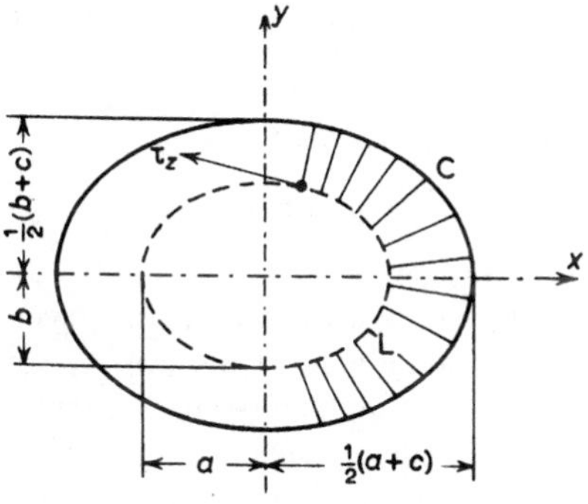


Fig. 60.

The solution for the elastic nucleus which satisfies the yield criterion $T = k$ on L is elementary

$$\tau_{xz} = -k\, y/b\,, \qquad \tau_{yz} = k\, x/a\,;$$

here the angle of torsion is

$$\omega = \frac{k}{2G}\,\frac{a+b}{ab}\,.$$

Suppose $x = -a \sin \psi$, $y = b \cos \psi$ are the parametric equations of the ellipse L; the tangential stresses then are

$$\tau_{xz} = -k \cos \psi\,, \qquad \tau_{yz} = -k \sin \psi\,.$$

The direction of the tangential stress $\boldsymbol{\tau}_z$ on the ellipse is determined from the relation

$$\tau_{yz}/\tau_{xz} = \tan \psi\,.$$

The equation of the straight line normal to the vector $\boldsymbol{\tau}_z$ and passing through points of L takes the form

$$y = -x \cot \psi + (b-a) \cos \psi\,.$$

For fixed $a$, $b$ this is the equation of a one-parameter family of straight slip lines. It is now necessary to construct the orthogonal trajectories of this family. The differential equation of these trajectories is:

$$\frac{\mathrm{d}y}{\mathrm{d}x} = \tan \psi = -\frac{x + a \sin \psi}{y - b \cos \psi}\,.$$

It is easy to show that the orthogonal trajectories have the parametric equations

$$2x = -\sin \psi\, [a + c + (a-b) \cos^2 \psi]\,,$$

$$2y = \cos \psi\, [b + c - (a-b) \sin^2 \psi]\,,$$

where $c$ is an arbitrary constant.

These equations define an oval with two axes of symmetry and semi-axes $\frac{1}{2}(a + c)$ and $\frac{1}{2}(b + c)$. The outlines of the oval do not differ significantly from the outlines of an ellipse with corresponding semi-axes. The solution is meaningful if ellipse L lies entirely inside the oval C, which is satisfied for sufficiently large angles of torsion $\omega$. With increasing angle $\omega$ the elastic nucleus (ellipse L) flattens, and in the limit degenerates to the line of rupture.

29.5. *Concluding remarks*

Another inverse method has been proposed by L.A. Galin [92]; in this method it is possible to establish the equations of the contours L and C, if the distribution of tangential stresses along L is given, and satisfies certain additional conditions. Using this result, L.A. Galin solved a number of elastic-plastic problems for rods with nearly polygonal cross-section. He also gave a method for solving the direct problem for rods of polygonal cross-section [93]. L.A. Galin's results agree well with Nadai's experiments.

We note in conclusion that a series of elastic-plastic problems (torsion of an angular profile, torsion of rods of square and triangular section) are soluble by numerical ("relaxation") methods [60].

The question of the existence of a solution to the elastic-plastic problem has been investigated by L.A. Galin and other authors.

One final point should be made. In the analysis of elastic-plastic torsion, it is tacitly assumed that with increasing torque (or with increasing angle of torsion $\omega$) loading takes place at all points of the plastic zone. But the boundary of the plastic zone is changing, and generally speaking, unloading can begin in certain parts of the zone. This question has been investigated in Hodge's work [183], which show that unloading does not occur with increase of torque in rods with a simply-connected cross-section. On the other hand, unloading can occur with a multiply-connected cross-section (for example, in a hollow cylinder). This greatly complicates the problem of elastic-plastic torsion of multiply-connected rods, since different equations are required in regions of unloading.

A generalization of the torsion problem for a straight rod is the problem of torsion of a circular annulus of unchanging cross-section, investigated by Freiberger and also by Wang and Prager (see [70]).

The stress state in plastic zones of a twisted circular rod of variable diameter, was studied by V.V. Sokolovskii [44]; the limit load for such a rod will be determined later (ch. 8).

Torsion of anisotropic and heterogeneous rods has also been studied (see [70, 71]).

The problem of elastic-plastic anti-plane deformation is mathematically similar to the problem of elastic-plastic torsion. Here too a state of pure shear is achieved, but the stresses on the contour of the body are given (see the works of G.P. Cherepanov [173]).

## §30. Torsion of hardening rods

30.1. *General remark*

With torsion of a rod of hardening material, simple loading does not take place; the form of the stress deviatoric is maintained, but the directions of the principal axes change. It is possible, however, to assume that these deviations are small, since a relatively simple stress state (pure shear) occurs and the directions of the principal axes are not substantially changed under torsion. In fact, the contour is one of the stress trajectories (§27) and the principal directions are obviously retained along it. The remaining stress trajectories act as if to "repeat" the outlines of the contour, therefore changes in these curves are relatively small under torsion. Changes in the directions of the principal axes which are connected with the rotation of the vector (tangential to the stress trajectories), can be assumed insignificant. Thus, it is possible to proceed approximately from the equations of deformation theory (see §15, paragraphs 15.1 and 15.4). Analysis of torsion of hardening rods on the basis of flow theory is associated with great difficulties and is not discussed here.

30.2. *Differential equation*

Substituting the strain components given by equations (14.23) of deformation theory

$$\gamma_{xz} = \bar{g}(T)\,\tau_{xz}\,, \qquad \gamma_{yz} = \bar{g}(T)\,\tau_{yz} \tag{30.1}$$

into the continuity condition (27.6) and introducing the stress function $F$, we obtain

$$\frac{\partial}{\partial x}\left[\bar{g}(T)\frac{\partial F}{\partial x}\right] + \frac{\partial}{\partial y}\left[\bar{g}(T)\frac{\partial F}{\partial y}\right] + 2\omega = 0\,, \tag{30.2}$$

where

$$T^2 = \left(\frac{\partial F}{\partial x}\right)^2 + \left(\frac{\partial F}{\partial y}\right)^2 .$$

On the contour $F$ = const. as before. The differential equation (30.2) is related to Monge-Ampere type equations; it is linear with respect to the second derivatives and by virtue of the properties of $\bar{g}(T)$ (see §12) it is of elliptic type. For $\bar{g}(T)$ = const. = $1/G$ (Hooke's elastic medium) we arrive at Poissons' equation (27.10).

30.3. *Solution for a circular section*

For a circular section the solution is elementary, since the cross-sections remain plane, i.e.

$$\gamma_{xz} = -\omega y\,, \qquad \gamma_{yz} = \omega x$$

and

$$\gamma_{\varphi z} = \Gamma = \omega r\,, \qquad \tau_{\varphi z} = T = g(\omega r)\,\omega r\,.$$

The angle of torsion $\omega$ is calculated from the condition of static equivalence

$$M = 2\pi\omega \int_0^a g(\omega r)\, r^3 \mathrm{d}r\,.$$

30.4. *Torsion of thin-walled rods*

First we consider torsion of open thin profiles. Our starting point is the problem of torsion of a long, thin rectangle (fig. 54a). Here it can be assumed that the stress function $F$ is independent of $x$; then from (30.2) we have:

$$\frac{\mathrm{d}}{\mathrm{d}y}\left[\bar{g}\left(\frac{\mathrm{d}F}{\mathrm{d}y}\right)\frac{\mathrm{d}F}{\mathrm{d}y}\right] + 2\omega = 0 \ ,$$

and so

$$\bar{g}\left(\frac{\mathrm{d}F}{\mathrm{d}y}\right)\frac{\mathrm{d}F}{\mathrm{d}y} = -2\omega y + \text{const.}$$

By virtue of the eveness of the stress function $\mathrm{d}F/\mathrm{d}y = 0$ when $y = 0$, and the arbitrary constant is equal to zero; now

$$\frac{\mathrm{d}F}{\mathrm{d}y} = -g(-2\omega y)\, 2\omega y ,$$

and since $F = 0$ on the contour,

$$F = F(h, \omega, y) = -2\omega \int_{y}^{\frac{1}{2}h} g(-2\omega y)\, y\mathrm{d}y \ .$$

For *open profiles* of arbitrary outline (fig. 54b)

$$M = 2 \int_0^l \int_{-\frac{1}{2}h(s)}^{\frac{1}{2}h(s)} F[h(s), \omega, y]\ \mathrm{d}y\mathrm{d}s \ ,$$

where $s$ is measured along the mean curve of the profile.

*Torsion of closed, thin-walled profiles* is discussed on the basis of a theorem concerning circulation of shear [16]. Consider the integral

$$I_* = \oint_{C_*} \gamma_{xz}\mathrm{d}x + \gamma_{yz}\mathrm{d}y \qquad (30.3)$$

along the closed contour $C_*$, lying entirely within the section. Introducing the strain components from (27.1) and using the condition of uniqueness of the displacement $u_z = w(x, y, \omega)$, we obtain:

$$I_* = 2\omega\Omega_* ,$$

where $\Omega_*$ is the area, bounded by the contour $C_*$ (fig. 61).

On the other hand, let us introduce in (30.3) the strain components given by (30.1), where stress components are expressed in terms of the stress func-

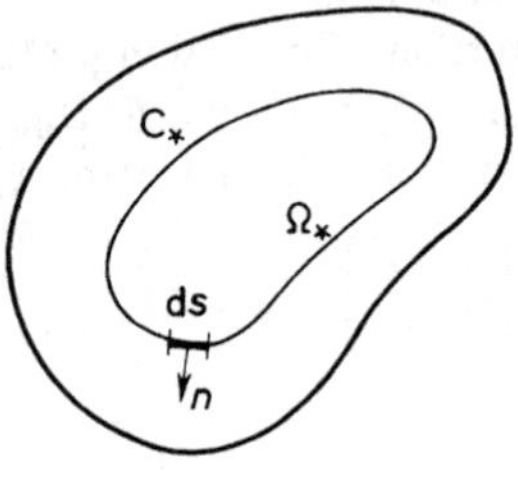


Fig. 61.

tion (27.7); since

$$\frac{\partial F}{\partial y}\,\mathrm{d}x - \frac{\partial F}{\partial x}\,\mathrm{d}y = -\frac{\partial F}{\partial n}\,\mathrm{d}s\,,$$

we find

$$I_* = -\oint_{C_*} \bar{g}(T)\frac{\partial F}{\partial n}\,\mathrm{d}s\,.$$

Thus,

$$\oint_{C_*} \bar{g}(T)\frac{\partial F}{\partial n}\,\mathrm{d}s = -2\omega\Omega_*\,. \tag{30.4}$$

When $\bar{g}(T) = \text{const.} = 1/G$ we obtain Brett's theorem on the circulation of tangential stress.

We consider now a *thin-walled tube*, whose section is bounded by the curves $C_0$, $C_1$ (fig. 62); here $C_*$ is the mean curve. On the contours $C_0$, $C_1$ the stress function takes certain constant values $F_0$, $F_1$; one of these can be set equal to zero (§27); let $F_0 = 0$.

Owing the small thickness of the tube $h(s)$, we can assume that $F$ changes linearly from $F = F_1$ on the inner contour to $F = F_0 = 0$ on the outer contour. From (27.9) we find:

$$M \approx 2F_1\Omega_*\,,$$

where $\Omega_*$ is the area, bounded by the curve $C_*$. Further,

$$\frac{\partial F}{\partial n} \approx -\frac{F_1}{h}\,, \qquad T \approx \frac{F_1}{h}\,.$$

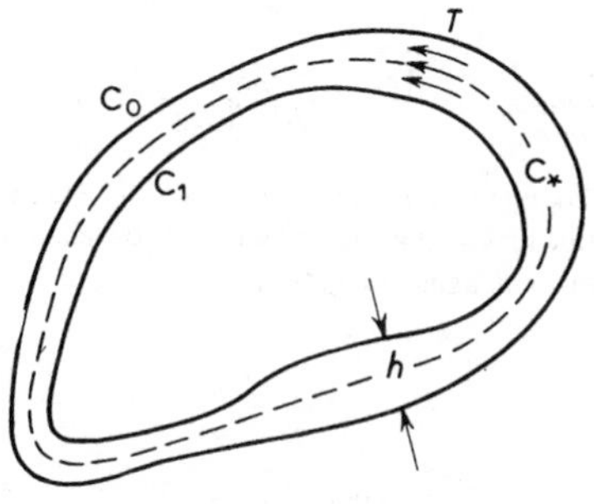


Fig. 62.

By the circulation theorem (30.4),

$$\oint_{C_*} \bar{g}\left(\frac{F_1}{h}\right) \frac{F_1}{h} ds = 2\omega\Omega_*, \tag{30.5}$$

from which the angle of torsion $\omega$ can be determined.

30.5. *Concluding remarks*

In particular problems integration of the differential equation (29.2) can be achieved by one or other method of successive approximations. There is a solution to the problem of concentration of stresses, due to a fine groove on the surface of a twisting rod [17]. In a series of cases it is possible to construct an approximate solution with the aid of variation methods (see §68).

Problems of the existence of solutions to the differential equation (30.2) and their properties have been studied.

## PROBLEMS

1. Calculate the limit turning moment for a rod of equilaterial triangular section.
2. Calculate the limit turning moment for a corner section.
3. Calculate the limit turning moment for a thin-walled ($h$ = const.) square tube.
4. Investigate the limit state of a circular (radius $a$) cylindrical rod under simultaneous torsion and extension (proceeding from the Saint Venant-von Mises equations of plasticity theory (13.12)); the cross-sections remain plane and rotate as a whole, only the stress components $\sigma_z$, $\tau_{\varphi z}$ being non zero); find the distribution of stress and the values of the axial forces and the torque.

*Answer*

$$\frac{\sigma_z}{\sigma_s} = \frac{1}{\sqrt{1+p^2\rho^2}}, \qquad \frac{\tau_{\varphi z}}{\tau_s} = \frac{p\rho}{\sqrt{1+p^2\rho^2}}, \qquad p = \frac{a\dot{\omega}}{\sqrt{3}\,\dot{\xi}_z}, \qquad \rho = \frac{r}{a}.$$

5. Use plastic flow theory to investigate the torsion and extension of a circular, cylindrical rod for the following loading path: the rod is stretched until it reaches the yield limit, then is twisted with fixed axial elongation.

*Answer*

$$\frac{\sigma_z}{\sigma_s} = \operatorname{sech} q\rho\,, \qquad \frac{\tau_{\varphi z}}{\tau_s} = \tanh q\rho\,, \qquad q = \frac{Ga\omega}{\tau_s}.$$

6. In the torsion of a circular rod of variable diameter, only the tangential displacement $u_\varphi = u_\varphi(r, z)$ is non-zero. Proceeding from the relations of deformation theory derive the differentual equation for $u_\varphi$ in the case of hardening.

# 5

# Plane Strain

## §31. Basic equations

### 31.1. *General concepts*

In plane strain the displacements of particles of the body are parallel to the $x$, $y$-plane, and are independent of $z$:

$$u_x = u_x(x, y)\,, \qquad u_y = u_y(x, y)\,, \qquad u_z = 0\,. \tag{31.1}$$

This kind of situation arises in thin prismatic bodies under loads which are normal to the lateral surface and independent of $z$.

As usual we suppose the body to be homogeneous and isotropic. In any section $z$ = const. there will be the same stress-strain configuration; the components of stress depend only on $x$, $y$, and $\tau_{xz}$, $\tau_{yz}$ are zero on account of the absence of the corresponding shears. Thus $\sigma_z$ is one of the principal stresses.

In elasticity theory the above conditions are known to be sufficient for formulating the plane strain problem. In plasticity theory, however, additional simplifications are needed, since otherwise it is not possible to derive an acceptable mathematical formulation of the problem.

In the sequel we shall use the plastic-rigid body model. As we have pointed out earlier (§23), this concept introduces an error which is difficult to estimate. On the other hand it is extremely difficult to undertake any systematic analysis of the plane strain problem without using the plastic-rigid model. In the problems we consider, the limit state is usually reached with some regions of the body still in an elastic state (e.g. the case of bending of a beam, §24). A different situation occurs in the torsion problem (ch. 4) or in the hollow sphere problem (§25), where in the limiting state every section of the rod (sphere) is affected by plastic deformation.

Thus it is really necessary to consider the elastic-plastic problem, but the difficulties of solving it are enormous. Complete neglect of the elastic regions deprives the formulation of determinacy and makes physical interpretation of solutions difficult.

It is far more expedient to proceed from the plastic-rigid model. This allows the stress field and the displacement field to be investigated simultaneously, the latter being related to the displacements of the rigid regions. In this way we can construct neaningful approximate solutions of elastic-plastic problems.

It should be understood that the error depends on the type of problem being considered. In technological problems, where large plastic deformations occur in definite parts of the body, the use of the plastic-rigid model can scarcely be disputed. Fig. 146 shows the deformation in a square network with the drawing of a strip through a hard conical matrix. It is obvious that the parts of the strip on the left and right of the matrix can be regarded as rigid, and that plastic deformation is localized in the neighbourhood of the contact surfaces. Technological problems of this type are related to plastic flow with large deformations (§49).

Another type of problem is that characterized by small deformations. These are problems on limiting loads, and are closely related to questions of strength. In this case the regions of plastic deformation for plastic-rigid and plastic-elastic bodies can be quite different. This assertion will be justified in ch. 8, where extremum principles are applied to plasticity theory.

To estimate the error it is desirable to have a collection of experimental data. As we shall see, recent tests are in good agreement with many of the results obtained from the plastic-rigid model.

31.2. *Basic equations*

It follows from (31.1) that $\epsilon_z = 0$. Using this condition we find, both from the deformation theory equations (14.3) and from the flow theory equations (13.5), that on neglecting the elastic strains [1]) we have

[1]) It is easily shown that (31.2) can be derived by assuming the incompressibility condition alone.

$$\sigma_z - \sigma = 0 , \tag{31.2}$$

and hence

$$\sigma = \tfrac{1}{2}(\sigma_x + \sigma_y) . \tag{31.3}$$

As already stated, $\sigma_z$ is one of the principal stresses. The other principal stresses $\sigma_i$ are the roots of the quadratic equation

$$\begin{vmatrix} \sigma_x - \sigma_i & \tau_{xy} \\ \tau_{xy} & \sigma_y - \sigma_i \end{vmatrix} = 0 .$$

This gives

$$\left.\begin{matrix} \sigma_{max} \\ \sigma_{min} \end{matrix}\right\} = \tfrac{1}{2}(\sigma_x + \sigma_y) \pm \tfrac{1}{2}\sqrt{(\sigma_x - \sigma_y)^2 + 4\tau_{xy}^2} . \tag{31.4}$$

It is clear that $\sigma_z$ is the mean principal stress, so that the maximum tangential stress will be

$$\tau_{max} = \tfrac{1}{2}(\sigma_{max} - \sigma_{min}) = \tfrac{1}{2}\sqrt{(\sigma_x - \sigma_y)^2 + 4\tau_{xy}^2} \equiv \tau .$$

The intensity of the tangential stresses is obviously also equal to $\tau_{max}$:

$$T = \tau . \tag{31.5}$$

Thus, the principal stresses are

$$\sigma_1 = \sigma + \tau , \qquad \sigma_z = \sigma , \qquad \sigma_2 = \sigma - \tau ,$$

i.e. *the stress state at every point is characterized by superposition of the hydrostatic pressure* $\sigma$ *on the pure shear stress* $\tau$ (fig. 63).

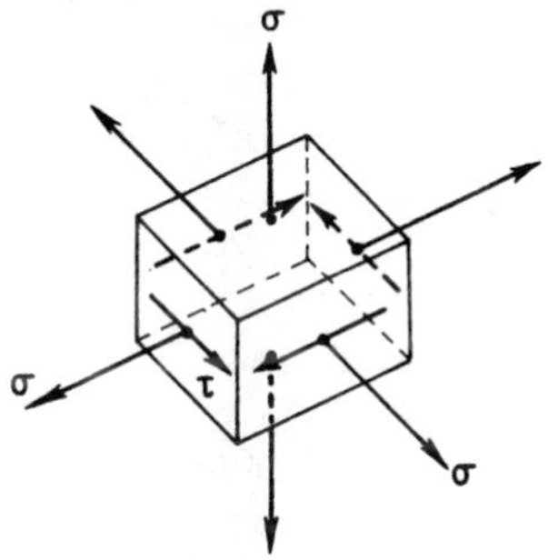


Fig. 63.

The values of the cosines which determine the first (taking $\sigma_1 \geqslant \sigma_2$) principal direction are found from the system

$$(\sigma_x - \sigma_1)\cos(1,x) + \tau_{xy}\cos(1,y) = 0\,,$$
$$\tau_{xy}\cos(1,x) + (\sigma_y - \sigma_1)\cos(1,y) = 0\,.$$

Eliminating $\sigma_1$ we obtain

$$\tan 2(1,x) = \frac{2\tau_{xy}}{\sigma_x - \sigma_y}\,. \tag{31.6}$$

The directions of the surfaces on which the maximum tangential stresses act make angles $\pm\frac{1}{4}\pi$ with the principal directions.

In the sequel an important concept will be that of *slip lines. A slip line is a line which is tangent at every point to the surface of maximum tangential stress*. It is obvious that there are two orthogonal families of slip lines, characterized by the equations

$$x = x(\alpha, \beta)\,, \qquad y = y(\alpha, \beta)\,,$$

where $\alpha$, $\beta$ are parameters. The lines of the first family (*$\alpha$-lines*) correspond to fixed values of the parameter $\beta$ ($\beta$ = const.); along *$\beta$-lines* the parameter $\alpha$ is constant. The $\alpha$-line is inclined to the right of the first principal direction at 45° (fig. 64); the $\beta$-line is inclined to the left of the first principal direction at the same angle.

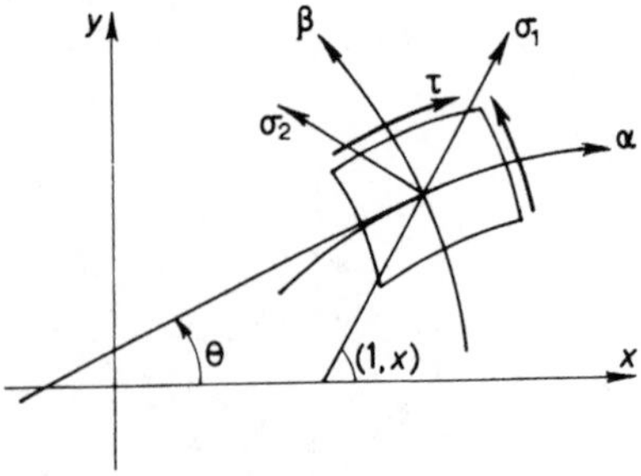


Fig. 64.

Let us fix the directions of the $\alpha$ and $\beta$ lines in such a way that they generate a right-handed coordinate system; the tangential stress $\tau$ is then positive [1]) (fig. 64). The angle of inclination of the tangent to the α-line, measured in the positive $x$-direction, will be denoted by $\theta$.

The differential equations of the α, β families are respectively

$$\frac{dy}{dx} = \tan\theta\,, \qquad \frac{dy}{dx} = -\cot\theta\,. \tag{31.7}$$

The slip lines cover the region with an orthogonal grid. An infinitesimal element cut out by slip lines experiences identical tension in the directions of the slip lines (fig. 65).

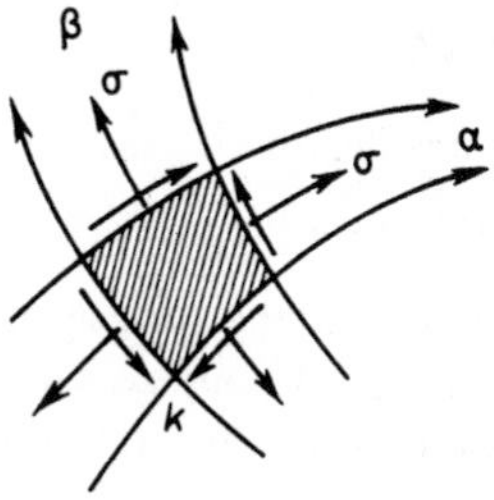


Fig. 65.

31.3. *Yield state*

Suppose the medium to be in a perfectly plastic state. Then we have the yield condition

$$\tau = \text{const.} = \tau_s$$

or

$$\sigma_{max} - \sigma_{min} = 2\tau_s\,.$$

Denoting $\tau_s$ by $k$, we obtain

$$(\sigma_x - \sigma_y)^2 + 4\tau_{xy}^2 = 4k^2\,. \tag{31.8}$$

To this we must add the two differential equations of equilibrium (body forces being absent):

[1]) Note that this condition remains true if the coordinate system is rotated through an angle $\pi$ with respect to that depicted in fig. 64.

$$\frac{\partial \sigma_x}{\partial x} + \frac{\partial \tau_{xy}}{\partial y} = 0 , \qquad \frac{\partial \tau_{xy}}{\partial x} + \frac{\partial \sigma_y}{\partial y} = 0 . \tag{31.9}$$

If the stresses are prescribed on the boundary of the body, then we have available a complete system of equations for the determination of the stress state (independent of the strain). Problems of this type are called *statically determinate.*

The above equations must be supplemented by relations which link the stress components with the increments in the strain components. Such relations are (13.7), with terms corresponding to elastic strain omitted, i.e. the relations of the Saint Venant-von Mises plasticity theory (13.12). For the case of plane strain we are left with only three relations (for $\xi_x$, $\xi_y$, $\eta_{xy}$), and from these we obtain the equation

$$\frac{\sigma_x - \sigma_y}{2\tau_{xy}} = \frac{\partial v_x/\partial x - \partial v_y/\partial y}{\partial v_x/\partial y + \partial v_y/\partial x} . \tag{31.10}$$

This states that the direction of the surface of maximum tangential stress coincides with the direction of the surface which experiences the maximum rate of shear strain. In addition the incompressibility condition must be satisfied:

$$\frac{\partial v_x}{\partial x} + \frac{\partial v_y}{\partial y} = 0 . \tag{31.11}$$

We now have five equations (31.8)–(31.11) for the five unknowns $\sigma_x$, $\sigma_y$, $\tau_{xy}$, $v_x$, $v_y$.

31.4. *Semi-inverse method*

If the problem is statically determinate, the stresses $\sigma_x$, $\sigma_y$, $\tau_{xy}$ can be found independently of the velocities $v_x$, $v_y$; once the stresses are known the velocities can be found from a *linear* system of equations (31.10), (31.11). With prescribed boundary conditions these can now be solved to give the velocity field.

If the problem is statically indeterminate, then the equations for stresses and velocities have to be solved in conjunction, and this we know to be extremely difficult. Such problems can sometimes be treated by a semi-inverse method; we attempt to choose a slip-line field such that the velocity distribution is in accord with the boundary conditions. To do this we have to prescribe to some extent the configuration of the plastic zone and to satisfy the boundary conditions for the stresses. Despite their obvious limitations, such methods have been used to find many important solutions. For this reason, analysis of the system of equations (31.8), (31.9) for stresses is very important. We therefore turn now to a detailed study of the properties of the solutions of this system. A more complex method of treating the combined stress and velocity equations will be discussed later (§51).

## §32. Slip lines and their properties

32.1. *Characteristics*

We consider the equations in stresses (31.8), (31.9).

We have the well-known formulae of stress theory:

$$\sigma_x = \tfrac{1}{2}(\sigma_1 + \sigma_2) + \tfrac{1}{2}(\sigma_1 - \sigma_2) \cos 2(1, x) ,$$
$$\sigma_y = \tfrac{1}{2}(\sigma_1 + \sigma_2) - \tfrac{1}{2}(\sigma_1 - \sigma_2) \cos 2(1, x) ,$$
$$\tau_{xy} = \tfrac{1}{2}(\sigma_1 - \sigma_2) \sin 2(1, x) .$$

In these equations we write $\sigma$ for half the sum of the principal stresses, $k$ for half their difference (as in the yield criterion), and transform to the angle $\theta = (1, x) - \frac{1}{4}\pi$. Then

$$\begin{aligned} \sigma_x &= \sigma - k \sin 2\theta , \\ \sigma_y &= \sigma + k \sin 2\theta , \\ \tau_{xy} &= k \cos 2\theta . \end{aligned} \tag{32.1}$$

It is clear that the yield criterion (31.8) is satisfied.

Substituting these quantities into the equilibrium equations, we obtain two non-linear partial differential equations of first order with respect to the unknown functions $\sigma(x, y)$, $\theta(x, y)$:

$$\begin{aligned} \frac{\partial \sigma}{\partial x} - 2k \left( \cos 2\theta \frac{\partial \theta}{\partial x} + \sin 2\theta \frac{\partial \theta}{\partial y} \right) &= 0 , \\ \frac{\partial \sigma}{\partial y} - 2k \left( \sin 2\theta \frac{\partial \theta}{\partial x} - \cos 2\theta \frac{\partial \theta}{\partial y} \right) &= 0 . \end{aligned} \tag{32.2}$$

Before we can construct solutions of this system and investigate their properties, we must first of all determine the type of the equations (cf. appendix). We show now that the system is of hyperbolic type.

To establish the hyperbolicity of the system it is necessary to show that there exist two distinct real families of characteristic curves (characteristics).

This can be done in various ways. The usual "determinantal" proof (cf. appendix) involves a good deal of algebra and is not very convenient. We shall therefore use an alternative, simpler method, as follows.

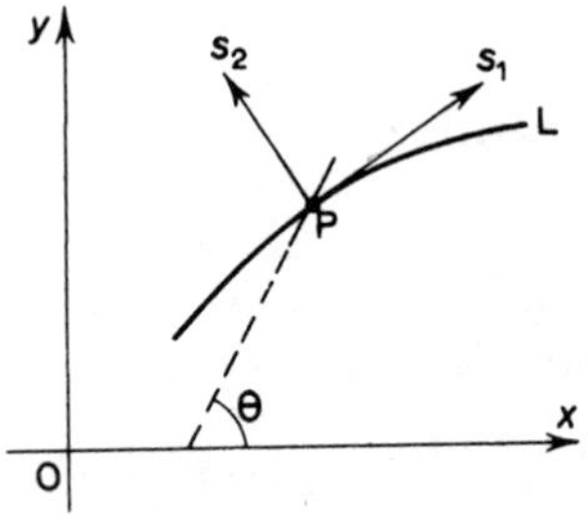


Fig. 66.

Suppose that along some line L in the $x, y$-plane (fig. 66), given by

$$x = x(s)\,, \qquad y = y(s)$$

the values of the required functions are known:

$$\sigma = \sigma(s)\,, \qquad \theta = \theta(s)\,.$$

We shall seek a solution $\sigma(x, y)$, $\theta(x, y)$ which assumes given values $\sigma(s)$, $\theta(s)$ along the line L. The problem of constructing this solution is called *Cauchy's problem.* In geometric terms it is the problem of finding an integral surface through a given curve.

If L is a characteristic, Cauchy's problem is insoluble, since in that case it is not possible to determine uniquely the first derivatives of the solution along the line L *from the differential equations* (geometrically – it is not possible to determine uniquely the tangent plane to the integral surface along L). The functions $\sigma$ and $\theta$ are known on the curve L. This means that their derivatives $\partial\sigma/\partial s_1$, $\partial\theta/\partial s_1$ are also known, if they are differentiable functions. Here $s_1$ and $s_2$ are measured in a local coordinate system, generated by the tangent and normal to L at any point P (fig. 66). Note that the equilibrium equations and the plasticity condition remain unchanged in transforming from the coordinate system $x$, $y$ to the system $s_1$, $s_2$. The differential equations (32.2) also retain their previous form:

$$\begin{aligned}
&\frac{\partial \sigma}{\partial s_1} - 2k\left(\cos 2\theta \frac{\partial \theta}{\partial s_1} + \sin 2\theta \frac{\partial \theta}{\partial s_2}\right) = 0\,,\\
&\frac{\partial \sigma}{\partial s_2} - 2k\left(\sin 2\theta \frac{\partial \theta}{\partial s_1} - \cos 2\theta \frac{\partial \theta}{\partial s_2}\right) = 0\,,
\end{aligned} \tag{32.3}$$

where the angle $\theta$, which defines the direction on the slip surface at the point P, is measured from the $s_1$-axis. If $\theta$ is different from 0 or $\frac{1}{2}\pi$, then (32.3) will give the derivatives $\partial\sigma/\partial s_1$, $\partial\theta/\partial s_1$ and solve Cauchy's problem, so long as the derivatives $\partial\sigma/\partial s_2$, $\partial\theta/\partial s_2$ are known on L.

If L coincides with a slip line, then $\theta = 0$ or $\frac{1}{2}\pi$, and these derivatives cannot be determined from the differential equations (32.3). In this case the line L is a characteristic.

Thus, *characteristics coincide with slip lines*; it is apparent that two distinct real families of characteristics exist, and, consequently, our original system is of hyperbolic type.

If the coordinate axes $s_1$, $s_2$ coincide in direction with the tangents to the slip lines, then the differential equations (32.3) take the simple form

$$\frac{\partial}{\partial s_\alpha}(\sigma - 2k\theta) = 0\,, \qquad \frac{\partial}{\partial s_\beta}(\sigma + 2k\theta) = 0\,, \tag{32.4}$$

where $\partial/\partial s_\alpha$, $\partial/\partial s_\beta$ are derivatives along the $\alpha$ and $\beta$ lines.

These equations have a simple mechanical interpretation: they are the differential equations of equilibrium of an infinitesimal element of a *plastic medium*, generated by a grid of slip lines (*slip element*, fig. 65), which has the character of a natural coordinate grid for the given problem.

Since P is an arbitrary point on the slip line, it follows that along slip lines of the $\alpha$ and $\beta$ families we have respectively

$$\left.\begin{aligned}
&\frac{dy}{dx} = \tan\theta\,,\\
&\frac{\sigma}{2k} - \theta = \text{const.} \equiv \xi\,,
\end{aligned}\right\} \qquad
\left.\begin{aligned}
&\frac{dy}{dx} = -\cot\theta\,,\\
&\frac{\sigma}{2k} + \theta = \text{const.} \equiv \eta\,.
\end{aligned}\right\} \tag{32.5}$$

These equations for plane problems of plasticity theory were first obtained by Hencky (1923). They were derived earlier in a slightly more general form by Kotter (1903), for a brittle medium.

In passing from one slip line of the family $\alpha$ to another, the parameter $\xi$ generally changes. Similarly the parameter $\eta$ changes in going from one line of the $\beta$ family to another. Thus, $\xi$ depends only on the parameter $\beta$, and $\eta$ only on $\alpha$, i.e.

$$\xi = \xi(\beta)\,, \qquad \eta = \eta(\alpha)\,.$$

If both the slip-line field and the values of the parameters $\xi, \eta$ on the lines are known, then $\sigma$, $\theta$ are determined at each point, i.e. the stress components $\sigma_x, \sigma_y, \tau_{xy}$ are known. Note that in the present problem the characteristics depend on the required solution – the stress field – which is in contrast with the situation in linear problems (such as the wave equation). In particular, any curve $y = y(x)$ can be a characteristic if a suitable stress state is attained along it (i.e. the corresponding angle $\theta$ is defined).

32.2. *Properties of slip lines*

Slip lines have a series of important properties (investigated principally by Hencky), which we now consider.

1. *Along a slip line the pressure changes in proportion to the angle between the line and the x-axis.* This property is obvious: along the $\alpha$-line $\sigma = 2k\theta$ + const., while $\sigma = -2k\theta$ + const. along the $\beta$-line.

2. *The change in the angle $\theta$ and the pressure $\sigma$ is the same for a transition from one slip line of the $\beta$-family to another along any slip line of the $\alpha$-family* (Hencky's first theorem).

From the relations

$$\sigma/2k - \theta = \xi\,, \qquad \sigma/2k + \theta = \eta \tag{32.6}$$

it follows that

$$\sigma = k(\xi + \eta)\,, \qquad \theta = \tfrac{1}{2}(\eta - \xi)\,. \tag{32.7}$$

Take two arbitrary slip lines $\beta = \beta_1, \beta = \beta_2$ of the $\beta$-family and two slip lines $\alpha = \alpha_1, \alpha = \alpha_2$ of the $a$-family (fig. 67). Along these lines we have respectively:

$$\xi = \xi_1\,, \qquad \xi = \xi_2\,; \qquad \eta = \eta_1\,, \qquad \eta = \eta_2\,.$$

Substituting these values into (32.7) for points of intersection $A_{11}, \ldots,$ $A_{22}$ we easily find

$$\varphi_1 = \theta_{A_{21}} - \theta_{A_{11}} = \tfrac{1}{2}(\eta_2 - \eta_1)\,, \qquad \varphi_2 = \theta_{A_{23}} - \theta_{A_{12}} = \tfrac{1}{2}(\eta_2 - \eta_1)\,,$$

i.e. $\varphi_1 = \varphi_2$. In the same way we obtain

$$\sigma_{A_{21}} - \sigma_{A_{11}} = \sigma_{A_{22}} - \sigma_{A_{12}}\,.$$

It is apparent that we arrive at analogous results if we consider a transition from one $\alpha$-line to another along a $\beta$-line.

(3) *If the value of $\sigma$ is known at an arbitrary point of a given slip grid, then it can be found everywhere in the field.*

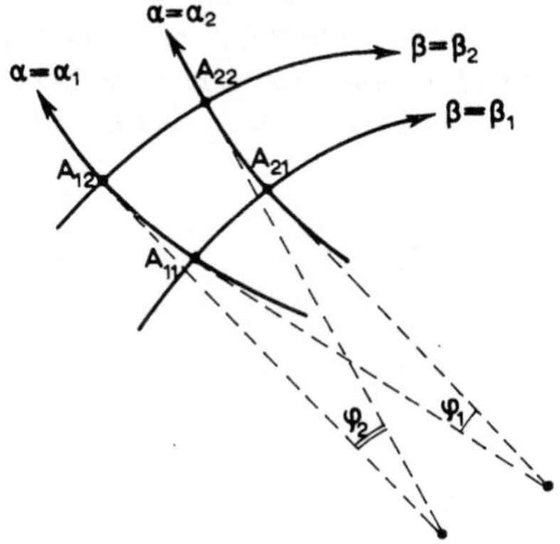


Fig. 67.

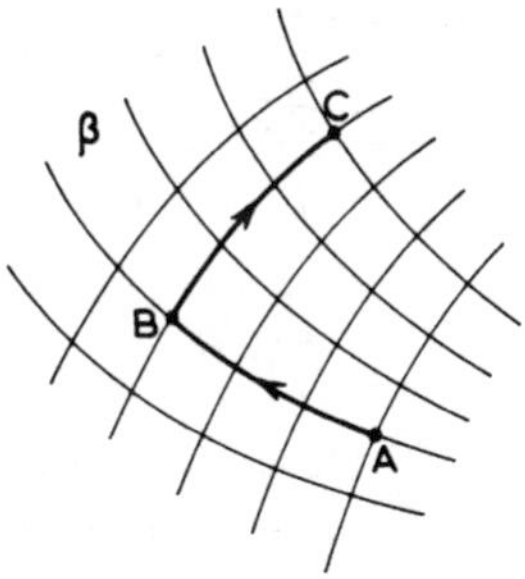


Fig. 68.

Suppose $\sigma_A$ is known at the point A (fig. 68); at this point we know $\theta_A$ and consequently we can immediately calculate the value of the parameter $\eta_1$ for the $\beta$ slip line passing through A.

Furthermore, we easily find $\sigma_B = 2k\ (\eta_1 - \theta_B)$ and $\xi_1 = \sigma_B/2k - \theta_B$ at the point B; the pressure $\sigma$ at the point C is then obtained from the formula $\sigma_C = 2k\ (\xi_1 + \theta_C)$.

4. *If some segment of a slip line is straight, then $\sigma, \theta$, the parameters $\xi$, $\eta$, and the stress components $\sigma_x, \sigma_y, \tau_{xy}$ are constant along it.* Suppose in fact that a segment of an $\alpha$-line is straight; then $\theta$ = const. along it and the parameter $\xi$ is constant. But then from (32.6) $\sigma$ = const. also. Consequently the parameter $\eta$ is also constant along the segment.

*If in some region both families of slip lines are straight, then the stress distribution is uniform and the parameters* $\xi$, $\eta$ *are constant in this region.*

5. *If some segment of a* $\beta$ *(or* $\alpha$*) slip line is straight, then all the corresponding segments of* $\beta$ ($\alpha$) *lines which are cut off by* $\alpha$ ($\beta$) *lines are also straight* (fig. 69).

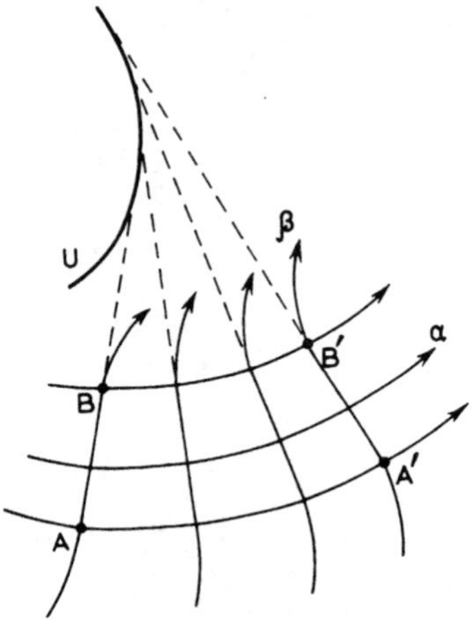


Fig. 69.

This conclusion follows from the second property, since the angle between respective tangents to any two slip lines remains constant as we move along the prescribed $\beta$-line.

In this region the stresses $\sigma_x$, $\sigma_y$, $\tau_{xy}$ are constant along each straight segment, but change from one segment to another. We shall call a stress state of this kind *simple.*

We have shown that both the parameters $\xi$, $\eta$ are constant along each straight segment; since the parameter $\xi$ has a constant value along each $\alpha$-line, it follows that $\xi$ = const. *in the whole region* ABB′A′.

6. *Straight segments cut off by slip lines of the other family all have the same length.* To show this, consider the slip lines AA′, BB′. The evolute (locus of the centre of curvature) of an arbitrary curve is the envelope of the family of normals to the curve. It is evident that the slip line AA′ and BB′ have the same evolute U. As we know, the original curve can be constructed by uncoiling a filament from the evolute. But then the filament will be shorter on the segment AB when the curve BB′ is traced than when the curve AA′ is traced.

7. Suppose that we move along some slip curve; then *the radii of curvature of the slip lines of the other family at the points of intersection change by the distance travelled* (Hencky's second theorem).

The radii of curvature $R_\alpha$, $R_\beta$ of the $\alpha$, $\beta$-lines are defined respectively by

$$\frac{1}{R_\alpha} = \frac{\partial\theta}{\partial s_\alpha}, \qquad \frac{1}{R_\beta} = -\frac{\partial\theta}{\partial s_\beta} \tag{32.8}$$

The radius of curvature $R_\alpha$ ($R_\beta$) is positive if the centre of curvature is located on the side of increasing $s_\beta$ (increasing $s_\alpha$). Consider neighbouring lines of the $\alpha$, $\beta$-families, bounding a slip element $\Delta s_\alpha \Delta s_\beta$ (fig. 70). It is apparent that

$$R_{\alpha\beta}\Delta\theta'' = \Delta s_\alpha\,, \qquad -R_\beta\Delta\theta' = \Delta s\,.$$

We evaluate the derivative of $\Delta s_\alpha$ along the $\beta$-line:

$$\frac{\partial}{\partial s_\beta}(R_\alpha\Delta\theta'') \sim \frac{(R_\alpha - \Delta s_\beta)\,\Delta\theta'' - R_\alpha\Delta\theta''}{\Delta s_\beta} = -\Delta\theta''\,.$$

As we have shown, the angle $\Delta\theta''$ between the two $\beta$-lines is constant, and consequently

$$\frac{\partial R_\alpha}{\partial s_\beta} = -1\,, \qquad \frac{\partial R_\beta}{\partial s_\alpha} = -1\,. \tag{32.9}$$

The second relation is derived just as the first.

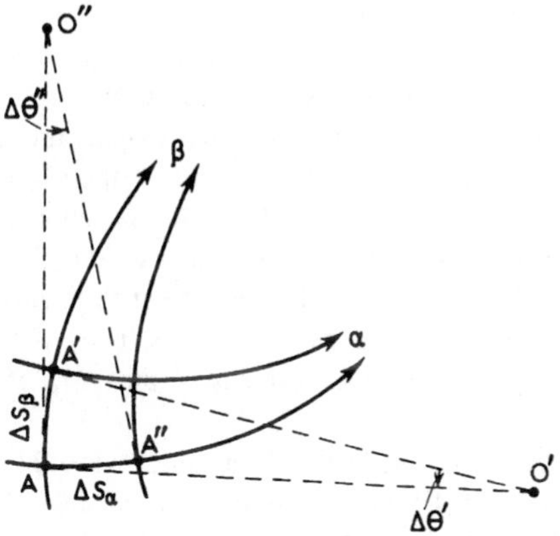


Fig. 70.

The points of intersection O′, O″ of the normals O′A, O′A′ and O″A, O″A″ are the centres of curvature respectively of the $\beta$, $\alpha$-lines at the point A.

The radius of curvature AP of the $\beta$-line at the point A is equal to the sum of the radius of curvature BQ of the $\beta$-line at the point B and the arc length AB (fig. 71).

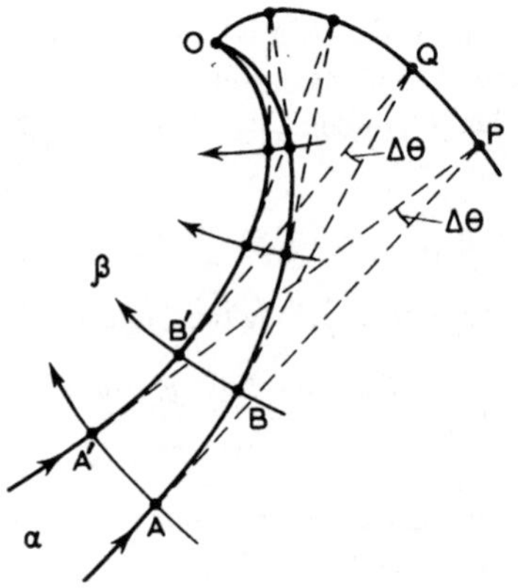


Fig. 71.

Hencky's theorem can also be presented in a different form (Prandtl); the centre of curvature of the $\beta$-lines at points of intersection with $\alpha$-lines generate the involute PO of the $\alpha$-line.

8. Hencky's theorem shows that *the radius of curvature of the $\beta$ slip lines diminishes with motion towards their concave side.*

If the plastic state is of sufficiently great extent, then the radius of curvature of the $\beta$-lines must tend to zero. This corresponds to the intersection of the involute OP with the slip line AO. In this case a line of the $\beta$-family has a cusp at the point O. It is clear from the construction (fig. 71), moreover, that the neighbouring slip lines AO, A′O converge at the point O. The point O belongs to the envelope of the family of slip lines. Thus, the envelope of the slip lines of one family is the geometric locus of the cusps of the slip lines of the other family.

Since they have a cusp at O the $\beta$ slip lines cannot intersect the envelope. In other words, the envelope is the boundary of an analytic solution.

Let AB be the envelope of $\alpha$-lines, and let $s_\alpha$, $s_\beta$ be a local coordinate system constructed at some point P of it (fig. 72). It follows from the relations

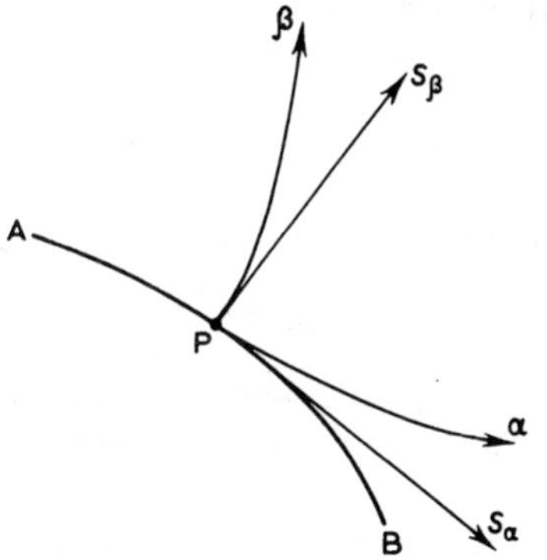


Fig. 72.

(32.8) that the derivative $\partial\theta/\partial s_\alpha$ is bounded at the point P, while $\partial\theta/\partial s_\beta$ tends to infinity, since the radius of curvature $R_\beta = 0$ on the envelope for $\beta$-lines. But then we see from the differential equations of equilibrium (32.4) that $\partial\sigma/\partial s_\alpha$ is bounded, and $\partial\sigma/\partial s_\beta \to \infty$. Thus, *the normal derivative of the mean pressure* $\sigma$ *tends to infinity along the envelope.*

9. *If derivatives of the stress components are discontinuous in moving through a slip line (for example, through some* $\beta$*-line), then the curvature of the slip lines of the second family* ($\alpha$) *is discontinuous along the* $\beta$*-line.*

In the local coordinate system $s_\alpha$, $s_\beta$ the normal stresses are equal to the mean pressure $\sigma$ (fig. 65), and the tangential stresses are constant.

The derivative $\partial\sigma/\partial s_\beta$ is continuous, while the derivative $\partial\sigma/\partial s_\alpha$ is discontinuous along a $\beta$-line.

On an $\alpha$-line we have $(\partial/\partial s_\alpha)\,(\sigma - 2k\theta) = 0$, and consequently in moving through a $\beta$-line the derivative

$$\partial\theta/\partial s_\alpha = 1/R_\alpha\,,$$

is discontinuous, i.e. the curvature also changes discontinuously.

Thus, an orthogonal grid of slip lines can be composed of pieces of distinct analytic curves; there is a continuously turning tangent in patches, and, in general, the curvature experiences discontinuities.

In conclusion we note that slip fields have a number of other interesting properties (cf., for example, [52, 54]), but we shall not dwell on them here, as they are not frequently used in solving problems of plasticity theory.

## §33. Linearization. Simple stress states

### 33.1. *Linearization*

As suggested by M. Levy, the original system of differential equations (32.2) can be linearized. We note first of all that the unknown functions can conveniently be replaced by the parameters $\xi, \eta$. Substituting in (32.2)

$$\sigma = k(\xi + \eta)\,, \qquad \theta = \tfrac{1}{2}(\eta - \xi)\,,$$

then multiplying the second of the equations so obtained by $\tan\theta$ and $-\cot\theta$ in turn, and adding to the first, we eventually obtain

$$\frac{\partial \xi}{\partial x} + \frac{\partial \xi}{\partial y}\tan\theta = 0\,, \qquad \frac{\partial \eta}{\partial x} - \frac{\partial \eta}{\partial y}\cot\theta = 0\,. \tag{33.1}$$

These are homogeneous non-linear equations, whose coefficients are functions of $\xi$, $\eta$ only. Such a system is called *reducible*, since by interchanging the dependent and independent variables we can reduce it to a linear system [1]). Thus let

$$x = x(\xi, \eta)\,, \qquad y = y(\xi, \eta)\,,$$

where in the region under consideration the Jacobian of the transformation is different from zero:

$$\Delta(\xi, \eta) \equiv \frac{D(\xi, \eta)}{D(x, y)} = \frac{\partial \xi}{\partial x}\frac{\partial \eta}{\partial y} - \frac{\partial \xi}{\partial y}\frac{\partial \eta}{\partial x} \neq 0\,.$$

Substituting the values of the partial derivatives

$$\frac{\partial \xi}{\partial x} = \Delta \frac{\partial y}{\partial \eta}\,, \qquad \frac{\partial \xi}{\partial y} = -\Delta \frac{\partial x}{\partial \eta}\,, \qquad \frac{\partial \eta}{\partial x} = -\Delta \frac{\partial y}{\partial \xi}\,, \qquad \frac{\partial \eta}{\partial y} = \Delta \frac{\partial x}{\partial \xi}$$

in the differential equations (32.1), and cancelling the multiplier $\Delta \neq 0$, we obtain

$$\frac{\partial y}{\partial \xi} + \frac{\partial x}{\partial \xi}\cot\theta = 0\,, \qquad \frac{\partial y}{\partial \eta} - \frac{\partial x}{\partial \eta}\tan\theta = 0\,. \tag{33.2}$$

This is a linear system with variable coefficients; it is called *canonical*, since in each of the equations we find derivatives with respect to one variable only.

[1]) Cf. appendix.

S.G. Mikhlin suggested the substitution

$$x = \bar{x}\cos\theta - \bar{y}\sin\theta\,, \qquad y = \bar{x}\sin\theta + \bar{y}\cos\theta\,,$$

where $\bar{x}$, $\bar{y}$ are new variables. The system (33.2) is then transformed into a system with constant coefficients:

$$\frac{\partial\bar{y}}{\partial\eta} - \frac{1}{2}\bar{x} = 0\,, \qquad \frac{\partial\bar{x}}{\partial\xi} - \frac{1}{2}\bar{y} = 0\,. \tag{33.3}$$

It is easy to see that each of the new variables ($\bar{x}$ and $\bar{y}$) satisfies the telegraph equation, for example

$$\frac{\partial^2\bar{x}}{\partial\xi\partial\eta} - \frac{1}{4}\bar{x} = 0\,.$$

33.2. *Integrals of the plane problem*

The system (33.2) is in general not equivalent to the original equations (32.2), since in the process of inversion we have lost those solutions for which the Jacobian $\Delta(\xi, \eta)$ is identically zero. However, these solutions ("integrals of the plane problem"), which are often encountered in applications and which have appeared earlier in a different way (§32), can be determined directly.

We shall follow the method adopted by S.A. Khristianovich [167] in investigating these solutions. With the aid of (33.1) we find that the condition $\Delta(\xi, \eta) = 0$ can be written in the form

$$\Delta(\xi,\eta) = \frac{2}{\sin 2\theta}\,\frac{\partial\xi}{\partial x}\,\frac{\partial\eta}{\partial x} = -\frac{2}{\sin 2\theta}\,\frac{\partial\xi}{\partial y}\,\frac{\partial\eta}{\partial y} = 0\,.$$

From this there arise three cases for which the solutions of (32.1) reduce the Jacobian $\Delta(\xi, \eta)$ to zero in some domain:

1) $\xi \equiv \text{const.} = \xi_0\,, \qquad \eta \equiv \text{const.} = \eta_0\,;$

2) $\eta \equiv \text{const.} = \eta_0\,;$

3) $\xi \equiv \text{const.} = \xi_0\,.$

The *first case* corresponds to a uniform stress state in some domain. The slip lines will here be two orthogonal families of straight lines (fig. 73a).

In the *second case* one of the equations (33.1) is satisfied. Since $\xi = -2\theta + \eta_0$, the other equation can be rewritten in the form

$$\frac{\partial\theta}{\partial x}\cos\theta + \frac{\partial\theta}{\partial y}\sin\theta = 0\,. \tag{33.4}$$

This is a quasilinear differential equation, whose integral surface consists of characteristics. The equations of the latter have the form:

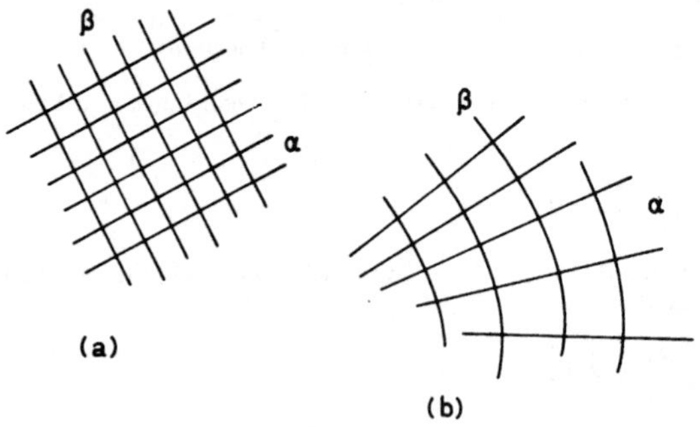


Fig. 73.

$$\frac{dx}{\cos\theta} = \frac{dy}{\sin\theta} = \frac{d\theta}{0}.$$

The solutions of this system of ordinary differential equations are clearly

$$\theta = \text{const.} = C_1, \qquad y - x\tan\theta = \text{const.} = C_2.$$

Thus, one family of slip lines is a family of straight lines depending on the two parameters $C_1$, $C_2$. Since $\sigma = 2k(\eta_0 - \theta)$ it follows that the stresses are constant along each such line (though they change from one line to another). That is, we have a *simple stress state* (§32). The general solution of equation (33.4) can be represented in the form

$$y - x\tan\theta = \Phi(\theta),$$

where $\Phi(\theta)$ is an arbitrary function.

The second family of slip lines can be constructed by standard methods as a family of orthogonal curves (fig. 73b). The differential equation of this family can be integrated in closed form [167].

The *third case* is similar to the second and can be investigated by repeating the preceding arguments.

33.3. *Mapping.*

The solution $\xi = \xi(x, y)$, $\eta = \eta(x, y)$ can be interpreted as a mapping of the "physical" plane $x, y$ on to the parameter plane $\xi, \eta$. The region in the $x, y$-plane in which the Jacobian of the transformation is different from zero is mapped on to some region (generally, multi-sheeted, cf. [56]) of the $\xi, \eta$-plane.

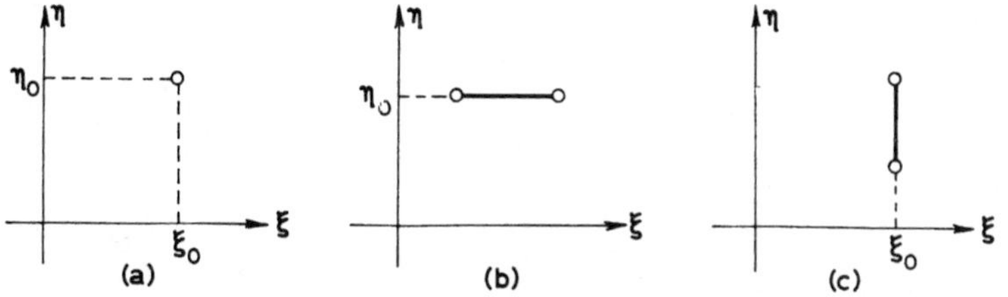


Fig. 74.

The integrals of the plane problem lead to a different mapping. Thus, in the first case $\xi = \xi_0$, $\eta = \eta_0$ (*uniform stress state*) some domain D in the $x, y$-plane is mapped into a point of the $\xi, \eta$-plane (fig. 74a).

In the second case $\eta = \eta_0$ (*simple stress state*) the domain D is mapped into a segment of the straight line $\eta = \eta_0$ (fig. 74b). A similar mapping occurs in the third case $\xi = \xi_0$ (fig. 74c).

33.4. *Simple stress states*

We shall consider in detail a number of solutions corresponding to simple stress states.

A special case of these solutions is the *uniform stress state*; in these regions the slip-line grid is generated by two orthogonal families of parallel straight lines (fig. 73a), and the parameters $\xi, \eta$ are constants ($\xi = \xi_0$, $\eta = \eta_0$, fig. 74a).

In the general simple-stress case one family of slip lines ($\alpha$, say) consists of straight lines, while the second family ($\beta$) is generated by orthogonal curves (fig. 73b). The parameter $\eta$ is here constant (fig. 74b). The picture is similar if the $\beta$-family consists of the straight lines (fig. 74c).

For a simple stress state the straight slip lines (for example, the $\beta$-lines, fig. 75) are tangential to the envelope of the family (cf. §32, fig. 69); this envelope is called the *limit curve*. In the present case the $\alpha$-family is generated by equidistant curves, which are involutes with respect to the limit curve.

An important case of simple stress is the *central* slip-line field, generated by a pencil of straight lines together with concentric circles (fig. 76). Here the envelope degenerates to a point – the centre O. In this example, when the $\alpha$-lines are straight, the parameter $\eta = \text{const.} = \eta_0$. The normal stresses on radial and circular surfaces are obviously equal to the mean pressure $\sigma = 2k\,(-\theta + \eta_0)$, i.e. they are linear functions of the angle of inclination of the line. The stresses are discontinuous at the centre O which is a singular point of the given stress field.

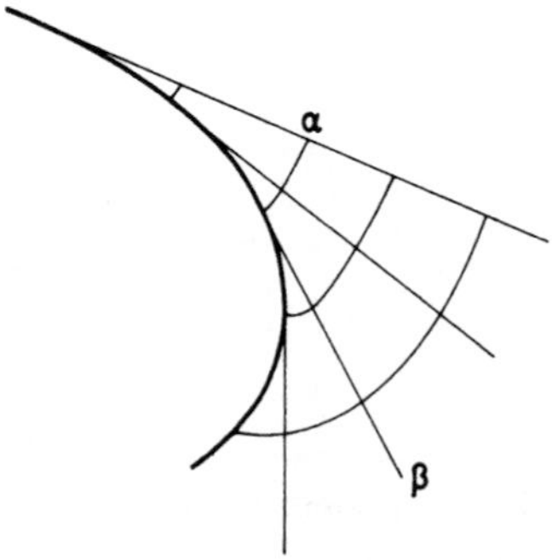


Fig. 75.

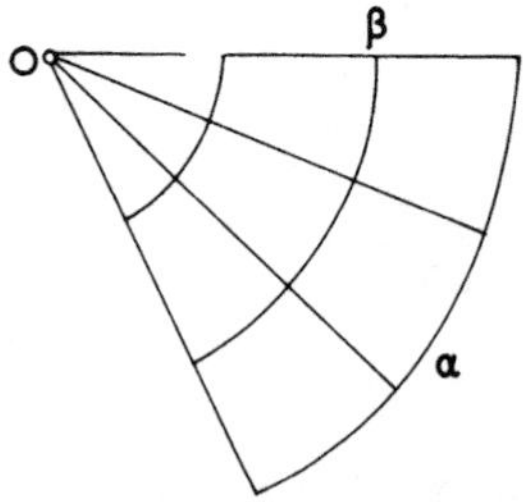


Fig. 76.

From the preceding remarks we can infer an important theorem:

*A region adjoining a region of uniform stress is always in a state of simple stress.*

Suppose the region A (fig. 77) is in a state of uniform stress, i.e. $\xi = \xi_0$, $\eta = \eta_0$. A segment of the slip line L which bounds the region A and belongs, say, to the $\beta$ slip-line family, is straight, so that $\xi = \xi_0, \eta = \eta_0$ on it as well. As we have shown earlier, the adjoining region B will have one family of slip lines ($\beta$) consisting of straight segments of equal length, and the parameter $\xi = \text{const.} = \xi_0$, since each of the $\alpha$-lines intersecting L bears the same constant value $\xi_0$.

In the $\xi$, $\eta$-plane we have the following configuration: the domain A is mapped into the point $\xi_0$, $\eta_0$, and the domain B into the segment of the line $\xi = \xi_0$ which originates from this point.

The given solution along the rectilinear boundary of the region can only be contiguous with simple stress states (in particular, uniform stress states).

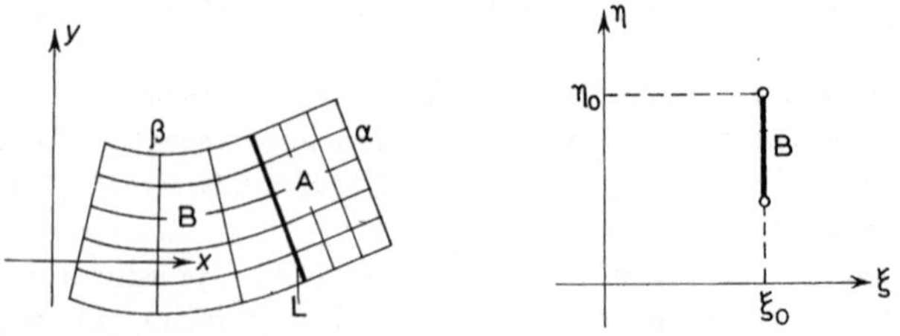


Fig. 77.

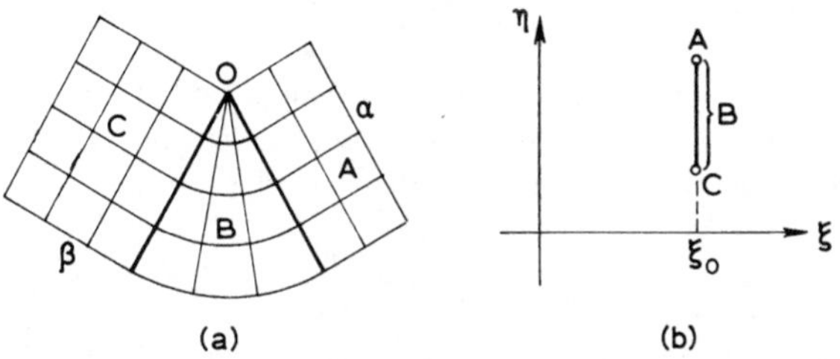


Fig. 78.

Regions of uniform stress can be linked in various ways through regions of simple stress. We shall illustrate this with simple examples.

Fig. 78a depicts a slip field consisting of two separate domains of uniform stress joined by a central field B. The stress field is continuous in the whole domain A + B + C (excluding the centre O): the parameter $\xi$ is constant ($\xi = \xi_0$). The mapping in the $\xi$, $\eta$-plane is shown on the right-hand side of fig. 78b.

A somewhat more complicated case is illustrated in fig. 79. Here the regions A, C, E of uniform stress are linked by two central fields B, D. The stresses are continuous, except at the point O. The mapping in the $\xi$, $\eta$-plane consists of two intersecting straight segments.

These kinds of constructions of slip fields are widely used in solving particular problems.

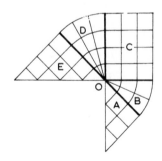

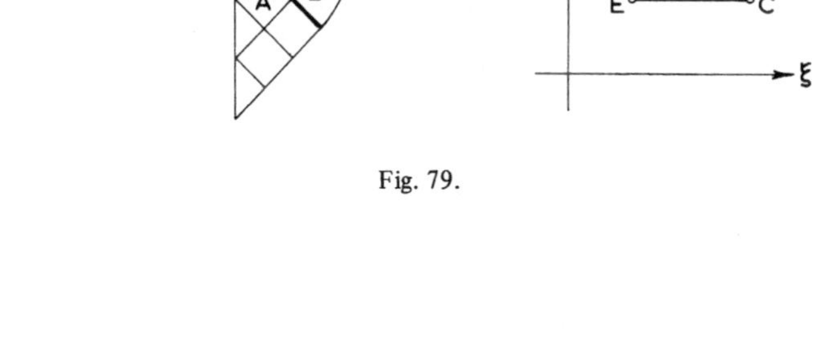

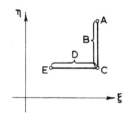

Fig. 79.

## §34. Axisymmetric field

It will be shown in §36 that an axisymmetric stress field is achieved in the neighbourhood of a *circular part* of a contour (either free or uniformly loaded). Such fields are often encountered in the solution of a variety of problems.

34.1. *The case* $\tau_{r\varphi} = 0$

Consider the slip field around a circular hole of radius $a$, loaded uniformly with a pressure $p$. Let $r$, $\varphi$ be polar coordinates. Since there is no tangential stress on the edge of the hole, the equilibrium condition gives $\tau_{r\varphi} = 0$. Then at every point of the field the principal planes have radial and circumferential directions. The slip line will be a curve which intersects at each of its points a ray, emerging from the centre, at an angle $\pm \frac{1}{4}\pi$. But this property is unique to the logarithmic spirals

$$\varphi - \ln(r/a) = \beta\,, \qquad \varphi + \ln(r/a) = \alpha\,, \tag{34.1}$$

which generate two orthogonal families (fig. 80). These lines have been observed in experiments (fig. 81).

If $\sigma_\varphi > 0$, $\sigma_r < 0$ in the neighbourhood of the boundary, the yield condition has the form $\sigma_\varphi - \sigma_r = 2k$, and the stress is determined by our previous

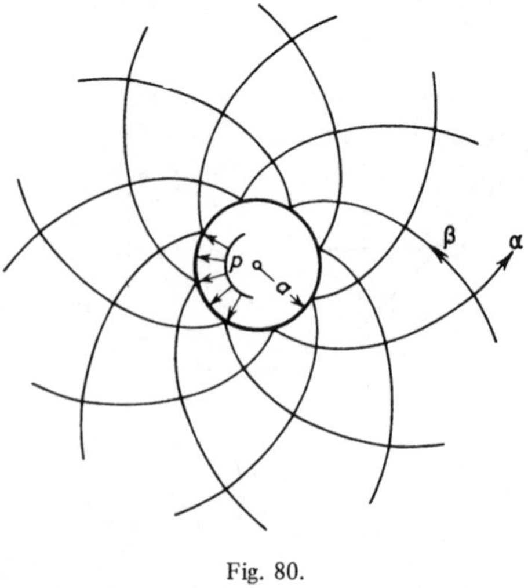


Fig. 80.

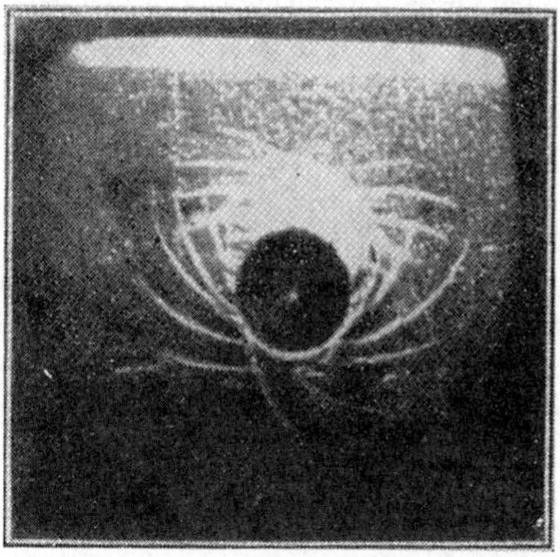

Fig. 81.

formulae (26.3):

$$\sigma_r = -p + 2k \ln (r/a)\,, \qquad \sigma_\varphi = \sigma_r + 2k\,. \tag{34.2}$$

It is easy to confirm that the relations (32.6) are fulfilled along the logarithmic spirals (34.1).

If the yield condition has the form $\sigma_\varphi - \sigma_r = -2k$, we have to substitute a minus sign in front of $2k$ in formulae (34.2).

34.2. *The general case*

The general case of an axisymmetric field with $\tau_{r\varphi} \neq 0$ was considered by S.G. Mikhlin. The yield condition now has the form

$$(\sigma_\varphi - \sigma_r)^2 + 4\tau_{r\varphi}^2 = 4k^2\,.$$

The differential equations of equilibrium can be written

$$\frac{\mathrm{d}\sigma_r}{\mathrm{d}r} + \frac{\sigma_r - \sigma_\varphi}{r} = 0\,, \qquad \frac{\mathrm{d}\tau_{r\varphi}}{\mathrm{d}r} + \frac{2\tau_{r\varphi}}{r} = 0\,. \tag{34.3}$$

Suppose the normal and tangential stress components are given on the boundary of the hole:

$$\text{when} \qquad r = a\,, \qquad \sigma_r = -p\,, \qquad \tau_{r\varphi} = q\,, \tag{34.4}$$

where $|q| \leqslant k$. From the second equilibrium equation and the boundary conditions we find

$$\tau_{r\varphi} = q(a/r)^2 . \tag{34.5}$$

The yield condition now gives

$$\sigma_\varphi - \sigma_r = \pm \sqrt{k^2 - q^2(a/r)^4} . \tag{34.6}$$

We substitute this in the equilibrium equation, integrate and determine the arbitrary constant from the first boundary condition. We then obtain

$$\sigma_r = -p \pm k\left[2 \ln \frac{\sqrt{r^2-A} + \sqrt{r^2+A}}{\sqrt{a^2-A} + \sqrt{a^2+A}} - \left(\frac{\sqrt{r^4-A^2}}{r^2} - \frac{\sqrt{a^4-A^2}}{a^2}\right)\right] , \tag{34.7}$$

where $A = a^2(q/k)$.

When $q \neq 0$ the slip lines are no longer logarithmic spirals.

## §35. Boundary conditions for stresses

Suppose that the normal and tangential components of stress are given on a contour C, and denote them by $\sigma_\mathrm{n}$, $\tau_\mathrm{n}$ respectively, with $|\tau_\mathrm{n}| \leqslant k$. From (1.3), (1.4) we have that $\sigma_\mathrm{n}$, $\tau_\mathrm{n}$ are related to the components $\sigma_x, \sigma_y, \tau_{xy}$ by the formulae

$$\begin{aligned} \sigma_\mathrm{n} &= \sigma_x \cos^2 \varphi + \sigma_y \sin^2 \varphi + \tau_{xy} \sin 2\varphi , \\ \tau_\mathrm{n} &= \tfrac{1}{2}(\sigma_y - \sigma_x) \sin 2\varphi + \tau_{xy} \cos 2\varphi . \end{aligned} \tag{35.1}$$

Here $\varphi$ denotes the angle between the normal to the contour C and the $x$-axis (fig. 82). Since the medium is in a plastic state, we substitute formulae (32.1) into (35.1) and obtain

$$\begin{aligned} \sigma_\mathrm{n} &= \sigma - k \sin 2(\theta - \varphi) , \\ \tau_\mathrm{n} &= k \cos 2(\theta - \varphi) . \end{aligned} \tag{35.2}$$

If $x = x(s)$, $y = y(s)$ are the equations of the contour, $\sigma_\mathrm{n}(s)$, $\tau_\mathrm{n}(s)$ are the prescribed stresses, then we can assume that $\sigma = \sigma(s)$, $\theta = \theta(s)$ are known on the contour, and consequently the parameters $\xi, \eta$ also. In particular, if a segment of the boundary is straight ($\varphi$ = const.) and the stresses $\sigma_\mathrm{n}$, $\tau_\mathrm{n}$ are constant on it, then $\sigma, \theta$, and also $\xi, \eta$, are constant on it.

Note that $\sigma$, $\theta$ (and, consequently, the stresses $\sigma_x, \sigma_y, \tau_{xy}$) are uniquely determined from (35.2):

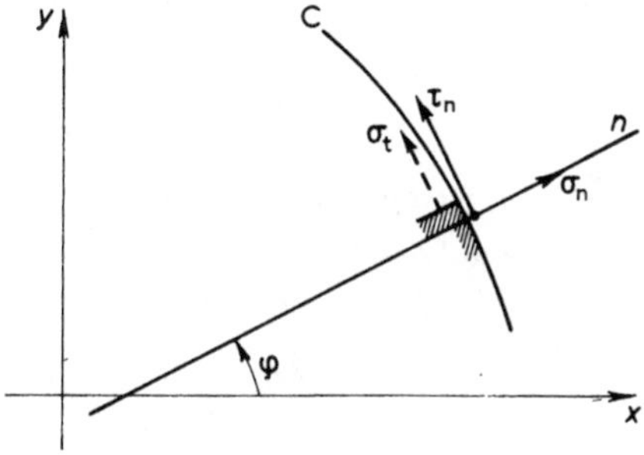


Fig. 82.

$$\theta = \varphi \pm \tfrac{1}{2} \arccos \frac{\tau_n}{k} + m\pi \, ,$$
$$\sigma = \sigma_n + k \sin 2(\theta - \varphi) \, , \qquad (35.3)$$

where by arccos we mean its principal value, and $m$ is an arbitrary integer. The presence of two solutions for $\sigma$, $\theta$ satisfying the yield condition is explained by the quadratic character of the latter.

To choose the sign we need to impose additional conditions, and these must be obtained from the mechanical formulation of the problem in each individual case. It is occasionally helpful to consider the normal stress $\sigma_t$ at the contour C (fig. 82):

$$\sigma_t = 2\sigma - \sigma_n \, .$$

The sign of $\sigma_t$ can sometimes be predetermined, and this enables the correct choice of solution to be made.

An important special case occurs when *there is no tangential stress* on the contour C ($\tau_n = 0$). Then formulae (35.3) take the simpler form

$$\theta = \varphi \pm \tfrac{1}{4}\pi + m\pi \, ,$$
$$\sigma = \sigma_n \pm k \, , \qquad (35.4)$$

and, correspondingly, $\sigma_t = \sigma_n \pm 2k$.

Consider the simplest example of a *free rectilinear boundary* $x = 0$ (fig. 83); on this boundary we have $\varphi = 0$, $\sigma_n = 0$, $\tau_n = 0$, and, consequently, $2\theta = \pm \frac{1}{2}\pi + 2m\pi$, $\sigma = \pm k$, $\sigma_x = 0$, $\sigma_y = \sigma_t = \pm 2k$ i.e. in the neighbourhood of the boundary there can only be either tension in the $y$ direction, or compression.

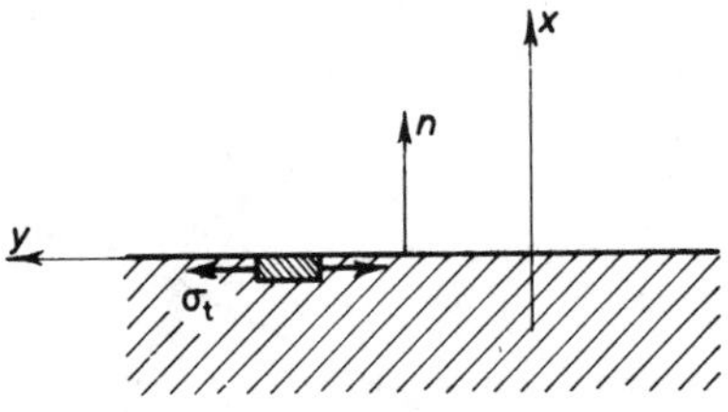


Fig. 83.

## §36. Fundamental boundary value problems

When considering particular problems it is necessary to find solutions of the hyperbolic equations (32.2) which satisfy certain boundary conditions. This usually involves the solution of a series of boundary value problems. We now give a brief account of the most important of these; more extensive information can be found in standard textbooks on mathematical physics.

36.1. *Cauchy's problem*

The most important is Cauchy's problem (the *initial value problem*). In the $x$, $y$-plane let AB (fig. 84), $x = x(s), y = y(s)$, where $s$ is some parameter, be a given smooth arc, which nowhere coincides with a characteristic and which intersects each characteristic once only [1]). Let the functions $\sigma = \sigma(s), \theta = \theta(s)$ be continuous together with their first and second derivatives on the arc AB. It is required to find a solution of equations (32.2) which assumes the prescribed values on AB. The required solution exists and is unique in a triangular region APB, bounded by the arc AB and the $\alpha, \beta$ slip lines (characteristics) which originate at the end points of the arc. In particular the functions $\sigma(x, y), \theta(x, y)$ are also determined on the sides AP, BP. The solution is continuous together with its derivatives up to second order.

A similar construction can be effected on the other side of the arc AB.

The solution at the point P depends only on the data along AB. The triangle APB is called the domain of dependence for the point P. *If the data are changed on the arc* AB, *then the solution changes only inside the triangle* APB. It follows that a fixed solution inside the triangle can, in general, be supplemented by different solutions along the slip lines. In other words solutions in adjoining regions can have different representations.

[1]) If the latter condition is not fulfilled Cauchy's problem is in general insoluble.

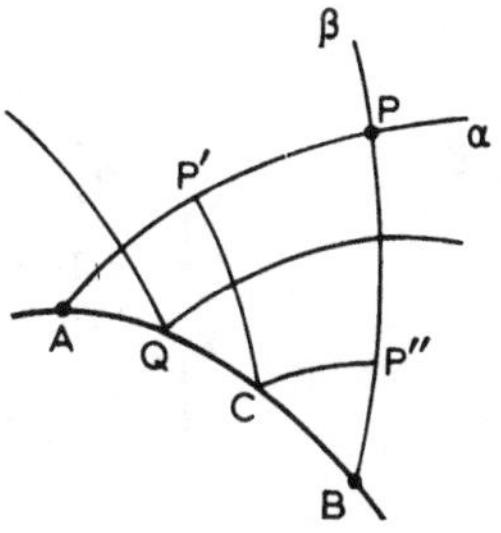


Fig. 84.

Further, prescribed values of $\sigma(s)$, $\theta(s)$ at an arbitrary point Q of the arc can influence the solution only at points lying within a "characteristic angle" generated by the slip lines originating at Q.

Existence and uniqueness of the solution apply if the requirements of smoothness of the arc and continuity of the initial data are satisfied. If derivatives of the initial data are discontinuous at some point C, then the above results will be valid only in triangular regions ACP′, BCP″. A solution in the remaining region CP′PP″ can also be constructed, but its derivatives will be discontinuous along the characteristic CP′, CP″. *Discontinuities of derivatives propagate only along characteristics, and cannot vanish along them.*

We note now some simple corollaries which are very useful in application.

*The stress field at a boundary, which is free of forces, depends only on the shape of the boundary.*

Since the tangential stress $\tau_n$ on the boundary is zero, the normal to the contour is one of the principal directions and the slip lines approach the contour at an angle of 45°. Consequently, the contour coincides nowhere with a characteristic direction, and we have Cauchy's problem, whose solution is unique.

In particular, a *rectilinear free boundary will always have a field of uniaxial uniform tension or compression, of magnitude* $2k$ *and parallel to the line of the boundary* (fig. 85a). For example, if the $x$-axis is parallel to the boundary AB, then in the domain APB, $\sigma_x = \pm 2k$, $\sigma_y = \tau_{xy} = 0$.

*At a circular free boundary* BA (fig. 85b) *the slip field is generated by logarithmic spirals, and the stresses are given by formulae* (34.2) with $p = 0$.

By (34.1) the equations of the slip lines BP, AP are respectively

$$\varphi - \ln\frac{r}{a} = -\ln\frac{r_1}{a}, \qquad \varphi + \ln\frac{r}{a} = +\ln\frac{r_1}{a},$$

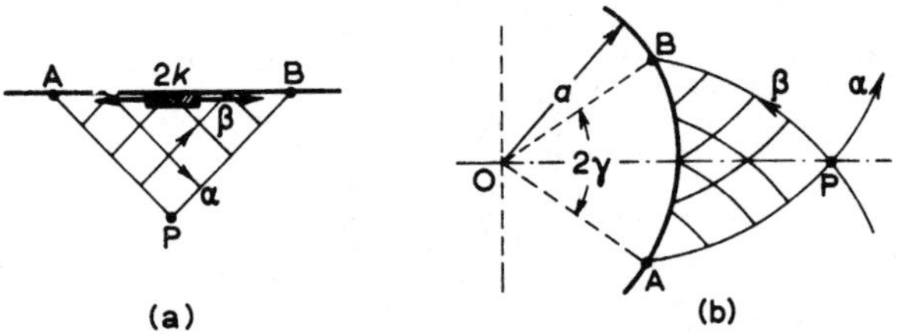


Fig. 85.

where $r_1$ is the distance of the point P from the centre. At the point B $\varphi = \gamma$; consequently $\ln (r_1/a) = \gamma$ and the stresses at P are

$$\sigma_r = 2k\gamma \; ; \qquad \sigma_\varphi = 2k(1 + \gamma) \, . \tag{36.1}$$

Note that if the yield condition has the form $\sigma_\varphi - \sigma_r = -2k$, a minus sign must be inserted in front of $2k$ in the preceding formulae.

These results continue to apply almost entirely if a uniform normal pressure $p$ is applied along the relevant portion of the contour. The geometry of the slip lines is just as before. At a rectilinear boundary we now have a uniform stress state (with the previous choice of axes):

$$\sigma_y = -p \, , \qquad \sigma_x = \pm\, 2k - p \, , \qquad \tau_{xy} = 0 \, .$$

At a circular boundary we have an axisymmetric stress field, given by formulae (34.2). This field is not influenced by initial data *outside the arc* AB. Just as before, this solution is independent of the shape of the boundary *outside the circular arc* AB. For example, if the boundary consists of a circular arc AB and a rectilinear segment BC (fig. 86), then the slip field near AB will be generated by logarithmic spirals, and that near BC will be a rectangular grid.

36.2. *Initial characteristic problem (Riemann problem)*

Suppose the values of the functions $\sigma, \theta$ are known on segments of the slip lines OA, OB (fig. 87). Since $\sigma, \theta$ satisfy on OA, OB the differential equations of equilibrium of a slip element (31.4), it follows that these values cannot be completely arbitrary: they are connected by the relations

$$\frac{\sigma}{2k} - \theta = \text{const.} = \xi \qquad \text{along OA} \, , \tag{36.2}$$

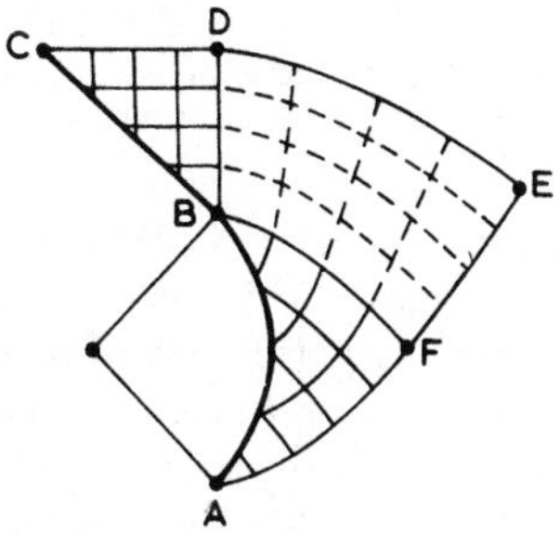


Fig. 86.

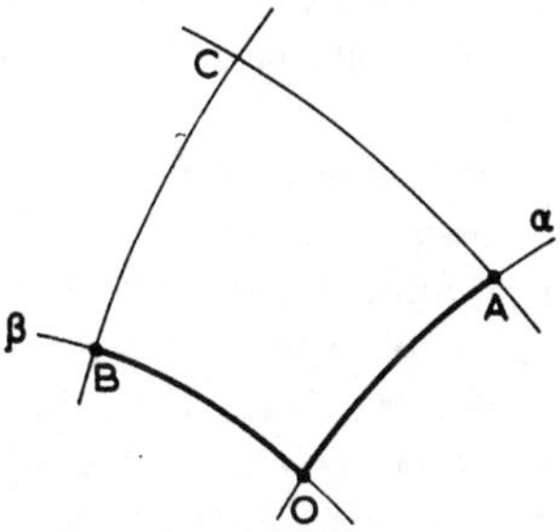


Fig. 87.

$$\frac{\sigma}{2k} + \theta = \text{const.} = \eta \qquad \text{along OB}, \tag{36.3}$$

The solution is thus determinate in the quadrilateral OACB. Note that the functions $\sigma$, $\theta$ on the segments OA, OB are usually determined from constructing solutions in adjoining domains, so that the above relations are automatically satisfied. Thus, in the example illustrated in fig. 86, an initial characteristic problem is to be solved in the domain BDEF. The values of $\sigma, \theta$ on the characteristics BD, BF are known from respective solutions of Cauchy problems in the domains BCD and ABF.

An important case is the *degenerate initial characteristic problem,* when the segment of the slip line OB (or OA) contracts on the point O, such that its radius of curvature becomes vanishingly small while the angle $\theta$ remains constant (fig. 88). All the $\alpha$ slip lines come together at the point O and the stress is discontinuous.

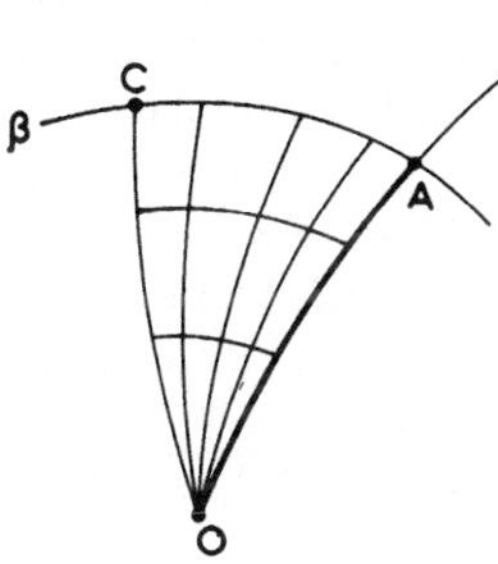


Fig. 88.

The solution can be determined in the triangle OAC for given angle of spread at the nodal point O and given values of $\sigma, \theta$ on the arc OA.

36.3. *Mixed boundary value problem*

The functions $\sigma, \theta$ satisfying the equilibrium condition (36.2), are known on a segment of the slip line OA (fig. 89). The segment OA joins a characteristic curve OB, along which the angle $\theta$ is prescribed. This kind of problem arises, for example, if OB is a friction-free boundary of the medium; the slip lines approach the curve OB at an angle $\frac{1}{4}\pi$, and, consequently, $\theta$ is known. It is assumed that the angle AOB is acute (i.e. lies inside the characteristic angle).

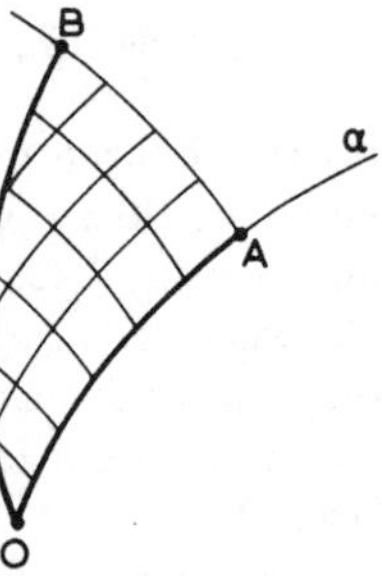


Fig. 89.

The solution to the mixed boundary value problem is determinate in the triangle OAB. The actual construction is different, depending on the magnitude of the angle $\theta$ defined on the curve OB at the point O. If this angle is equal to the angle $\theta$ on OA at the point O, then the slip line field has the form shown in fig. 89. In particular, if OB is frictionless, the angle between the curves OA, OB must be equal to $\frac{1}{4}\pi$ at the point O.

If an $\alpha$-line, originating from the point O, lies inside the region BOA (fig. 90), then the latter is divided into two parts, BOA′ and A′OA. The condition of the previous case will apply to the first of these if we can find the values of $\sigma, \theta$ on the slip line OA′. But these values can be determined by solving an initial characteristic problem for the domain AOA′, since the angle of spread of the characteristic pencil AOA′ is known.

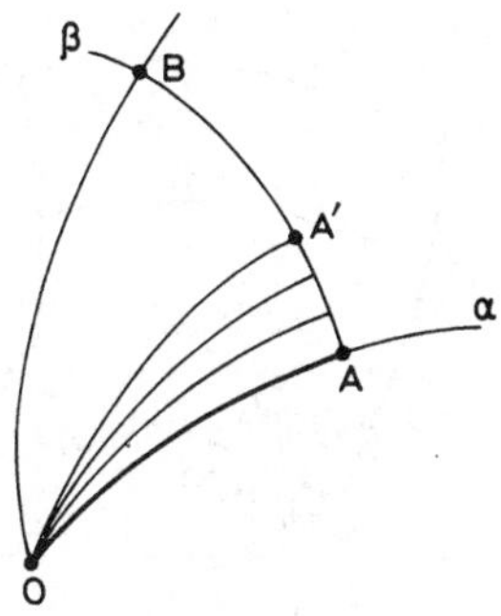


Fig. 90.

## §37. Numerical methods of solution

The boundary value problems we have been discussing can also be solved by other methods. In particular, the solutions of the Cauchy problem and initial characteristic problem for the linearized equations (33.2) can be represented in closed form by means of the Riemann function [54]. But the use of these solutions requires a considerable amount of calculation.

The apparatus of so-called meta-cylindrical functions, introduced by L.S. Agamirzyan [83] can be used to construct an analytical solution of various boundary value problems encountered in plane plasticity problems. With the aid of tables of these functions the amount of calculation can be considerably reduced.

Much simpler, however, are approximate methods of constructing slip fields based on the idea of transforming to finite-difference equations and utilizing certain properties of the slip lines (note that the method in its general form was developed by Masso (1899), cf. [52]). Different versions of this technique have been used by Sokolovskii [44], Hill [54], Prager and Hodge [31], and other authors.

We shall now present an outline of some methods of solving numerically the fundamental boundary value problems.

### 37.1. *Initial characteristic problem*

The segments OA, OB of slip lines (fig. 91) are divided into small intervals by the points

$$(1,0),(2,0),\ldots,(m,0),\ldots,\quad (0,1),(0,2),\ldots,(0,n),\ldots$$

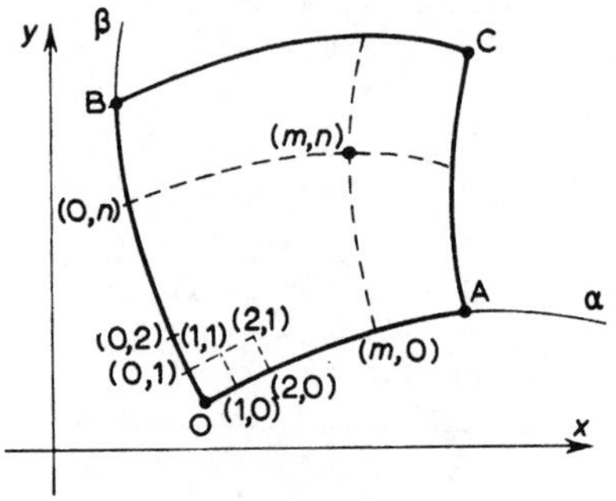


Fig. 91.

The intersections of the slip lines passing through these points will be called the nodal points of the grid and will be denoted by $(m, n)$. The functions $\sigma$, $\theta$ are known on the sides OA, OB; by Hencky's first theorem we can find the values of these functions at the node $(m, n)$:

$$\theta_{m,n} = \theta_{m,0} + \theta_{0,n} - \theta_{0,0}\,, \tag{37.1}$$

$$\sigma_{m,n} = \sigma_{m,0} + \sigma_{0,n} - \sigma_{0,0}\,. \tag{37.2}$$

The coordinates of the nodal points are calculated step by step. Suppose we know the coordinates of the nodal points $(m-1, n)$, $(m, n-1)$, and the angle $\theta$ at these points. The position of the point $(m, n)$ is determined by the intersection of small arcs; we replace the latter by chords, whose inclinations are equal to the mean value of the inclinations at the initial and final points [1]). The differential equations of the slip lines,

$$\frac{dy}{dx} = \tan\theta\,, \qquad \frac{dy}{dx} = -\cot\theta$$

are replaced by the differences

$$y_{m,n} - y_{m-1,n} = (x_{m,n} - x_{m-1,n}) \tan \tfrac{1}{2}(\theta_{m,n} + \theta_{m-1,n})\,, \tag{37.3}$$

$$y_{m,n} - y_{m,n-1} = -(x_{m,n} - x_{m,n-1}) \cot \tfrac{1}{2}(\theta_{m,n} + \theta_{m,n-1}) \tag{37.4}$$

and we determine $x_{m,n}$, $y_{m,n}$ from these. It is always necessary to begin at the point (1.1).

[1]) The inclination of the chord is often taken equal to the inclination at the initial point; this leads to somewhat inferior results.

The last relations give simple formulae for calculation of $x_{m,n}$, $y_{m,n}$. The results of the calculation can conveniently be written in a table (fig. 92). In the squares of the shaded strips we insert the known values of $x$, $y$; $\sigma$, $\theta$ at the selected points of the slip lines OA, OB. Then we calculate consecutively the values of $x$, $y$; $\sigma$, $\theta$ at the nodes and insert them in the appropriate squares of the table. We then plot the coordinates of the nodal points $x_{m,n}$, $y_{m,n}$ which have been found. Joining one line of points, corresponding to one row of the table, we obtain an $\alpha$-line; points corresponding to a column give a $\beta$-line.

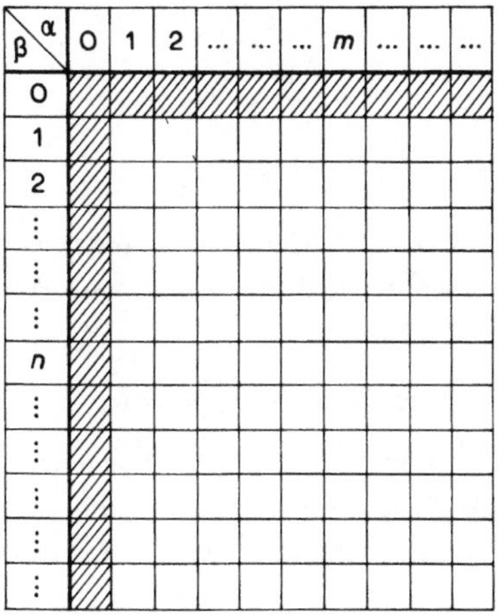


Fig. 92.

*In the degenerate case* $\sigma, \theta$ are known on the segment OA (fig. 93), as well as the change in the angle $\theta$ at the vertex O (i.e. the angle AOC). We divide this angle into a number of small sectors $\theta_{0,0}, \theta_{0,1}, \ldots, \theta_{0,n}, \ldots$ where $\theta_{0,n}$ is the angle between the slip lines O$n$ and OA at the vertex O. The value of $\theta_{m,n}$ at the node $(m, n)$ is found from formula (37.1). The pressure $\sigma$ is discontinuous at the point O and cannot be directly computed from formula (37.2). We first determine $\sigma_{1,n}$ at the points $(1, n)$; the values $\sigma_{1,0}$, $\theta_{1,0}$ are given at the point (1, 0), and consequently we know the parameter $\eta_1 = \sigma_{1,0}/2k + \theta_{1,0}$ which is constant along the $\beta$-line passing through (1, 0); then $\sigma_{1,n} = 2k(\eta_1 - \theta_{1,n})$. For the rest we can use (37.2) replacing $\sigma_{0,0}$ by the values of $\sigma_{1,n}$. The coordinate of the nodal points are computed as before.

### 37.2. *Cauchy's problem*

We divide the arc AB into small intervals with the points (0, 0), (1, 1), . . . ,

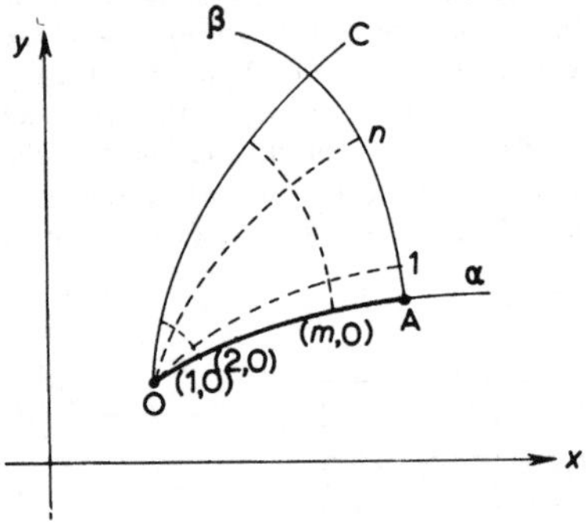


Fig. 93.

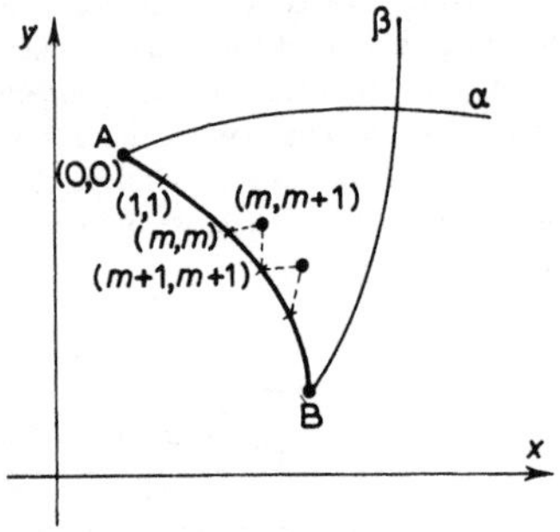


Fig. 94.

$(m, m), \ldots$ The values of $\sigma, \theta$ at nodal points in the neighbourhood of the arc (fig. 94), are found from the condition of constancy of the parameters $\xi$, $\eta$ on the $\alpha, \beta$-lines:

$$\sigma_{m,m+1} - 2k\theta_{m,m+1} = \sigma_{m,m} - 2k\theta_{m,m} \,,$$

$$\sigma_{m,m+1} + 2k\theta_{m,m+1} = \sigma_{m+1,m+1} + 2k\theta_{m+1,m+1} \,.$$

The coordinates of the nodes are determined by the previous formulae. The slip grid is then calculated in the same way as for the initial characteristic problem.

The results of the calculation can be shown in the cells of a square table (fig. 95). The known values of $x$, $y$; $\sigma$, $\theta$ are inserted in the shaded cells along the principal diagonal. The values of $x$, $y$; $\sigma$, $\theta$ obtained for the nodal points of the grid complete the table on one side of the diagonal.

### 37.3. *Mixed boundary value problem*

We consider the general case of the mixed problem shown in fig. 90.

In the domain OAA′ the solution can be constructed just as for the degenerate case of the initial characteristic problem. Next we turn to the region OA′B. We subdivide OA′ into small intervals by points (1, 0), (2, 0), . . . (fig. 96). The values of $\sigma$, $\theta$ are known on OA′. We begin the construction at the point (1, 0), and draw a straight line from this point towards the $\beta$-line (i.e. along the normal to the $\alpha$-line); this locates a point P′ on OB; the value of $\theta$ at P′ is given from the condition on OB (§36). We calculate the mean value of the angle $\theta$ at the points (1, 0), P′, and then again draw a straight line from the point (1, 0); this locates a point P″ on OB, etc., until the difference between consequtive positions of the points P becomes small. This determines

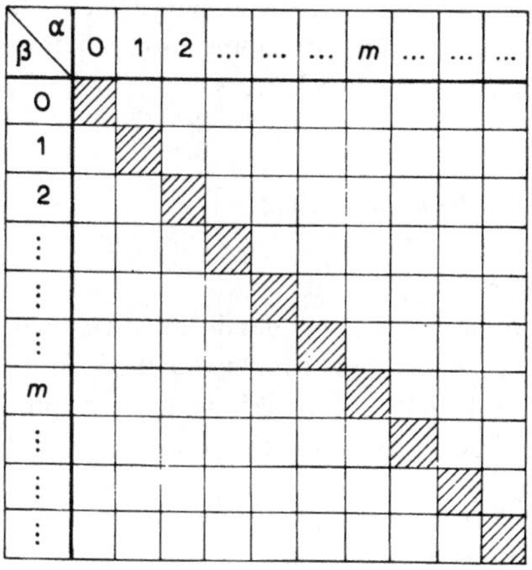


Fig. 95.

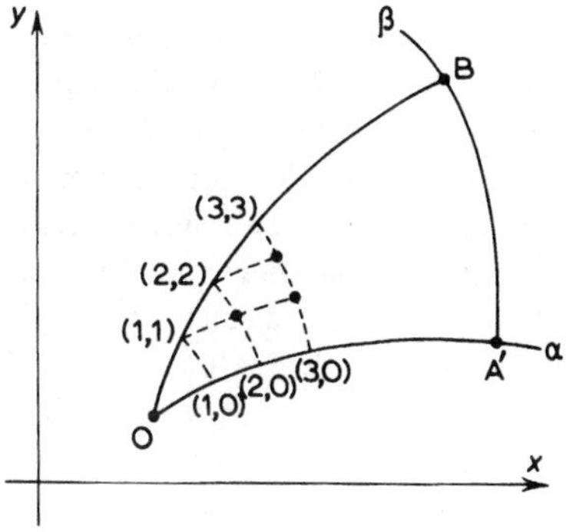


Fig. 96.

the point (1, 1). The points (2, 1), (3, 1), . . . are computed as in the initial characteristic problem. To determine the point (2, 2) we have to repeat the preceding successive approximation method.

### 37.4. *Concluding remarks*

When the slip line grid has been constructed, the values of $\sigma, \theta$ are known at its nodal points, and, therefore, so are the stress components $\sigma_x$, $\sigma_y$, $\tau_{xy}$. With a sufficiently dense grid of slip lines the plastic state can be determined to any desired degree of accuracy.

We note that useful and convenient graphical methods of solution have also been suggested [8, 29]; but their application requires the use of a large-scale diagram and involves substantial errors.

## §38. Determination of the velocity field

### 38.1. *General relations*

We have been considering in some detail the stress field, and we now turn to the remaining equations of plane strain (31.10), (31.11), which contain the components of the velocity vector. Taking into account the substitution (32.1), we can write these equations in the form

$$\left(\frac{\partial v_x}{\partial y} + \frac{\partial v_y}{\partial x}\right) \tan 2\theta + \left(\frac{\partial v_x}{\partial x} - \frac{\partial v_y}{\partial y}\right) = 0 \,, \tag{38.1}$$

$$\frac{\partial v_x}{\partial x} + \frac{\partial v_y}{\partial y} = 0 \,. \tag{38.2}$$

Once the stresses have been found, the angle $\theta$ is known and the problem of determining the velocities is linear. *This system of equations is of hyperbolic type, and its characteristics coincide with the slip lines.* Let the velocities $v_x$, $v_y$ be continuous, and let their values be prescribed on some line L. The form of the equations remains unchanged if we transform (cf. §32) to a local coordinate system $x$, $y$, generated by the tangent and normal to the line L. Since the velocities are given on L, we can calculate their derivatives along the tangent $\partial v_x/\partial x$, $\partial v_y/\partial x$. Then we can always find from (38.1), (38.2) the derivatives along the normal $\partial v_x/\partial y$, $\partial v_y/\partial y$ as well, provided $\theta \neq 0, \frac{1}{2}\pi$, i.e. provided the line L does not coincide with a slip line. If the latter is the case (as before we denote the respective "characteristic" coordinates by $s_\alpha$, $s_\beta$ and the components of the velocity vector along these directions by $u$, $v$) the normal stresses are equal to the mean pressure, $\sigma_\alpha = \sigma_\beta = \sigma$, and the tangential stress is $\tau_{\alpha\beta} = k$. Then it follows from the system (38.1), (38.2) that

$$\partial u/\partial s_\alpha = 0 \,, \qquad \partial v/\partial s_\beta = 0 \,, \tag{38.3}$$

and the derivative $\partial u/\partial s_\beta$ is indeterminate.

Thus, *the rates of relative elongation along slip lines are zero*; just as equations (32.4) express conditions of equilibrium of a slip element, so equations (38.3) characterize the singular deformation of a slip element. We write these equations in a different, somewhat more convenient, form.

Consider an infinitesimal segment $ds_\alpha$ of an $\alpha$-line (fig. 97). The rate of

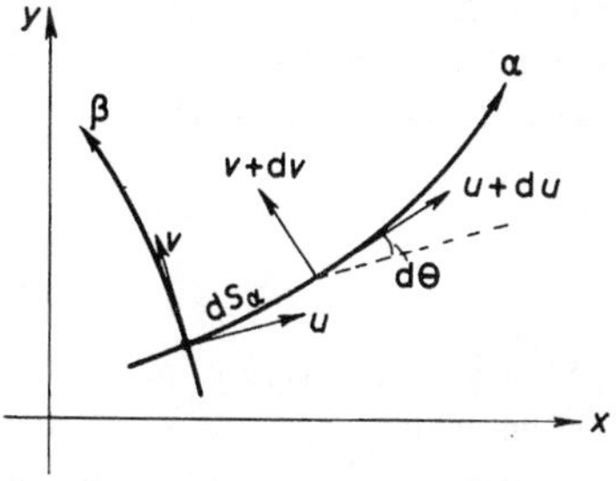


Fig. 97.

elongation in the $\alpha$-direction is $(u + \mathrm{d}u - v\mathrm{d}\theta) - u$, neglecting second order quantities. From (38.3) we have along the $\alpha$-line

$$\mathrm{d}u - v\mathrm{d}\theta = 0\,. \tag{38.4}$$

Similarly we obtain for the $\beta$-line

$$\mathrm{d}v + u\mathrm{d}\theta = 0\,. \tag{38.5}$$

These relations were found by Geiringer, and are called *the equations for the velocities along the slip lines.*

38.2. *Condition of positive dissipation*

The velocity field is determined from the above equations and supplementary boundary conditions. In this situation *the dissipation must be positive in plastic zones* ($\sigma_{ij}\xi_{ij} > 0$). This is a condition of compatibility of the stress and velocity fields, and imposes limitations on the choice of the solution constructions. It is verified *a posteriori.*

38.3. *Velocity fields for simple stress states*

The velocity field which corresponds to a simple stress state has a series of elementary properties. Thus, if the stress is uniform in some region, then $\theta$ = const. everywhere, and, consequently, we find from (38.4), (38.5),

$$u = u(\beta)\,, \qquad v = v(\alpha)\,, \tag{38.6}$$

where $u(\beta)$, $v(\alpha)$ are arbitrary functions.

The case $u = u(\beta)$, $v = 0$ defines a shear flow in the $\alpha$-direction, and the other case $u = 0$, $v = v(\alpha)$ gives a shear flow in the $\beta$-direction. The general case (38.6) is obtained by superposing two arbitrary shear flows in the given directions.

In the case of a *central field* the angle $\theta$ is constant along radial lines, for example $\theta$ = const. along $\alpha$-lines (fig. 76). Then $u$ = const. along the $\alpha$-line, i.e. $u = u(\theta)$. From (38.4), (38.5) it follows that

$$\begin{aligned} v &= \varphi(\theta) + \psi(\rho)\,, \\ u &= -\varphi'(\theta)\,, \end{aligned} \tag{38.7}$$

where $\varphi(\theta)$, $\psi(\rho)$ are arbitrary functions ($\rho$ is the distance from the centre O, fig. 76; the prime denotes derivative).

If $\varphi(\theta) = 0$, then formulae (38.7) describe rotational motion (shear flow with the streamlines in the form of concentric circles).

*In the general case of simple shear* $\theta$ = const. *along straight slip lines, and, consequently, the velocity component along each straight line is constant.*

It follows from (38.1), (38.2) that a *uniform velocity field* $v_x$ = const., $v_y$ = const. is possible in a plastic region, i.e. the plastic region is displaced like a rigid body. These regions can be interpreted as regions of negligible plastic deformation. Such fields are encountered, for example, in the problem of indentation by a flat die (§45).

38.4. *Numerical construction of the velocity field*

We consider the general case when neither of the slip-line families consists of straight lines. In this case the velocity field cannot be determined by elementary means. As for the stress field (§37), the simplest method is that of finite-differences. We shall outline here the solution of several boundary value problems. We shall not give a detailed exposition of this method of construction, nor shall we investigate other types of boundary value problems, since these are all very similar to their counterparts in the stress field case (§37).

*Initial characteristic problem.* In fig. 91, let normal components of velocity be prescribed on the segments OA, OB of the slip line ($v$ on OA and $u$ on OB, so that the tangential components are determined from (38.4), (38.5)). Both components can be prescribed if they satisfy equations (38.4), (38.5).

From (38.4), (38.5) we obtain the tangential components in the form

$$u = \int v\,\mathrm{d}\theta + C_1 \qquad \text{along } \alpha\text{-line}\,,$$

$$v = -\int u\,\mathrm{d}\theta + C_2 \qquad \text{along } \beta\text{-line}\,.$$

The constants are found from continuity conditions at the point O.

Denote by $u_{m-1,n}$, $v_{m-1,n}$, $u_{m,n-1}$, $v_{m,n-1}$ the values of the velocities $u$, $v$ at the nodal points of the slip grid $(m-1, n)$, $(m, n-1)$. If in (38.4), (38.5) we replace the infinitesimal increments by finite ones, we obtain formulae for calculating $u$, $v$ at the nodal point $(m, n)$:

$$\begin{aligned} u_{m,n}-u_{m-1,n} &= \tfrac{1}{2}(v_{m,n}+v_{m-1,n})(\theta_{m,n}-\theta_{m-1,n}) , \\ v_{m,n}-v_{m,n-1} &= -\tfrac{1}{2}(u_{m,n}+u_{m,n-1})(\theta_{m,n}-\theta_{m,n-1}) . \end{aligned} \tag{38.8}$$

In order to enhance the accuracy for the functions $u$, $v$, we have taken the arithmetic mean of the values of $u$, $v$ at neighbouring points. The construction begins at the node (1, 1).

*Cauchy's problem.* Let $u$, $v$ be given on some arc AB (fig. 94), which is not a slip line. The velocity field can then be constructed with the aid of relations (38.8).

*Mixed problem.* The normal component of the velocity $v$ is given along the segment OA of the $\alpha$-line (fig. 90), and the relation between the components of the velocity vector is known on the curve OB:

$$au + v = 0 ,$$

where $a$ is a constant. We restrict ourselves to consideration of the slip field for the case, shown in fig. 96, where the angle $\theta$, given on OB at the point O, is equal to the angle of inclination of the slip line at the same point. The slip line grid is known (fig. 96); the values $u_{1,1}$, $v_{1,1}$ at the node (1, 1) are computed from the formulae

$$\begin{aligned} v_{1,1}-v_{1,0} &= -\tfrac{1}{2}(u_{1,1}+u_{1,0})(\theta_{1,1}-\theta_{1,0}) , \\ au_{1,1}-v_{1,1} &= 0 . \end{aligned} \tag{38.9}$$

The relations (38.8) determine the values of the velocities $u$, $v$ at consecutive points (2, 1), (3, 1), . . . . For the point (2, 2) we have to begin again with equations similar to (38.9), and so on.

### 38.5. *The line separating plastic and rigid regions is a slip line or the envelope of slip lines*

We shall assume that the rigid region is at rest; this can always be done by superposing a velocity field corresponding to a rigid translation of the body. In the plastic zone there is some non-zero field.

We first assume that *the velocities are continuous on the line of separation* L. Then $u = 0$, $v = 0$ on L. If the boundary under consideration has nowhere a characteristic direction (i.e. nowhere coincides with a slip line), then we have Cauchy's problem for determination of the velocities $u$, $v$ in the plastic zone. Since the initial values are zero, and since the equations are homogeneous, the velocities must be zero in the plastic region, which contradicts the original assumption. If the boundary L coincides with some slip line, the vanishing on it of $u$, $v$ does not imply the vanishing of the velocity field in the plastic zone.

We now suppose that *discontinuities can occur on the line of separation.* The discontinuity can only be in the component of velocity $v_t$ tangential to L, since discontinuity in the normal component $v_n$ is connected with the appearance of a "crack". The boundary L can thus be thought of as the limiting position of a thin plastic layer (fig. 98; here $n$ is the normal, $t$ the tangent), in which the tangential velocity $v_t$ changes rapidly through its thickness, while the normal component $v_n$ is almost constant. It is evident that as the thickness of the layer decreases the shear velocity $\eta_{tn}$ increases without limit, while at the same time the other components of velocity will be almost invariant. But then the Saint Venant-von Mises relations (13.12)

$$\frac{\sigma_n - \sigma}{2k} = \frac{\xi_n}{H}, \ldots, \quad \frac{\tau_{nt}}{k} = \frac{\eta_{nt}}{H},$$

written in coordinates $t$, $n$, imply that

$$\sigma_n \to \sigma, \qquad \sigma_t \to \sigma, \qquad |\tau_{nt}| \to k,$$

since the intensity $H \to |\eta_{nt}| \to \infty$. Thus the boundary L will be a slip line or an envelope of slip lines. Note that the line of separation is often at the same time a line of velocity discontinuity (§39).

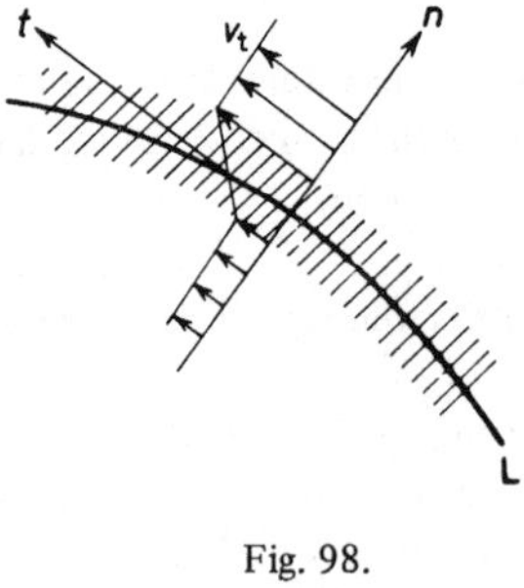


Fig. 98.

### 38.6. *Geiringer's equation*

We transform equations (38.1), (38.2) to new independent variables – the characteristic parameters $\xi$, $\eta$. Let

$$x = x(\xi, \eta), \qquad y = y(\xi, \eta)$$

subject to the condition that the Jacobian $\Delta(\xi, \eta) \neq 0$ in the region under consideration. Then using (33.1), we easily obtain the system

$$\frac{\partial v_y}{\partial \xi} - \tan\theta \frac{\partial v_x}{\partial \xi} = 0 \; ; \qquad \frac{\partial v_y}{\partial \eta} + \cot\theta \frac{\partial v_x}{\partial \eta} = 0 \; .$$

We transform to the velocity components $u$, $v$ in the $\alpha, \beta$ directions using

$$v_x = u\cos\theta - v\sin\theta \; , \qquad v_y = v\cos\theta + u\sin\theta \; ,$$

and after some simplification obtain

$$\frac{\partial u}{\partial \eta} - \tfrac{1}{2}v = 0 \; , \qquad \frac{\partial v}{\partial \xi} - \tfrac{1}{2}u = 0 \; . \tag{38.10}$$

Hence it follows that each velocity component satisfies the telegraph equation.

## §39. Lines of discontinuity in stress and velocity

### 39.1. *General remarks*

In the preceding sections we have already touched on the question of discontinuities in the *derivatives* of the stress (or velocity). Such discontinuities, called *weak*, develop along slip lines and are a consequence of discontinuities in derivatives of the initial data.

There are various cases where it is impossible to construct solutions with continuous stresses or velocities. At the same time there exist solutions with discontinuous stresses (velocities) which satisfy the boundary conditions (these discontinuities are called *strong*).

We consider a few simple examples.

In the problem of bending of a beam (§24) the stress $\sigma_x$ in the limiting state experiences a jump from $+\sigma_s$ to $-\sigma_s$ in passing through the neutral plane. Again, problems of pure plastic torsion are characterised by the appearance of lines of discontinuity, along which the tangential stress is discontinuous (in direction). In both cases the lines (or surfaces) of discontinuity are limit positions of elastic regions.

It is natural that in the plane strain problem under consideration discontinuous solutions are also possible. Nevertheless the significance of discontinuous solutions in the plane problem has escaped the attention of researchers, and has only comparatively recently been emphasized in the work of Prager (cf. [31]).

Discontinuous stress and velocity fields are also of interest because they can be used to derive simple approximate solutions on the basis of extremum principles. This question will be considered in ch. 8.

39.2. *Relations on a line of stress discontinuity*

Along a line of discontinuity it is necessary to satisfy some simple relations, which ensue from the equilibrium equations and the plasticity condition. Let L be a line of discontinuity (fig. 99), and consider an infinitesimal element lying on L. We suppose the thickness of this element to be vanishingly small. Normal stresses $\sigma_n$, $\sigma_t$ and a tangential stress $\tau_n$ act on the faces of the element.

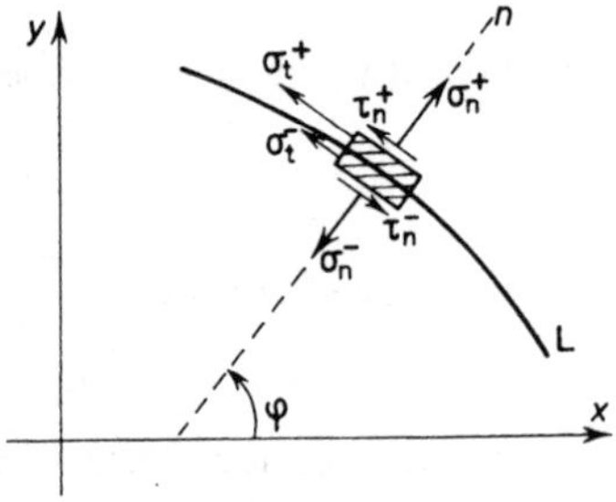


Fig. 99.

The values of the stress components on the two sides of the line of discontinuity will be distinguished by the indices +, −. By virtue of the equilibrium equations for the element, we must have (recalling that the thickness of the element tends to zero)

$$\sigma_n^+ = \sigma_n^- = \sigma_n\,, \qquad \tau_n^+ = \tau_n^- = \tau_n\,.$$

It follows that a discontinuity is possible only for the normal stress $\sigma_t$. The plasticity condition (31.8), which is valid on both sides of L, can be solved with respect to $\sigma_t$:

$$\sigma_t = \sigma_n \pm 2\sqrt{k^2 - \tau_n^2}\,. \tag{39.1}$$

Since by hypothesis L is a line of discontinuity, the upper and lower signs respectively in this formula should be taken for the values of $\sigma_t^+$, $\sigma_t^-$; then the jump in $\sigma_t$ will be $4\sqrt{k^2-\tau_n^2}$. The jump in the mean pressure $\sigma = \frac{1}{2}(\sigma_n + \sigma_t)$ will be equal to $2\sqrt{k^2-\tau_n^2}$.

*On a line of discontinuity the angle of inclination $\theta$ of a slip line exper-*

*iences a jump discontinuity.* The normal and tangential stresses $\sigma_n$ and $\tau_n$ on the two sides of L can be expressed with the aid of formula (35.2). Then the continuity conditions for $\sigma_n, \tau_n$ can be written in the following form:

$$\sigma^+ - k \sin 2(\theta^+ - \varphi) = \sigma^- - k \sin 2(\theta^- - \varphi) ,$$

$$k \cos 2(\theta^+ - \varphi) = k \cos 2(\theta^- - \varphi) .$$

Hence

$$\theta^+ - \varphi = \pm (\theta^- - \varphi) \pm m\pi ,$$

where $m$ is an integer. If we choose the + sign in front of the term in brackets on the right-hand side, we can easily show that the stress distribution will be continuous in the neighbourhood of L. We therefore choose the – sign, and then

$$\theta^- = -\theta^+ + 2\varphi + m\pi , \tag{39.2}$$

$$\sigma^- = \sigma^+ - 2k \sin 2(\theta^+ - \varphi) . \tag{39.3}$$

The first of these expressions gives that *at each of its points the line of discontinuity* L *is the bisector of the angle generated by slip lines of the same family* which approach L from opposite sides. This is easily seen if we take the $x$-axis to coincide with the tangent to the line L. Fig. 100 shows a slip element intersected by the line L, and the four directions of the slip lines $\alpha^+, \beta^+, \alpha^-, \beta^-$ which meet at each point of L.

Thus, the slip-line configuration experiences reflection with respect to the line of discontinuity L.

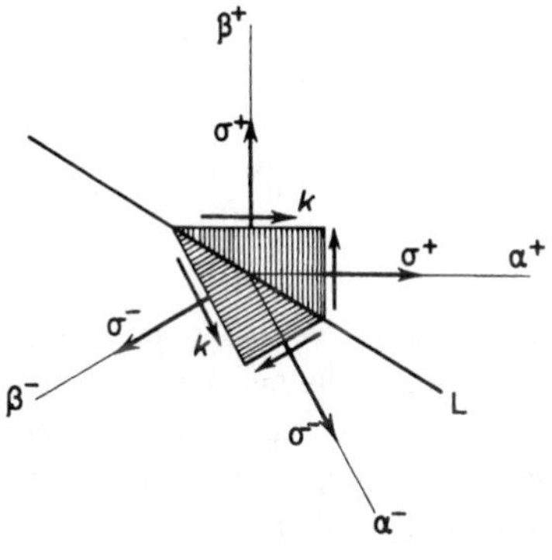


Fig. 100.

It follows from the preceding analysis that *stress discontinuities on slip lines are impossible* (since $\tau_n = k$ on slip lines, and then the normal stresses $\sigma_t = \sigma_n$ are continuous).

We note one further property of the stress field in the neighbourhood of the line of discontinuity L.

*The curvature of a slip line undergoes a jump in passing through a line of stress discontinuity* [54].

39.3. *Continuity of velocity near a line of stress discontinuity*

Since we have excluded the possibility of cracks, a discontinuity in the components of the velocity vector normal to the line L cannot occur, and we need only consider discontinuities in the tangential components.

It is not difficult to show that *the tangential component is also continuous on* L. Being the limiting position of the elastic region, the line of discontinuity can be replaced by a thin elastic strip. From the equilibrium equations for an element of the strip, it follows that the stresses $\sigma_n$, $\tau_n$ are almost constant in this strip; the tangential stress $\sigma_t$ changes very rapidly over the width of the strip (from $\sigma_t^+$ to $\sigma_t^-$, fig. 101), which means, among other things, that the narrow strip under consideration must be elastic (since with almost constant $\sigma_n$, $\tau_n$ and rapidly varying stress $\sigma_t$ the yield condition cannot be fulfilled).

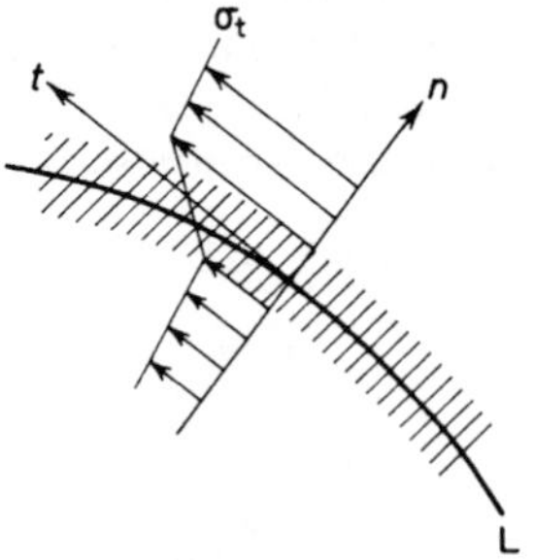


Fig. 101.

The velocity $v_n$ is continuous; we assume now that the component $v_t$ is discontinuous. Then the derivative $\partial v_t/\partial n$ is unbounded in the limit, and, consequently, so is the shear velocity $\eta_{nt}$; the other derivatives, and hence the other strain-rate components, are bounded. Repeating the argument used at the end of §38, we can establish that

$$\sigma_n \to \sigma\,, \qquad \sigma_t \to \sigma\,, \qquad |\tau_n| \to k\,.$$

This implies that the directions $n$, $t$ coincide with the directions of the slip lines. But then formula (39.1) with $\tau_n = k$ yields the continuity of the stresses along L, which contradicts the original hypothesis.

Thus, *a discontinuity in the component $v_t$ is also impossible.* Along L $v_t^+ = v_t^-$ and hence $\xi_t^+ = \xi_t^-$; therefore, by virtue of the von Mises equation (13.11), we have:

$$\lambda'_+(\sigma_t^+ - \sigma^+) = \lambda'_-(\sigma_t^- - \sigma^-) .$$

Using the expression (31.3) for the mean pressure, we have

$$\sigma_t^+ - \sigma^+ = \tfrac{1}{2}(\sigma_t^+ - \sigma_n^+) , \qquad \sigma_t^- - \sigma^- = \tfrac{1}{2}(\sigma_t^- - \sigma_n^-) .$$

By (39.1) these quantities are different in sign, and consequently, $\lambda'_+ = -\lambda'_-$. Since $\lambda' \geqslant 0$ we must have $\lambda' = 0$ at every point of the line of discontinuity L, i.e. the strain-rate components are zero along L:

$$\xi_n = 0 , \qquad \xi_t = 0 , \qquad \eta_{nt} = 0 .$$

Thus, *a line of stress discontinuity is not extended.* This result is natural enough if we note that a line of discontinuity is the trace of an elastic strip, where we are neglecting elastic strains.

39.4. *Lines of velocity discontinuity*

Suppose that along some line L the stresses are continuous but the velocity vector is discontinuous. At an arbitrary point of L we construct a coordinate system $n$, $t$, with the $t$-axis directed along the tangent to L. A discontinuity in the normal velocity component $v_n$ is impossible, and therefore we need only consider the tangential component $v_t$. We repeat the argument given at the end of the preceding section (fig. 98). The line of discontinuity L is the limit position of the layer, in which the velocity $v_n$ is almost constant, while the velocity $v_t$ is changing rapidly through the thickness of the layer (from $v_t^+$ to $v_t^-$). As the thickness of the layer decreases, the shear velocity $\eta_{nt}$ increases without limit, and at the same time the other velocity components undergo little change. This means that, in the limit, the direction of the line of discontinuity must coincide with the direction of the slip line. Thus, *a line of discontinuity in the velocity vector must be either a slip line or the envelope of slip lines.* In what follows we shall replace $v_n$, $v_t$ by $u$, $v$ (the components of the velocity vector in the direction of the slip lines $\alpha$, $\beta$; cf. §38). *The velocity $u$ can be discontinuous on an $\alpha$-line, and $v$ on a $\beta$-line.* From (38.4), (38.5) we obtain

$$u = \int v \mathrm{d}\theta + \text{const. along } \alpha\text{-line} ,$$

$$v = \int u \mathrm{d}\theta + \text{const. along } \beta\text{-line} .$$

Since $v$ is continuous on an $\alpha$-line and $u$ on a $\beta$-line, we easily see that *the jump in u (or v) is constant along an $\alpha$ ($\beta$) line of discontinuity.*

The magnitude of the tangential stress along a line of discontinuity is equal to $k$. In passing through such a line, an element experiences a finite shear in the direction in which the tangential stresses act, and changes its direction. Thus, for example, the jump in the velocity $u$ and the direction of the tangential stress $\tau$ are related by the condition that the dissipation be positive:

$$\tau(u^{+}-u^{-}) > 0 .$$

If the jump $[u] > 0$ then $\tau = +k$; if $[u] < 0$ then $\tau = -k$.

39.5. In conclusion we note that questions in the general theory of discontinuities in a plastic medium have been discussed in the books of Thomas [50] and Ivlev [11].

## §40. Non-uniqueness of the velocity field. Criterion for selection. Complete solution

The construction of the slip field depends on the choice of the plastic and rigid domains. Since the stresses are indeterminate in the rigid zones, this choice is to a large extent arbitrary. For this reason the plastic-rigid model is characterized by the *non-uniqueness of the stress and velocity fields* [1]). By way of illustration we consider a simple example.

### 40.1. *Extension of a strip with a hole*

We shall examine the problem of the extension of a strip with a sufficiently large [2]) circular hole (fig. 102a).

Since the circular contour is stress-free, it can be adjoined by an axisymmetric field of logarithmic spirals (§34). On the other hand, near the free rectilinear boundary of the strip there can be a field of uniform uniaxial tension (§36). Let these fields meet at some point C (fig. 102a). The stresses in the region ABC are determined from the formulae derived in §34:

$$\sigma_r = 2k \ln (r/a) , \qquad \sigma_\varphi = \sigma_r + 2k . \tag{40.1}$$

[1]) Note that the limit load is unique (cf. §65).

[2]) This condition guarantees the development of plastic deformation in weakened sections.

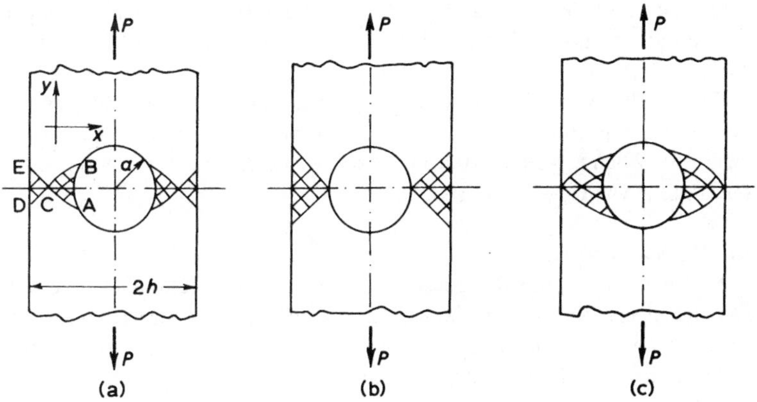


Fig. 102.

In a cartesian coordinate system $x, y$ the stresses in the region CDE have the form

$$\sigma_x = \tau_{xy} = 0\,, \qquad \sigma_y = 2k\,. \tag{40.2}$$

The limit load is

$$P_* = 2\int_a^{r_c} \sigma_\varphi \mathrm{d}r + 2\int_{r_c}^{h} \sigma_y \mathrm{d}x\,.$$

Substituting here for $\sigma_\varphi$, $\sigma_y$ we obtain

$$P_* = 4k(h-a) + 4k\int_a^{r_c} \ln\frac{r}{a}\,\mathrm{d}r\,.$$

The second term is easily evaluated, but this is now unnecessary. We need only observe that the integral term is non-negative and is a monotonically increasing function of $r_c$.

Consider now the velocity field. In the limit state the rigid parts of the body move respectively up and down with velocity $V$. The normal velocity components on the boundaries BC, AC; CD, CE are continuous and are easy to calculate since these boundaries are known. The tangential velocity components are discontinuous along lines of separation. The velocity fields in the plastic regions ABC, CDE are uniquely determined from the solution of the initial characteristic problem. Thus, the stress and velocity fields are compatible (it can be shown that the dissipation is positive at every point of the field).

We see that we can have whatever solution we wish, depending on the choice of the arbitrary point C; to each solution there corresponds some limiting load. When $r_c = a$ (fig. 102b) the load $P_* = 4k(h-a)$ is a minimum, and is a maximum when $r_c = h$ (fig. 102c).

40.2. *Selection criterion*

It is natural to pose the question, which of these solutions is preferred? The answer can be given on the basis of theorems concerning extremal properties of the limit load, which will be treated later (cf. §65). Anticipating somewhat, we shall formulate, without proof, a selection criterion ensuing from these theorems.

The solution obtained in the present section defines throughout the whole of the body (i.e. in both plastic and rigid zones) a velocity field compatible with the boundary conditions. Such a field is called *kinematically possible.* It will be shown below (§65) that *every kinematically possible velocity field results in a lower bound of the limiting load.*

Consequently, *the most appropriate solution corresponds to the lowest value of the load.* We shall call this proposition the *selection criterion.*

We now assume that *in the whole body* (i.e. in both plastic and rigid zones) a stress field $\sigma_x, \sigma_y, \tau_{xy}$ has been constructed which

1. satisfies the differential equations of equilibrium;
2. satisfies given boundary conditions for stresses for some value of the load parameter;
3. lies inside or on the yield circle, i.e.

$$(\sigma_x - \sigma_y)^2 + 4\tau_{xy}^2 \leqslant 4k^2 .$$

everywhere.

A stress field of this kind is called a *statically possible plastic field.*

From the second theorem (§65) *every statically possible plastic stress field results in a lower bound of the limiting load.*

The solution to the problem obtained earlier does not give such a field, since the stress state is unknown in the rigid zones. In accordance with the selection criterion it is necessary to focus on the solution shown in fig. 102b. This solution corresponds to the minimum load

$$P_* = 4k(h-a) \tag{40.3}$$

and is supported by observations. Fig. 103 reproduces photographs [79] of

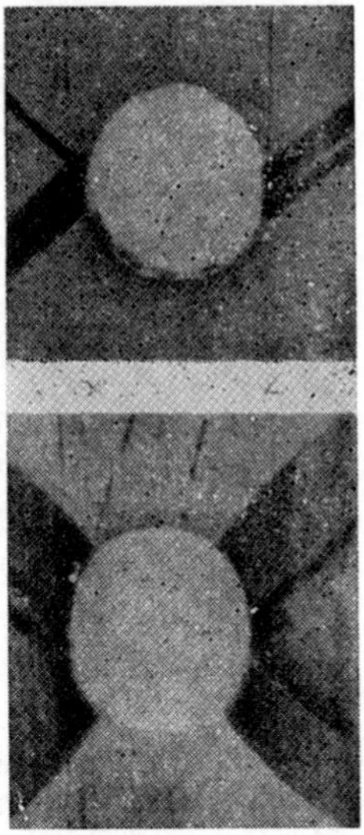

Fig. 103.

slip lines in initial and subsequent stages of plastic flow. The plastic zones are observed after special treatment (grinding, etching) of deforming steel specimens.

*Remark.* The selection criterion is not always sufficient for estimating the slip field. Cases can arise when different slip fields lead to the same limit load. This situation occurs, for example, in solving the problem of indentation of a flat die (§45). In these cases it is necessary to adduce supplementary mechanical considerations.

40.3. *Complete solution*

In the example we have been discussing the construction of an appropriate statically possible plastic stress field is easy; it is shown in fig. 104. The stresses are equal to zero in the shaded strip, while in the side strips,

$$\sigma_x = \tau_{xy} = 0 , \qquad \sigma_y = 2k . \tag{40.4}$$

It is obvious that the boundary conditions on the contour of the circular hole and on the lateral edges are satisfied. The corresponding limiting load is clearly equal to the previous value (40.3). Since the upper and lower bounds of the limit load coincide, the value obtained for the latter is exact. Al-

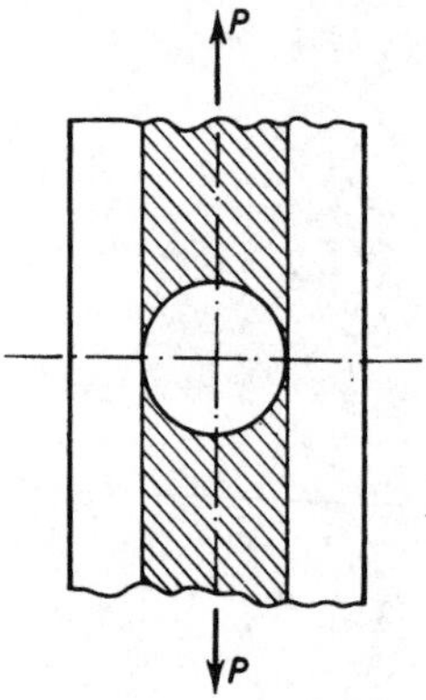


Fig. 104.

though the construction of the field in this particular case is elementary, the general case of field construction involves considerable difficulties.

One further remark should be made. Suppose a kinematically possible solution has been constructed (for example, the solution shown in fig. 102b). If it is possible to extend the stress state in the plastic zones into the rigid zones such that the yield criterion is nowhere exceeded, then the stress field so constructed in the whole of the body will be a statically possible plastic state. It is evident that in this case the upper and lower bounds of the limiting load will coincide. These solutions are called *complete,* since they result in the exact value of the limiting load.

The solution of the problem discussed in this section (fig. 102b) is complete; it can easily be extended through the whole body, as shown in fig. 104.

40.4. *Problems of finding limit loads*

We have repeatedly emphasized the significance of limit loads in determining a practical safety factor.

For an elastic-plastic body deformations usually develop gradually as the load increases; initially the elastic regions restrict the deformations of the body, the restricting effect diminishes as these regions contract until, finally, unrestricted plastic flow, corresponding to a limiting state, sets in. We have shown in a series of examples that loads close to the limit can be achieved with comparatively small deformations. As a rule the plastic deformations are localized and increase rapidly, while at the same time the elastic deformations change very little; as a consequence the latter can be neglected. This enables us to use a plastic-rigid model for the evaluation of the limit loads. The conditions which have to be satisfied by solutions in the plastic-rigid model have been discussed earlier (§23). In particular, it is necessary that the yield criterion should not be exceeded in the rigid zones. As we have already observed, this is not amenable to verification; but the construction throughout the body of a statically possible plastic field enables a lower bound of the limit load to be obtained.

Another requirement is that *the dissipation be positive everywhere in the slip field.* This is a condition of compatibility of stress and velocity fields, and can be verified, though not always simply.

The moment when the limit load is attained is characterized by an instantaneous movement, corresponding to the transition of the body from a rigid state to a "plastic mechanism". It is evident here that all changes in the exterior dimensions of the body can be neglected; the instantaneous movement is associated with a given ("instantaneous") combination of loads.

For a real elastic-plastic body a finite combination of loads can be achieved in various ways, and the question arises of the dependence of the limiting state on the loading path. In the sequel it is assumed that *the limit state is independent of the loading path.* This hypothesis is supported by experimental data, which are in good agreement with solutions obtained from the plastic-rigid model. Another important feature, which was discussed earlier (§15), is the asymptotic convergence of the stresses (with developing strain in a fixed direction) to a value independent of the strain path. As a rule, approach to the limit state of a body is accompanied by rapid development of strain in directions along which given external loads act; it can be assumed that a stress state is then established which is practically independent of the loading path.

A full evaluation of this hypothesis requires further investigation.

## §41. Extension of a strip weakened by notches

We consider the problem of the extension of a strip weakened by symmetric, deep notches of various shapes. It is assumed that the strip is sufficiently long so that the manner in which the ends are fixed has no influence on the plastic flow in the weakened section. As the following discussion will show, the form of the sides of the strip has no significance in the case of deep notches.

### 41.1. *Strip with ideal (infinitely thin) cuts*

The extension of a strip with ideal cuts (fig. 105) is the simplest problem of this type. In the limit state the strip expands in the $y$ direction along both sides of the mean section with velocity $V$. The slip field, shown in fig. 105, consists of four equivalent regions. Along the stress-free boundary of separation OA we have simple uniform compression or tension $\pm\, 2k$ in $\Delta$OAB; we shall suppose that in $\Delta$OAB we in fact have tension (as regards the other possibility, cf. below). Adjoining the region OAB there is a central field OCB, and then another triangular region OCD of uniform stress. The boundary of the plastic region is the $\beta$-line DCBA; the parameter $\eta$ = const. in the whole domain. But in $\Delta$OAB $\sigma = k$, $\theta = -\frac{1}{4}\pi$ and $\eta = \frac{1}{2}-\frac{1}{4}\pi$; in $\Delta$ODC $\sigma$ is unknown, $\theta = -\frac{3}{4}\pi$ and $\eta = \sigma/2k-\frac{3}{4}\pi$. Comparing the values of $\eta$, we see that the mean pressure in $\Delta$ ODC is $\sigma = k(1+\pi)$. From formulae (32.1) the stress components in $\Delta$ ODC are

$$\sigma_x = k\pi\,, \qquad \sigma_y = k(2+\pi)\,,$$

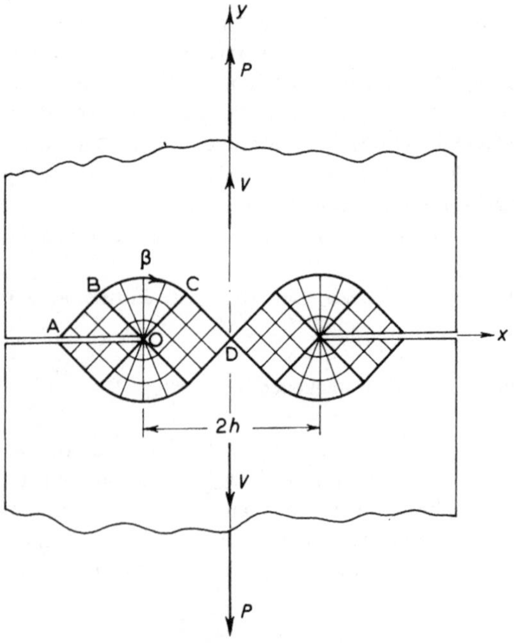


Fig. 105.

and consequently, the limit load is

$$P_*/P_*^0 = 1 + \tfrac{1}{2}\pi\,, \tag{41.1}$$

where $P_*^0 = 4kh$ is the limit extending force for a smooth strip of width $2h$.

We now turn to the determination of the velocities. By symmetry the velocity normal to OD is zero. On the boundary CD the normal component of velocity is continuous, since it is constant and equal to $V/\sqrt{2}$ on this boundary. Since $u = u(\beta)$, $v = v(\alpha)$ in $\Delta$ ODC (cf. §38), then $u = -V/\sqrt{2}$. Projecting on the vertical axis the velocity of the particles on OD, we have $u/\sqrt{2} + v/\sqrt{2} = 0$, and consequently $v = \text{const.} = V/\sqrt{2}$. Thus when the limit load is achieved the triangle ODC begins to move like a rigid body with velocity $V$ in the direction OD.

In the region OBC, $u$ is constant along each of the straight $\alpha$-lines; since the normal velocity component on the boundary BC is continuous, therefore $u = V \sin\theta$ on BC and everywhere in the region under consideration. Integrating now the relation (38.5), $dv + V \sin\theta \; d\theta = 0$, along $\beta$-lines, we obtain $v = V\cos\theta + \text{const.}$ But along OC $\theta = -\tfrac{3}{4}\pi$, $v = V/\sqrt{2}$, and so $v = V(\cos\theta + \sqrt{2})$. Thus $\Delta$ OAB moves like a rigid body with velocities $v_x = V$, $v_y = 2V$. The tangential velocity component along the line ABCD experiences a discontinuity.

The foregoing construction cannot be used in the case of strips with shallow notches.

It is easy to see that the load $P_*^0 = 4kh$ is the lower bound. For, assuming that in the middle part of the strip of width $2h$ there is a uniaxial tension ($\sigma_x = \tau_{xy} = 0$, $\sigma_y = 2k$), and that in the lateral zones $|x| > h$ the stresses are zero, we obtain a statically possible plastic stress field; to this there corresponds the load $P_*^0$.

In conclusion we note that the choice of the solution $-2k$ in the region OAB corresponds to compression of the strip with cuts.

41.2. *Strip with angular notches (fig. 106)*

In this case a solution can be constructed in a way similar to that of the preceding problem. Omitting the details of the calculation, which are easy enough, we find for the value of the limiting load

$$P_*/P_*^0 = 1 + \tfrac{1}{2}\pi - \gamma\,. \tag{41.2}$$

41.3. *Strip with notches with a circular base (fig. 107)*

Here the construction of the slip field depends on the ratio of the length of the section $2h$ to the radius of curvature $a$. If $h/a \leqslant 3.81$ the

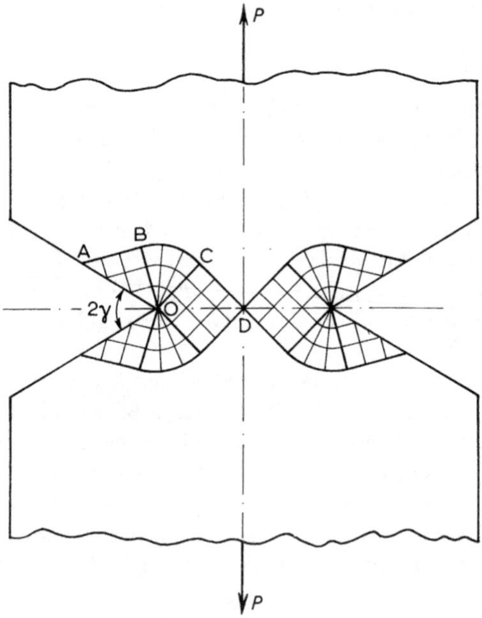


Fig. 106.

slip field adjoins the circular base only and is completely determined by the shape of the latter; this field is generated by logarithmic spirals (§36). The angle $\gamma$ is related to the distance $h$ by the expression

$$\gamma = \ln(1 + h/a)\,.$$

The stress distribution $\sigma_y$ along the segment AB is given by the formula

$$\sigma_\varphi = 2k(1 + \ln r/a)\,, \tag{41.3}$$

where $r$ is distance measured from the centre O. The limit load is determined from the formula

$$\frac{P_*}{P_*^0} = \left(1 + \frac{a}{h}\right) \ln\left(1 + \frac{h}{a}\right). \tag{41.4}$$

Since $\gamma \leqslant \frac{1}{2}\pi$ this solution is only of use when $\frac{1}{2}h \leqslant e^{\frac{1}{2}\pi} - 1 = 3.81$.

The velocity field can be determined from the solution of a characteristic problem for the region CC′B, with normal velocity components known and continuous for transition through BC.

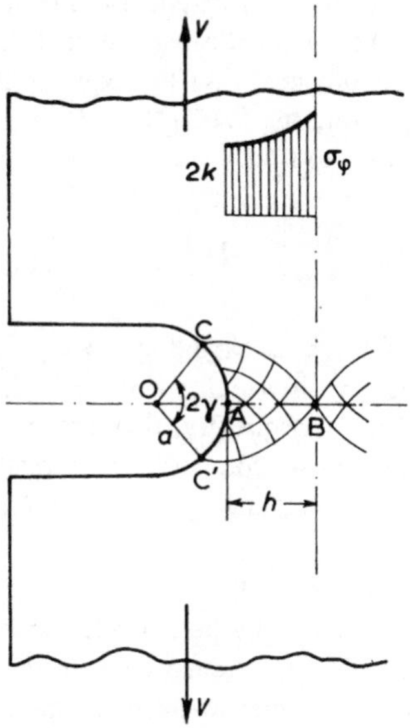


Fig. 107.

For large values of $h/a$ the construction is somewhat more complicated. Here the logarithmic field is bounded by the value $\gamma = \frac{1}{2}\pi$, and the segment AB (fig. 108) is equal to $3.81a$. To the right of this region we have the triangle BDE of uniform stress, and adjoining it the quadrilateral BEFC. In the lat-

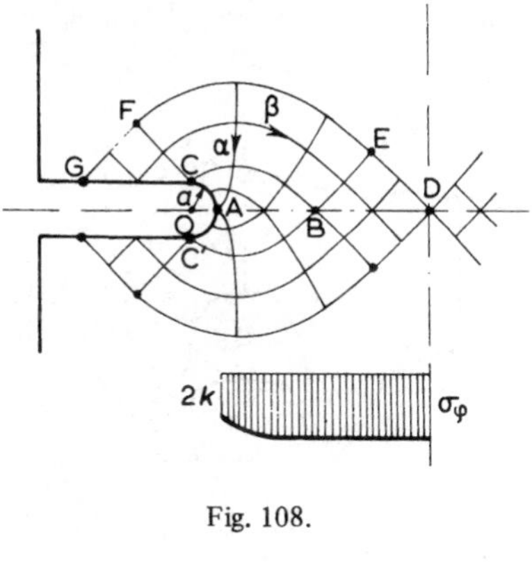


Fig. 108.

ter $\eta$ = const. and the $\alpha$-slip lines are equal segments of straight lines, perpendicular to BC, while the $\beta$-lines are parallel to BC. Along FC this region is linked to a triangle of uniform stress CFG, equal to BDE. Along AB the stress distribution is given by formula (41.3). In $\Delta$ BDE we have $\sigma_y = 2k(1 + \frac{1}{2}\pi)$. The limit load is now easy to find:

$$\frac{P_*}{P_*^0} = 1 + \tfrac{1}{2}\pi - \frac{a}{h}\,(e^{\frac{1}{2}\pi} - 1 - \tfrac{1}{2}\pi)\,. \tag{41.5}$$

The velocity distribution is as follows. We find as before that $\Delta$ BDE moves like a rigid body, with velocity $V$ to the right. The velocities $u$, $v$ are then constant along BE; consequently, they are constant along each $\alpha$-line in the quadrilateral BECF and are compatible with rigid-body motion of $\Delta$ FCG. The computation is completely analogous to that of par. 1.

41.4. *Other types of notches*

The same technique can easily be extended to investigate other types of weakening (angular notch with circular base, rectangular notch, etc.). Stress distributions with more complex weakening can be studied by the use of numerical construction of slip fields. The solutions we have been discussing are not applicable to shallow notches. The lower bound for the width of the strip is determined by the possibility of constructing the slip field (for example, the distance AD, fig. 108). However, the width of the strip must obviously be fairly large. Solutions of elastic-plastic problems obtained by numerical integration, as well as experimental observations, show that in the presence of shallow notches the plastic zones break through to the axis of the strip as the load increases, not along the weakened section, but above and below it.

The shaded area in fig. 109 shows the plastic region in a strip under ten-

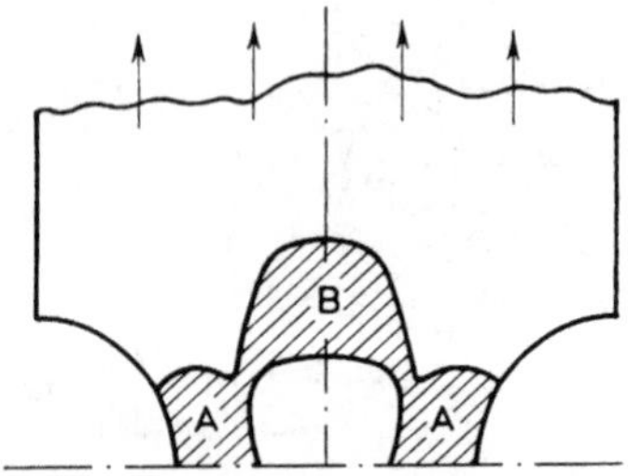


Fig. 109.

sion, weakened by shallow semi-circular notches. Initially, as the force increases, the regions A develop and expand; as the load approaches the limit, a new plastic zone arises on the axis of the strip. This latter region, B, quickly grows and merges with regions A as the limiting state is attained (fig. 109).

## §42. Bending of a strip weakened by notches

Following Green [104] we consider the problem of pure bending of a strip (fig. 110) weakened by notches of various shapes. In the construction of solutions it will be assumed that the plastic region takes hold of the most weakened section. This will be true for sufficiently deep notches; for shallow notches an increase in load may result in the plastic zones breaking through to the axis of the strip, away from the weakest section. This possibility is supported by experimental data and a number of numerical solutions of elastic-plastic problems.

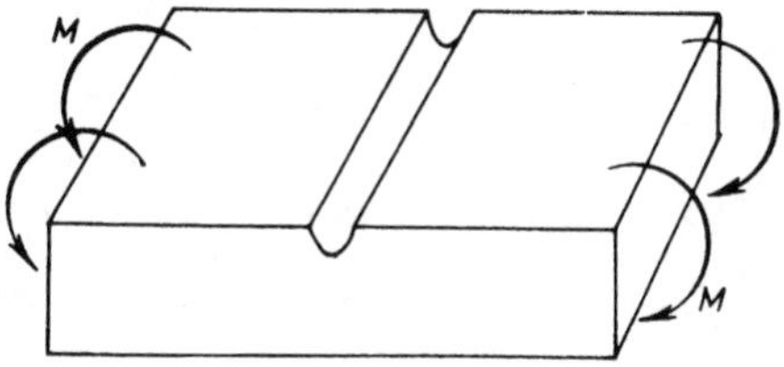


Fig. 110.

42.1. *One-sided deep notch with a circular base*

The notch is shown in fig. 111. As we shall see below, the method of extension of the circular arc is not applicable. Along the lower, stress-free boundary, there is in the triangle AAC a uniform stress, namely the compressive stress $-2k$, parallel to the base. The circular arc BB (of radius $a$) is also stress free, and, consequently, the boundary conditions are independent of the polar angle. In the neighbourhood of the arc there will be an axisymmetric slip field (§36), where

$$\sigma_r = 2k \ln (r/a) , \qquad \sigma_\varphi = 2k (1 + \ln (r/a)) . \tag{42.1}$$

Here $r$ is the radius vector originating from the centre O; the sign is chosen so that $\sigma_\varphi$ represents tension.

In the limit state the regions AAC and BBC are joined at the point C,

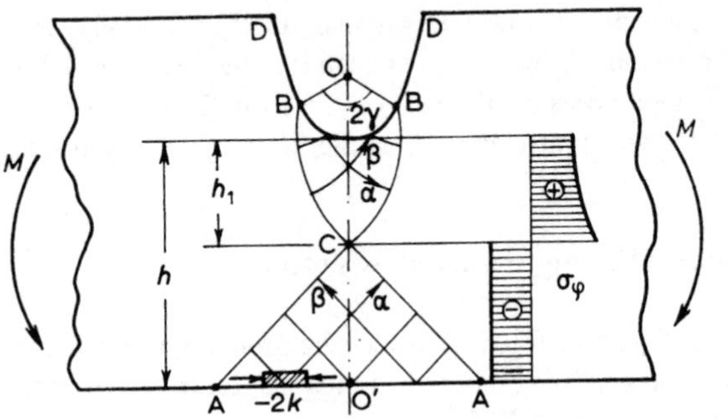


Fig. 111.

whose position is determined by the condition that the principal stress vector along the section OO′ is zero:

$$\int_a^{a+h} \sigma_\varphi \mathrm{d}r = 0 \,. \tag{42.2}$$

Substituting in this the stress

$$\sigma_\varphi = \begin{cases} 2k\,(1 + \ln\,(r/a)) & \text{for} \quad a \leqslant r < a + h_1 \,, \\ -2k & \text{for} \quad a + h_1 < r \leqslant a + h \,, \end{cases}$$

we obtain the equation

$$\rho_1(1 + \ln \rho_1) = \rho \,,$$

where

$$\rho_1 = 1 + h_1/a \,, \qquad \rho = 1 + h/a \,.$$

The stress distribution on the segment OO′ is shown in fig. 111; the stress $\sigma_\varphi$ is discontinuous at the point C.

Evaluating the limit moment $M_*$ (for unit width of the strip) from the equilibrium condition,

$$M_* = -\int_a^{a+h} \sigma_\varphi r \mathrm{d}r \,, \tag{42.3}$$

we find

$$\frac{M_*}{M_*^0} = -[(\rho+\rho_1)^2+(1-2\rho)^2-7\rho^2+8\rho_1]\,(\rho-1)^2\,,$$

where $M_*^0 = \frac{1}{2}kh^2$ is the limit bending moment for a smooth strip of height $h$. This relation is depicted by the dotted line in fig. 112.

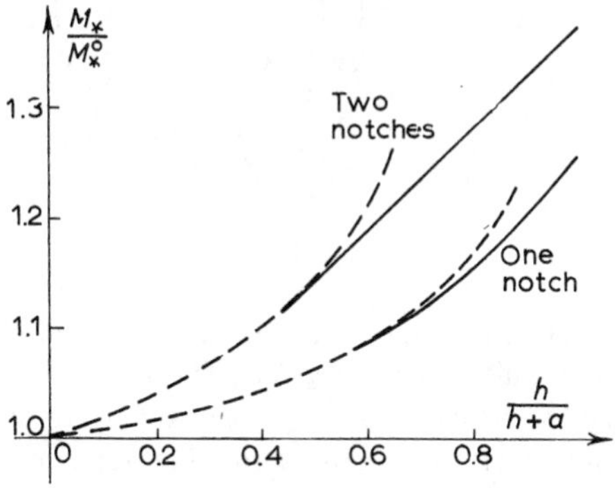


Fig. 112.

Note that the quantity $M_*^0$ *is the lower bound of the limit load.* Thus, the moment $M_*^0$ corresponds to the elementary solution of the bending problem (cf. §24) for a smooth strip of height $h$. We add this solution to a zero stress field lying above this smooth strip. It is obvious that we now have in the whole of the body a statically possible plastic stress state. This, of course, does not allow us to assert that the solution shown in fig. 111 is complete (since the solution is not extended throughout the body).

The presence of elastic (in our case, rigid) regions near the weakened section restricts the development of plastic deformation and increases the limit load by comparison with a smooth strip of height $h$. The ratio $M_*/M_*^0$ is sometimes called the *strength coefficient.*

The rigid zones are rotated about the point C. The normal velocity components along the slip lines BC, CA are known, and the velocity fields inside the plastic zones AAC and BBC are determined from the solutions of characteristic problems.

Consider now a different construction of slip field, shown in fig. 113. Adjoining the lower edge of the strip there is, as before, a region of uniform compression $-2k$, and near the circular boundary the solution is again described by the formulae (42.1). But the outline of these domains is now differ-

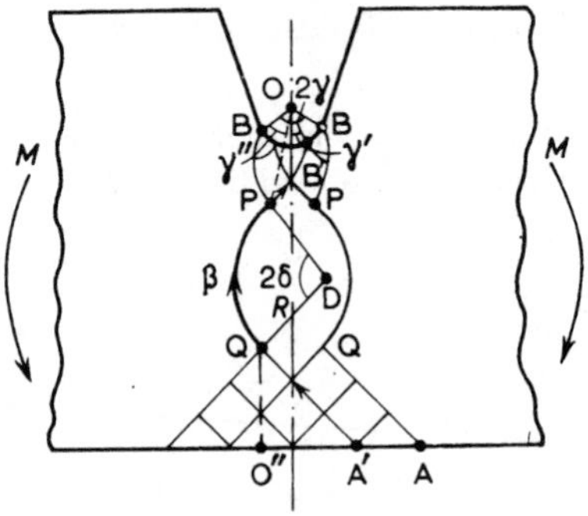


Fig. 113.

ent; they are less developed, contain re-entrant angles and are joined by two circular arcs – the so-called *isolated slip lines* PQ. The material inside PPQQ is in a rigid state, and so is that outside the lines BPQA. In the limit state the outer rigid portions experience complete rotation, for example the left-hand portion is rotated with respect to the centre of arc D. The material inside the nucleus PPQQ is stationary, so that the nucleus behaves like a journal around which instantaneous rotation along the slip lines PQ takes place. The line B′PQA′ is a slip line with a continuous tangent, consisting of a straight segment A′Q, a circular arc QP and a segment B′P of a logarithmic spiral. We denote ∠ BOP by $\gamma''$, the angle of spread of the arc PQ by $2\delta$, its radius by $R$, and the angle BOB by $2\gamma$. The mean pressure at the point P is equal to $\sigma = 2k(\gamma'' + \frac{1}{2})$ (cf. §36). The values of the parameter $\eta$ at the points P, Q respectively are

$$\eta_P = \tfrac{1}{2} + \gamma'' + \tfrac{1}{4}\pi - 2\delta\ , \qquad \eta_Q = -\tfrac{1}{2} + \tfrac{1}{4}\pi\ .$$

Along the $\beta$-slip line A′QPB′ $\eta$ = const., and consequently $\eta_P = \eta_Q$. Hence

$$\gamma'' = 2\delta - 1\ .$$

Further, at the point B′ we have $\sigma = k$, $\theta = \gamma' - \frac{1}{4}\pi$ and from the condition $\eta = \text{const.} = \eta_{B'} = \eta_Q$ we obtain

$$\gamma' = \tfrac{1}{2}\pi - 1\ .$$

Finally, it is obvious that

$$\gamma = 2\gamma'' - \gamma' = 4\delta - \tfrac{1}{2}\pi - 1\ .$$

To find the unknown parameters $\delta$, $R$ we have the two equations of equilibrium.

In the first place, the sum of the projections on the $x$-axis of the stresses acting in any cross-section (for example, in the section OPQO″) is equal to zero. Note that along OP only the tensile stress $\sigma_\varphi$ acts; along PQ the tangential stress $k$ and the normal stress $\sigma$ (varying on the arc PQ like a linear function of the angle $\theta$) act; and along QO″ we have the normal compressive stress $-2k$.

In the second place, the sum of the projections on the vertical $y$-axis of the stresses acting in the cross-section is also equal to zero.

Next, the limit moment $M_*$ is found as the moment of the stresses (in the same section OPQO″ with respect to the point D, say).

The results of the calculations, the details of which are omitted, are shown by the solid lines in fig. 112. Since $R$ must be non-negative it follows that $\delta \geqslant \frac{1}{4}\pi$ which in its turn corresponds to $h/(h+a) \geqslant 0.64$. When $\delta = \frac{1}{4}\pi$, $R = 0$ and the two constructions coincide. Beginning with $h/(h+a) > 0.64$, we find that the limit moment in the second solution is less than the first.

The velocity field is determined by consecutive solution of characteristic problems.

It is not possible to assert that the slip fields which have been constructed correspond to statically possible plastic fields (§40), since the stress state in the rigid zones is unknown. However, *the corresponding velocity fields are kinematically possible, and, consequently, both solutions give an upper bound for the limit moment* (§40). *It is therefore necessary to take the solution which leads to the smaller value of* $M_*$.

42.2. *Two-sided deep notches with circular base (fig. 114)*

It is assumed that the notches are symmetric.

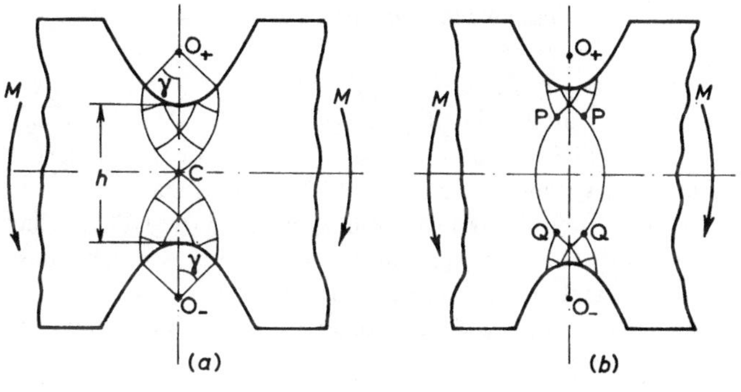


Fig. 114.

The simplest slip field consists of two symmetric fields (generated by logarithmic spirals) joined at the point C, which is a point of discontinuity of the stress. The latter is given by the expressions

$$\sigma_r = \pm\, 2k \ln (r/a)\,, \qquad \sigma_\varphi = \pm\, 2k\,(1 + \ln (r/a))\,,$$

where the plus sign relates to the upper field, the minus sign to the lower. The radius vector $r$ is measured from the respective centre ($O_+$ or $O_-$).

When $h > 2a$ the angle $\gamma$ is given for the formula

$$\gamma = \ln (h/2a)\,.$$

The equilibrium conditions reduce to one equation for the moments about the point C:

$$M_* - 2 \int_0^{\frac{1}{2}h} \sigma_\varphi y \,\mathrm{d}y = 0$$

(where $y$ is the distance from C on $O_+O_-$). This determines the value of the limit bending moment $M_*$.

Putting $y = a + \frac{1}{2}h - r$, substituting $\sigma_\varphi$ in the moment equation and integrating, we obtain the dependence of the ratio $M_*/M_*^0$ on the same parameter $h/(h + a)$; this relationship is shown by the dotted line in fig. 112.

If we begin with $h/(h + a) = 0.398$, another slip field is possible, whose construction is similar to that for the second solution in the case of one notch. Here the two axisymmetric fields are joined by circular arcs PQ (fig. 114b), along which slip takes place – rotation of the outer rigid portions. The calculations are as in the previous case; the solid lines in fig. 112 illustrate the respective dependence of the limiting moment on the geometric parameter; this is to be used when $h/(h + a) \geqslant 0.398$.

42.3. *Deep angular notch (fig. 115)*

The construction of the slip field is clear from fig. 115. Since the free boundaries are rectilinear, it follows that uniform stress states exist in the triangles OBD, CAA (tension + $2k$ in $\Delta$ OBD, compression $-2k$ in $\Delta$ CAA). Central fields ODE join the triangle OBD with the square OECE, which experiences uniform stress. It is apparent from symmetry considerations that, in OECE, only the tension $q$ acts along vertical sections.

Along the $\beta$-line CB (and $\beta$-lines parallel to it) the parameter $\eta$ is constant. Hence $\eta_C = \eta_B$, i.e.

$$\sigma/2k - \tfrac{1}{4}\pi = \tfrac{1}{2} + \tfrac{1}{4}\pi - \gamma\,.$$

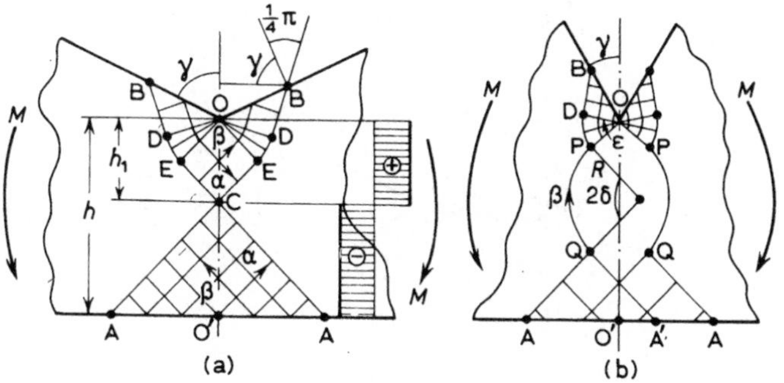


Fig. 115.

But $q = \sigma + k$ in the square OECE, and consequently we obtain $q = k(2 + \pi - 2\gamma)$. The position of the point of discontinuity C, which determines the whole construction, is found from the equilibrium condition: $qh_1 - 2k(h-h_1) = 0$.

The limit moment is

$$M_* = \tfrac{1}{2}qh_1^2 + k(h-h_1)^2 . \tag{42.4}$$

Substituting here the values of $q$ and $h_1$, we obtain the relation

$$\frac{M_*}{M_*^0} = 1 + \frac{\pi - 2\gamma}{4 + \pi - 2\gamma} , \tag{42.5}$$

which is shown by the dotted line in fig. 116. As before, the rigid portions experience rotation about the point C; the velocity field is determined by consecutive solution of characteristic problems.

Consider now the other version of the slip field (fig. 115b); here the rotation takes place along the circular slip line PQ around the solid, stationary "journal" OPQQP. In Δ OBD there is a uniform tension + $2k$, while in the domain adjoining the lower boundary there is a uniform compression $-2k$. The triangular region OBD is contiguous with the central field ODP, which is linked to the plastic zone on the lower part of the circular slip lines PQ. Along A′QPO $\eta$ = const., and hence $\eta_P = \eta_Q$; since

$$\eta_P = \frac{\sigma_P}{2k} + \tfrac{1}{4}\pi - 2\delta , \qquad \eta_Q = -\tfrac{1}{2} + \tfrac{1}{4}\pi ,$$

we have $\sigma_P = k(4\delta - 1)$; it is clear from geometric considerations that $\delta + \epsilon +$

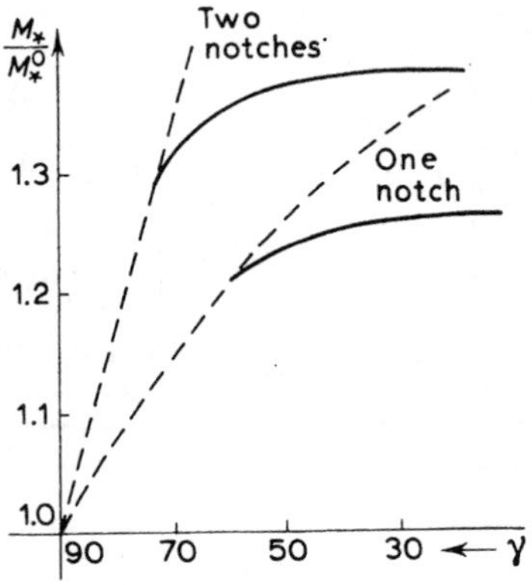


Fig. 116.

$+\frac{1}{4}\pi+\gamma=\pi$; moreover, $\xi=$ const. in the regions OPD, ODB, from which it follows that $3\delta=1-\gamma+\frac{3}{4}\pi$.

The construction is now defined by two parameters – the distance O′A′ and the radius of curvature $R$. As before, these are found from two equilibrium equations. The calculations are straightforward, through somewhat involved, and we pass directly to the results. When $\gamma=1$, $R$ tends to zero and the second type of field merges into the first. When $\gamma<1$ the second solution gives a smaller value for the limit moment. The corresponding relationship is represented by the solid lines in fig. 116. For small angles $\gamma$ the limiting moment is practically independent of $\gamma$; in this case the limit load will be the same as for a notch with circular base when the radius of curvature tends to zero.

42.4. *Two-sided deep angular cuts, symmetrically distributed*

This case can be treated in a similar way. The dependence of the limit moment on the angle of the cut is shown in fig. 116.

42.5. *Experimental data*

With the aid of special processes (grinding, etching) it is possible to observe the plastic zones during the deformation of steel specimens. Fig. 117 shows photographs obtained by Handy for the cases $2\gamma=60^\circ$ and $2\gamma=140^\circ$ [79]. Experimental data are in good agreement with the theoretical conclusions regarding the two different types of slip field (shown on the left in fig. 117).

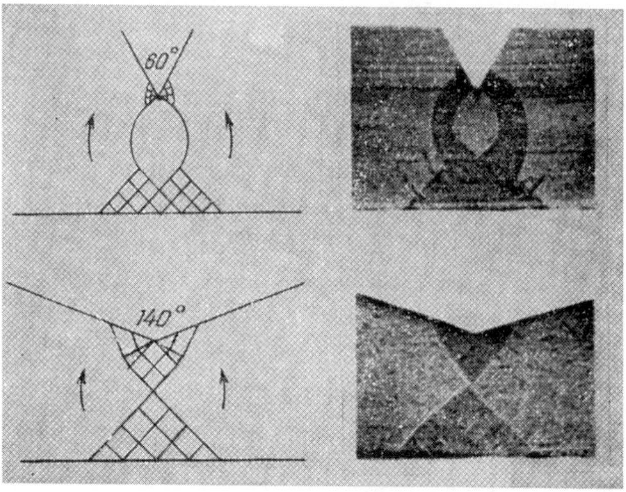


Fig. 117.

## §43. Bending of a short cantilever

### 43.1. *Formulation of the problem*

In §24 we considered the problem of the plastic bending of a beam by a couple. However, just as in the case of an elastic material, the results of that analysis can also be applied to bending by transverse loads, provided the beam is sufficiently long (for then the effect of tangential stresses is negligible).

For short beams neglect of tangential stresses can lead to substantial errors.

Let a cantilever beam of constant rectangular cross-section be bent by a force $P$ (per unit width), applied at the end (fig. 118); the left-hand end of the beam is rigidly clamped. The width $b$ in the horizontal direction is constant and much greater (at least six times [104]) than the height $2h$. In these circumstances we can assume plane strain. As usual, we proceed from a plastic-rigid model of the material.

### 43.2. *First type of slip field*

In this problem a possible slip field is that shown in fig. 119. In the triangles ABC, A′B′C′ the stresses are determined by the free boundaries, so that

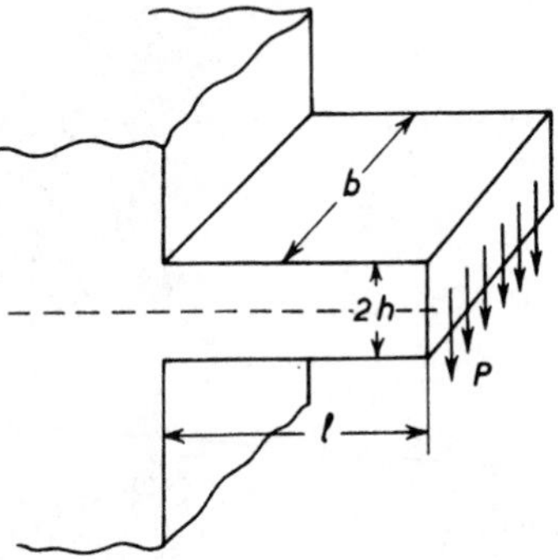


Fig. 118.

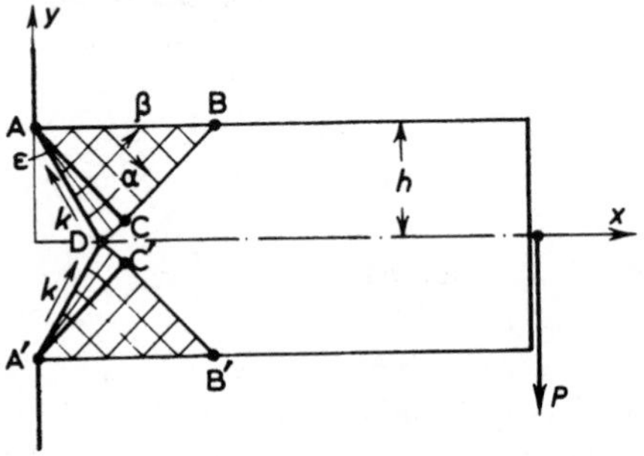


Fig. 119.

we have uniaxial tension + $2k$ in ABC, and compression $-2k$ in A'B'C'. In $\Delta$ ABC $\theta = -\frac{1}{4}\pi$, $\sigma = k$, $\xi_1 = \frac{1}{2} + \frac{1}{4}\pi$, $\eta_1 = \frac{1}{2} - \frac{1}{4}\pi$. Adjoining this triangle is the central field ACD, in which $\eta$ = const. = $\eta_1$. Let $d$ denote the length AD (the point D lies on the axis of the beam), and $\epsilon$ the angle DAC. Then obviously

$$d \cos\left(\tfrac{1}{4}\pi - \epsilon\right) = h\,.$$

Along AD $\theta = -\frac{1}{4}\pi - \epsilon$, and hence the mean pressure, equal on the slip line AD to the normal stress, is $\sigma_1 = k(1 + 2\epsilon)$; the tangential stress on AD is clearly equal to $k$. Along A'D the mean pressure is $\sigma_1' = -k(1 + 2\epsilon)$. At the point D the stresses are discontinuous.

Consider now the equilibrium of the part of the cantilever to the right of the line ADA'. We have the condition that the sum of the projections on the

$y$-axis of the forces is equal to zero. This leads to the equation

$$\cos(\tfrac{1}{4}\pi-\epsilon)-(1+2\epsilon)\sin(\tfrac{1}{4}\pi-\epsilon)=\frac{P}{2kd},$$

which determines the angle $\epsilon$. The length of the cantilever $l$ cannot be arbitrary, and is found from the requirement that the sum of the moments with respect to D are equal to zero:

$$k(1+2\epsilon)\,d^2=P[l-d\sin(\tfrac{1}{4}\pi-\epsilon)]\ .$$

These equations establish the relationship between $P/2kh$ and $l/2h$.

In the limiting state a rotation of the rigid part of the cantilever (to the right of BDB′) takes place with respect to the point D. It is easy to show that the velocity field in the plastic zones is compatible with the velocities on the boundaries.

43.3. *Second type of slip field (fig. 120)*

Here again we have tension and compression respectively in the regions ABC and A′B′C′. Adjacent to these are the central fields ADC, A′D′C′, which are linked by the isolated circular slip line DD′ of radius $R$. The right-hand part of the cantilever slides along this arc in the limit state.

Let $d$ be the length of AD, $\epsilon$ the angle DAC, and $2\delta$ the angle subtended by the arc DD′. In $\Delta$ ABC $\sigma=k$, $\theta=-\frac{1}{4}\pi$, $\xi_1=\frac{1}{2}+\frac{1}{4}\pi$, $\eta_1=\frac{1}{2}-\frac{1}{4}\pi$. In the sector ACD $\eta=\eta_1$ and it is easy to see that along AD $\sigma_1=k(1+2\epsilon)$ and $\xi_{AD}=$ $=(1+2\epsilon)/\epsilon+\frac{1}{4}\pi+\epsilon$. Similarly we have along A′D′ $\sigma_1'=-k(1+2\epsilon)$, $\xi_{A'D'}=$ $=-(1+2\epsilon)/\epsilon+\frac{3}{4}\pi-\epsilon$. It is clear from geometrical considerations that $\delta=$ $=\frac{1}{4}\pi-\epsilon$. The curve ADD′A′ is a continuous slip line, with $\xi=$ const. on it, and

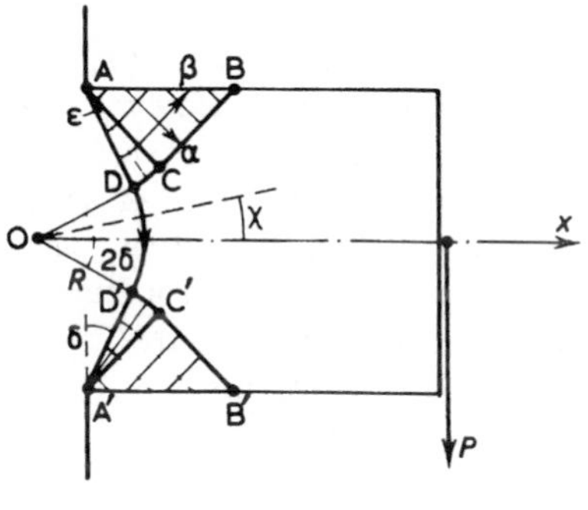


Fig. 120.

so $\xi_{AD} = \xi_{A'D'}$. Hence

$$2\epsilon = \tfrac{1}{4}\pi - \tfrac{1}{2} = 16°20' .$$

Thus $2\delta = 73°40'$. The length $d$ is found from the geometrical relation

$$R \sin \delta + d \cos \delta = h .$$

The unknown radius $R$ is determined from the condition that the sum of the projections on the vertical of all the forces acting on the portion of the beam to the right of the curve ADD′A′ is zero. Along the arc DD′ the tangential stress is zero, and the normal stress is computed from the condition $\xi = \text{const.} = \xi_{AD}$; since $\xi = \sigma/2k - (\chi - \frac{1}{2}\pi)$ we have on DD′ $\sigma = 2k\chi$, where the angle $\chi$ is measured from the horizontal. Thus,

$$kd \cos \delta - \sigma_1 d \sin \delta + kR \int_0^\delta \cos \chi d\chi - 2kR \int_0^\delta \chi \sin \chi d\chi = \tfrac{1}{2}P$$

or

$$0.03d + 0.43R = \tfrac{1}{2}P/k .$$

Next we form the momental equation relative to the point O:

$$kdR + \tfrac{1}{2}\sigma_1 d^2 + k\delta R^2 = \tfrac{1}{2}Pl' ,$$

where $l'$ is the distance of the loading end from the centre O:

$$l' = l + R \cos \delta - d \sin \delta .$$

As the force increases, the radius $R$ increases, and the regions ABCD contract rapidly. The plastic deformation is thus localized essentially along the isolated circular slip line. The above equations establish the relationship between $P/2kh$ and $l/2h$.

When $2\epsilon < \frac{1}{4}\pi - \frac{1}{2}$ only the first solution is valid; when $2\epsilon = \frac{1}{4}\pi - \frac{1}{2}$ the second solution leads to a smaller, and therefore more appropriate, value of the limiting load $P_*/2kh$ (since these solutions, being kinematically possible, give an upper bound for the load). The first type of field arises with long beams $l/2h \geqslant 13.73$; when $l/2h = 13.73$ the two fields coincide, since $R = 0$. Observations on the deformation of beams [79] confirm the presence of two types of slip fields. Fig. 121 shows the theoretical field and a picture of the bending of a short cantilever obtained in experiment.

The elementary solution, based on ignoring the tangential stresses (§24), leads to the relation

$$P_*^0/2kh = h/l .$$

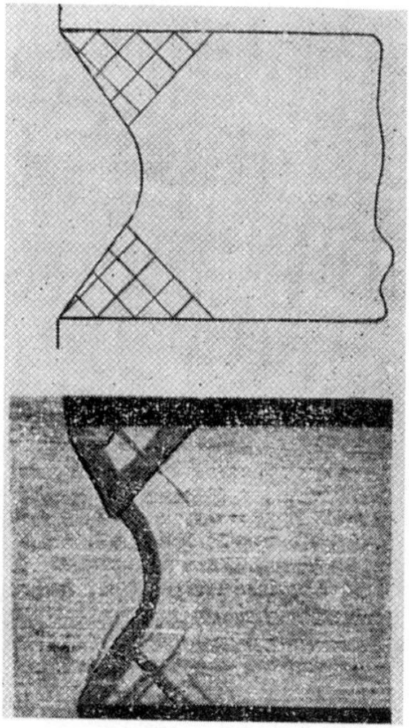

Fig. 121.

The dependence of the limit loads $P_*/P_*^0$ on the parameter $l/2h$ are shown in fig. 122. It is easy to see that in the case of plane strain the approximate load $P_*^0$ obtained from the elementary solution is about 5% below the exact load $P_*^0$ when $l/2h = 10$.

For short beams the divergence increases, but does not exceed 13% for $l/2h \geqslant 1$. As the length increases $P_* \to P_*^0$.

### 43.4. *Concluding remarks*

Methods similar to the above enable us to investigate the bending of a wedge-shaped cantilever and cantilevers bounded by circular arcs. In the latter case the possible slip field (type II) is shown on the left of fig. 123. Because of the variable height of the beam the region of plastic deformation begins at some distance from the fixed end. Experimental observations (cf. the photograph on the right of fig. 123) are in good agreement with theoretical predictions. Another case which has been examined is that of the bending of a cantilever by a uniformly distributed load [184].

We note in conclusion that the distribution of plastic deformation is strongly influenced by the method of fixing the end [184].

The lower bound of the limit load has been given in [170].

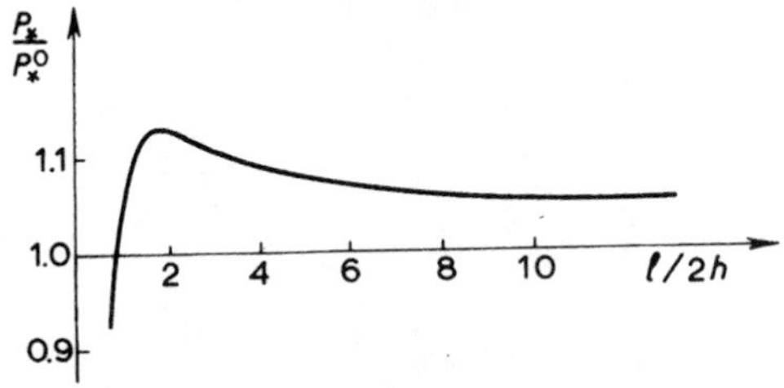


Fig. 122.

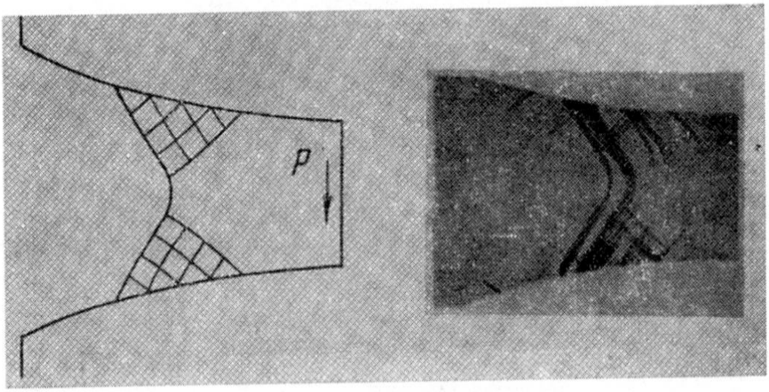


Fig. 123.

## §44. Rectangular neck

We have been considering the effects of tension and bending on a strip weakened by notches. A.P. Green [102] investigated a series of problems on the limit states when the weakening operated in conditions of shear and pressure. Problems of this type arise, in particular, in the analysis of the behaviour of dry friction of metals, which is connected with plastic deformations at uneven contact surfaces.

We shall examine one such case here, which is typical of a wider class.

Let the massive portions I and II be linked by a neck of height $h$ and length $l$ (fig. 124); the upper portion I moves to the left with speed $V$, while the lower portion moves to the right with the same speed. It is required to find the limit load and the plastic zones.

Fig. 125 shows a possible construction of the slip field, symmetric with

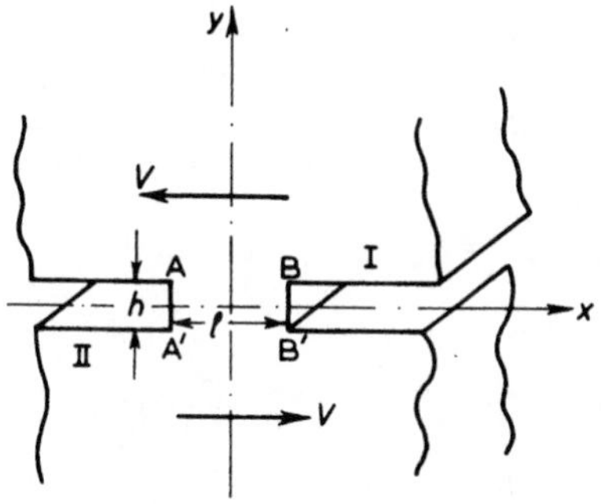


Fig. 124.

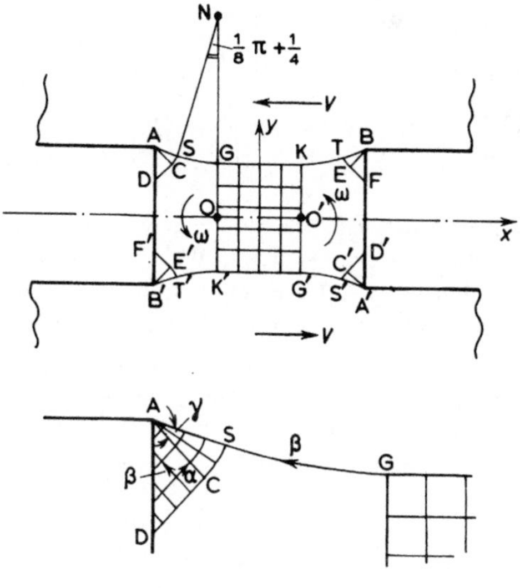


Fig. 125.

respect to the $x$, $y$-axes. There are plastic zones at the corners A, A′, B, B′. These zones contain the triangles ACD, A′C′D′ of uniform uniaxial tension $+2k$ and the triangles BEF, B′E′F′ of uniaxial compression $-2k$, which adjoin the free boundaries AB′ and BA′. These triangles are adjacent to the central fields CAS, EBT, . . . with angle of spread $\gamma$. The central zone GKG′K′ is a region of uniform shear with zero pressure $\sigma$, and it is linked to the plastic zones by the circular slip lines SG, KT, S′G′, K′T′ of radius $R$. Denote the lengths of the segments AC and GK′ by $a$, $b$ respectively. The regions DCSGK′T′E′F′ and KTEFD′C′S′G′ remain rigid (i.e. experience no plastic deformation). The lower part of fig. 125 shows, on a larger scale, a representation of the slip field near one of the corners. In $\Delta$ ACD

$$\theta = \tfrac{1}{4}\pi\,, \qquad \sigma = k \quad \text{and} \quad \xi = \text{const.} = \tfrac{1}{2} - \tfrac{1}{4}\pi \equiv \xi_1\,;$$

and in $\Delta$ BEF

$$\theta = \tfrac{3}{4}\pi\,, \qquad \sigma = -k \quad \text{and} \quad \xi = \text{const.} = -\tfrac{1}{2} - \tfrac{3}{4}\pi \equiv \xi_2\,.$$

Along AS $\theta = \frac{1}{4}\pi + \gamma$, while along TB $\theta = \frac{3}{4}\pi - \gamma$. It is evident that in each of the regions ASCD, BTEF the parameter $\xi$ is a constant, equal to $\xi_1$, $\xi_2$ respectively; these conditions enable us to find the mean pressure on the sides AS and TB: $\sigma_1 = k(1 + 2\gamma)$, $\sigma_2 = -k(1 + 2\gamma)$. We next calculate the values of the parameter $\eta$ on the sides AS, TB:

$$\eta_1 = \tfrac{1}{2} + \tfrac{1}{4}\pi + 2\gamma\,, \qquad \eta_2 = -\tfrac{1}{2} + \tfrac{3}{4}\pi - 2\gamma\,.$$

Since ASGKTB is a continuous $\beta$-line, $\eta$ = const. along it; i.e. $\eta_1 = \eta_2$ from which we find $\gamma = \frac{1}{8}\pi - \frac{1}{4}$. It is clear from geometrical considerations that $\angle\,\text{SNG} = \frac{1}{8}\pi + \frac{1}{4}$.

The unknown quantities $R$, $a$, $b$ can be determined from the following conditions. First, the vertical projection of the curve ASG is equal to $\frac{1}{2}(h-b)$. Next, the rigid region DCSGK′T′E′F′ must be in equilibrium; its boundary F′D is free and a tangential stress equal to $k$ in magnitude acts along the segment DSGK′T′F′. The normal stress is zero on GK′, $k$ on DC and $-k$ on F′E′. On the arc CS it ranges in value from $k$ to $k(1 + 2\gamma)$ as a linear function of the angle of inclination of the slip line, while on the arc SG it decreases from the latter value to zero. The normal stress along the line E′T′K′ has a similar variation.

By symmetry, the sum of the projections of the stresses on the $x$-axis is equal to zero ($\Sigma\, X = 0$), and only two conditions remain: $\Sigma\, Y = 0$ and $\Sigma$ mom $= 0$. After some straightforward but tedious computation we can now derive the values of the parameters:

$$a/h = 0.052\,, \qquad R/h = 1.076\,, \qquad b/h = 0.739\,.$$

Here the width of the segment GK depends only on the ratio $l/h$, and, as is easily seen, tends to zero when $l/h = 0.68$. Thus the slip field under consideration can be achieved when $l/h \geqslant 0.68$.

Calculating the sum of the horizontal projections of the stresses acting along the slip line ASGKTB we obtain the limit shearing force

$$Q_* = Q_*^0(1-0.249\,h/l)\,, \tag{44.1}$$

where $Q_*^0 = kl$ is the shearing force from an elementary result of strength of materials. Obviously, $Q_* \to Q_*^0$ as $h/l \to 0$.

We now turn to the kinematic picture.

Denote by $U$ the discontinuities (equal in magnitude) of the tangential velocity components along the upper and lower $\beta$-lines (AGKB and A′G′K′B′ respectively). The central zone experiences a uniform shear; points of the line OO′ are stationary, and the velocity of shear is equal to $(2/b)\,(V-U)$. The rigid regions rotate with respect to the points O (O′) with angular speed $\omega$ in a positive direction. On the other hand, these regions undergo rotation with respect to the lower and upper portions around the centres of the corresponding circular arcs (SG, KT, . . .) with some angular speed $\omega'$.

From the continuity condition for the normal velocity components we can easily find that

$$V-U = \tfrac{1}{2}\omega b\,, \qquad R\omega' = U\,, \qquad (R+\tfrac{1}{2}b)\,\omega' = V\,,$$

and hence

$$\omega = U/R\,, \qquad V = (1+b/2R)\,U\,.$$

Fig. 126 is a photograph of the plastic deformation fields observed in a neck after etching; the region of uniform shear in the central portion of the neck is very clear.

In conclusion we note that analogous solutions can be constructed [102] for symmetric necks of different shapes (e.g. with circular sides).

## §45. Indentation by a flat die

We consider the problem of the occurrence of plastic flow due to indentation by a rigid die having a flat base (fig. 127). The plastic medium is bounded by a plane, and friction at the contact surface is neglected. In the limiting state the die moves downwards with velocity $V$. The deformations are assumed small, so that changes in the shape of the free surface can be ignored.

### 45.1. *Prandtl's solution*

Prandtl's solution is an early approach to the plane problem. Suppose that in the limiting state the distribution of pressure under the die is uniform.

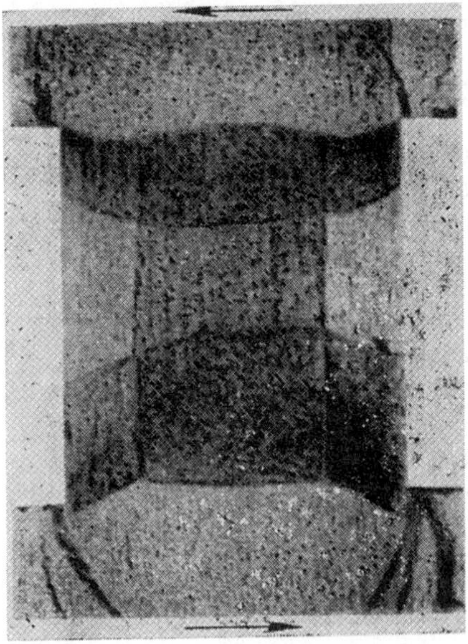

Fig. 126.

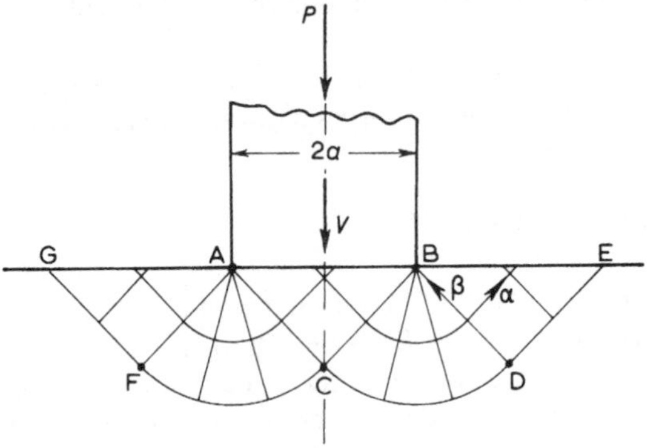


Fig. 127.

Then the slip field (fig. 127) can be constructed as follows: under the die and bordering it there will be regions of uniform stress; in particular, the triangles BDE and AFG will experience simple compression, parallel to the boundary. In the triangle ABC the pressure $\sigma$ is unknown, and

$$\theta = \tfrac{1}{4}k\,, \qquad \xi = \text{const.} = \sigma/2k + \tfrac{1}{4}\pi \equiv \xi_1\,,$$

while in $\Delta$ BDE

$$\sigma = -k\,, \qquad \theta = \tfrac{1}{4}\pi\,, \qquad \xi_2 = -\tfrac{1}{2} - \tfrac{1}{4}\pi\,.$$

The triangular domains are joined by central fields. The parameter $\xi$ is constant along an $\alpha$-line, and hence $\xi_1 = \xi_2$, and $\sigma = -k(1+\pi)$. From (32.1) we find the stresses in $\Delta$ ABC:

$$\sigma_x = -k\pi\,, \qquad \sigma_y = -k(2+\pi)\,.$$

The limit load is

$$P_* = 2ak(2+\pi)\,. \tag{45.1}$$

We next find the velocity distribution: $\Delta$ ABC moves downwards like a rigid body, with the velocity $V$ of the die. Along BC the tangential velocity component is discontinuous, while the normal component is equal to $V/\sqrt{2}$. Along CD the tangential velocity component is discontinuous, and the normal component is zero. Then, from (38.7), $v = 0$ and $u = V/\sqrt{2}$ in the central field. Finally, the region BDE slips like a rigid body, with velocity $V/\sqrt{2}$ in the direction DE. A similar field can be constructed on the left-hand side of fig. 127.

### 45.2. *Hill's solution*

A different solution, proposed comparatively recently by Hill, is shown in fig. 128. Here too it is assumed that uniform pressure acts along the contact line AB. Then we have in the region OCDEB the same stress field as in the

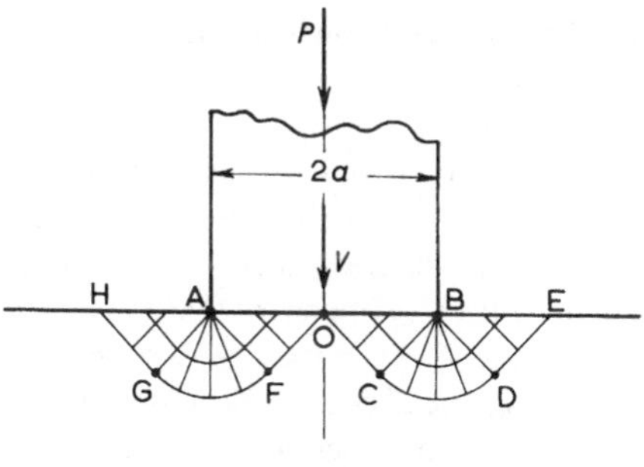


Fig. 128.

region ACDEB of Prandtl's solution (fig. 127). Again, the stress field in OFGHA of fig. 128 is the same as in BCFGA of fig. 127. It is obvious that the same uniform pressure $\sigma_y = -k(2 + \pi)$ acts along the line AOB. The limit load $P_*$ has its previous value (45.1). However, the slip field and the kinematic picture are different (fig. 128). In this case OCB slips like a rigid body with velocity $\sqrt{2}V$ along OC. The velocity on BC is continuous, and in the central field BCD $v = 0$, $u = \sqrt{2}V$. The triangular region BDE moves in the direction DE with velocity $\sqrt{2}V$. In contrast with Prandtl's solution, the velocity field in the plastic zones is continuous.

As Prager has observed, it is possible to construct a solution in the form of a combination of the Prandtl and Hill solutions, and containing an arbitrary parameter which characterizes the overlap of the regions OBC and OAF.

This problem illustrates the non-uniqueness of solutions based on the plastic-rigid body model. For this reason the construction of possible slip and velocity fields requires the introduction of various additional considerations and the utilization of experimental results.

In particular, it is helpful to attempt to use elasticity theory or some other considerations to represent the nature of the origin and development of plastic zones. From this point of view Hill's solution gives a more correct picture, since plastic zones, if they arise from solution of the corresponding problem of the pressure of a rigid die on an elastic half-plane, originate in the neighbourhood of the corners A, B and later spread towards the centre.

A different approach to the question of choosing the solution is discussed in the work of Ya. Rykhlevskii [151].

## §46. Wedge under the action of a one-sided pressure

We attempt to find the limit load for a wedge acted on by a uniform pressure $p$ applied to its right-hand boundary (fig. 129).

### 46.1. *Case of an obtuse wedge*

We first consider the case of an obtuse-angled wedge, with $2\gamma > \frac{1}{2}\pi$. In the triangular region OCD, generated by slip lines originating at the ends of the segment OD, there is a uniform stress. It is evident that uniform stress is also present in OAB, adjoining the free boundary OA.

By virtue of (35.4), $\sigma_t = -p \pm 2k$; the sign is chosen from some additional consideration (or by trying out all possibilities). We suppose that the wedge is "bent" as a result of the action of one-sided pressure; then we must expect a tensile stress on the side OD and a compressive stress on the side OA. On this

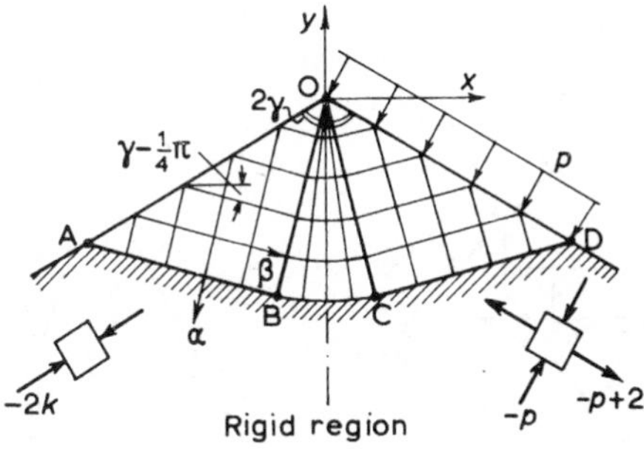


Fig. 129.

basis we assume that $\sigma_t = -p + 2k$ in $\Delta$ OCD, and that $\sigma_t = -2k$ in $\Delta$ OAB (fig. 129). These two regions are joined by a central field (§33) OBC; in the latter, stresses are constant along every ray and vary (as linear functions of the angle of inclination of a ray) from their values on the slip line OC to their values on the slip line OB. The line of separation ABCD, being a slip line, supports a tangential stress $\tau = k$. The normal stress is $\sigma = -p + k$ on the segment CD and $\sigma = -k$ on the segment AB; on the segment BC it has a linear variation.

These regions can be in equilibrium only for a definite value of the pressure $p = p_*$, called the *limit* value. This quantity can be calculated from the condition that the parameter $\eta$ is constant along $\beta$-lines:

$$\text{in } \Delta\ \text{OCD} \qquad \theta = \gamma - \tfrac{3}{4}\pi\,, \qquad \eta = \frac{-p+k}{2k} + \gamma - \tfrac{3}{4}\pi\,,$$

$$\text{in } \Delta\ \text{OAB} \qquad \theta = -(\gamma + \tfrac{1}{4}\pi)\,, \qquad \eta = -\tfrac{1}{2} - \gamma - \tfrac{1}{4}\pi\,.$$

Equating the values of $\eta$, we obtain Prandtl's formula

$$p_* = 2k(1 + 2\gamma - \tfrac{1}{2}\pi)\,. \tag{46.1}$$

In the $\xi$, $\eta$-plane, the solution is represented in the form of a segment of the straight line $\eta$ = const. When $\gamma = \frac{1}{4}\pi$ (rectangular wedge) the central field degenerates to a straight line, and the solution has a simple form – uniform uniaxial compression everywhere (fig. 130). Suppose the normal component is given along OD; the base of the wedge is stationary, and the normal velocity component on the line ABCD is zero. It is easy to see that the velocity field can be determined by consecutive solution of boundary-value problems for the regions OCD, BCO, BAO.

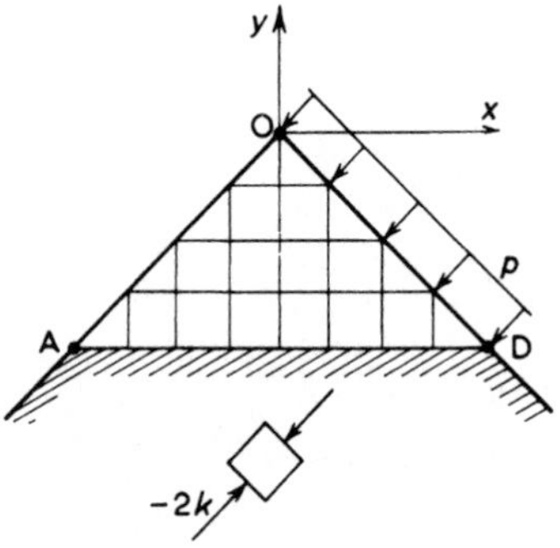


Fig. 130.

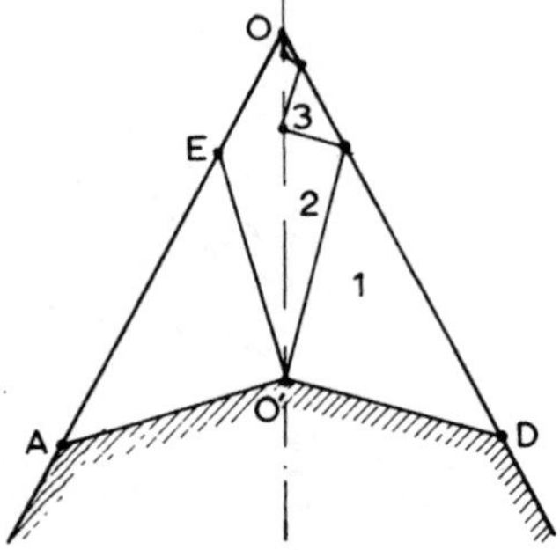


Fig. 131.

46.2. *Case of an acute wedge*

When $\gamma < \frac{1}{4}\pi$ the previous construction is no longer meaningful, since the triangles OAB and OCD overlap, and this leads to non-uniqueness of the stress state. In this case a continuous stress field is impossible, and the solution is characterized by a discontinuous field, as shown in fig. 131. To the right and left of the line of discontinuity we have, as before, a uniform stress state:

$$\text{in } \Delta\, \mathrm{OO'D} \qquad \theta^+ = \gamma - \tfrac{3}{4}\pi\,, \qquad \sigma^+ = -p + k\,,$$

$$\text{in } \Delta\, \mathrm{OO'A} \qquad \theta^- = -(\gamma + \tfrac{1}{4}\pi)\,, \qquad \sigma^- = -k\,.$$

On the line of discontinuity we obtain from (39.3)

$$p_* = 2k(1 - \cos 2\gamma) \tag{46.2}$$

(for $\varphi = 0$).

It is worth noting that a discontinuous stress field can also be constructed for an obtuse wedge, but, as we shall show, this leads to a contradiction. In order to establish the validity of a solution it is necessary to verify that the fields satisfy all outstanding conditions.

We consider briefly the velocity field for an acute wedge. The lines of separation AO′, O′D (fig. 131) are slip lines, and the normal velocity component $v_n = 0$ on them. Along the line of discontinuity OO′, $v_y = \text{const.}$ (§39); since the base of the wedge is stationary, $v_y = 0$ at the point O′, and, consequently, $v_y = 0$ everywhere on the line of discontinuity. Suppose the normal velocity component on the boundary OD is given. We partition the right-hand half of the wedge by slip lines into an infinite sequence of decreasing triangles (fig. 131); in each of these the velocity field can be found by solving consecutive mixed boundary-value problems (§38), passing from triangle 1 to triangle 2, and so on. In this way the velocity component $v_x$ on the line of discontinuity OO′ can be found.

We now turn to the construction of the velocity field in the left-hand half of the wedge. In OO′E the velocities can be determined by solving Cauchy's problem with known values of $v_x$, $v_y$ on the straight line OO′. For O′EA we have a characteristic problem.

An analogous construction of the velocity field for a discontinuous solution to the obtuse wedge leads to a contradiction. In Δ ODD′ we have a mixed problem (fig. 132 – normal velocity components are given on the line OD and on the characteristic DD′), which determines the velocity field in this triangle. Next, we have to solve a characteristic problem for the triangle OO′D′, and this will then give the velocities on the line of discontinuity OO′. It is now not possible to insist that $v_y = 0$ on OO′, which is a consequence of the assumption that a discontinuous solution exists, and therefore this construction is invalid.

In conclusion we note that the elastic-plastic problem for a wedge was also examined by Shapiro and Naghdi; the equilibrium of a wedge of hardening material was considered by Sokolovskii [44]. We mention also the recent work of Nayar, Rykhlevskii and G.S. Shapiro [138].

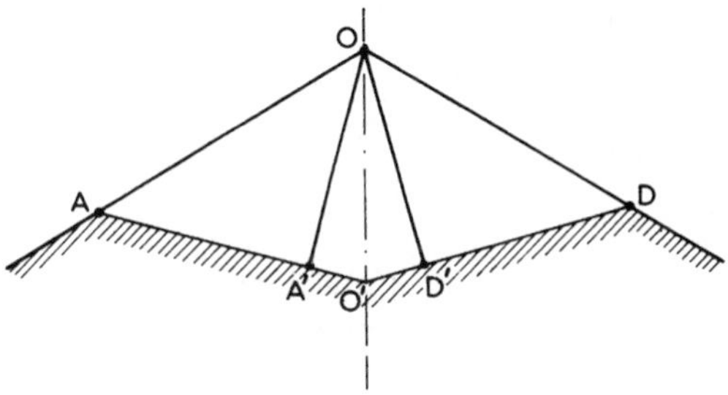


Fig. 132.

## §47. Compression of a layer between rigid plates

We consider the problem of compression of a plastic layer between parallel, rigid, rough plates (fig. 133). The plastic layer is compressed sideways and flows from the centre to the edges; large tangential stresses arise at the con-

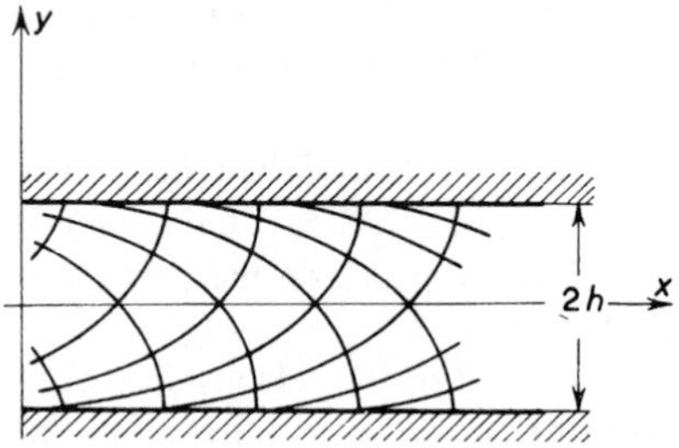


Fig. 133.

tact surfaces. For developing plastic deformations it is necessary to assume that these tangential stresses attain a maximum value $k$.

47.1. *Prandtl's solution for a thin layer*

Let the thickness $2h$ of the layer be substantially less than its breadth $2l$. Then the balancing loads at the end sections of the layer cannot influence significantly the state of the layer at some distance from the ends. In these circumstances we can reasonably consider solutions which do not exactly satisfy the boundary conditions at the ends of the layer.

It is easy to see that the stresses

$$\begin{aligned} \sigma_x &= -p-k(x/h-2\sqrt{1-y^2/h^2})\,, \\ \sigma_y &= -p-k\,x/h\,, \\ \tau_{xy} &= k\,y/h \end{aligned} \tag{47.1}$$

satisfy the differential equations of equilibrium (31.9) and the plasticity condition (31.8) for any value of the arbitrary constant $p$. The components of the velocity vector

$$\begin{aligned} v_x &= V + c(x/h-2\sqrt{1-y^2/h^2})\,, \\ v_y &= -c\,y/h \end{aligned} \tag{47.2}$$

satisfy the incompressibility condition (31.11) and equation (31.10), for arbitrary values of the constants $c$, $V$. It follows from (47.2) that each of the plates moves on the layer with speed $c$.

We next find the slip lines. Comparing the last of formulae (32.1) with the formula $\tau_{xy} = k\,y/h$, we obtain

$$y = h\cos 2\theta\ .$$

Hence

$$\frac{\mathrm{d}y}{\mathrm{d}x} = -2h \sin 2\theta \frac{\mathrm{d}\theta}{\mathrm{d}x} .$$

With the aid of (31.7) we now find the differential equations of the slip lines:

$$2h \sin 2\theta \frac{\mathrm{d}\theta}{\mathrm{d}x} = -\tan\theta , \qquad 2h \sin 2\theta \frac{\mathrm{d}\theta}{\mathrm{d}x} = \cot\theta .$$

Separating the variables and integrating, we obtain the parametric equations of the families of slip lines:

$$x = -h(2\theta + \sin 2\theta) + \text{const.} , \qquad y = h \cos 2\theta ; \qquad (\alpha)$$

$$x = h(2\theta - \sin 2\theta) + \text{const.} , \qquad y = h \cos 2\theta . \qquad (\beta)$$

The slip-line field is generated by two orthogonal families of cycloids, where the radius of the generating circle is equal to $h$. The straight lines $y = \pm h$ are the envelopes of these families of cycloids, and, consequently, they are lines of discontinuity. It is easy to see that the derivatives $\partial\sigma_x/\partial y$ and $\partial v_x/\partial y$ tend to infinity along the latter. The shear rate $\eta_{xy}$ is also unbounded on the lines $y = \pm h$.

The conditions on the free edge $x = 0$ are satisfied in the sense of Saint Venant, i.e. we require that when $x = 0$

$$\int_{-h}^{h} \sigma_x \mathrm{d}y = 0 .$$

Substituting in this the stress components, we obtain $p = \frac{1}{2}k\pi$. The pressure distribution $(\sigma_y)_{y=h}$ is linear. The limit compressive force (denoted by $2P$) is easily calculated:

$$2P = 2\int_0^l \sigma_y \mathrm{d}x = -kl(l/h + \pi) . \qquad (47.3)$$

The parameters $c$ and $V$ are related through the incompressibility condition; the flux of material through the section $x = 0$ must be equal to the amount of material which extrudes in unit time and length $l$ as the plates move together:

$$-\int_0^h v_x \mathrm{d}y = lc .$$

Replacing $v_x$ from (47.2) with $x = 0$, we obtain

$$V = c(\tfrac{1}{2}\pi - l/h)\ .$$

The normal stress $\sigma_y$ is constant through the thickness of the layer, and is a linear function of $x$. *Away from the free boundary* the stress $\sigma_x$ differs from $\sigma_y$ only by a small quantity, of order $h/l$ compared with unity; to this degree of accuracy the flow velocity in the $x$-direction is constant through the layer. The tangential stresses are small compared with the normal stresses.

Prandtl's solution is unsatisfactory near the edges (when $x = 0$ the boundary condition is satisfied only in the sense of Saint Venant) and in the middle portion (near $x = 1$), since the tangential stresses must vanish on the axis of symmetry. It is necessary to assume that there is a rigid region in the middle part of the layer, and that the material extrudes from it on both sides (fig. 134). Nevertheless, Prandtl's solution is a good approximation for a thin layer.

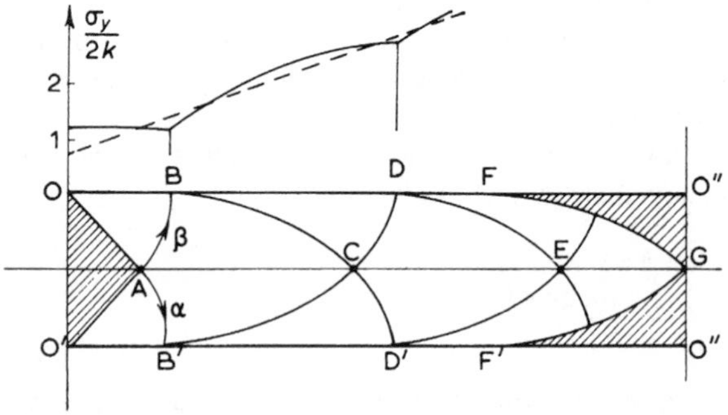


Fig. 134.

### 47.2. *Moderately thick layer*

For a layer of finite thickness it is not possible to neglect the effect of the conditions at the ends of the layer and in its central portion; the solution must satisfy all the boundary conditions for stresses and velocities. The construction of the slip field for this case was also carried out by Prandtl; subsequently this problem was examined by Sokolovskii [44], Hill, Lee and Tupper [54] and others.

We assume that the $\alpha$-slip line which separates the plastic and rigid regions is a straight line OA (fig. 134); the justification of this will be given later. By symmetry the tangential stresses are zero on the axis of the layer, and therefore the line OA is inclined at an angle $\frac{1}{4}\pi$. At the point O the stress field has a singularity, and the solution in the region OAB is represented by a central field which, in accordance with the condition on the contact surface, is bounded by an $\alpha$-slip line coinciding with the boundary. We consider the forces acting on the rigid portion OAO$'$. At the ends OA, O$'$A there are uniformly distributed tangential stresses, of magnitude $k$ and directed towards the vertex A; the normal stress is equal to the mean pressure. From the equilibrium condition for the rigid portion OAO$'$ it is easy to see that the mean pressure along the segment OA of the $\alpha$-line must be equal to $-k$. By Hencky's theorem $\sigma = -2k\theta$ + const. along the circular $\beta$-line; determining this constant from the values of $\sigma$, $\theta$ on OA, and then passing to the contact curve OB, for which $\theta = 0$, we find that the pressure is constant and equal to $-k(1 + \frac{1}{2}\pi)$ along OB. Next, the values of $\xi$, $\eta$ are known on the segments AB, AB$'$ of the characteristics $\alpha$, $\beta$, and the stress state in the region ABCB$'$ can be determined by the solution of a characteristic problem. In the region BCD we have a mixed problem (the angle $\theta = 0$ is given along BD, since $\tau_{xy} = k$ on the contact curve and the slip surfaces coincide with the boundaries).

We can proceed with this construction so long as we do not reach the axis O$''$O$''$. By symmetry the tangential stresses are zero on the line O$''$O$''$, and therefore the condition $\tau_{xy}$ = const. $= k$ on the contact curve cannot be satisfied near the middle of the layer. A rigid zone occurs here, bounded by the contact curve and the slip line FG, extending to the point G. The pressure distribution on the portion FO$''$ remains undetermined, and it is only possible to calculate the mean pressure for the stresses acting along the line of separation FG. Our construction is possible if the point C does not get to the other side of the axis of symmetry O$''$O$''$. The calculations show that this is the case when $l/h > 3.64$. The solution to the right of AB is obtained by numerical methods.

An example of a numerically-calculated slip grid is shown in fig. 135 (the thickness of the layer has been taken equal to two, i.e. $h = 1$). In the sector OAB the solution is known. We divide the arc AB into 10 equal parts by the points (0, 0), (0, 1), . . . (0, 10). The value of $\theta$ at each of these points is equal to the inclination of the corresponding $\alpha$-ray, for example $\theta = -\frac{1}{4}\pi$ at the point (0, 0), $\theta = -\frac{1}{8}\pi$ at the point (0, 5), etc. The mean pressure on the arc AB is equal to $\sigma = -k(1 + \frac{1}{2}\pi + 2\theta)$.

The arc AB$'$ is also divided into 10 equal parts by the points (0, 0), (1, 0), . . . , (10, 0). The pressure $\sigma$ at symmetric points assumes its previous values; the angle $\theta$ is easily found. Next, we calculate the coordinates of the nodes $x_{0,1}$, $y_{0,1}$, $x_{0,2}$, $y_{0,2}$, etc., and we write all the data on the arcs AB, AB$'$ in the top row and left-hand column of the table respectively. The coordinates of the node $(m, n)$ and the values of the unknown functions $\theta_{m,n}$, $\sigma_{m,n}$ at this point are calculated from formulae (37.1), (37.2), (37.3), (37.4). By symmetry it is sufficient to calculate the field above AC (we note that $\theta$ =

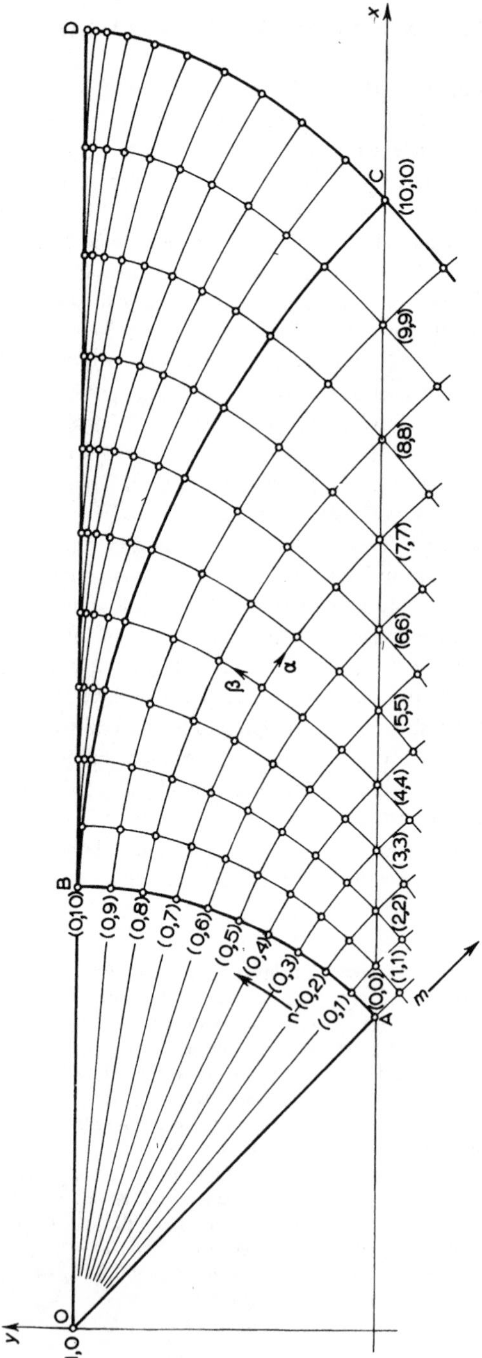

x
y
O
1,0
A
B
C
D
(0,0)
(0,1)
(0,2)
(0,3)
(0,4)
(0,5)
(0,6)
(0,7)
(0,8)
(0,9)
(0,10)
(1,1)
(2,2)
(3,3)
(4,4)
(5,5)
(6,6)
(7,7)
(8,8)
(9,9)
(10,10)
α
β
m
n


Fig. 135.

Compression of a plastic layer between rigid plates

| | $m \backslash n$ | 0 | 1 | 2 | 3 | 4 | 5 | 6 | 7 | 8 | 9 | 10 |
|---|---|---|---|---|---|---|---|---|---|---|---|---|
| $-\theta_{0,n}$ | 0 | 0.785 | 0.706 | 0.628 | 0.550 | 0.471 | 0.392 | 0.314 | 0.235 | 0.157 | 0.078 | 0.00 |
| $-\sigma_{0,n}/k$ | | 1.00 | 1.16 | 1.31 | 1.47 | 1.63 | 1.78 | 1.94 | 2.10 | 2.26 | 2.41 | 2.57 |
| $x_{0,n}$ | | 1.00 | 1.07 | 1.14 | 1.21 | 1.26 | 1.31 | 1.34 | 1.37 | 1.40 | 1.41 | 1.41 |
| $y_{0,n}$ | | 0.00 | 0.08 | 0.17 | 0.26 | 0.36 | 0.46 | 0.56 | 0.67 | 0.78 | 0.89 | 1.00 |
| $-\theta_{1,n}$ | 1 | | 0.785 | 0.706 | 0.628 | 0.550 | 0.417 | 0.392 | 0.314 | 0.235 | 0.157 | 0.078 |
| $-\sigma_{1,n}/k$ | | | 1.31 | 1.47 | 1.63 | 1.78 | 1.94 | 2.10 | 2.26 | 2.41 | 2.57 | 2.73 |
| $x_{1,n}$ | | | 1.16 | 1.24 | 1.32 | 1.39 | 1.45 | 1.50 | 1.54 | 1.57 | 1.60 | 1.61 |
| $y_{1,n}$ | | | 0.00 | 0.09 | 0.18 | 0.29 | 0.39 | 0.51 | 0.62 | 0.74 | 0.86 | 0.99 |
| $-\theta_{2,n}$ | 2 | | | 0.785 | 0.706 | 0.628 | 0.550 | 0.417 | 0.392 | 0.314 | 0.235 | 0.157 |
| $-\sigma_{2,n}/k$ | | | | 1.63 | 1.78 | 1.94 | 2.10 | 2.26 | 2.41 | 2.57 | 2.73 | 2.88 |
| $x_{2,n}$ | | | | 1.34 | 1.43 | 1.51 | 1.59 | 1.66 | 1.71 | 1.76 | 1.80 | 1.82 |
| $y_{2,n}$ | | | | 0.00 | 0.10 | 0.20 | 0.32 | 0.43 | 0.56 | 0.69 | 0.82 | 0.96 |
| $-\theta_{3,n}$ | 3 | | | | 0.785 | 0.706 | 0.628 | 0.550 | 0.471 | 0.392 | 0.314 | 0.235 |
| $-\sigma_{3,n}/k$ | | | | | 1.94 | 2.10 | 2.26 | 2.41 | 2.57 | 2.73 | 2.88 | 3.04 |
| $x_{3,n}$ | | | | | 1.53 | 1.63 | 1.73 | 1.81 | 1.88 | 1.95 | 2.00 | 2.04 |
| $y_{3,n}$ | | | | | 0.00 | 0.11 | 0.22 | 0.35 | 0.48 | 0.62 | 0.77 | 0.92 |
| $-\theta_{4,n}$ | 4 | | | | | 0.785 | 0.706 | 0.628 | 0.550 | 0.471 | 0.392 | 0.314 |
| $-\sigma_{4,n}/k$ | | | | | | 2.26 | 2.41 | 2.57 | 2.73 | 2.88 | 3.04 | 3.20 |
| $x_{4,n}$ | | | | | | 1.75 | 1.86 | 1.96 | 2.05 | 2.14 | 2.21 | 2.26 |
| $y_{4,n}$ | | | | | | 0.00 | 0.12 | 0.25 | 0.39 | 0.53 | 0.69 | 0.85 |
| $-\theta_{5,n}$ | 5 | | | | | | 0.785 | 0.706 | 0.628 | 0.550 | 0.471 | 0.392 |
| $-\sigma_{5,n}/k$ | | | | | | | 2.57 | 2.73 | 2.88 | 3.04 | 3.20 | 3.36 |
| $x_{5,n}$ | | | | | | | 1.98 | 2.10 | 2.12 | 2.32 | 2.42 | 2.49 |
| $y_{5,n}$ | | | | | | | 0.00 | 0.13 | 0.27 | 0.43 | 0.59 | 0.77 |
| $-\theta_{6,n}$ | 6 | | | | | | | 0.785 | 0.706 | 0.628 | 0.550 | 0.471 |
| $-\sigma_{6,n}/k$ | | | | | | | | 2.88 | 3.04 | 3.20 | 3.36 | 3.51 |
| $x_{6,n}$ | | | | | | | | 2.24 | 2.38 | 2.50 | 2.62 | 2.72 |
| $y_{6,n}$ | | | | | | | | 0.00 | 0.15 | 0.31 | 0.48 | 0.66 |
| $-\theta_{7,n}$ | 7 | | | | | | | | 0.785 | 0.706 | 0.628 | 0.550 |
| $-\sigma_{7,n}/k$ | | | | | | | | | 3.20 | 3.36 | 3.51 | 3.67 |
| $x_{7,n}$ | | | | | | | | | 2.53 | 2.68 | 2.82 | 2.95 |
| $y_{7,n}$ | | | | | | | | | 0.00 | 0.17 | 0.34 | 0.53 |
| $-\theta_{8,n}$ | 8 | | | | | | | | | 0.785 | 0.706 | 0.628 |
| $-\sigma_{8,n}/k$ | | | | | | | | | | 3.51 | 3.67 | 3.83 |
| $x_{8,n}$ | | | | | | | | | | 2.86 | 3.02 | 3.18 |
| $y_{8,n}$ | | | | | | | | | | 0.00 | 0.18 | 0.38 |
| $-\theta_{9,n}$ | 9 | | | | | | | | | | 0.785 | 0.706 |
| $-\sigma_{9,n}/k$ | | | | | | | | | | | 3.83 | 3.98 |
| $x_{9,n}$ | | | | | | | | | | | 3.22 | 3.41 |
| $y_{9,n}$ | | | | | | | | | | | 0.00 | 0.20 |
| $-\theta_{10,n}$ | 10 | | | | | | | | | | | 0.785 |
| $-\sigma_{10,n}/k$ | | | | | | | | | | | | 4.14 |
| $x_{10,n}$ | | | | | | | | | | | | 3.62 |
| $y_{10,n}$ | | | | | | | | | | | | 0.00 |

$= -\frac{1}{4}\pi$ on AC; using this we can solve the mixed problem for the domain ABC). We find consecutively the points $(1, n)$ (i.e. we fill out the second row of the table), $(2, n)$, etc.

Joining the nodes by straight lines we obtain a grid of slip lines (fig. 135). In the region BCD the field is determined from the solution to the mixed problem (§37), since at the nodes $(m, 10)$ on BC we now know $\sigma$, $\theta$, as well as having $\theta = 0$ on the line $y = 1$.

The calculated pressure distribution on the contact surface is shown by solid lines in fig. 134. The dotted line shows the pressure according to Prandtl's solution. It is evident that Prandtl's solution is a good approximation when $l \gg h$.

The slip field constructed in this way must be consistent with the velocity field which corresponds to it. We now turn to this question.

By symmetry $u = v$ on AG; on OF we have the boundary condition $v = -c$ (fig. 134). Since the rigid zone in the middle region is displaced in a vertical direction with velocity $c$ and the normal velocity component $v$ is continuous on the boundary FG, then $v = -c \cos\theta$ on the latter. Integrating now equation (38.4) and determining the arbitrary constant from the condition $u = v$ at the point G, for which $\theta = -\frac{1}{4}\pi$, we find the second velocity component on the boundary FG:

$$u = -c(\sqrt{2} + \sin\theta)\,.$$

In transition through the line of separation FG the tangential velocity component $u$ experiences a jump of magnitude $c\sqrt{2}$, which implies an infinite velocity of shear. From calculation of the values $u$, $v$ we construct the velocity field, advancing consecutively from the right to the left, until the line AB is reached; thus the velocities on AB are determined in a unique way. Next we find the velocities in the sector OAB from the data on the segments AB and OB. The velocity $u$ is constant along each of the rays, i.e. $u = u(\theta)$; then along the $\beta$-lines (cf. §38)

$$v = -\int u(\theta)\,\mathrm{d}\theta + \psi(\rho)\,.$$

Since $v$ is constant on OB, $\psi(\rho) = \text{const.}$ and so $v = v(\theta)$. Thus the velocities $u$, $v$ are constant along the line of separation OA. The condition that the normal velocity component $v$ is constant along AO corresponds to the required motion of the rigid part of the layer OAO$'$.

For a real *plastic-elastic layer* the deformation picture is shown schematically in fig. 136. The shaded zones are those in which elastic and plastic deformations are of the same order.

Note that in our solution $h$ can be regarded as the thickness at a given moment of time, and hence the solution is valid for finite deformations. Here the form of the extruded portion of the layer is easily determined from the in-

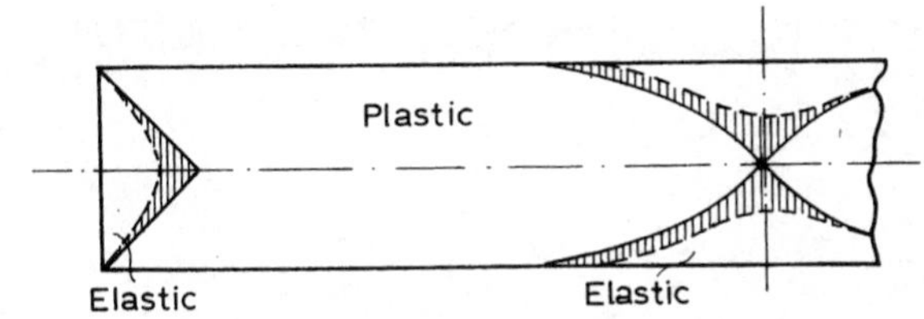


Fig. 136.

compressibility condition. Thus, suppose the plate to have been displaced by an amount $dh$; then $h dx + l dh = 0$, where $dx$ is the displacement of the extruded part of the layer (fig. 137). Hence

$$h = h_0 e^{-x/l},$$

where $h_0$ is the initial thickness of the layer, and $x$ is measured from its end.

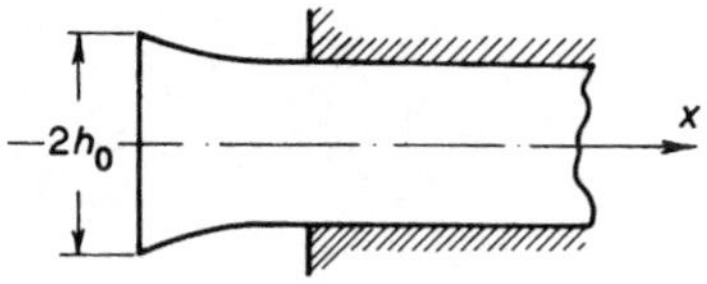


Fig. 137.

Calculations [54] have shown that the velocities undergo significant variations only in a narrow strip near the contact curve, and change very little in the rest of the layer. These results are confirmed by observations on an initially square layer, compressed between rough plates. In fig. 138 the "rigid" zones are clearly visible.

### 47.3. *Short layer*

For a short layer ($l/h < 3.64$) the above solution is unsuitable. In this case, when $1 \leqslant l/h \leqslant 3.64$ the slip field has the form shown in fig. 139. The straight slip lines OA intersect the horizontal axis at an angle $\frac{1}{4}\pi$. The angle of spread $\gamma$ of the central field is determined from the condition on the slip line OBC at the centre of the strip. The shaded regions remain rigid; the pressure distribution along the contact curve is indeterminate and we can only indicate its mean value. When $h/l = 1$ the points A, C coincide; when $h/l < 1$ the slip field is constructed in a different way (cf. [54]).

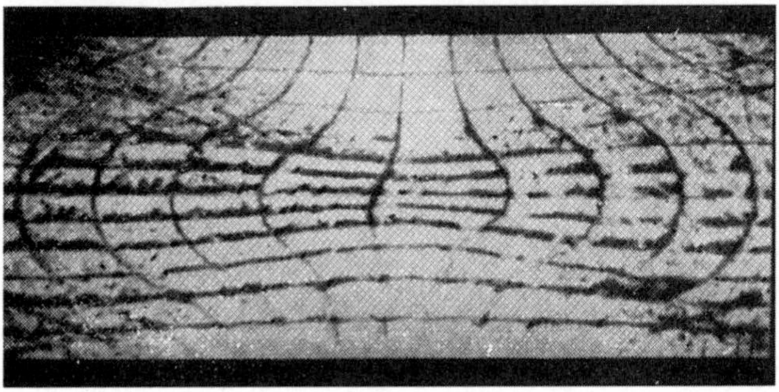

Fig. 138.

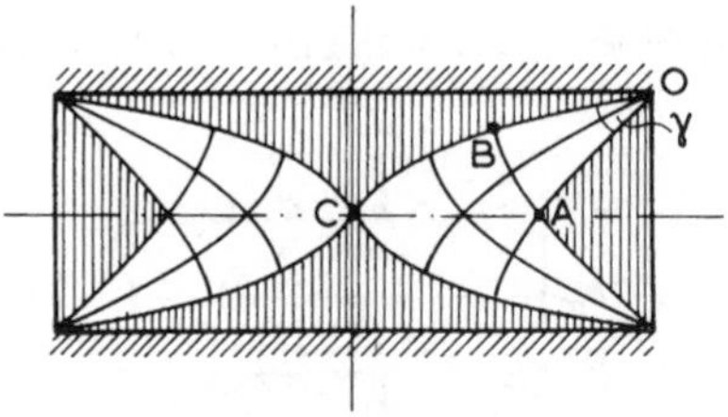


Fig. 139

47.4. *Concluding remarks*

It has been assumed above that the maximum tangential stress $\tau_{xy} = k$ develops along the contact surface. The case when the tangential stress is constant but less than $k$ has been studied by Sokolovskii [44].

Approximate methods of calculation have been developed on the basis of the preceding solutions. Thus, the work of Meyerhof and Chaplin [187] gives approximate solutions for the compression of layers of different planforms (circular, rectangular, etc.), and indicates experimental confirmations. A.A. Il'yushin [111] has considered plastic flow of a layer between two non-deforming surfaces.

The stress state in a thin plastic layer with rigid attachments is a problem of considerable practical interest. Problems of this kind arise, for example, in examining the working of a seam (soldered joint), of welded joints, etc.

The solution for a thin layer is relevant to the *final stage* of plastic flow, when tangential stresses equal to the yield limit arise on a contact surface. However, the stress distribution in these layers changes, depending on the load, from simple uniaxial compression (tension) to the ultimate complex stress state discussed above. An approximate analysis of the development of stress in a thin layer is given in [120]; see also §60.

We note finally that the presence of a force bending the plate substantially lowers the limiting compressive force $2P$. This question is briefly discussed below (§66).

## §48. Elastic-plastic extension of a plane with a circular hole

### 48.1. *Formulation of the problem*

We consider the extension of a plane having a circular hole of radius $a$ and experiencing at infinity different tensions in the directions of the $x$, $y$-axes (fig. 140); i.e. $\sigma_x \to p$, $\sigma_y \to q$, $\tau_{xy} \to 0$ as $r \to \infty$, with $q \geqslant p$; the hole is stress-free.

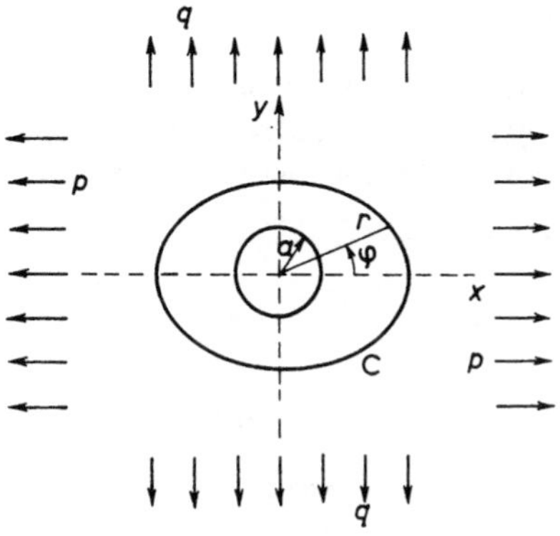


Fig. 140.

Plastic regions will develop for sufficiently large loads $p$ and $q$. Ignoring the history of the gradual development of these regions (which may be accompanied by unloading in some portions of the plastic zones) we assume that the plastic zone completely surrounds the hole. We suppose further that the material is incompressible, so that the plasticity condition (31.8) holds. Then in the plastic zone the stress components in polar coordinates are (§34)

$$\sigma_r = 2k \ln (r/a) , \qquad \sigma_\varphi = 2k(1 + \ln (r/a)) . \tag{48.1}$$

To construct the solution in the elastic zone we utilize the stress function $F = F(x, y)$:

$$\sigma_x = \frac{\partial^2 F}{\partial y^2} , \qquad \sigma_y = \frac{\partial^2 F}{\partial x^2} , \qquad \tau_{xy} = -\frac{\partial^2 F}{\partial x \partial y} .$$

In polar coordinates the stress components are expressed in terms of the stress function $F = F(x, y)$ by the formulae .

$$\sigma_r = \frac{1}{r}\frac{\partial F}{\partial r} + \frac{1}{r^2}\frac{\partial^2 F}{\partial \varphi^2}, \qquad \sigma_\varphi = \frac{\partial^2 F}{\partial r^2}, \qquad \tau_{r\varphi} = \frac{1}{r^2}\frac{\partial F}{\partial \varphi} - \frac{1}{r}\frac{\partial^2 F}{\partial r \partial \varphi}. \tag{48.2}$$

It is easy to see that there is a "plastic" stress function $F_\mathrm{p}$ corresponding to the solution (48.1):

$$\frac{1}{k}F_\mathrm{p} = -\tfrac{1}{2}r^2 + r^2 \ln\frac{r}{a}. \tag{48.3}$$

In the elastic zone the stress function $F_\mathrm{e}$ satisfies the biharmonic equation. Every biharmonic function can be represented by Goursat's formula

$$F = \mathrm{Re}\,[\bar{z}\Phi_*(z) + \Psi_*(z)], \tag{48.4}$$

where $\Phi_*(z)$, $\Psi_*(z)$ are analytic functions of the complex variable $z = x + \mathrm{i}y$. The line above a symbol denotes complex conjugate, as usual. The stress components can be found from the Kolosov-Muskhelishvili formula

$$\begin{aligned} \sigma_x + \sigma_y &= 4\,\mathrm{Re}\,\Phi_1(z), \\ \sigma_y - \sigma_x + 2\mathrm{i}\tau_{xy} &= 2[\bar{z}\Phi_1'(z) + \Psi_1(z)], \end{aligned} \tag{48.5}$$

where we have put

$$\Phi_1(z) = \Phi_*'(z), \qquad \Psi_1(z) = \Psi_*''(z).$$

For simplicity of notation we shall henceforth use the operators $L$ and $M$, corresponding to the expressions (48.5), namely,

$$L(F) = \frac{\partial^2 F}{\partial x^2} + \frac{\partial^2 F}{\partial y^2},$$

$$M(F) = \frac{\partial^2 F}{\partial x^2} - \frac{\partial^2 F}{\partial y^2} - 2\mathrm{i}\frac{\partial^2 F}{\partial x \partial y}.$$

Determination of the analytic functions $\Phi_1$ and $\Psi_1$ requires knowledge of the boundary conditions. In particular, the stresses must be continuous on the unknown boundary C between the elastic and plastic zones.

48.2. *Solution of L.A. Galin [91]*

It is easy to see that the plastic stress function $F_\mathrm{p}$ described above satisfies the biharmonic equation for an axisymmetric field. This property permits the construction of an elegant, closed solution with the aid of the complex representation (48.4).

We introduce a new biharmonic function $F = F_e - F_p$. Here $F_p$ is given by formula (48.3) and can be represented (recall that $z = re^{i\varphi}$) in the form

$$F_p = k \operatorname{Re} [z\bar{z}(\ln (z/a) - \tfrac{1}{2})] \ . \qquad (48.6)$$

It is evident that

$$L(F) = L(F_e) - L(F_p) \ , \qquad M(F) = M(F_e) - M(F_p) \ .$$

We formulate the boundary conditions for the function $F$. First of all we observe that at infinity

$$L(F_e) = q + p \ , \qquad M(F_e) = q - p \ ,$$

$$L(F_p) = \left(\frac{d^2}{dr^2} + \frac{1}{r}\frac{d}{dr}\right) F_p = 2k \left(1 + 2 \ln \frac{r}{a}\right),$$

$$M(F_p) = 2k\frac{\bar{z}}{z} = 2k e^{-2i\varphi} \ .$$

Consequently, $L(F)$ and $M(F)$ tend at infinity to the expressions

$$L(F) = q + p - 2k(1 + 2 \ln (r/a)) \ ,$$

$$M(F) = q - p - 2ke^{-2i\varphi} \ .$$

Continuity of stress gives, on the unknown boundary C,

$$L(F) = 0 \ , \qquad M(F) = 0 \ . \qquad (48.7)$$

We now transform the elastic region in the $z$-plane (i.e. the exterior of C) on to the exterior of the unit circle $\gamma$ in the $\zeta$-plane. Let the point at infinity in the $z$-plane map into the point at infinity in the $\zeta$-plane. Then the mapping function $\omega(\zeta)$ has the form

$$z = \omega(\zeta) = c\zeta + g(\zeta) \ ,$$

where $c$ is a real positive constant, and $g(\zeta)$ is a function analytic outside the circle $\gamma$; it can be assumed that $g(\infty) = 0$.

The mapping function can be represented as a Laurent series

$$\omega(\zeta) = c\zeta + \sum_{n=1}^{\infty} \frac{a_n}{\zeta^n}, \qquad (48.8)$$

where the coefficients $a_n$ are real because of the symmetry of the contour C with respect to the $x$-axis.

We introduce the notation

$$\Phi(\zeta) = \Phi_1[\omega(\zeta)]\,, \qquad \Psi(\zeta) = \Psi_1[\omega(\zeta)]\,.$$

Then formulae (48.5) can be written in the form

$$L(F) = 4\,\mathrm{Re}\,\Phi(\zeta)\,,$$
$$M(F) = 2\left[\frac{\overline{\omega}(\zeta)}{\omega'(\zeta)}\,\Phi'(\zeta) + \Psi(\zeta)\right]. \tag{48.9}$$

Taking into account the behaviour of $L(F)$ and $M(F)$ at infinity and on the contour $\gamma$, as given by (48.7), we arrive at the following conditions for determination of the functions $\Phi(\zeta)$, $\Psi(\zeta)$, $\omega(\zeta)$:

$$4\,\mathrm{Re}\,\Phi(\zeta) = \begin{cases} 0 & \text{on } \gamma\,, \\ p+q-4k(\frac{1}{2}+\ln(c/a)-\ln|\zeta|) & \text{for } \zeta\to\infty\,, \end{cases} \tag{48.10}$$

$$2\left[\frac{\overline{\omega}(\zeta)}{\omega'(\zeta)}\,\Phi'(\zeta) + \Psi(\zeta)\right] = \begin{cases} 0 & \text{on } \gamma\,, \\ q-p-2k\mathrm{e}^{-2\mathrm{i}\varphi_1}\,, & \varphi_1 = \arg\zeta\,,\ \text{for } \zeta\to\infty\,. \end{cases} \tag{48.11}$$

By virtue of (48.10) the function $\Phi(\zeta)$ has the form

$$\Phi(\zeta) = -k\ln\zeta + h(\zeta)\,, \tag{48.12}$$

where $h(\zeta)$ is a function analytic outside the unit circle $\gamma$ and bounded at infinity; its real part is equal to zero on $\gamma$, and consequently $h(\zeta)$ is necessarily equal to zero everywhere. Then comparing (48.12) with the condition at infinity we arrive at the requirement

$$p+q-4k(\tfrac{1}{2}+\ln(c/a)) = 0\,,$$

from which it follows that

$$c = a\exp\left(\frac{p+q}{4k}-\frac{1}{2}\right). \tag{48.13}$$

Moreover, when $\zeta\to\infty$ we have $\omega(\zeta) = c\zeta$, $\omega'(\zeta) = c$. Substituting these values in (48.11) we find that as $\zeta\to\infty$

$$2[-k\,\overline{\zeta}/\zeta + \Psi(\zeta)] = q-p-2k\mathrm{e}^{-2\mathrm{i}\varphi_1}\,.$$

Since $2k\overline{\zeta}/\zeta = 2k\mathrm{e}^{-2\mathrm{i}\varphi_1}$, we have at infinity

$$\Psi(\zeta) = \tfrac{1}{2}(q-p)\,. \tag{48.14}$$

The condition on the unit circle gives

$$\Psi(\zeta) = \frac{\bar{\omega}(\zeta)}{\omega'(\zeta)} \frac{k}{\zeta}. \tag{48.15}$$

Since $\bar{\zeta} = 1/\zeta$ on the unit circle, it follows that on $\gamma$

$$\bar{\omega}(\zeta) = \frac{c}{\zeta} + \sum_{n=1}^{\infty} a_n \zeta^n .$$

In consequence, condition (48.15) can be written in the form

$$\frac{c}{\zeta} + \sum_{n=1}^{\infty} a_n \zeta^n = \frac{1}{k} \omega'(\zeta)\, \Psi(\zeta)\, \zeta . \tag{48.16}$$

The right-hand side is a function analytic outside the unit circle everywhere except at the point at infinity where it has a first-order pole, since, as $\zeta \to \infty$,

$$\frac{1}{k} \omega'(\zeta)\, \Psi(\zeta)\, \zeta = \kappa c \zeta , \qquad \kappa = \frac{q-p}{2k} . \tag{48.17}$$

To determine the coefficients $a_n$ of the series we multiply (48.16) by $\zeta^{-m-1}$ ($m = 1, 2, 3, \ldots$) and integrate along the contour of the unit circle:

$$\int_\gamma \left[ \frac{c}{\zeta} + \sum_{n=1}^{\infty} a_n \zeta^n \right] \zeta^{-m-1} d\zeta = \int_\gamma \left[ \frac{1}{k} \omega'(\zeta)\, \Psi(\zeta)\, \zeta \right] \zeta^{-m-1} d\zeta .$$

From Cauchy's residue theorem we have

$$\int_\gamma \left[ \frac{c}{\zeta} + \sum_{n=1}^{\infty} a_n \zeta^n \right] \zeta^{-m-1} d\zeta = 2\pi i a_m .$$

On the other hand, the quantity $\omega'(\zeta)\, \Psi(\zeta)\, \zeta^{-m}$ is an analytic function in the whole plane exterior to $\gamma$, and therefore the integral around $\gamma$ must equal the integral around a circle of larger radius. Thus

$$a_m = \frac{1}{2\pi i} \int_{|\zeta| = \infty} \frac{1}{k} \omega'(\zeta)\, \Psi(\zeta)\, \zeta^{-m} d\zeta .$$

Evaluating this integral we find that

$$a_m = \begin{cases} \kappa c & \text{when} \quad m = 1 \,, \\ 0 & \text{when} \quad m > 1 \,. \end{cases}$$

Hence

$$\omega(\zeta) = c(\zeta + \kappa/\zeta) \,. \tag{48.18}$$

As is well-known, this function is the mapping for the exterior of an ellipse, and consequently the boundary C will be elliptic (fig. 141). Using (48.15) to evaluate the function $\Psi$ on the contour of the unit circle, we obtain

$$\Psi(\zeta) = k \, \frac{1 + \kappa\zeta^2}{\zeta^2 - \kappa} \,.$$

When $\kappa = 1$ the stress $\tau_{max}$ at infinity is equal to $k$ and the whole plane is in a plastic state. It is therefore necessary to assume that $\kappa < 1$, and then the poles of $\Psi$ lie inside $\gamma$ and $\Psi(\zeta)$ is in fact regular outside $\gamma$.

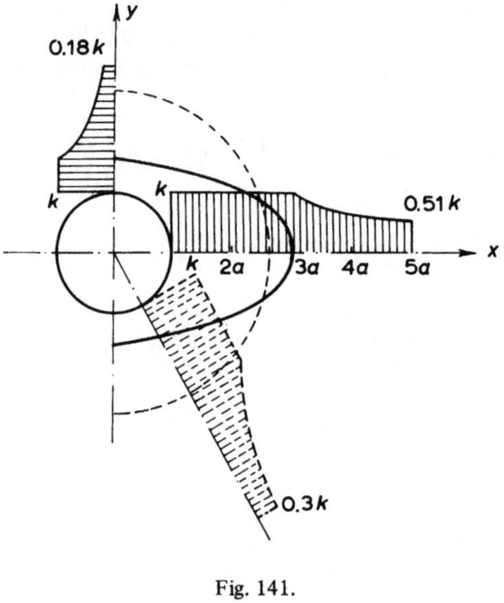


Fig. 141.

The equation of the ellipse C has the form

$$\frac{x^2}{c^2(1+\kappa)^2}+\frac{y^2}{c^2(1-\kappa)^2}=1\,.$$

The solution is possible if the plastic zone completely surrounds the hole. This requirement is satisfied if

$$c(1-\kappa)>a\,.$$

From this it follows that the stresses $p$ and $q$ at infinity cannot be very different. The stresses in the elastic region can be found once the potentials $\Phi$ and $\Psi$ are known. We refer the reader to the article by Galin, which contains detailed analysis, and merely state some results of the calculations for the stress field in the case $p = 2.4k$, $q = 3.0k$. The semi-axes of the ellipse are here equal to $3.04a$, $1.64a$ respectively. The solid lines in fig. 141 show the curves of the distribution of the intensity of tangential stresses along the $x$, $y$-axes. For comparison, the dotted line shows the circle of radius $2.72a$, which is the line of separation in the axisymmetric elastic-plastic problem (when $p = q = 3k$); the distribution of the intensity of tangential stresses along the radius vector is also shown by a dotted line.

48.3. *Concluding remarks*

In developing the solution we have made great use of the fact that the stress function in the plastic zone adjoining the hole is biharmonic. With certain additional conditions Galin's solution has been generalized to the case of a plastically inhomogeneous medium (A.I. Kuznetsov) and to the case of a non-uniform thermal field (V.L. Fomin). An approximate method for solving the elastic-plastic problem for a plane with a hole was developed by P.I. Perlin in an inverse formulation (the plastic zone is given). Recently G.P. Cherepanov has utilized methods from the theory of functions of a complex variable; here the biharmonic condition on the function $F_{\mathrm{p}}$ is not needed.

## §49. Steady plastic flow. Drawing of a strip

49.1. *Steady plastic flow*

We have been considering (§§40–47) problems of determining bearing capacity, which are related to questions of strength of structures. In this we have been able to limit ourselves to the examination of small deformations.

Another important area of application of plasticity theory is related to the analysis of continuous technological machining processes of metals under pressure (rolling, drawing, indentation, cutting, etc.), which are widely used in industry. Here the most interesting question is the prediction of the forces required to carry out the given machining process, and the analysis of the deformations which take place. In problems of this type it is natural to assume that the stresses and velocities do not change at each fixed point of space.

As an illustration, we consider the drawing of a strip (fig. 142). Here the strip, of initial thickness $H$, is pulled with constant velocity $U$ through a rigid tapering slit. As a result of the plastic deformation experienced by the strip in passing through the slit, its thickness decreases to a value $h$, and its length increases accordingly. The part of the strip to the left of the slit moves with constant velocity $V < U$. The velocity and stress fields do not change with time (stationary); finally, the parts of the strip which are far from the slit are assumed to be undeformed.

An example of pressing (or extrusion) of a sheet from a container through a slit is shown in fig. 143a. In contrast with the previous case the force (pressure) is now applied over a large part of the sheet. Metal of thickness $h$

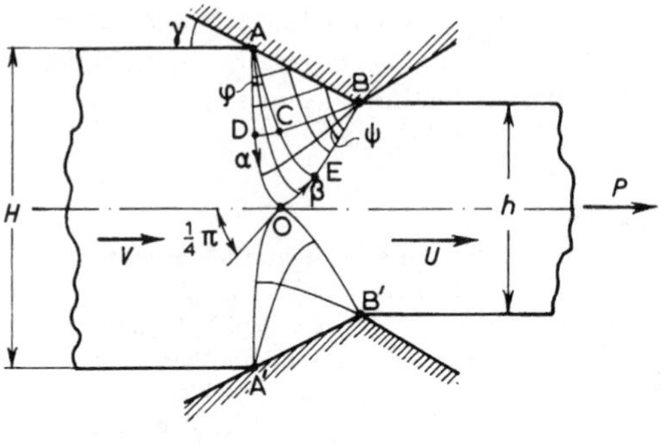


Fig. 142.

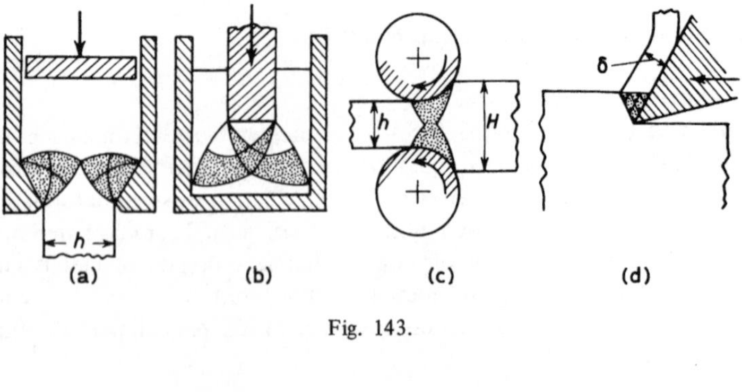


Fig. 143.

emerges from the slit. The shaded regions in fig. 143a are the zones of plastic deformation.

The process of pressing has many variations. In fig. 143b we show an example of inverse pressing (inverted extrusion). A rigid tool is indented into the metal in the container. The metal extrudes upwards on both sides of the die.

Another important process is that of rolling, different versions of which are widely used. The simplest version is shown in fig. 143c. The cylindrical balls turn in different directions, engage the sheet of thickness $H$ and reduce it to thickness $h$.

Various cutting processes for metals (turning, piercing, etc.) can also be regarded as steady plastic flows. Fig. 143d shows schematically the machining of a plane surface. A chip of thickness $\delta$ is cut off along the front edge of the tool. A zone of plastic deformation accumulates locally at the edge of the tool (shaded region in the diagram). Distant parts and the metal chips can be assumed rigid.

An analysis of various technological processes on the basis of plane strain solutions plays an important role in the monographs of Hill [54], Sokolovskii [44] and Prager [29], as well as in many articles.

Problems of steady plastic flow are closely related to the particular features of technological processes, and require special treatment. We note that significant developments have been made in approximate ("one-dimensional") schemes for calculating continuous processes. We refer here to the books by Hoffman and Sachs [10] and Tomlenov [51], which present calculations of a number of technological processes of metal machining under pressure. Further references to the literature can also be found in those books.

Turning now to the general case, we observe that the region occupied by the medium divides into rigid and plastic zones; since plastic deformations are large, the plastic-rigid model is completely acceptable. Large deformations are connected with the development of hardening; nevertheless one usually proceeds from the perfect yield model, taking some mean value for the constant $k$. When hardening is insignificant, comparison between theory and experiment for cold machining processes gives good agreement, and the differences between calculated and observed forces do not exceed 10%.

In hot machining processes changes in temperature are found to have a substantial influence, and in this case experimental results can differ significantly from theoretical predictions.

In the majority of technological processes involving steady plastic flow contact boundary conditions are encountered. As a rule the distribution of pressure on contact curves is difficult to estimate; it can only be done in the simplest problems (in particular when the curves of contact with the tool are straight lines). In general it is necessary to consider simultaneously the equations for stresses and velocities (cf. §51).

We shall consider below the drawing of a strip through a smooth conical slit; a semi-inverse method turns out to be effective for this problem.

### 49.2. *Drawing of a strip*

The strip (of initial thickness $H$) is pulled with velocity $U$ through a rigid, smooth, tapering slit; the strip experiences plastic deformations in the region adjoining the slit and its thickness is reduced to the value $h$. The angle between the planes of the slit is $2\gamma$ (fig. 142). At some distance from the slit the portions of the strip move like a rigid body with velocities $U$ and $V$. Incompressibility of the material implies that the velocity $V = (h/H)U$.

For simplicity we neglect friction (though it is not difficult to allow for a constant frictional force), and therefore the contact stress is normal to AB. We assume that a uniform pressure $p$ acts along AB, and show that all the conditions are then satisfied.

In $\Delta$ ABC we now have a uniform stress state; adjoining AC and BC we have the central fields ACD and BCE, where the angles $\varphi$ and $\psi$ are so far unknown. The stress distribution in these regions depends on the pressure $p$. For the quadrilateral CDOE we have an initial characteristic problem with data on the slip lines CD, CE. By symmetry, the point O lies on the axial line of the strip, and the slip lines intersect the axis at an angle of 45°. These conditions determine the values of the angles $\varphi$, $\psi$. In particular it follows from the second condition that

$$\gamma = \psi - \varphi \, . \tag{49.1}$$

Now in $\Delta$ ABC, $\theta = -\frac{1}{4}\pi - \gamma$, and we denote the magnitude of the mean pressure by $\sigma'$; the parameters $\xi, \eta$ are constant and have the respective values $\xi' = \sigma'/2k + \frac{1}{4}\pi + \gamma$, $\eta' = \sigma'/2k - \frac{1}{4}\pi - \gamma$. Furthermore, $\eta = \text{const.} = \eta'$ in the region ADC; along AD $\theta = -\frac{1}{4}\pi - \gamma - \varphi$ and the mean pressure is constant; we denote it by $\sigma''$. Equating the corresponding value of $\eta$ with the above value of $\eta'$, we find $\sigma''/2k = \sigma'/2k - \varphi$. At the point D we have $\xi_D = \sigma'/2k + \varphi + \frac{1}{4}\pi + {} + \gamma + \varphi$.

Similarly $\xi = \text{const.} = \xi'$ for the region BCE, the mean pressure $\sigma'''/2k = {} = \sigma'/2k + \psi$ on BE, and $\eta = \eta_E = \sigma'/2k + \psi - \frac{1}{4}\pi - \gamma + \psi$ at the point E. Finally, $\theta = -\frac{1}{4}\pi$ at the point O, and we denote the unknown pressure by $\sigma_0$. The corresponding values of the parameters $\xi$, $\eta$ at O will be $\xi_0 = \sigma_0/2k + \frac{1}{4}\pi$, $\eta_0 = \sigma_0/2k - \frac{1}{4}\pi$. But $\xi = \text{const.}$ along ADO and $\eta = \text{const.}$ along BEO; consequently $\xi_0 = \xi_D$, $\eta_0 = \eta_E$ and hence equation (49.1) follows immediately.

On any section of the strip to the right of BOB′ the sum of the horizontal stress components is equal to the unknown drawing force $P$, while to the left of AOA′ this sum is zero, since the left-hand part of the strip does not experience the action of internal forces. It is easy to see that

$$P = p(H - h)\,.$$

The required pressure $p$ and one of the angles (say, $\varphi$) can be found from the conditions that the sum of the horizontal stress components in the section AOA′ is zero, and that the point O lies on the axis. This requires numerical calculations, since the solution for the quadrilateral ODCE is obtained by numerical integration. In the work of Hill and Tupper results have been calculated for different angles $\gamma$ and ratios $h/H$. One of the results (for $\gamma = 15°$) is shown in fig. 144. The drawing force increases with large compression of the strip.

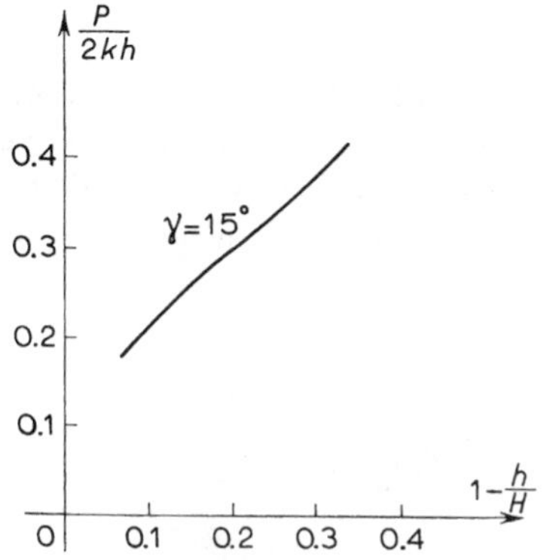


Fig. 144.

We now show that the velocity field is consistent with the stress field. Along ADO and BEO the normal velocity components are continuous and therefore known, since the velocities of motion of the rigid portions are given. From these data on DO, EO we can determine the velocity field in the quadrilateral ODCE. Next we can find the velocities in the central fields, and, finally, in Δ ABC. The normal velocity components are obviously constant along the straight lines AD, BE; then because of (38.7) the velocity components are constant along every straight slip line in the central fields. Consequently $u$, $v$ are constant on AC, BC; but then by (38.6) the velocities $u$, $v$ are constant everywhere in Δ ABC.

The equations of the velocity field are based on the incompressibility condition; by virtue of the relation $VH = Uh$ between the velocities in the right and left parts of the strip, the flux of mass through ADO is equal to its flux through BEO; hence the flux through AB must be equal to zero. Since the velocity is constant in ABC, its direction is along the contact line AB [1]) which is a necessary condition for the velocity field to be correct.

Note that the tangential velocity component is discontinuous along the slip lines ADO, BEO.

Our solution will be valid if for each $\gamma$ the compression ratio $h/H$ does not exceed some value. It is clear from the geometrical picture that the maximum is attained for $\varphi = 0$, when the quadrilateral ODCE degenerates to the point O (fig. 145). In this case the solution is elementary and

$$\frac{P}{2kh} = \frac{2(1+\gamma)\sin\gamma}{1+2\sin\gamma} .$$

Fig. 146 shows the distortion of an initially square grid, evaluated at $\varphi = 0$, $\gamma = 15°$, $h/H = 0.66$. The quantity $2kh$ is equal to the limiting load for uniaxial tension of a flat strip of width $h$. Drawing is possible if $P < 2kh$ (otherwise there is fracture in the right-hand part of the strip); hence $\gamma \sin\gamma < \frac{1}{2}$ i.e. $\gamma < 42°27' \equiv \gamma_1$.

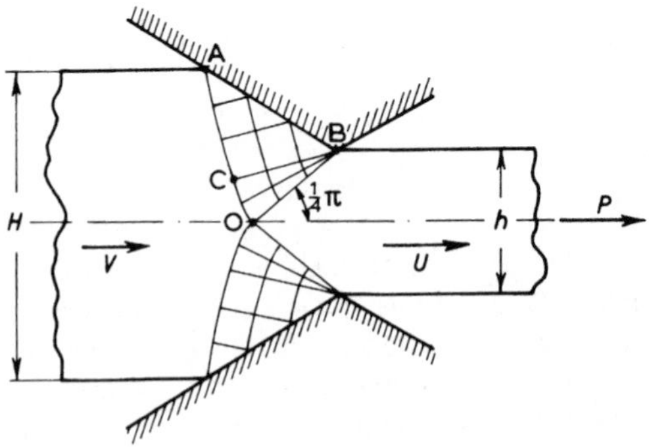


Fig. 145.

1) It need not be assumed that the pressure is uniform; it is easy to see that the condition for the velocity on AB is satisfied only for rectilinear AC, BC, but then the pressure is constant on AB.

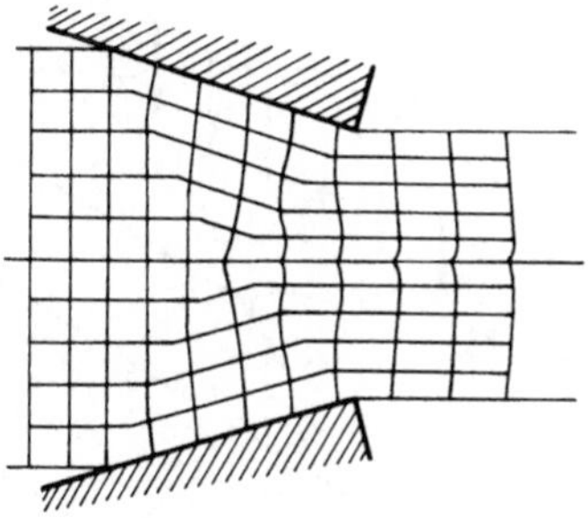

Fig. 146.

Note that for the field shown in fig. 145,

$$\frac{h}{H} = \frac{1}{1 + 2 \sin \gamma}.$$

If $\gamma < \gamma_1$ but $h/H$ exceeds the limiting value, then the construction discussed in this section is not possible; the solution is then of a different form.

## §50. Non-steady plastic flow with geometric similarity. Indentation by a wedge

### 50.1. *Non-steady plastic flow with geometric similarity*

Following the work of Hill, Lee and Tupper [54], we consider one class of problem in non-steady plastic flow, which lends itself to relatively simple analysis. This is concerned with problems in which the plastic region changes in such a way that its configuration always retains geometrical similarity to some initial state. The simplest examples are problems of the expansion of cylindrical and spherical cavities in unbounded space, beginning from zero radius. We shall give below the solution to the problem of indentation by a wedge.

In problems of this type deformation begins at a point or on a line, and the medium is unbounded.

### 50.2. *Indentation by a rigid wedge*

We consider the problem of the indentation of a semi-infinite plastic-rigid medium by a symmetric, rigid (non-deformable) wedge of angle $2\gamma$. Friction at the contact surface is neglected (the surface is lubricated).

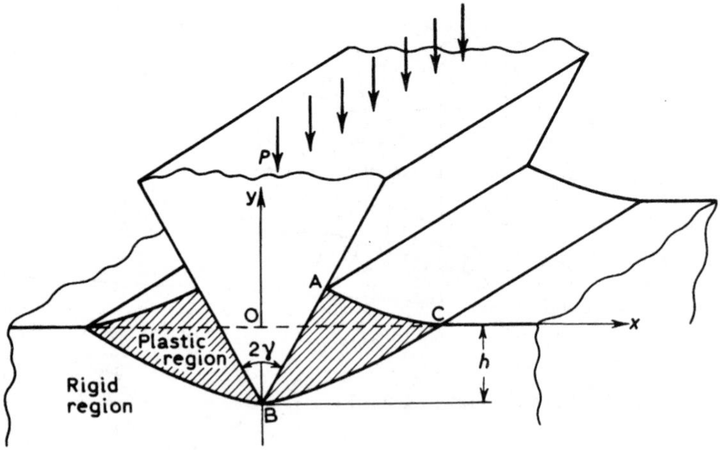


Fig. 147.

With indentation by a wedge the material extrudes along both its sides, and the deformation has the form shown schematically in fig. 147. The shaded region ABC is in a plastic state, and its boundary AC is, to a good approximation, a straight line. The slip field can then be constructed in the following way (fig. 148). We suppose that the contact pressure $p$ is constant along AB; as subsequent analysis will show, this will satisfy all the conditions of the problem. In the triangles ABD and AEC we then have uniform stress

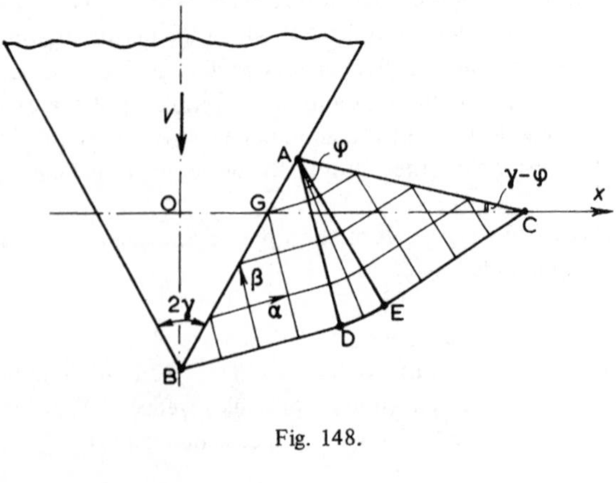


Fig. 148.

distributions. We denote the depth of the indentation OB by $h$, and the length AC (which equals AB because of the equality of the triangles ABD and ADE) by $l$; the pressure $p$ and the length $l$ are unknown. The regions of uniform stress are joined by a central field ADE of angle $\varphi$.

In $\Delta$ ABD $\theta = \frac{1}{4}\pi - \gamma$, and we denote the mean pressure by $\sigma'$; the parameter $\xi$ is here constant and equal to $\xi' = \sigma'/2k - \frac{1}{4}\pi + \gamma$. In $\Delta$ ACE, $\theta = \frac{1}{4}\pi - \gamma + \varphi$, $\sigma = -k$ and the parameter $\xi$ is constant and equal to $\xi'' = -\frac{1}{2} - \frac{1}{4}\pi + \gamma - \varphi$. The parameter $\xi$ = const. throughout the whole plastic region, and, consequently, $\xi' = \xi''$. Hence we obtain

$$\sigma' = -k(1 + 2\varphi)\,. \tag{50.1}$$

The line AC makes an angle $\gamma - \varphi$ with the horizontal axis; from the diagram,

$$l \cos \gamma - h = l \sin (\gamma - \varphi)\,. \tag{50.2}$$

Here $\gamma$, $h$ are given, and therefore (50.2) establishes the relation between $l$ and $\varphi$. Finally, by virtue of the incompressibility of the medium, the areas of the triangles OBG and ACG are equal, i.e.

$$h^2 \tan \gamma = (l \cos \gamma - h)\,[l \cos (\gamma - \varphi) + (l \cos \gamma - h) \tan \gamma]\,. \tag{50.3}$$

Eliminating $l/h$ from (50.2) and (50.3), and simplifying, we obtain the expression

$$2\gamma = \varphi + \text{arc cos} \tan \left(\tfrac{1}{4}\pi - \tfrac{1}{2}\varphi\right),$$

which determines $\varphi$.

The pressure $p$ is one of the principal stresses, and therefore (§31) it is equal to $\sigma' - k$ i.e.

$$p = -2k(1 + \varphi)\,. \tag{50.4}$$

The total force on unit length of the wedge in the direction of the $z$-axis is

$$P = 2pl \sin \gamma$$

and is a function of the angle $\gamma$ and the indentation depth $h$. The graph of the variation of $p$ with $\gamma$ is shown in fig. 149.

We now turn to the velocity field. As we know (§38), the velocity $v$ along each of the straight $\beta$-lines is constant; but on the line of separation BDEC the normal velocity component is zero (the rigid zone is at rest), and therefore $v = 0$ everywhere in the plastic region. The other velocity component is $u = c + \psi(\rho)$ in the central field, and $u = u(\beta)$ in the triangles of uniform stress. It is obvious that the velocity $u$ is constant on each of the $\alpha$-lines (consisting of straight segments and circular arcs). Let the velocity of indentation of the

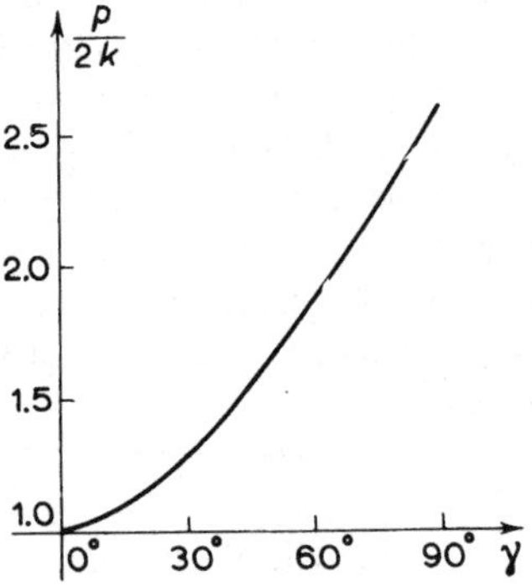


Fig. 149.

wedge be $V$; then projecting the velocities $V$ and $u$ along the normal to the contact line AB we find $u = \sqrt{2}\, V \sin\gamma$, i.e. the velocity component $u$ is constant everywhere in the plastic region. Since $v = 0$ the modulus of the velocity vector is constant and equal to $\sqrt{2}\, V \sin\gamma$.

Direct determination of the trajectories of particles of the material in the above velocity field is difficult, since this field is not fixed (in contrast with the steady case), and it is necessary to calculate the continuous expansion of the plastic region and the associated variation in the velocity field.

This difficulty can be overcome with the aid of a simple transformation which utilizes the similarity condition for the plastic region.

Let $\mathbf{r}$ be the radius vector of some particle M, with respect to a coordinate origin O, for indentation depth $h$. Consider the plane $\Pi^*$, on which the point $M^*$ corresponds to the point M, the former being defined by the radius vector

$$\mathbf{r}^* = \mathbf{r}/h\,. \tag{50.5}$$

By virtue of the similarity condition the region of plastic deformation in the plane $\Pi^*$ does not change with increasing $h$, and the indentation depth of the wedge always remains equal to unity. For this reason the fixed configuration in the plane $\Pi^*$ is called the *unit diagram* (fig. 150). As the wedge penetrates the medium the particle M undergoes a certain displacement. Because the whole configuration is determined by the increasing indentation depth $h$, the latter can be regarded as "time". Then the "velocity" of the point M will be equal to $\mathbf{v} = \mathrm{d}\mathbf{r}/\mathrm{d}h$.

As the point M moves in the physical plane the image point $M^*$ moves on the unit diagram with velocity $\mathbf{v}^* = \mathrm{d}\mathbf{r}^*/\mathrm{d}h$. For example, if the particle M is

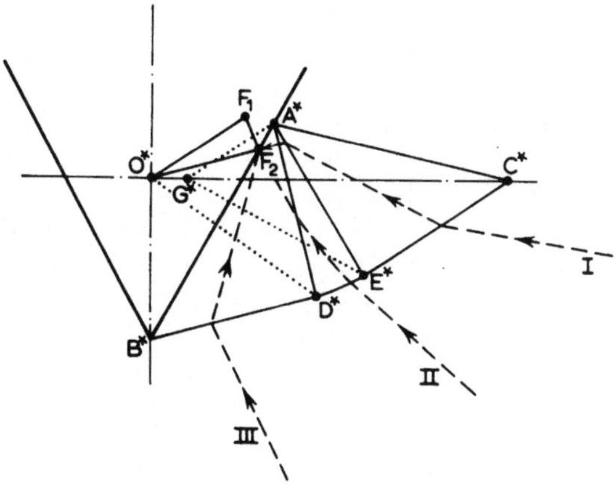


Fig. 150.

stationary ($\mathbf{r}$ = const.) in the physical plane, then $\mathbf{r}^*$ decreases in inverse proportion to $h$ in the unit diagram.

Differentiating (50.5) with respect to the "time" $h$, we obtain

$$\mathbf{v}^* = -\frac{1}{h}(\mathbf{r}^*-\mathbf{v})\,. \tag{50.6}$$

Thus, the velocity of the point $M^*$ is directed from $M^*$ to the point whose radius vector is $\mathbf{v}$; we shall call the latter point the focus. The magnitude of the velocity $\mathbf{v}^*$ is given by the ratio of the distance of the point $M^*$ from the focus ("focal distance") to the indentation depth $h$.

As we have already pointed out, in this problem the velocity vector has constant length $\sqrt{2}\ V \sin\gamma$. In the sequel we shall assume for simplicity that $V = 1$; if $V \neq 1$ the eventual results change proportionately. The foci then lie on the circle of radius $\sqrt{2}\sin\gamma$ and centre $O^*$. Since the direction of the velocity vector $\mathbf{v}$ does not change significantly – from the direction $B^*D^*$ in $\Delta\ A^*B^*D^*$ to the direction $E^*C^*$ in $\Delta\ A^*C^*E^*$ – it follows that on the unit diagram we obtain the circular arc $F_1F_2$, where the radii $O^*F_1$, $O^*F_2$ are parallel to $E^*C^*$, $B^*D^*$ respectively. The segment $O^*F_2$, equal to $\sqrt{2}\sin\gamma$, makes an angle $\frac{1}{4}\pi$ with the line $A^*B^*$.

We consider the mapping of the trajectory of the particle M on the unit diagram. So long as the moving line of separation BDEC has not reached M, the velocity $\mathbf{v} = 0$ and, by (50.6), the image point $M^*$ moves towards the centre $O^*$ along a straight line until the intersection with the boundary

$B^*D^*E^*C^*$. The nature of the subsequent motion of the image point depends on the location of this intersection. It is necessary to distinguish three cases.

In the first case (dotted line I) the intersection is on the segment $E^*C^*$ of the boundary; the velocity **v** is constant in the region ECA, and consequently the image point moves along a straight line towards the focus $F_1$ in $E^*C^*A^*$ until it intersects the line $A^*E^*$. In the sector $A^*E^*D^*$ the velocity **v** is variable, the focus is displaced along an arc of the circle from $F_1$ to $F_2$, and the trajectory is distorted. After the intersection with the line $A^*D^*$ the image point again moves along a straight line, but this time towards the focus $F_2$.

In the second case (line II) there is an intersection with the circular arc $D^*E^*$; the trajectory is bent in the sector $A^*E^*D^*$, and after intersection with the line $A^*D^*$ the image point moves in a straight line towards the focus $F_2$.

Finally, in the third case (line III) when the segment $B^*D^*$ is intersected, we have simple rectilinear motion towards the focus $F_2$.

In $D^*B^*F_2$ the material is displaced in the direction $O^*F_2$, having initially occupied the region $B^*D^*O^*$. Similarly, the material in $\Delta\ A^*E^*C^*$ is displaced in the direction $O^*F_1$, its initial position having been in $\Delta\ G^*E^*C^*$ (note that $A^*G^* \parallel O^*F_1$). In these regions we have pure shearing deformation, parallel to $B^*D^*$ and $E^*C^*$ respectively.

The material which initially occupied the region $E^*D^*O^*G^*$ experiences complex deformation and is transformed into the quadrilateral $E^*D^*A^*F_2$. The distortion of the initially square grid can be calculated using the unit diagram. It is necessary to find the final position (for indentation depth $h$) of an initial right angle, characterized in the initial state by the radius vector $\mathbf{r}_0$. Let the boundary of the plastic zone reach the point $\mathbf{r}_0$ at some value $h_0 < h$; here the corresponding image point $M_0^*$ is defined by the radius vector $\mathbf{r}_0^* = \mathbf{r}_0/h_0$. Let $s$ be the distance from the point $M_0^*$ which has been traversed along the trajectory when the wedge reaches a depth $h$, and let $F(s)$ be the focal dis-

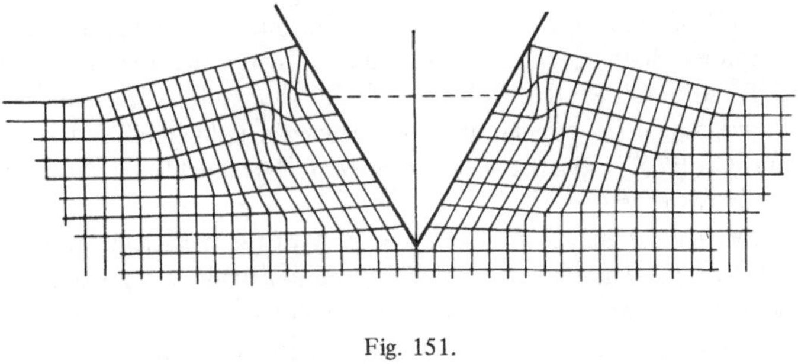

Fig. 151.

tance. Then by (50.6)

$$\ln \frac{h}{h_0} = \int_0^s \frac{ds}{F(s)},$$

since the velocity of motion of the image point is equal to $ds/dh$.

This equation determines $s$ as a function of $h$ and, consequently, the radius vector $\mathbf{r} = h\mathbf{r}^*$ of the point M. The integral can easily be evaluated in the triangles $D^*B^*F_2$, $A^*C^*E^*$, since in these $F(s) = d-s$, where $d$ is the distance from the point $M_0^*$ to the foci $F_2$, $F_1$ respectively. Thus

$$h/h_0 = d/(d-s) .$$

In the quadrilateral $E^*D^*O^*G^*$ the integral is evaluated numerically. Fig. 151 shows the distortion of an initially square grid, calculated for $\gamma = 30°$; the three deformation zones considered above are clearly visible. The photograph (fig. 152) shows the deformation of a grid for the indentation of lead by a lubricated steel wedge; all three zones can be distinguished. Experimentally-derived points confirm the theoretical variation (solid line in fig. 153) of the indentation parameter $kh/P$ with the angle $\gamma$.

A number of other problems of non-steady flow with geometric similarity (oblique indentation by a rigid wedge, compression of a wedge by a rigid plane, etc.) have been studied by Hill and other authors [54].

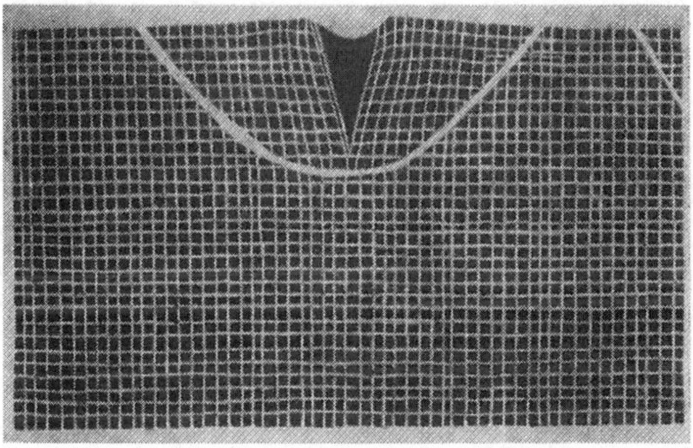

Fig. 152.

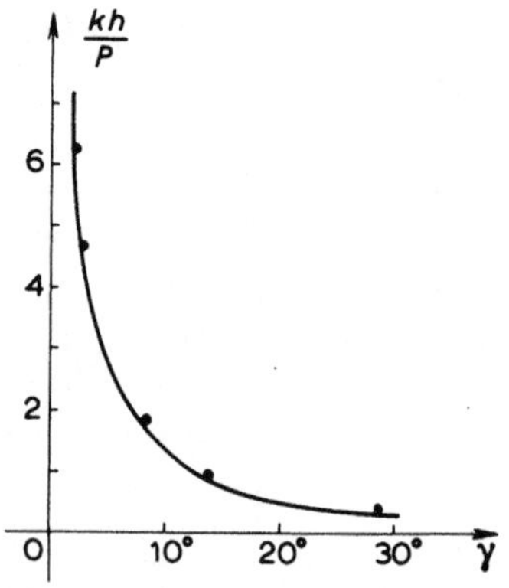


Fig. 153.

## §51. Construction of compatible stress and velocity fields

### 51.1. *Semi-inverse method*

We have made use of the semi-inverse method in solving plastic-rigid problems. First, the stress field is determined, with guesses made for the unattained boundary conditions. In general it is difficult to effect this scheme for contact problems. If the contact lines are straight and simple boundary conditions are prescribed on them (such as zero friction or constant tangential stress), the method can be carried through.

Examples of contact problems of this type were given in §45 (indentation by a flat die; pressure assumed constant), in §47 (compression of a layer between plates; pressure on the segment OB, fig. 134, assumed constant), and in §49 (drawing of a strip; pressure on the surface of the tool assumed constant).

If contact takes place along a curved line, the semi-inverse method is ineffective, since it is virtually impossible to guess the correct distribution of contact stresses. These stresses can sometimes be calculated by solving a mixed boundary-value problem. But in general it is necessary to use the method of compatible construction of stress and velocity fields, described in the works of Druyanov [109] and Sokolovskii [159].

51.2. *Construction of compatible stress and velocity fields*

It is assumed that the structure of the slip field in the physical $x, y$-plane can be indicated, and that some configuration corresponds to this field in the plane of the characteristics $\xi, \eta$. First we construct the velocity field. The velocity components satisfy equations (38.10):

$$\frac{\partial u}{\partial \eta} - \frac{1}{2}v = 0\,, \qquad \frac{\partial v}{\partial \xi} - \frac{1}{2}u = 0\,. \tag{51.1}$$

The lines which divide the plastic region from the rigid regions are characteristics. Since the velocities of motion of the rigid portions are prescribed, and since the angle of inclination of the normal to the line of separation can be calculated from $\theta = \frac{1}{2}(\eta - \xi)$, it follows that the normal velocity components ($u$ or $v$) are known along the line of separation. Although the slip-line grid is unknown in the $x, y$-plane, the above data enables us to find the functions $u = u(\xi, \eta)$, $v = v(\xi, \eta)$. Here the boundary conditions for velocities on the contact line are superfluous, and we are able to find the transformation of the contact line in the $\xi, \eta$-plane. Along this line $x, y$ are known and, consequently, so are the "coordinates" $\bar{x}, \bar{y}$ (cf. §33).

For $\bar{x}, \bar{y}$ we have the differential equations (33.3)

$$\frac{\partial \bar{y}}{\partial \eta} - \frac{1}{2}\bar{x} = 0\,, \qquad \frac{\partial \bar{x}}{\partial \xi} - \frac{1}{2}\bar{y} = 0\,. \tag{51.2}$$

Knowledge of the boundary values of $\bar{x}, \bar{y}$ along the transform of the contact line allows us eventually to find the functions $\bar{x}, \bar{y}$ and thus $x, y$ also, i.e. the slip-line grid in the physical plane.

As an example, we consider the procedure for solving the problem of indentation by a smooth convex die of a plastic layer, lying on a smooth base [159]. The convexity of the surface of the layer in the neighbourhood of the die is neglected. The die is impressed with velocity $V$; the rigid lateral regions move out with velocity $U$. The assumed structure of the slip field is shown in fig. 154. The mapping of this field on the characteristic plane $\xi, \eta$ is given in fig. 155.

Along the separation line $OA_0B_0C$ (or $OA_0'B_0'C'$) the normal velocity component is continuous, and can be calculated from $U$ and the angle $\theta = \frac{1}{2}(\eta - \xi)$. Thus, the velocity component $u$ (or $v$) is known along the characteristic segment $OA_0B_0C_0$ (or $OA_0'B_0'C_0'$). Solving consecutive characteristic problems

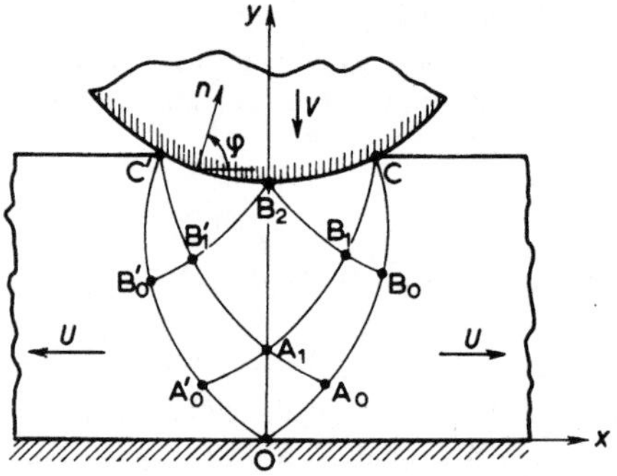


Fig. 154.

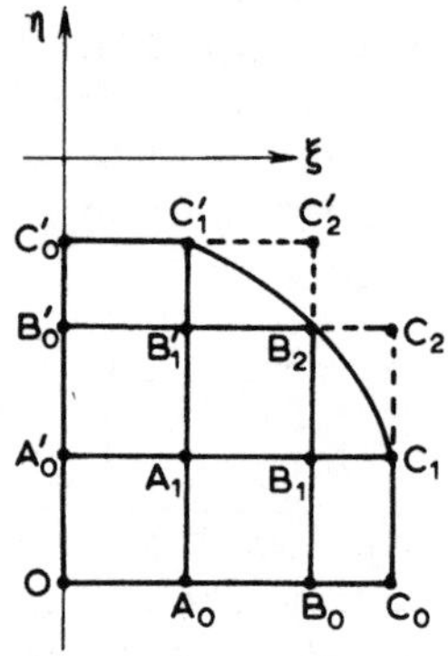


Fig. 155.

for equations (51.1), we find $u = u(\xi, \eta)$, $v = v(\xi, \eta)$ in the domain $OC_0C_2B_2C_2'C_0'O$ (fig. 155).

Along the contact line C′C the normal velocity component is $v_n = -V \sin\varphi$ or

$$u - v = V \sin\varphi \,. \tag{51.3}$$

But on the contact line the tangential stress is zero, and hence the slip lines meet the contour at an angle $\pm\frac{1}{4}\pi$ and $\varphi = \theta - \frac{3}{4}\pi = \frac{1}{2}(\eta - \xi) - \frac{3}{4}\pi$. The condition (51.3) then determines the image of the contact line in the characteristic plane: the curve $C_1'B_2C_1$; the points C′, C are transformed into the segments $C_0'C_1'$ and $C_0C_1$. Since the equations of the contact line $x = x(\varphi)$, $y = y(\varphi)$ are given, the boundary values of the functions $\bar{x}$, $\bar{y}$ can be found along $C_0'C_1'B_2C_1C_0$ (cf. §33). The functions $\bar{x}$, $\bar{y}$ can be determined in the region $C_1'A_2C_1$ by solving Cauchy's problem with data on the arc $C_1'B_2C_1$. For the remaining regions we have characteristic problems.

Note that by using conditions on the axis of symmetry O$y$ we can construct a solution only in the region $OA_1B_2CB_0A_0O$. The solution of the boundary value problem for equations (51.1) and (51.2) can be obtained by various methods, the simplest of which is the finite-difference method of Masso. The functions $u$, $v$; $\bar{x}$, $\bar{y}$ satisfy the telegraph equation. The solutions to the corresponding boundary-value problems can also be obtained analytically by using Riemann's formula.

The reference cited earlier give details of the calculations.

# PROBLEMS

1. Let $\eta$ = const. = $\eta_0$ in some domain (simple stress); one family of slip lines is a family of straight lines. Find the equation of the second family of slip lines (as a family of orthogonal lines).

2. Write down the formulae for calculating the slip and velocity fields, assuming the inclination of the chord to have the value $\theta$ at the initial point.

3. Find the limit load for a symmetric wedge ($\angle 2\gamma$) with vertex cut off, subject to a uniform pressure on the plane of the cut.

*Answer.* $p_* = 2k(1 + \frac{1}{2}\pi - \gamma)$ .

4. Find the limit bending moment for a strip weakened by symmetrically distributed angular notches (§42.4), with the first type of field.

*Answer.* $M_* = M_*^0(1 + \frac{1}{2}\pi - \gamma)$ .

5. Find the limit moment for the bending of a short, wedge-shaped cantilever by a force (§43.4); consider the first type of slip field only.

6. Derive formula (41.2) for the limit load in the extension of a strip with angular notches.

7. Derive formula (41.5).

8. Find the limit load for the extension of a strip having deep symmetric angular notches with round bases.

9. Find the limit load for the extension of a strip with rectangular notches.

10. In the problem of compression of a layer, construct the slip grid in the region ABCB′ (fig. 134) by dividing the arc AB into a small (4–5) number of parts.

11. In the strip-drawing problem, calculate the slip grid for the case $1-h/H = 0.2$, $\gamma = 15°$. Divide the circular arcs DC, CE into a small number (3–5) of parts.

12. Find the limit load for the bending of a cantilever bounded by arcs of a circle of radius $R$ (fig. 123), for the second type of slip field.

13. The same, but with one boundary of the cantilever rectilinear.

14. Find the limit load for the bending of a cantilever (fig. 118) by a pressure distributed uniformly on its upper boundary (the fields will be asymmetric with respect to the $x$-axis and similar to figs. 119, 120).

15. Find the limit load for a circular semi-annulus ($0 \leqslant \varphi \leqslant \pi$), bent by forces $P$ tangential to the ends $\varphi = 0, \varphi = \pi$.

# 6

# Plane Stress

## §52. Equations of plane stress

### 52.1. *Plane stress*

This is the case in which the stress components $\sigma_z$, $\tau_{xz}$, $\tau_{yz}$ (in a cartesian coordinate system) are equal to zero, while the components $\sigma_x$, $\sigma_y$, $\tau_{xy}$ are independent of $z$.

A plane stress state is approximately achieved in a thin lamina deformed under the action of forces which lie in its median plane. The stress components $\sigma_z$, $\tau_{xz}$, $\tau_{yz}$ are small by comparison with the other components, since the bases $z = \pm \frac{1}{2}h$ of the lamina are load-free, and the thickness $h$ is small compared with the transverse dimensions. For this reason the stresses $\sigma_x$, $\sigma_y$, $\tau_{xy}$ change very little through the thickness of the lamina. In the sequel we shall understand by $\sigma_x$, $\sigma_y$, $\tau_{xy}$ the values of the respective components averaged through the lamina, while the components $\sigma_z$, $\tau_{xz}$, $\tau_{yz}$ will be assumed zero.

In these circumstances the differential equations of equilibrium of the element $h\mathrm{d}x\mathrm{d}y$ of the lamina (fig. 156), with constant thickness $h$ and in the absence of body forces, have the form

$$\frac{\partial \sigma_x}{\partial x} + \frac{\partial \tau_{xy}}{\partial y} = 0\,, \qquad \frac{\partial \tau_{xy}}{\partial x} + \frac{\partial \sigma_y}{\partial y} = 0\,. \tag{52.1}$$

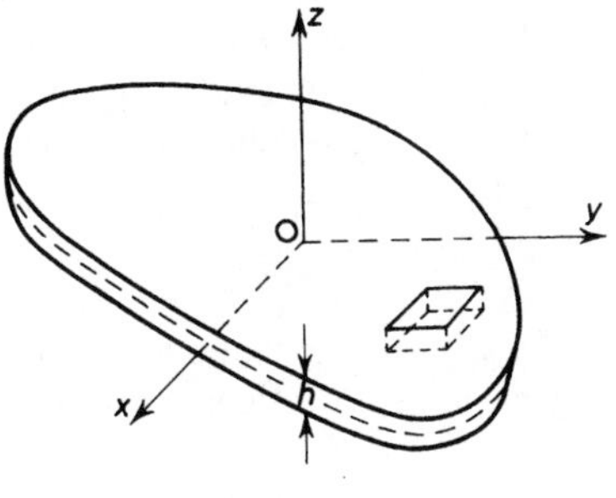


Fig. 156.

52.2. *Equations of plane stress with von Mises yield criterion*

The von Mises yield criterion in the present case has the form

$$\sigma_x^2 + \sigma_y^2 - \sigma_x \sigma_y + 3\tau_{xy}^2 = \sigma_s^2 = 3k^2 , \tag{52.2}$$

or, in principal axes,

$$\sigma_1^2 + \sigma_2^2 - \sigma_1 \sigma_2 = \sigma_s^2 .$$

The constant $k$ is the yield limit with pure shear, $\tau_s$.

The last equation represents an ellipse in the $\sigma_1$, $\sigma_2$-plane, inclined at an angle $\frac{1}{4}\pi$ to the coordinate axes (fig. 157) and cutting off segments $\sigma_s$ on the latter; the principal stresses cannot here exceed $(2/\sqrt{3})\,\sigma_s = 2k$ in magnitude. The semi-axes of the ellipse are respectively $\sqrt{2}\sigma_s$ and $\sqrt{2}\tau_s$.

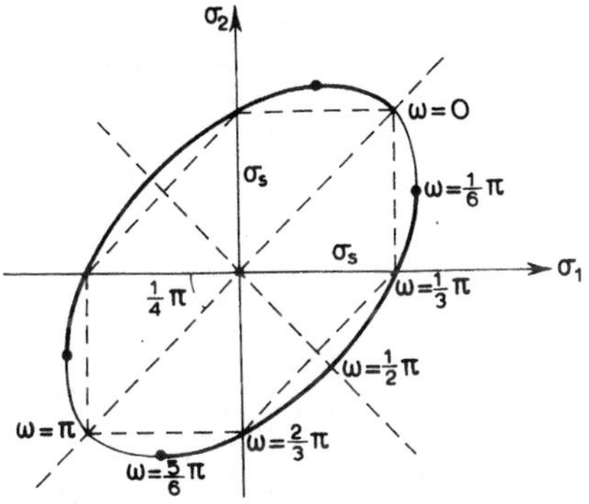


Fig. 157.

With developing plastic deformation we can neglect the elastic strain components in the plastic flow equations, and proceed from the Saint Venant-von Mises relations (13.2) for a plastic rigid body. In our problem these relations can be written in the form ($v_x$, $v_y$ being independent of $z$)

$$\frac{\partial v_x/\partial x}{2\sigma_x-\sigma_y} = \frac{\partial v_y/\partial y}{2\sigma_y-\sigma_x} = \frac{\partial v_x/\partial y + \partial v_y/\partial x}{6\tau_{xy}} . \tag{52.3}$$

We now have, together with the equilibrium equation (52.1) and the yield criterion (52.2), a system of five equations for the five unknown functions $\sigma_x, \sigma_y, \tau_{xy}, v_x, v_y$.

52.3. *Equations of plane stress with Tresca-Saint Venant yield criterion*

Depending on the sign of the principal stresses $\sigma_1$, $\sigma_2$, the maximum tangential stresses develop along different surface elements. If $\sigma_1$, $\sigma_2$ are of different sign, then, as in the case of plane strain, the maximum tangential stress is

$$\tau_{max} = \tfrac{1}{2}|\sigma_1-\sigma_2| = \tfrac{1}{2}\sqrt{(\sigma_x-\sigma_y)^2 + 4\tau_{xy}^2} ,$$

and acts along a surface, normal to the $x$, $y$-plane and bisecting the angle between the principal axes $\sigma_1$, $\sigma_2$ (fig. 158a). In this case there are two orthogonal families of slip lines in the $x$, $y$-plane.

If $\sigma_1$, $\sigma_2$ have the same sign (for example, $\sigma_1 > 0$, $\sigma_2 > 0$, with $\sigma_1 > \sigma_2$), then the maximum tangential stress is

$$\tau_{max} = \tfrac{1}{2}\sigma_1$$

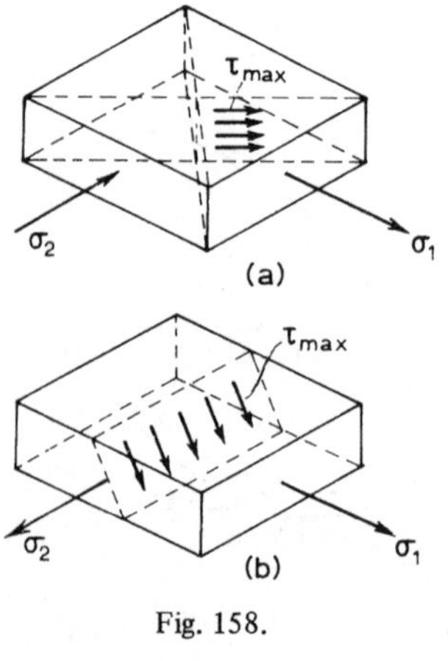


Fig. 158.

and it acts along surface elements parallel to the $\sigma_2$-axis and inclined at an angle $\frac{1}{4}\pi$ to the $x$, $y$-plane (fig. 158b). On the latter the slip elements have a single trace, i.e. one family of lines which are also called slip lines. The direction of one of these slip lines coincides with the principal direction $\sigma_2$. As a consequence, the Tresca-Saint Venant yield criterion takes the form

$$\begin{aligned} &\sigma_1 - \sigma_2 = \pm\,\sigma_s\,, \qquad \text{if} \qquad \sigma_1\sigma_2 \leqslant 0\,; \\ &\sigma_1 = \sigma_s \qquad \text{or} \qquad \sigma_2 = \pm\,\sigma_s\,, \qquad \text{if} \qquad \sigma_1\sigma_2 \geqslant 0\,. \end{aligned} \tag{52.4}$$

These equations represent a hexagon, inscribed in the von Mises ellipse, in the $\sigma_1$, $\sigma_2$-plane (see fig. 157 – dotted line).

The question of the relation between the stresses and the strain-rates with the Tresca-Saint Venant yield criterion has been discussed in §16. For plane stress $\sigma_3 = \sigma_z = 0$; the section on the plane $\sigma_3 = 0$ of the right hexahedral prism, which represents the Tresca-Saint Venant yield criterion in the stress space $\sigma_1$, $\sigma_2$, $\sigma_3$, is the hexagon considered earlier. The plane of the diagram does not contain the normal to the prism, but does contain the projection of the normal perpendicular to the side of the hexagon (fig. 159). In consequence, the ratio of the principal strain rates is equal to the ratio of the direction cosines of the normal to the hexagon at the point under consideration. Using the incompressibility condition

$$\xi_1 + \xi_2 + \xi_3 = 0\,, \tag{52.5}$$

we can find the principal strain rate $\xi_3$.

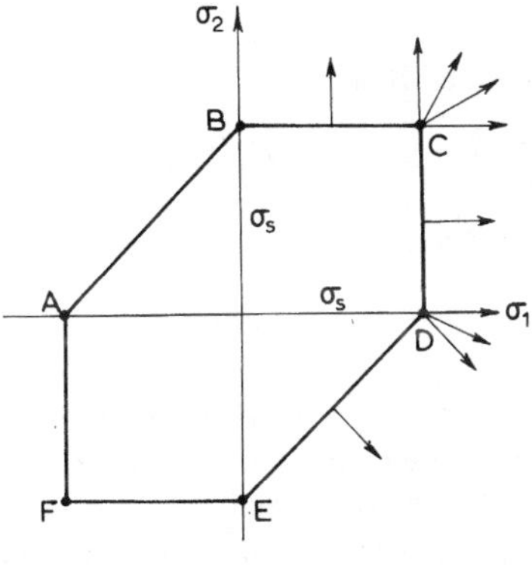


Fig. 159.

Internal points of the segments AB, BC, . . . will be called respectively the *regimes* AB, BC, . . . , while the vertices of the hexagon A, B, . . . will be called the *regimes* A, B, . . . . We consider in greater detail some typical cases.

*Case* $\sigma_1\sigma_2 < 0$ corresponds to inclined edges AB, DE (regimes AB, DE). We concentrate for definiteness on the *regime* DE. Then $\sigma_1 > 0$, $\sigma_2 < 0$, and the yield criterion has the form

$$\sigma_1 - \sigma_2 = \sigma_s \,. \tag{52.6}$$

From the associated flow law (16.10) we obtain

$$\xi_1 = \lambda \,, \qquad \xi_2 = -\lambda \,, \tag{52.7}$$

i.e. $\xi_1 = -\xi_2$.

It now follows from the incompressibility condition that

$$\xi_3 = 0 \,,$$

i.e. the thickness of the lamina remains constant. The quantity $\lambda \geqslant 0$ is an unknown function, determined in the solution of each particular problem. The above equations also hold in the case of plane strain (§31).

In fact, the slip lines introduced in §31 retain their significance. Suppose the first principal direction makes an angle $\varphi$ with the $x$-axis. Then from well-known transformation formulae we have

$$\left.\begin{array}{l}\xi_x \\ \xi_y\end{array}\right\} = \tfrac{1}{2}(\xi_1 + \xi_2) \pm \tfrac{1}{2}(\xi_1 - \xi_2)\cos 2\varphi \,, \qquad \eta_{xy} = (\xi_1 - \xi_2)\sin 2\varphi \,. \tag{52.8}$$

In our case

$$\xi_x = \lambda \cos 2\varphi \,, \qquad \xi_y = -\lambda \cos 2\varphi \,, \qquad \eta_{xy} = 2\lambda \sin 2\varphi \,.$$

Eliminating $\lambda$ we arrive at equation (31.10)

$$\frac{\partial v_x/\partial x - \partial v_y/\partial y}{\partial v_x/\partial y + \partial v_y/\partial x} = -\cot 2\varphi = \frac{\sigma_y - \sigma_x}{2\tau_{xy}}$$

and the previous incompressibility condition

$$\frac{\partial v_x}{\partial x} + \frac{\partial v_y}{\partial y} = 0 \,.$$

Thus, for the regime DE the system of equations for stresses and velocities coincides with the system of equations in the case of plane strain.

A similar conclusion holds for the regime AB.

*Case* $\sigma_1\sigma_2 > 0$ corresponds to vertical and horizontal edges of the hexagon (fig. 159). For definiteness let $\sigma_1 > \sigma_2 > 0$, which relates to the *regime* CD.

The yield criterion is

$$\sigma_1 = \sigma_s . \tag{52.9}$$

By the associated law, flow occurs only in the first principal direction, i.e.

$$\xi_1 = \lambda , \qquad \xi_2 = 0 . \tag{52.10}$$

From the incompressibility condition it follows that

$$\xi_3 = -\xi_1 , \tag{52.11}$$

If we consider the *regime* BC, then

$$\sigma_2 = \sigma_s . \tag{52.12}$$

$$\xi_1 = 0 , \qquad \xi_2 = \lambda , \tag{52.13}$$

where

$$\xi_3 = -\xi_2 . \tag{52.14}$$

*Regime* C. We now turn to the corner point C. Here

$$\sigma_1 = \sigma_s , \qquad \sigma_2 = \sigma_s , \tag{52.15}$$

and the rates of strain are linear combinations, with non-negative coefficients $\lambda_1, \lambda_2$, of the flows in the adjacent regimes CD and BC, i.e.

$$\xi_1 = \lambda_1 , \qquad \xi_2 = \lambda_2 , \qquad \xi_3 = -\lambda_1 - \lambda_2 . \tag{52.16}$$

The coefficients $\lambda_1, \lambda_2$ are unknown functions, determined in the solution of each specific problem; an additional arbitrary function is introduced because of "two yield conditions" (52.15) on the edge of the prism. Owing to the equality of the principal stresses we have:

$$\sigma_x = \sigma_y = \sigma_s , \qquad \tau_{xy} = 0 .$$

A similar flow is achieved in the regime F.

We consider, finally, uniaxial tension, corresponding to the *regime* D:

$$\sigma_1 = \sigma_s , \qquad \sigma_2 = 0 . \tag{52.17}$$

The rates of strain will be linear combinations, with non-negative coefficients $\lambda_1, \lambda_2$, of the flows in adjacent regimes CD and DE, i.e.

$$\xi_1 = \lambda_1 + \lambda_2 , \qquad \xi_2 = -\lambda_1 , \qquad \xi_3 = -\lambda_2 . \tag{52.18}$$

The flows have a similar form in the other uniaxial regimes A, B, E.

52.4. *Concluding remarks*

It is easy to generalize the equations obtained above to the case of a plate of variable thickness $h = h(x, y)$ provided that the latter changes slowly. The differential equations of equilibrium then are:

$$\frac{\partial h\sigma_x}{\partial x} + \frac{\partial h\tau_{xy}}{\partial y} = 0\,, \qquad \frac{\partial h\tau_{xy}}{\partial x} + \frac{\partial h\sigma_y}{\partial y} = 0\,.$$

We note further, that a plane stress state can be realized in thin, zero-moment shells. In shells and plates which experience bending we can have plane stress which varies with the thickness (the non-zero stress components $\sigma_x$, $\sigma_y$, $\tau_{xy}$ depend on $z$, where $z$ is measured along the normal to the mean surface).

## §53. Solutions with von Mises yield criterion. Discontinuous solutions

53.1. *General remarks*

Simultaneous solution of the non-linear system of equations (52.1), (52.2), (52.3) presents great difficulties. However, as in the case of plane strain, a semi-inverse method is often useful. That is, we can attempt to investigate in sequence the solution of the equations for stresses (52.1), (52.2) and the equations for velocities (52.3). If the stresses are known, then the system of equations for the velocities will be linear; it is necessary to construct a velocity field consistent with the stress field which has been determined.

53.2. *Equations for stresses*

We turn to the system of equations for stresses (52.1), (52.2), investigated by Sokolovskii.

It is possible to satisfy the yield condition (52.2) in principal axes by putting

$$\sigma_1 = 2k\cos\left(\omega - \tfrac{1}{6}\pi\right), \qquad \sigma_2 = 2k\cos\left(\omega + \tfrac{1}{6}\pi\right), \tag{53.1}$$

where $\omega = \omega(x, y)$ is a new unknown function, specifying the position of the point on the ellipse (fig. 157). When $\sigma_1 \geqslant \sigma_2$ the angle $\omega$ changes between the limits $0 \leqslant \omega \leqslant \pi$. It is easy to see that angle $\omega$ is connected with the value of the mean pressure $\sigma = \frac{1}{3}(\sigma_1 + \sigma_2)$, namely:

$$\cos\omega = \frac{\sqrt{3}\sigma}{2k}\,. \tag{53.2}$$

Now, with the help of the well-known formulae

$$\left.\begin{matrix}\sigma_x\\ \sigma_y\end{matrix}\right\} = \tfrac{1}{2}(\sigma_1 + \sigma_2) \pm \tfrac{1}{2}(\sigma_1 - \sigma_2)\cos 2\varphi\,, \qquad \tau_{xy} = \tfrac{1}{2}(\sigma_1 - \sigma_2)\sin 2\varphi\,, \tag{53.3}$$

where $\varphi$ is the angle between the first principal direction and the $x$-axis, it is possible to express the stress components $\sigma_x$, $\sigma_y$, $\tau_{xy}$ in terms of the functions $\omega, \varphi$:

$$\left.\begin{matrix}\sigma_x\\ \sigma_y\end{matrix}\right\} = k(\sqrt{3}\cos\omega \pm \sin\omega\cos 2\varphi)\,, \qquad \tau_{xy} = k\sin\omega\sin 2\varphi\,. \tag{53.4}$$

From this it follows, that (in contrast to the case of plane strain) the components of stress are bounded:

$$|\sigma_x| \leqslant 2k\,, \qquad |\sigma_y| \leqslant 2k\,, \qquad |\tau_{xy}| \leqslant k\,.$$

Substituting the stress components into the differential equations of equilibrium, and carrying out some simple transformations, we obtain a system of two equations for the two unknown functions $\varphi(x, y)$, $\omega(x, y)$:

$$\begin{aligned}&(\sqrt{3}\sin\omega\cos 2\varphi - \cos\omega)\frac{\partial\omega}{\partial x} + \sqrt{3}\sin\omega\sin 2\varphi\frac{\partial\omega}{\partial y} - 2\sin\omega\frac{\partial\varphi}{\partial y} = 0\,,\\ &\sqrt{3}\sin\omega\sin 2\varphi\frac{\partial\omega}{\partial x} - (\sqrt{3}\sin\omega\cos 2\varphi + \cos\omega)\frac{\partial\omega}{\partial y} + 2\sin\omega\frac{\partial\varphi}{\partial x} = 0\,.\end{aligned} \tag{53.5}$$

We shall ascertain the type of this system using the "determinantal method" (cf. appendix). Suppose that along some line $x = x(s), y = y(s)$, the functions $\varphi = \varphi(s)$, $\omega = \omega(s)$ are given. For the integral surface passing through L, we have:

$$\frac{\partial\varphi}{\partial x}\,dx + \frac{\partial\varphi}{\partial y}\,dy = d\varphi\,, \qquad \frac{\partial\omega}{\partial x}\,dx + \frac{\partial\omega}{\partial y}\,dy = d\omega\,. \tag{53.6}$$

Along L the tangent plane to the integral surface is defined by the partial derivatives $\partial\varphi/\partial x$, $\partial\varphi/\partial y$; $\partial\omega/\partial x$, $\partial\omega/\partial y$, which can be found from (53.3), (53.6), since on L these will be linear algebraic equations with respect to the derivatives.

If the line L is a characteristic of equations (53.5), then the derivatives are indeterminate along it; consequently, the determinant of the above algebraic system and the appropriate numerators vanish. Equating to zero the determinant of the system, we find the differential equations of the characteristic lines:

$$\frac{dy}{dx} = \frac{\sqrt{3}\sin\omega\sin 2\varphi \pm \Sigma(\omega)}{\sqrt{3}\sin\omega\cos 2\varphi - \cos\omega}\,, \tag{53.7}$$

where we have introduced the notation

$$\Sigma(\omega) = \sqrt{3 - 4\cos^2\omega}\,.$$

Equating the numerators to zero, we obtain (after some transformations and integrations) the relations between the unknown functions $\varphi$, $\omega$, satisfied along the characteristics

$$\Omega \pm \varphi = \text{const.}\,, \tag{53.8}$$

where we have put

$$\Omega = -\frac{1}{2} \int_{\frac{1}{6}\pi}^{\omega} \frac{\Sigma(\omega)}{\sin \omega} \mathrm{d}\omega\,. \tag{53.9}$$

The original system (53.5) will have two distinct families of real characteristics (i.e. will be *hyperbolic*) if $3\text{–}4 \cos^2 \omega > 0$, i.e.

$$\tfrac{1}{6}\pi < \omega < \tfrac{5}{6}\pi\,.$$

The points of hyperbolicity are marked by heavy lines on the ellipse (fig. 157), and fill a major part of its perimeter.

The system (53.5) will have only one family of real characteristics (i.e. will be *parabolic*) in the case $\Sigma(\omega) = 0$, i.e. if the function $\omega$ assumes either of the values $\frac{1}{6}\pi$, $\frac{5}{6}\pi$. Finally, when $3\text{–}4 \cos^2 \omega < 0$ there are no real characteristics and the system (53.5) is *elliptic*. The interior points of the arcs of the ellipse which are marked by light lines correspond to this type.

It is easy to see that in the region of hyperbolicity $|\sigma| < \tau_{\max}$, at the parabolic points $|\sigma| = \tau_{\max}$, and in the region of ellipticity $|\sigma| > \tau_{\max}$. The type of the system is determined by the relationship between the magnitude of the mean pressure and the maximum tangential stress.

Thus it is possible to have hyperbolic, parabolic and elliptic regions in the solution with the transition boundary not known *a priori*. This fact substantially complicates the solution of plane stress problems by comparison with corresponding problems in plane stress.

## 53.3. *Hyperbolic case. Properties of the characteristics*

We consider in greater detail the properties of the solutions in the hyperbolic region. Here $\Sigma(\omega) > 0$, and the function $\Omega$ is easily evaluated:

$$\Omega = -\tfrac{1}{4}\pi + \arcsin \frac{2 \cos \omega}{\sqrt{3}} - \tfrac{1}{4}\arctan \frac{4 \cos \omega + 3}{\Sigma(\omega)} - \tfrac{1}{4}\arctan \frac{4 \cos \omega - 3}{\Sigma(\omega)} < 0\,. \tag{53.10}$$

The principal values of the trigonometrical functions are understood in the

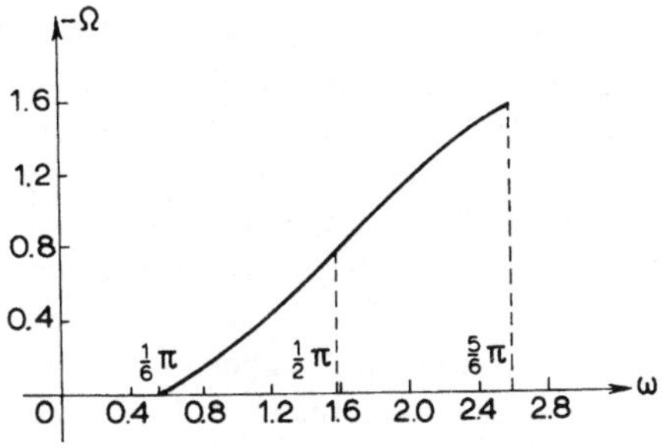


Fig. 160.

above expression (in the interval $-\frac{1}{2}\pi, \frac{1}{2}\pi$). The graph of the function $\Omega$ is shown in fig. 160. We introduce a new unknown function $\psi(x, y)$, equal to

$$\psi = \tfrac{1}{2}\pi - \tfrac{1}{2}\arccos \frac{\cot \omega}{\sqrt{3}} . \tag{53.11}$$

Then the equations of the characteristics take the form

$$(\alpha) \quad \frac{dy}{dx} = \tan(\varphi - \psi), \quad \Omega - \varphi = \text{const.} \equiv \xi ; \qquad (\beta) \quad \frac{dy}{dx} = \tan(\varphi + \psi), \quad \Omega + \varphi = \text{const.} \equiv \eta . \tag{53.12}$$

The characteristics intersect at an angle $2\psi$ (fig. 161) and generate a non-orthogonal network of curves which, obviously, does not coincide with the slip-line grid. The principal directions bisect the angles between the character-

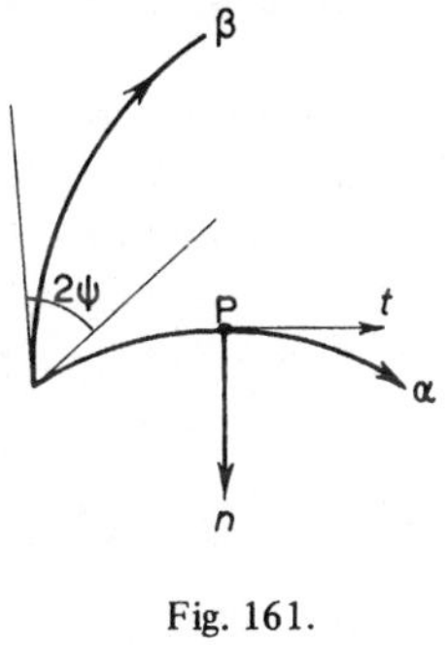


Fig. 161.

istics. We shall distinguish the characteristics of these two families by the parameters $\alpha$, $\beta$, as before (these parameters have a different meaning than in ch. 5). The characteristics of the first family (*$\alpha$-characteristics*) correspond to fixed values of the parameter $\beta$, while $\alpha$ is constant along the $\beta$-characteristics. In general the angle of intersection $2\psi$ varies from point to point. The parameter $\xi$ varies in passing from one characteristic of the $\alpha$-family to another; similarly the parameter $\eta$ varies in passing from one member of the $\beta$-family to another.

The stress components obey simple relations along the characteristics (we recall that in the case of plane strain the normal stresses in the characteristic directions are equal to the mean pressure $\sigma$, cf. §31).

Suppose that continuous stress components are prescribed on some line L. At an arbitrary point P of the line L we construct a system of coordinates $t$, $n$, the $t$-axis being in the direction of the tangent to L, and the $n$-axis the normal. The differential equations of equilibrium (52.1) and the yield criterion (52.2) retain their previous form in the new coordinate system. We calculate the derivatives $\partial\sigma_t/\partial t$, $\partial\sigma_n/\partial t$, $\partial\tau_{nt}/\partial t$ along L at the point P. Then the normal derivatives $\partial\sigma_n/\partial n$, $\partial\tau_{nt}/\partial n$ can be determined from the differential equations. The derivative $\partial\sigma_t/\partial n$ can also be found from differentiating the yield criterion with respect to $n$:

$$\frac{\partial\sigma_t}{\partial n}(2\sigma_t-\sigma_n)+\frac{\partial\sigma_n}{\partial n}(2\sigma_n-\sigma_t)+6\tau_{nt}\frac{\partial\tau_{nt}}{\partial n}=0\,,$$

if $2\sigma_t-\sigma_n\neq 0$. If the condition

$$2\sigma_t-\sigma_n=0$$

is fulfilled, Cauchy's problem cannot be solved, and the line L is characteristic.

The same condition is fulfilled along the characteristics of the second family.

The characteristics have a series of properties similar to certain properties of slip lines in the problem of plane strain (§32). We shall present these without detailed proof (it is easy for the reader to verify them).

1. *If we cross from one characteristic of the $\beta$-family to another along any characteristic of the $\alpha$-family, then the angle of inclination of the first principal direction $\varphi$ and the function $\Omega$ will change by the same amount* (analog of Hencky's first theorem).

For a proof it is sufficient to use the relations

$$2\Omega=\xi+\eta\,,\qquad 2\varphi=\eta-\xi\,. \tag{53.13}$$

2. *If some segment of a characteristic is a straight line, then along it* $\omega$, $\varphi$, *the angle of intersection* $\psi$, *the parameters* $\xi$, $\eta$ *and the stress components* $\sigma_x$, $\sigma_y$, $\tau_{xy}$, *are all constant.*

3. *If some segment of a characteristic (for example of the α-family) is straight, then all the corresponding segments of the characteristics of that family are straight; in such a region a simple stress state is realized and the parameter* $\eta$ *is constant.*

Indeed, let us take a construction analogous to fig. 67; owing to the rectilinearity of the characteristic $A_{21}A_{22}$ it follows from the first property that $\varphi_{11} = \varphi_{12}$, $\Omega_{11} = \Omega_{12}$; but then $\omega_{11} = \omega_{12}$, $\psi_{11} = \psi_{12}$. All β-characteristics intersect a straight α-line at the same angle (changing, of course, from one α-line to another).

4. *If both families of characteristics are rectilinear then in such a region the stresses are uniformly distributed and the parameters* $\xi$, $\eta$ *are constant.*

The system of differential equations (53.5) is reducible and can be linearized by the method of inversion of the variables (analogously to the equations of plane strain). Simple and uniform stress states correspond to the integrals

1. $\xi = \text{const.}$,
2. $\eta = \text{const.}$,
3. $\xi = \text{const.}$, $\eta = \text{const.}$

As in the case of plane strain, it is necessary to consider various boundary value problems. In the general case the field of characteristics is constructed by numerical (or graphical) methods, similar to those discussed in the previous chapter, Here the initial relations are the equations of characteristics (53.12).

For simple stress states the boundary value problems have an elementary solution.

### 53.4. *Simple stress states*

We consider in more detail simple stress states, which play an important role in applications.

The simplest solution of this type is the *uniform stress*. In this region the characteristic grid is generated by two non-orthogonal families of parallel lines (fig. 162), and the parameters $\xi$, $\eta$ are constant ($\xi = \xi_0$, $\eta = \eta_0$). Then $\Omega = \frac{1}{2}(\xi_0 + \eta_0)$, $\varphi = \frac{1}{2}(\eta_0 - \xi_0)$, and $\omega$ and the angle $\psi$ can be calculated subsequently.

*At a rectilinear free boundary there will always be a field of uniform uniaxial tension or compression of magnitude* $\sqrt{3}k$, parallel to the boundary (fig. 163). For example, if the $x$-axis is parallel to the boundary, then in the adjoining region $\sigma_x = \pm\sqrt{3}k$, $\sigma_y = \tau_{xy} = 0$. In fact, the normal to the free

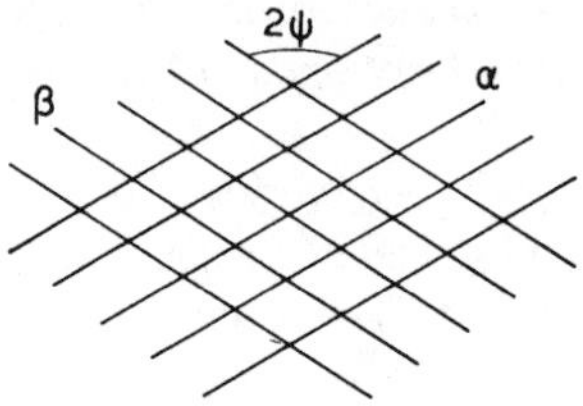


Fig. 162.

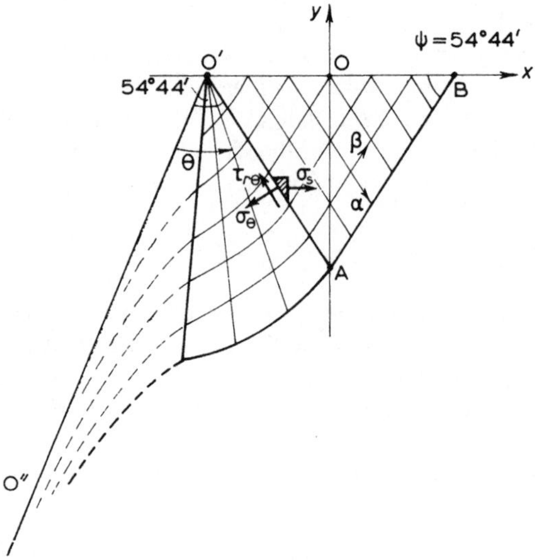


Fig. 163.

boundary coincides with one of the principal directions, $\varphi = 0$, and the boundary points experience either uniaxial tension or uniaxial compression. We consider the case of tension; then on the boundary $\omega = \frac{1}{3}\pi$ (fig. 157), i.e. the representative point lies in the hyperbolic region. By virtue of (53.11) we find that

$$\psi = 54°44' \equiv \psi_1 \, .$$

Thus, the contour is not characteristic, and we have a Cauchy problem for finding the stress field near the boundary. The solution can be determined in the triangular domain O′AB.

Suppose a uniform normal stress $p$ acts on the rectangular boundary. The second principal direction can be calculated from the yield condition (52.2). If the stress on the boundary corresponds to points of hyperbolicity on the ellipse (fig. 157), then the solution near the boundary can easily be constructed as in the previous case (when $p = 0$).

In simple stress only one of the families of characteristics consists of straight lines. As in the case of plane strain (§33), the following theorem holds:

*Only a region of simple stress can adjoin a region of uniform stress.*

Let O′A (fig. 163) be a boundary characteristic of a region of uniform stress. If this region adjoins along O′A a region having a different solution, then by property 3 of the previous section all the characteristics of the family to which O′A belongs will be straight lines.

We consider in more detail an important special case, when rectilinear characteristics originate at a point (*central field*). Let this field be adjacent to a region of uniform uniaxial tension O′AB. We introduce an auxiliary polar system of coordinates $r$, $\theta$ with pole at the point O′ and polar axis O′O″; the position of the latter will be chosen later.

The differential equations of equilibrium in polar coordinates are

$$\frac{\partial \sigma_r}{\partial r} + \frac{1}{r}\frac{\partial \tau_{r\theta}}{\partial \theta} + \frac{\sigma_r - \sigma_\theta}{r} = 0\,,$$

$$\frac{\partial \tau_{r\theta}}{\partial r} + \frac{1}{r}\frac{\partial \sigma_\theta}{\partial \theta} + \frac{2\tau_{r\theta}}{r} = 0\,, \qquad (53.14)$$

and the yield criterion is

$$\sigma_r^2 - \sigma_r\sigma_\theta + \sigma_\theta^2 + 3\tau_{r\theta}^2 = 3k^2\,. \qquad (53.15)$$

These are satisfied if we take

$$\sigma_r = k\cos\theta\,, \qquad \sigma_\theta = 2k\cos\theta\,, \qquad \tau_{r\theta} = k\sin\theta\,. \qquad (53.16)$$

It is obvious that the stress is constant along the radius vector. Along the transition line O′A the stress components $\sigma_\theta$, $\tau_{r\theta}$ must be continuous (from the condition of equilibrium of an element of the line O′A, cf. §39). Considering the equilibrium of the shaded triangular element adjoining the boundary O′A (fig. 163), we easily obtain $\sin\theta = \sqrt{\frac{2}{3}}$, i.e. the polar axis O′O″ must be inclined at an angle 54°44′ to the line O′A. In this coordinate system the equation of the characteristics of the second family has the simple form

$$r^2 \sin\theta = \text{const.} \qquad (53.17)$$

Curvilinear characteristics tend asymptotically to O′O″, for which $\theta = 0$; this is shown by the dotted line in fig. 163. Along O′O″ the two families of characteristics converge into one, and $\sigma_\theta = 2k$, $\sigma_r = k$ (the parabolic point $\omega = \frac{1}{6}\pi$ on the von Mises ellipse, fig. 157).

## 53.5. *Axisymmetric field*

We consider an axisymmetric field of stresses with the condition $\tau_{r\theta} = 0$ (zero torsion). Then the stresses $\sigma_r$, $\sigma_\theta$ will be principal stresses, and from (53.1) we have

$$\sigma_r = 2k \cos(\omega + \tfrac{1}{6}\pi), \qquad \sigma_\theta = 2k \cos(\omega - \tfrac{1}{6}\pi). \tag{53.18}$$

Here we have chosen a variant of formula (53.1) for the case $\sigma_\theta > \sigma_r$. Substituting (53.18) in the equilibrium equation

$$\frac{\mathrm{d}\sigma_r}{\mathrm{d}r} + \frac{\sigma_r - \sigma_\theta}{r} = 0, \tag{53.19}$$

we obtain the differential equation

$$(\sqrt{3} + \cot\omega)\,\mathrm{d}\omega + 2\frac{\mathrm{d}r}{r} = 0,$$

whose solution has the form

$$r^2 = \frac{C}{\sin\omega}\,\mathrm{e}^{-\sqrt{3}\omega}, \tag{53.20}$$

where $C$ is an arbitrary constant. Suppose, for example, the boundary of a hole $r = a$ (fig. 164) to be free, i.e. $\sigma_r = 0$ on it. Then $\omega = \frac{1}{3}\pi$ when $r = a$. It is

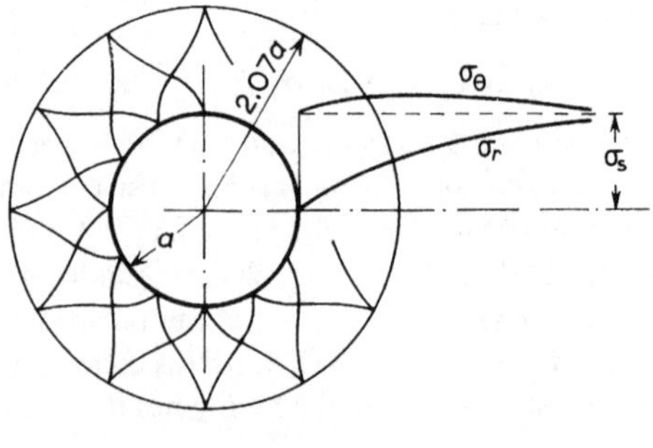


Fig. 164.

easy to see that

$$\left(\frac{r}{a}\right)^2 = \frac{\sqrt{3}}{2}\,\frac{1}{\sin\omega}\,e^{\sqrt{3}(\frac{1}{3}\pi-\omega)}\,. \tag{53.21}$$

Since the axes $r$, $\theta$ are principal, the relations (53.8) have the form

$$\Omega \pm \theta = \text{const.}\,,$$

and the equations of the characteristics in parametric form can be calculated from (53.21). Away from the boundary $r = a$ the angle between the characteristics decreases and when $\omega = \frac{1}{6}\pi$ (here $r/a = 2.07$) the two families converge into one (fig. 164).

When $r/a > 2.07$ the solution is again determined from (53.18) and (53.20), but there are no real characteristics (region of ellipticity).

The stress distribution $\sigma_r$, $\sigma_\theta$ is shown in fig. 164. As $r \to \infty$, $\omega \to 0$, and $\sigma_r \to \sigma_s$, $\sigma_\theta \to \sigma_s$, i.e. the lamina experiences uniform stress at infinity.

### 53.6. *Parabolic and elliptic cases*

*At parabolic points* $\Sigma(\omega) = 0$, so that $\omega$ has a constant value $\frac{1}{6}\pi$ or $\frac{5}{6}\pi$ (fig. 157). The two families of characteristics converge into one ($\psi = 0$). It follows from the system of differential equations (53.5) that $\varphi = $ const. The family of characteristics is a family of parallel straight lines. Thus, this solution leads to a uniform stress state of a particular type, for example, when $\omega = \frac{1}{6}\pi$, $\sigma_1 = 2k$, $\sigma_2 = k$, and only the principal direction is arbitrary.

*In the elliptic case* there are considerable difficulties in the construction of the solutions to the non-linear equations (53.5); there are no general methods, and solutions have been found only for axisymmetric problems.

### 53.7. *Determination of the velocity field*

We consider next the system (52.3) for velocities, assuming, as usual, that the stress state is known. The system (52.3) is then linear with variable coefficients. We write it in the following form:

$$\begin{aligned} (2\sigma_x-\sigma_y)\frac{\partial v_y}{\partial y} &= (2\sigma_y-\sigma_x)\frac{\partial v_x}{\partial x}\,, \\ (2\sigma_x-\sigma_y)\left(\frac{\partial v_x}{\partial y}+\frac{\partial v_y}{\partial x}\right) &= 6\tau_{xy}\frac{\partial v_x}{\partial x}\,. \end{aligned} \tag{53.22}$$

In the region where the equations for the stresses are hyperbolic *the equations for the velocities are also hyperbolic, with the characteristics of the two systems coinciding.*

Thus, let the velocity be given on a line L, which is not a line of velocity discontinuity. We construct at an arbitrary point P of L a system of coordinates $t$, $n$, the $t$-axis being tangent to L. Equations (53.22) retain their form in the new coordinate system. At the point P the derivatives $\partial v_t/\partial t$, $\partial v_n/\partial t$ are known. The derivatives $\partial v_t/\partial n$, $\partial v_n/\partial n$ are bounded and uniquely determined from equations (53.22), except in the case when

$$2\sigma_t - \sigma_n = 0 \ .$$

But the condition $2\sigma_t = \sigma_n$ is fulfilled along a stress characteristic (cf. above, par. 3).

It follows that no solution of Cauchy's problem for the velocities is possible along a characteristic of the stress state. *The characteristic equations for velocities coincide with the characteristic equations for stresses.*

Next, it emerges from (53.22) that along a characteristic

$$\partial v_t/\partial t = 0 \ , \tag{53.23}$$

i.e. *the velocity of relative elongation is zero along a characteristic* (just as in the case of plane strain). The same condition is satisfied along characteristics of the second family. These conditions can be expressed in the form of differential equations, similar to the Geiringer equations – see Hill [54].

### 53.8. *Lines of velocity discontinuity*

In plane stress, as in plane strain, it is important to consider discontinuous solutions, which can arise in hyperbolic and parabolic regions. Consider some line of velocity discontinuity L. In contrast with the plane strain case, it is necessary in plane stress to allow the possibility of a jump not only in the tangential, but also in the normal, component of velocity. This kind of jump leads to abrupt thinning (fig. 165a) or thickening (fig. 165b) of the lamina along the line of discontinuity. This line is a mathematical idealization of local development of a neck, which is observed in experiments. We shall therefore call this line of discontinuity a "neck"; it should be regarded as the limiting position of a band of intense strain, where in keeping with our plane stress model we assume the strain-rate in the neck to be uniform.

The jump in velocity cannot be arbitrary, since it is connected by definite conditions with the stress distribution. We consider these conditions.

Let one side (+) of the neck L (fig. 165c) be displaced relative to the other (–) with velocity **v**. Denote by $b$ the width of the neck (in the limit $b \to 0$). The vector **v** is inclined to the line L at an angle $\gamma$, i.e. the tangential and normal velocity components are discontinuous. At an arbitrary point M of the neck we construct a local coordinate system $n$, $t$, with the $t$-axis directed

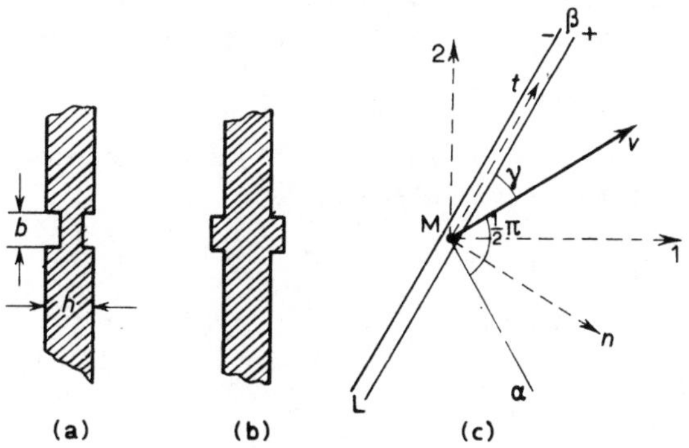


Fig. 165.

along the tangent. Then the strain-rate components in the neck will be

$$\xi_n = \frac{v_n}{b} = \frac{v \sin \gamma}{b}, \qquad \xi_t = 0, \qquad \eta_{nt} = \frac{v_t}{b} = \frac{v}{b} \cos \gamma, \tag{53.24}$$

where $v$ is the speed.

The rate of extension in the $\alpha$-direction, perpendicular to the velocity vectoı, is also equal to zero: $\xi_\alpha = 0$. Consequently the $t$ and $\alpha$-directions are characteristic, and the angle between them is $2\psi$. Thus, the velocity discontinuity vector $\mathbf{v}$ is inclined to the neck at an angle

$$\gamma = 2\psi - \tfrac{1}{2}\pi; \tag{53.25}$$

the line of discontinuity traverses the characteristic.

When $\gamma = 0$ the characteristics are orthogonal, thinning does not occur and only relative slip takes place. This happens when the stress state is a pure shear ($\omega = \frac{1}{2}\pi$).

Proceeding from (53.24), it is easy to find by the usual method the principal rates of extension in the neck:

$$\xi_1 = \frac{v}{2b}(1 + \sin \gamma), \qquad \xi_2 = -\frac{v}{2b}(1 - \sin \gamma). \tag{53.26}$$

The third principal strain rate (in the direction perpendicular to the $n$, $t$-plane) is determined from the incompressibility equation

$$\xi_z = -\frac{v}{b} \sin \gamma .$$

The principal directions 1, 2 bisect the angles between the characteristics. The rate of work of plastic deformation per unit length of the neck is

$$bh\dot{A}_\mathrm{p} = bh(\sigma_1 \xi_1 + \sigma_2 \xi_2) \,.$$

Substituting the principal strain-rates in the Saint Venant-von Mises relations (13.12), we obtain the principal stresses

$$\frac{\sigma_1}{k} = \frac{1 + 3\sin\gamma}{\sqrt{1 + 3\sin^2\gamma}} \,, \qquad \frac{\sigma_2}{k} = -\frac{1 - 3\sin\gamma}{\sqrt{1 + 3\sin^2\gamma}} \,. \tag{53.27}$$

It is easy to see that

$$\sin\gamma = \frac{\sigma_1 + \sigma_2}{3(\sigma_1 - \sigma_2)} \,, \qquad bhA_\mathrm{p} = kvh\sqrt{1 + 3\sin^2\gamma} \,. \tag{53.28}$$

*In the parabolic case* the two characteristics merge and coincide along the neck; the latter supports a stress state corresponding to one of the parabolic points on the ellipse (fig. 157).

For the point $\omega = \frac{1}{6}\pi$ the stresses are $\sigma_1 = 2k$, $\sigma_2 = k$, the angle $\gamma = \frac{1}{2}\pi$; consequently, *there is no slip jump, only the normal velocity component has a discontinuity*, and thinning occurs along the line of discontinuity. For the point $\omega = \frac{5}{6}\pi$ the stresses are $\sigma_1 = -k$, $\sigma_2 = -2k$, the angle $\gamma = -\frac{1}{2}\pi$; again there is no jump in the slip, only the normal component is discontinuous, and thinning takes place along the line of discontinuity.

As an example, consider the extension of a plane specimen (fig. 166), assuming that the neck occupies the whole of its section. Here $\sigma_2 = 0$, $\gamma = \arcsin\frac{1}{3} = 19°28'$, $\psi = 54°44'$. The velocity discontinuity vector is inclined to the axis of the specimen at an angle $35°16'$. Along both sides of the neck the material is "rigid". Under certain circumstances the development of this kind of neck is observed in experiments, with the angle of inclination to the direction of extension being 55–60°. This is discussed extensively in Nadai's monograph [25] from which the photograph (fig. 167) has been taken.

### 53.9. *Lines of stress discontinuity*

Stress discontinuities must satisfy the equilibrium and plasticity conditions. We use the diagram introduced in the previous chapter (fig. 99). Consider an element of the line of discontinuity L on which stress components $\sigma_n$, $\tau_n$, $\sigma_t^+$, $\sigma_t^-$ act; by the equilibrium condition the components $\sigma_n$, $\tau_n$ are continuous, and a discontinuity can occur only in the normal component $\sigma_t$.

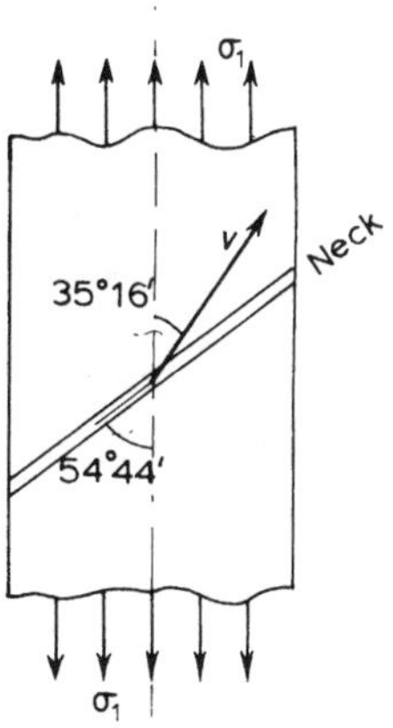


Fig. 166.

Fig. 167.

From the plasticity condition (52.2) we find

$$\left.\begin{matrix}\sigma_t^+\\ \\ \sigma_t^-\end{matrix}\right\} = \tfrac{1}{2}\sigma_n \pm \sqrt{3(k^2 - \tau_n^2 - \tfrac{1}{4}\sigma_n^2)}\,. \tag{53.29}$$

The jump in the component $\sigma_t$ is

$$[\sigma_t] = \sigma_t^+ - \sigma_t^- = 2\sqrt{3(k^2 - \tau_n^2 - \tfrac{1}{4}\sigma_n^2)}\,.$$

The angle of inclination of the characteristics to the line of stress discontinuity changes discontinuously. Let the position of the principal axes of stress be defined by the angles $\delta^+$ and $\delta^-$ (fig. 168). The components $\sigma_t$, $\sigma_n$, $\tau_{nt}$ can be evaluated with the aid of (53.3), with $\varphi$ replaced by $\delta$. The conditions of continuity of $\sigma_n$, $\tau_{nt}$ lead to the expressions

$$p^+ + \tau^+ \cos 2\delta^+ = p^- - \tau^- \cos 2\delta^-\,,$$

$$\tau^+ \sin 2\delta^+ = \tau^- \sin 2\delta^-\,,$$

where half the sum and half the difference of the principal stresses are denoted by $p$ and $\tau$ respectively. From the yield condition (52.2) we have $p^2 + 3\tau^2 = 3k^2$.

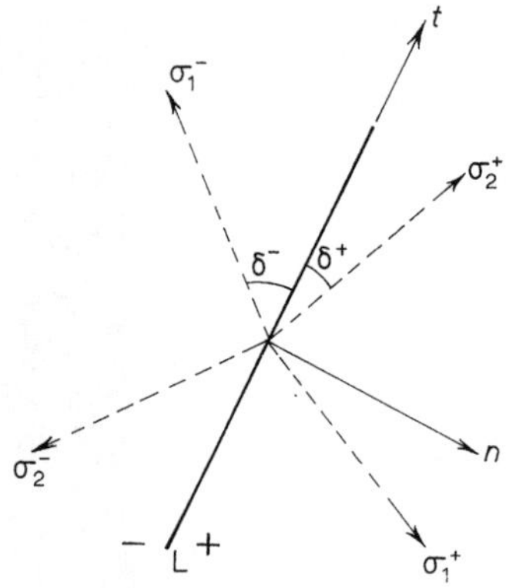


Fig. 168.

It follows from the above relations that

$$\tan 2\delta^- = \frac{2 \sin 2\delta^+}{\cos 2\delta^+ + p^+/\tau^+},$$

$$p^- = \tfrac{1}{2}p^+ - \tfrac{3}{2}\tau^+ \cos 2\delta^+ .$$

The quantity $\tau^-$ is evaluated in terms $p^-$ from the yield criterion.

*On a line of velocity discontinuity the stress components are continuous,* since the stress on this line is determined by formulae (53.27).

Conversely, *the velocity field is continuous on a line of stress discontinuity* (since otherwise stress discontinuity would not be possible). We show next that *a line of stress discontinuity does not experience extension.* In fact we have from the Saint Venant-von Mises equations (13.12)

$$\xi_t^+ = \lambda^+(\sigma_t^+ - \sigma^+), \qquad \xi_t^- = \lambda^-(\sigma_t^- - \sigma^-).$$

It follows from the continuity of the velocities that $\xi_t^+ = \xi_t^-$ along the line of discontinuity, i.e. $\lambda^+(2\sigma_t^+ - \sigma_n) = \lambda^-(2\sigma_t^- - \sigma_n)$.

By virtue of (53.29) the quantities in brackets are of opposite sign, and since $\lambda$ is non-negative, we have that $\lambda^+ = \lambda^- = 0$ i.e. $\xi_t = 0$.

### 53.10. *Plastic-rigid boundary*

Let L be a line which separates a plastic region from a rigid region; we shall assume that the latter is at rest (cf. §38.5).

If the velocity is discontinuous on the separation line L, then, as was shown above, this line traverses a characteristic.

If the velocity is continuous it is equal to zero on the line L; if L is not a characteristic, then the characteristic triangle adjoining L will be at rest (since the velocity field in it is determined from a Cauchy problem). If deformation takes place in the plastic zone, then this is possible only in the case where L is a characteristic.

Thus, *a plastic-rigid boundary traverses a characteristic (or the envelope of characteristics).*

## §54. Solutions with Tresca-Saint Venant yield criterion. Discontinuous solutions

54.1. *General remarks*

In §52 we formulated the equations of plane stress with the Tresca-Saint Venant yield criterion; these equations are different in different regimes. The solution of specific problems usually requires consideration of the flows in the different regimes, which are realized in one or other part of the plastic zone. In this regard it is not difficult to make an error and to choose an incorrect arrangement of the different regions of stress; for the solution demands an accurate construction of a compatible velocity field.

The equations for the stresses in different regimes were examined by Sokolovskii [44]. Consecutive construction of the velocity field became possible only after the introduction of the associated flow law. The analysis of discontinuous solutions was undertaken by Hill [182].

54.2. *Regimes* DE *and* AB *($\sigma_1 \sigma_2 < 0$)*

We noted in §52 that the system of equations for stresses and velocities in this regime coincided with the corresponding system for the case of plane strain. In these regions the characteristics are orthogonal and coincide with the slip lines. The results obtained in the preceding chapter carry over completely to the case of the lamina with $\sigma_1 \sigma_2 < 0$. For the lamina only the magnitudes of the normal stresses are bounded (for example, for DE $\sigma_1 < \sigma_s$, $|\sigma_2| < \sigma_s$). The lamina does not experience thinning ($\xi_z = 0$).

The impossibility of thinning excludes discontinuities in the normal velocity components; it follows from the condition $\xi_z = 0$ (cf. (53.24)) that $\xi_n = (v/b) \sin \gamma = 0$, i.e. $\gamma = 0$. *As in the case of plane strain, relative slip takes place along a line of velocity discontinuity which traverses a characteristic. A "neck" cannot occur.*

As regards lines of stress discontinuity, the results of §39 remain fully valid. A line of stress discontinuity is the bisector of the angle between like slip lines.

54.3. *Regimes* CD *and* EF

For the regime CD (cf. §52) $\sigma_1 = \sigma_s$, $0 < \sigma_2 < \sigma_s$. We put

$$\sigma_1 - \sigma_2 = 2\sigma_s \chi \,, \tag{54.1}$$

where $\chi = \chi(x, y)$ is an unknown function. Then

$$\sigma_1 + \sigma_2 = 2\sigma_s \kappa (1 - \chi) \,, \tag{54.2}$$

where $\kappa$ = sign $\sigma_1$ = sign $\sigma_2$. For the regime CD $\kappa = +1$; the quantity $\kappa$ is introduced so that the solution we obtain can be carried over to the regime EF, where $\kappa = -1$. With the aid of formula (53.3) we find

$$\left.\begin{matrix}\sigma_x \\ \sigma_y\end{matrix}\right\} = \sigma_s[\kappa(1-\chi) \pm \chi \cos 2\varphi] \,, \qquad \tau_{xy} = \sigma_s \chi \sin 2\varphi \,, \tag{54.3}$$

where $\varphi = \varphi(x, y)$ is the unknown angle between the first principal direction and the $x$-axis. Substituting these values in the differential equations of equilibrium, multiplying respectively by $\sin\varphi$, $\cos\varphi$, then adding and subtracting these equations, we obtain, after some simplification,

$$\begin{gathered} \sin 2\varphi \frac{\partial \varphi}{\partial x} - (\kappa + \cos 2\varphi) \frac{\partial \varphi}{\partial y} = 0 \,, \\ \sin 2\varphi \frac{\partial \ln \chi}{\partial x} - (\kappa + \cos 2\varphi) \frac{\partial \ln \chi}{\partial y} + 2\kappa \frac{\partial \varphi}{\partial y} = 0 \,. \end{gathered} \tag{54.4}$$

It is not difficult to construct the general solution of this system. The characteristic equation of the first equation has the form

$$\frac{dx}{\sin 2\varphi} = \frac{dy}{-(\kappa + \cos 2\varphi)} = \frac{d\varphi}{0}$$

and it can be integrated easily:

$$\varphi = \text{const.} = C_1 \,, \qquad y = x \tan\left[\varphi + \tfrac{1}{4}(\kappa + 1)\,\pi\right] + C_2 \,. \tag{54.5}$$

The characteristics are straight lines, inclined at an angle $[\varphi + \frac{1}{4}(\kappa + 1)\,\pi]$ to the $x$-axis, and, consequently, coincident with the slip lines, i.e. with the straight trajectories of the principal stresses (§52.3).

For the regime CD the characteristics coincide with the rectilinear trajectories of the numerically smaller principal stress $\sigma_2$ (for the regime EF $\sigma_2 = -\sigma_s$, $\kappa = -1$ and the characteristics traverse the rectilinear trajectories of the principal stress $\sigma_1$). By the associated law $\xi_2 = 0$ in the regime CD, i.e. the characteristic is not elongated.

The general solution of equation (54.4) can be represented in the form

$$y = x \tan\left[\varphi + \tfrac{1}{4}(\kappa + 1)\pi\right] + \Phi(\varphi)\,, \tag{54.6}$$

where $\Phi(\varphi)$ is an arbitrary function determined from the given boundary conditions.

We now construct the characteristic equation of the second equation of (54.4):

$$\frac{\mathrm{d}x}{\sin 2\varphi} = \frac{\mathrm{d}y}{-(\kappa + \cos 2\varphi)} = \frac{\mathrm{d} \ln \chi}{-2\kappa\, \partial\varphi/\partial x}\,.$$

This equation has the same family of characteristic lines, i.e. the system (54.4) is parabolic.

Along a characteristic line

$$\mathrm{d} \ln \chi = -2\kappa \frac{\partial\varphi}{\partial x} \frac{\mathrm{d}x}{\sin 2\varphi}\,.$$

Differentiating the solution (54.6) with respect to $x$ and calculating the derivative $\partial\varphi/\partial x$, we substitute it in the last equation and integrate. Then the general solution of the second of (54.4) is

$$\chi = \frac{\Psi(\varphi)}{2x + (1-\kappa \cos 2\varphi)\,\Phi'(\varphi)}\,, \tag{54.7}$$

where $\Psi(\varphi)$ is an arbitrary function determined from the boundary conditions.

We note the obvious solution

$$\varphi = \text{const.}\,, \qquad \chi = \text{const.}\,,$$

which describes a *uniform stress distribution.*

To find the arbitrary functions it is necessary to prescribe $\varphi$ and $\chi$ along some curve C in the $x$, $y$-plane. The solution of this Cauchy problem is indeterminate if C is itself a characteristic.

We consider a *velocity discontinuity along some line* L (fig. 165). Let the right-hand side (+) of the neck be displaced relative to the left (−) with velocity **v**. In the local coordinate system $n$, $t$ we have the previous formulae (53.24) for the strain-rate components. Since $\xi_t = 0$, *the line of discontinuity traverses a characteristic (recall that $\xi_2 = 0$ along it), i.e. it is rectilinear and coincides with the trajectory of the numerically smaller principal stress.* The directions of $n$, $t$ are consequently principal, and $\eta_{nt} = 0$ i.e. $\gamma = \frac{1}{2}\pi$. Thus, *a discontinuity can occur only in the normal velocity component.*

We note, further, that *the numerically smaller principal stress can be dis-*

*continuous along a characteristic, with the magnitude of the discontinuity arbitrary in the limits* $(0, \sigma_s)$. For example, $\sigma_2$ can be discontinuous along a characteristic for the regime CD, with $0 < |[\sigma_2]| < \sigma_s$.

54.4. *Regimes* C *and* F

For the singular regimes C and F we have

$$\sigma_1 = \sigma_2 = \kappa\sigma_s\,, \tag{54.8}$$

where $\kappa = +1$ for the regime C, and $\kappa = -1$ for the regime F. It follows from (53.3) that for any cartesian coordinate system $x, y$ we have

$$\sigma_x = \sigma_y = \kappa\sigma_s\,, \qquad \tau_{xy} = 0\,. \tag{54.9}$$

Obviously the differential equations of equilibrium (52.1) are satisfied. Thus, the regimes under consideration correspond to a *uniform hydrostatic (in the x, y-plane) stress state.* Any direction is principal for the stresses. The strain rates for the regime C are $\xi_1 = \lambda_1 \geqslant 0$, $\xi_2 = \lambda_2 \geqslant 0$.

In this region we consider a *velocity discontinuity along some line* L (fig. 165). Let the first principal direction for the strain-rates make an angle $\varphi$ with the $n$-axis. From formulae (52.8) and (53.24) we have

$$\xi_t = \lambda_1 \sin^2\varphi + \lambda_2 \cos^2\varphi = 0\,.$$

Consequently $\varphi = 0$, $\lambda_2 = 0$ and the axes $n, t$ are principal; thus $\eta_{nt} = 0$ i.e. $\gamma = \frac{1}{2}\pi$. Hence *the discontinuity vector* **v** *is normal to the line* L *and there is no relative slip.* The orientation of the line of discontinuity with respect to the stress field is arbitrary.

Clearly a discontinuity in the stresses in these fields is excluded (because of the condition $\sigma_n^+ = \sigma_n^-$).

54.5. *Regimes* D, A *(uniaxial tension or compression)*

Here $\sigma_1 = \kappa\sigma_s$, $\sigma_2 = 0$. Substituting the stresses $\sigma_x, \sigma_y, \tau_{xy}$ from formula (53.3) in the differential equations of equilibrium (52.1) we find that the angle of inclination of the first principal direction $\varphi =$ const., which corresponds to a field of homogeneous tension (or compression). The strain rates are given by formulae (52.18):

$$\xi_1 = \lambda_1 + \lambda_2\,, \qquad \xi_2 = -\lambda_1\,, \qquad \xi_3 = -\lambda_2\,,$$

where $\lambda_1 \geqslant 0$, $\lambda_2 \geqslant 0$.

We consider next *a velocity discontinuity along some line* L (fig. 165c). As before, let the first principal direction make an angle $\varphi$ with the $n$-axis. The principal rates of extension in the neck are given by equations (53.26), and hence

$$\frac{v}{2b}(1+\sin\gamma)=\lambda_1+\lambda_2\,, \qquad -\frac{v}{2b}(1-\sin\gamma)=-\lambda_1\,.$$

The rate of extension along the neck is zero, and therefore

$$\xi_t=(\lambda_1+\lambda_2)\sin^2\varphi-\lambda_1\cos^2\varphi=0\,. \tag{54.10}$$

Eliminating $v/b$, $\lambda_1$ and $\lambda_2$ from these equations we find that $\sin\gamma =$ $= \cos 2\varphi$; hence

$$\gamma=\tfrac{1}{2}\pi-2\varphi\,, \tag{54.11}$$

i.e. the principal stress $\sigma_1 = \sigma_s$ acts in the direction 1, the bisector of the angle between the normal to the line of discontinuity and the velocity vector **v**.

Note that because of (54.10) $\tan^2\varphi \leqslant 1$ i.e. $|\varphi| \leqslant \frac{1}{4}\pi$. In these limits the direction of the line of velocity discontinuity with respect to the tensile stress is arbitrary [1]). The orientation of the vector **v** is given by (54.11). *In the general case slipping and thinning proceed simultaneously.* If the line of discontinuity is normal to the direction of tension ($\varphi = 0$) there is no slip ($\gamma = \frac{1}{2}\pi$), and only the neck is formed.

*The stress components are continuous on a line of velocity discontinuity.*

Moreover, it is easy to see that the line of stress discontinuity is parallel to the direction of the tension, where $\sigma_1^+ = \sigma_s$ on its right, and $\sigma_1^- = -\sigma_s$ on its left (cf. fig. 39). *The velocity field is continuous along a line of stress discontinuity.*

These results (with obvious changes in notation) apply to the other two uniaxial regimes B and E.

### 54.6. *Axisymmetric field*

We consider an axisymmetric stress field in the absence of torsion (i.e. with $\tau_{r\theta} = 0$). The stress components $\sigma_r$, $\sigma_\theta$ are principal. The solution depends on which regime is attained. For definiteness we concentrate on an infinite lamina with a circular hole (fig. 169), which we considered in the preceding section under the von Mises yield criterion.

If an internal pressure acts on the contour of the cut $r = a$, and if the stresses are zero at infinity, then $\sigma_r < 0$, $\sigma_\theta > 0$ near the hole and the solution will be the same as in the case of plane strain.

[1]) This result does not agree with observations (cf. §53.8) and we are again reminded of the tentativeness of the flow model in a singular regime.

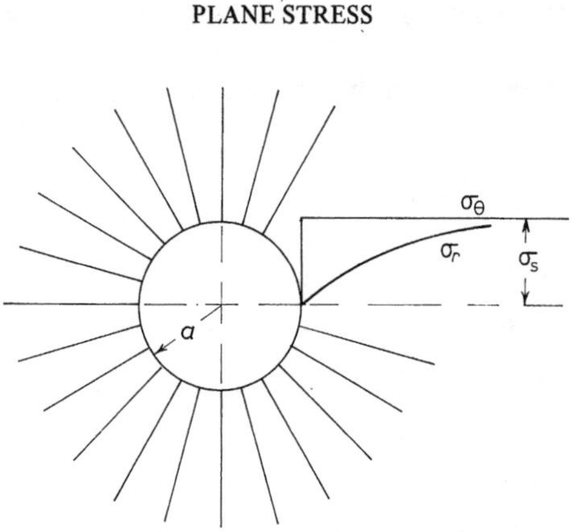


Fig. 169.

If the hole is free and the lamina experiences uniform tension at infinity, then the stresses $\sigma_r$, $\sigma_\theta$ are of the same sign (this follows from the analysis of the elastic problem), and the yield criterion has the form

$$\sigma_\theta = \text{const.} = \sigma_s \, .$$

The boundary condition is

$$\text{when} \quad r = a \qquad \sigma_r = 0 \, .$$

We thus obtain from the differential equation of equilibrium with this boundary condition:

$$\sigma_r = \sigma_s(1 - a/r) \, .$$

The problem is of parabolic type, and the single family of characteristics is a pencil of straight lines originating from the centre (fig. 169). In the same figure we have shown the stress distribution; it does not differ significantly from the stress field with the von Mises yield criterion.

54.7. *Remarks on plastic-rigid boundaries and discontinuities*

A plastic-rigid boundary which adjoins a hyperbolic zone (the regimes DE, AB) traverses a characteristic (in the present case, a slip line). If the line of separation borders a parabolic zone (regimes CD, EF, . . .) it also traverses a characteristic. The proof is similar to that at the end of the preceding section.

The cases considered above do not exhaust the possible discontinuities. In principle any flow regime can be achieved on either side of a line of discontinuity, and it is necessary to consider all possible variations. We shall not do this here; discontinuities with the Saint Venant plasticity condition have been studied in detail in the book [11].

## §55. Elastic-plastic equilibrium of a lamina with a circular hole under uniform pressure

As a simple example we consider the axisymmetric elastic-plastic problem for an infinite lamina with a circular hole. Along the edge of the hole $r = a$ a uniform pressure $p$ is applied (fig. 170).

55.1. *Elastic state*

When the pressure $p$ is small the lamina is in an elastic state and the stresses (in polar coordinates $r$, $\theta$, where $\theta$ is the polar angle) are

$$\sigma_r = -p(a/r)^2 , \qquad \sigma_\theta = +p(a/r)^2 . \tag{55.1}$$

Thus, with elastic deformations the lamina achieves a state of pure shear. The maximum tangential stress and the intensity of tangential stresses are equal to $T = p(a/r)^2$, and plastic deformation first appears at the edge of the hole when the pressure $p = \tau_s = k$. With increasing pressure plastic deformations spread through an annulus $a \leqslant r \leqslant c$, whose radius $c$ has to be determined. It is evident that in the elastic region $r \geqslant c$ the stresses are

$$\sigma_r = -k(c/r)^2 , \qquad \sigma_\theta = k(c/r)^2 . \tag{55.2}$$

55.2. *Elastic-plastic equilibrium with von Mises yield criterion*

The solution to the axisymmetric problem in the plastic zone was obtained in §53.5. On the separation boundary $r = c$ the stresses are continuous and correspond to pure shear, i.e.

$$\text{when} \qquad r = c \qquad \omega = \tfrac{1}{2}\pi .$$

The arbitrary constant in the solution (53.19) is determined from this condition, and we obtain

$$(c/r)^2 = \mathrm{e}^{-\sqrt{3}(\frac{1}{2}\pi - \omega)} \sin \omega . \tag{55.3}$$

Putting $r = a$ here, we find the value $\omega_a$ which corresponds to a given $c$; the pressure along the edge of the hole is determined from equation (53.18) for $\sigma_r$ with $\omega = \omega_a$. It is easy to see that $\omega_a \geqslant \frac{1}{2}\pi$ and increases with $(c/a)$. The pressure $p$ increases, reaching a maximum value $2k$ when $\omega_a = \frac{5}{6}\pi$. Further increase in pressure and expansion of the plastic zone are not possible; at this maximum pressure $p = 2k$ thickening of the lamina at the edge of the hole takes place freely (for small deformations, of course). To show this we neglect elastic strains at the edge of the hole, and then by (52.3) we have

$$\xi_r + \frac{\sigma_\theta - 2\sigma_r}{2\sigma_\theta - \sigma_r}\,\xi_\theta = 0\ .$$

Hence, using the incompressibility condition,

$$\xi_r + \xi_\theta + \xi_z = 0\ ,$$

we find the rate of relative thickening of the plate:

$$\xi_z = -\frac{\sigma_r + \sigma_\theta}{2\sigma_\theta - \sigma_r}\,\xi_\theta \qquad \left(\xi_\theta = \frac{v_r}{r}\right)\ .$$

At the maximum pressure $p = 2k$ the stresses at the edge of the hole $r = a$ are (cf. fig. 157): $\sigma_r = -2k$, $\sigma_\theta = -k$, and, obviously, $\xi_z \to \infty$. From (55.3) with $r = a$, $\omega = \frac{5}{6}\pi$ we obtain the maximum radius of the plastic zone:

$$(c/a)_{\max} \approx 1.75\ .$$

The equations in the plastic zone are hyperbolic; at maximum pressure $p = 2k$ the characteristics of the $\alpha, \beta$ families are mutually tangent ($\psi = 0$) on the circle $r = a$. This corresponds to the parabolic point $\omega = \frac{5}{6}\pi$ in fig. 157.

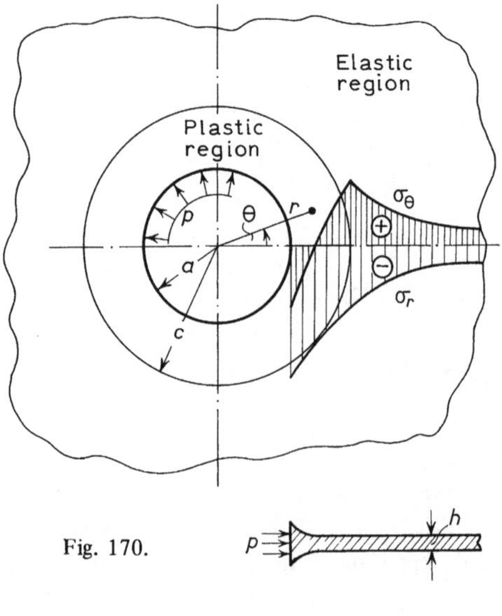


Fig. 170.

On the separation boundary $r = a$ the characteristics are orthogonal. The stress distribution is shown in fig. 170.

### 55.3. *Elastic-plastic equilibrium with Tresca-Saint Venant yield criterion*

In this case the regime DE is attained at the edge of the hole, and the yield criterion has the form

$$\sigma_\theta - \sigma_r = 2k ,$$

where $2k = \sigma_s$. From the differential equation of equilibrium (53.14) and the continuity condition at the circle $r = c$ we obtain

$$\sigma_r = -k(1 + 2 \ln (c/r)) , \qquad \sigma_\theta = k(1 - 2 \ln (c/r)) . \tag{55.4}$$

By virtue of the yield criterion, the radial stress $\sigma_r$ cannot exceed $\sigma_s$ in magnitude (cf. fig. 159); the maximum radius of the plastic zone is thus found from (55.4):

$$(c/a)_{\max} \approx 1.65 .$$

Fig. 171 shows the elastic (dotted line) and elastic-plastic (solid line, $c/a = 1.40$) stress distributions; the same figure shows the residual stresses $\sigma_r^0$, $\sigma_\theta^0$ which arise when the pressure is removed.

Note that the radial residual stress $\sigma_r^0$ is compressible. This fact is widely

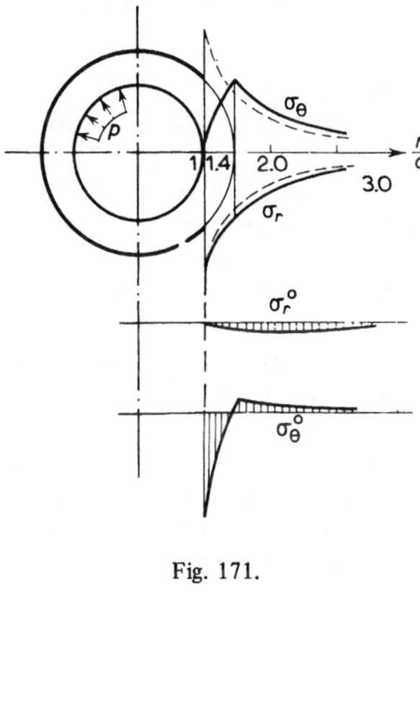


Fig. 171.

used in practical applications for clamping tubes in plane (or curved) metal sheets (plates) by the method of flaring. This process consists of an expansion from within the tube 1 (fig. 172), inserted in the hole in the sheet 2. After the pressure is removed the tube is seen to be firmly fixed in the sheet (cf. [25], ch. 33).

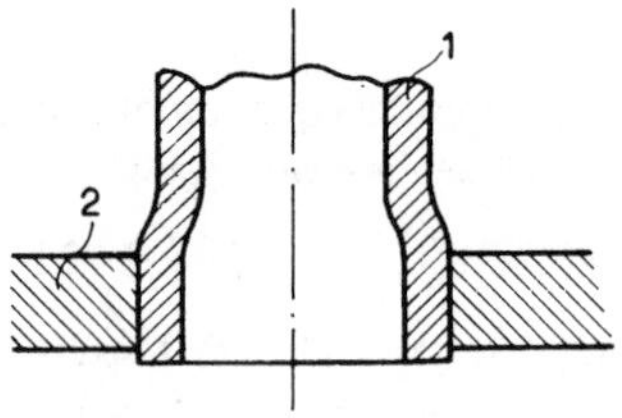


Fig. 172.

## §56. Extension of a strip weakened by notches

We consider two problems of the extension of a strip weakened by notches; we use the von Mises yield criterion. The solutions to these problems were given by Hill [182]; it is assumed that the notches are sufficiently deep.

56.1. *Extension of a strip with circular notches*

Let the strip be weakened by symmetric notches with a circular base of radius $a$ (fig. 173). Near the circular part of the contour an axisymmetric stress field develops (owing to the hyperbolicity of the equations in these regions the stress fields are completely determined by the form of the free boundary). In consequence, the stresses in these zones are given by equations (53.18), where the distance $r$ from the centre O and the function $\omega$ are related through (53.21). It is evident that these fields can be extended on either side to a distance not greater than $r = 2.07a = c_1$; in this limiting case the angle of spread of the circular portion of the arc AB must not be less than $38°56'$.

Let $h < 1.07a$; then the fields spreading out from each notch meet at the centre C (fig. 173a). To effect the construction, the arc AB must subtend a sufficiently large angle (in particular, the construction is always possible if the angle $\geqslant 38°56'$). Using the differential equation of equilibrium (53.19) and

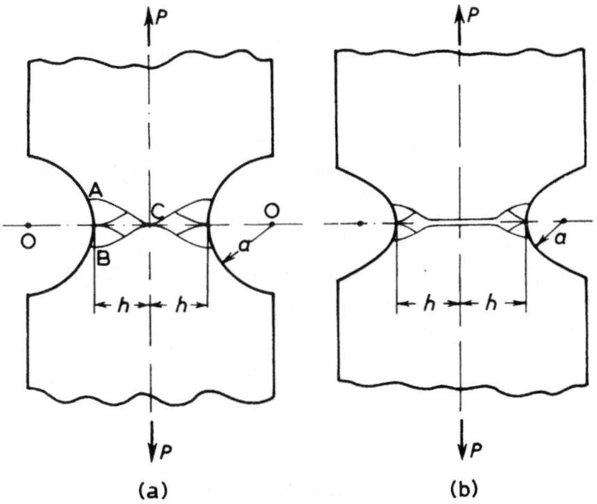


Fig. 173.

integrating by parts, we obtain the limit load

$$P_* = 2 \int_a^{a+h} \sigma_\theta \, dr = 2(a+h)\,(\sigma_r)_{r=a+h} \, . \tag{56.1}$$

The elementary limit load is $P_*^0 = 2\sigma_s h = 2\sqrt{3}kh$. The ratio $P_*/P_*^0$ (*strength coefficient*) can be represented by the approximate formula

$$\frac{P_*}{P_*^0} \approx 1 + 0.23 \frac{h}{h+a} \qquad \left(0 \leqslant \frac{h}{a} \leqslant 1.07\right) . \tag{56.2}$$

When $h = 1.07a$ the characteristics converge at the centre C.

When $h > 1.07a$ the axisymmetric plastic regions do not change, being joined along the $x$-axis by a neck (fig. 173b) which corresponds to the parabolic point on the yield ellipse (fig. 157). Along the neck

$$\sigma_r = k \, , \qquad \sigma_\theta = 2k \, .$$

The limit load is easily calculated with the aid of (56.1):

$$P_* = 2 \int_a^{a+h} \sigma_\theta \, dr = 2k(a + 2h - c_1)$$

and the strength coefficient is

$$P_*/P_*^0 \approx 1.15-0.04\, a/h \qquad (h > 1.07a)\,. \tag{56.3}$$

We conclude with two remarks.

Comparing these results with the corresponding strength coefficient (41.4) in the case of plane strain, we observe that *the strength coefficient is somewhat smaller in the case of plane stress.*

The upper and lower portions of the strip move rigidly in the direction of action of the loads. Along the line of separation traversing the characteristics the velocity is discontinuous and necks develop.

Handy's observations [79] on the extension of a copper strip weakened by circular notches (fig. 174) give a qualitative confirmation of this conclusion. The neck is clearly visible in the stage of deformation preceding fracture.

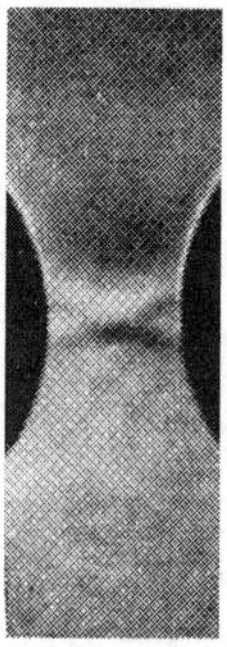

Fig. 174.

56.2. *Extension of a strip with angular notches*

The strip is weakened by symmetric angular notches (fig. 175). The free rectilinear sides of the notches can only be adjoined by regions of uniform uniaxial tension, with $\psi = \psi_1 = 54°44'$. We consider the scheme of solution shown in fig. 175a. Adjacent to the region of uniform tension there is the central field (53.16). By the stress continuity condition the angle $\theta$ should be measured from the polar axis O′O″, in accordance with fig. 163. In the rhomboidal region ABCD there is a uniform stress state, where, by symmetry, the tangential stresses are zero along the line AB. But then $\varphi$ is the angle between the principal direction and the $r$-axis, traversing the line BD; the latter is determined by the angle $\theta = \theta_1$.

Equating the stresses from (53.3) and the stresses given by (53.16) with

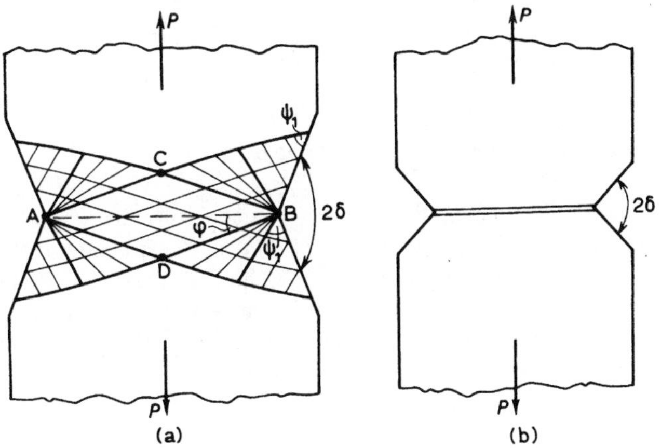


Fig. 175.

$\theta = \theta_1$, we find

$$\tan 2\varphi = 2 \tan \theta_1 \, . \tag{56.4}$$

From the construction shown in fig. 175a it follows that

$$\varphi = \theta_1 - (\delta + 2\psi_1 - \pi) \, .$$

Thus, the angle $\theta_1$ is determined from the equation

$$2 \tan \theta_1 + \tan 2(\delta + 2\psi_1 - \theta_1) = 0 \, . \tag{56.5}$$

When $\theta_1$ is known we can easily find the principal stresses in the region ABCD from (31.4) and (53.16):

$$\sigma_1, \sigma_2 = \tfrac{1}{2}k\left(3 \cos \theta_1 \pm \sqrt{1 + 3 \sin^2 \theta_1}\right) . \tag{56.6}$$

The construction is possible if $\varphi \geqslant 0$. When $\varphi = 0$ it follows from (56.4) that $\theta_1 = 0$. Thus, $\delta \geqslant \pi - 2\psi_1 = 70°32'$. The strength coefficient is now

$$\frac{P_*}{P_*^0} = \frac{1}{2\sqrt{3}}\left(3 \cos \theta_1 + \sqrt{1 + 3 \sin^2 \theta_1}\right) , \qquad (\tfrac{1}{2}\pi - 2\psi_1 \leqslant \delta \leqslant \tfrac{1}{2}\pi) \, . \tag{56.7}$$

When $\theta_1 = 0$ the whole construction degenerates into a line of discontinuity (neck) along AB; the two families of characteristics merge into one (cf. fig. 163). This solution, shown in fig. 175b, remains valid for more acute notches, with $\delta < 70°32'$. Along AB $\sigma_\theta = 2k$, $\sigma_r = k$ and the strength coefficient is now

$$P_*/P_*^0 = 2/\sqrt{3}\,, \qquad \delta < 70°32'\,. \tag{56.8}$$

## §57. Bending of a strip with a one-sided notch

We consider the bending of a strip with a one-sided notch (fig. 176), and restrict ourselves to the case of an angular notch. The structure of the field of characteristics is shown in fig. 176a. Adjoining the lower rectangular boundary is a triangle of uniform uniaxial compression $-\sqrt{3}k$. At the notch the field is the same as in the preceding problem (fig. 175a). By (56.6) the normal stress on the line AB is $\sigma_1$, while the angle $\theta_1$ is found from (56.5).

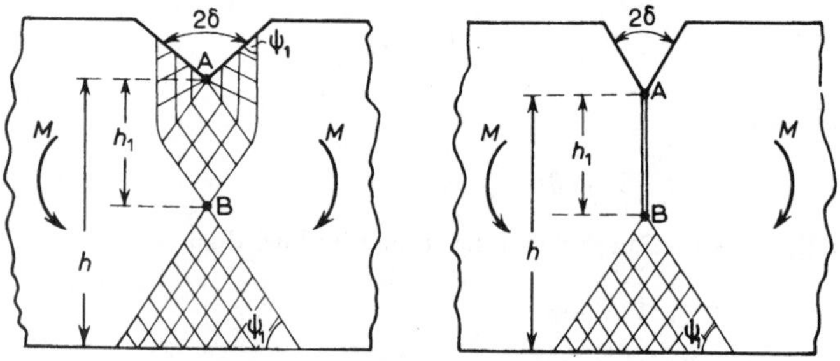


(a) (b)

Fig. 176.

The position of the point B (characterised by the segment $h_1$) is determined from the condition that the principal stress vector acting in the smallest cross-section is equal to zero. That is,

$$\sigma_1 h_1 - \sigma_s(h-h_1) = 0 \qquad (\sigma_s = \sqrt{3}k)\,.$$

The limit bending moment is

$$M_* = \tfrac{1}{2}\sigma_1 h_1^2 + \tfrac{1}{2}\sigma_s(h-h_1)^2\,.$$

For a flat strip of height $h$ the limit moment is $M_*^0 = \frac{1}{4}\delta_s h^2$. The strength coefficient is equal to

$$\frac{M_*}{M_*^0} = \frac{2\sigma_1}{\sigma_1 + \sigma_s}\,. \tag{57.1}$$

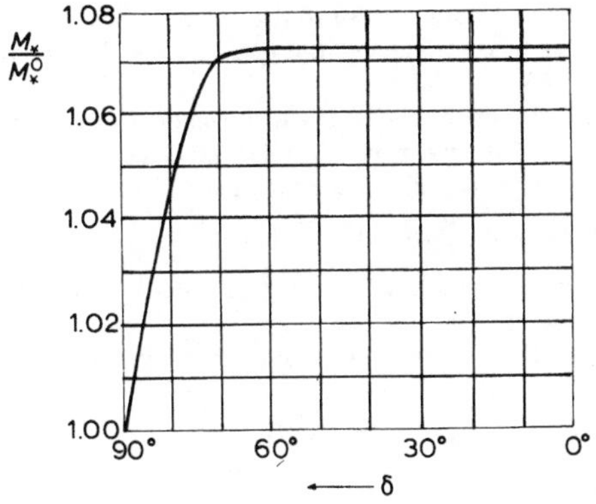


Fig. 177.

As in the preceding section, the solution is valid for $\delta \geqslant \pi - 2\psi_1 = 70°32'$. When $\delta = \pi - 2\psi_1$ the whole construction above the point B degenerates into a line of discontinuity (neck) along AB. For $\delta \leqslant \pi - 2\psi_1$ the solution follows the pattern shown in fig. 176b. The stresses in the neck are $\sigma_1 = 2k$, $\sigma_2 = k$. It is easy to see that

$$\frac{h_1}{h} = \frac{\sqrt{3}}{2 + \sqrt{3}},$$

and the strength coefficient is constant:

$$\frac{M_*}{M_*^0} = \frac{4}{2 + \sqrt{3}} = 1.072\,. \qquad (57.2)$$

The graph of the strength coefficient as a function of the angle of the notch is shown in fig. 177. In the limit state the rigid portions of the strip experience rotation with respect to the point B.

A somewhat more complex problem is that of the bending of a strip with a circular notch (cf. [182]).

## PROBLEMS

1. Consider the elastic-plastic state of a rapidly rotating disc of constant thickness, subject to the Tresca-Saint Venant yield criterion.

At what angular velocity is a pure plastic state attained?

2. Consider (in conditions of plane stress) the bending of a short cantilever by a force (fig. 119) with the first type of velocity field (i.e. for "long" cantilevers). Use the solutions indicated in fig. 163.

3. By examining solutions of the differential equations of equilibrium and the von Mises yield criterion which are independent of the radius vector $r$ show that the function $w = s^2 + (ds/d\theta)^2$, where $s = \sigma_\theta$, satisfies the equation

$$(dw/ds)^2 + 12w - 48k^2 = 0 . \qquad (*)$$

4. Show that the particular integral $w = 4k^2$ of equation (*) leads to Hill's solution with one arbitrary constant.

5. Show that integration of equation (*) leads to a solution with two arbitrary constants $B$, $\omega$:

$$\sigma_\theta = A + B \cos 2(\theta + \omega) , \qquad \sigma_r = A - B \cos 2(\theta + \omega) , \qquad \tau_{r\theta} = B \sin 2(\theta + \omega) ,$$

where $A^2 = 3(k^2 - B^2)$.

6. Consider the problem of the bending of a strip with two angular notches (cf. §57, fig. 176).

7. Consider the problem of the bending of a strip with one circular notch in the case where the compression field (fig. 176a) joins the axisymmetric field at some point B.

8. Find the limit load for the bending of a cantilever by a pressure uniformly distributed along its upper boundary (fig. 119).

9. The same for a beam on two supports.

# 7

# Axisymmetric Strain

## §58. Equations of axisymmetric strain with von Mises yield criterion

58.1. *General remarks*

Axisymmetric problems in plasticity theory are of great significance from the viewpoint of applications. Consider a body of revolution whose axis coincides with the axis of a cylindrical coordinate system $r, \varphi, z$; suppose also that prescribed loads (or displacements) have axial symmetry. Then the components of stress and displacement are independent of the polar angle $\varphi$.

We shall not consider the problem of torsion of the body; the twisting of a circular rod of variable diameter is considered briefly in ch. 8 (§66). We can then assume that the azimuthal velocity component is

$$v_\varphi = 0$$

and the stress components are

$$\tau_{r\varphi} = \tau_{z\varphi} = 0 \, .$$

The non-zero stress components $\sigma_r$, $\sigma_\varphi$, $\sigma_z$, $\tau_{rz}$ (fig. 178) and the velocity components $v_r$, $v_z$ (in the rest of this chapter these will be written $u$, $w$ respectively) are functions of the coordinates $r$, $z$.

In the absence of body forces the differential equations of equilibrium

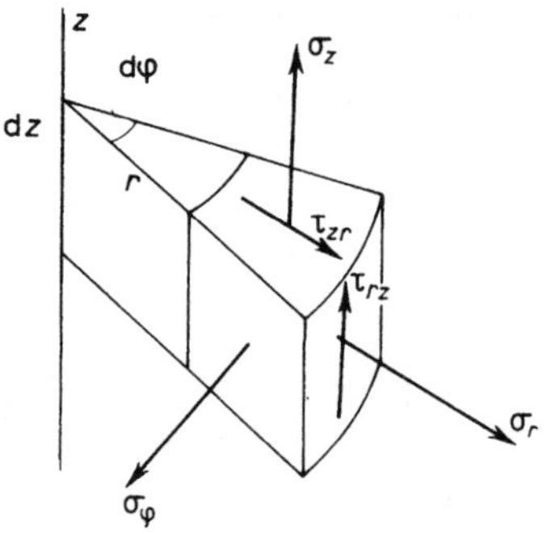


Fig. 178.

have the form

$$\frac{\partial \sigma_r}{\partial r}+\frac{\partial \tau_{rz}}{\partial z}+\frac{\sigma_r-\sigma_\varphi}{r}=0\,,$$
$$\frac{\partial \tau_{rz}}{\partial r}+\frac{\tau_{rz}}{r}+\frac{\partial \sigma_z}{\partial z}=0\,. \tag{58.1}$$

The strain-rate components are determined by the expressions

$$\xi_r=\frac{\partial u}{\partial r}\,,\qquad \xi_\varphi=\frac{u}{r}\,,\qquad \xi_z=\frac{\partial w}{\partial z}\,,\qquad \eta_{rz}=\frac{\partial u}{\partial z}+\frac{\partial w}{\partial r}\,. \tag{58.2}$$

To these must be added the equations of the plastic state. Although this is easy to do, solutions are very difficult, and we shall therefore limit ourselves to an analysis of the equations for a plastic-rigid body.

"One-dimensional" axisymmetric problems, in which the stress-strain distribution depends on one independent variable (the radius $r$) only, are relatively simple, although even these sometimes require the application of numerical methods. Problems of this type have been touched on earlier (hollow sphere and cylindrical tube under pressure, axisymmetric equilibrium of a thin plate, etc.). In these problems the elastic strains, the hardening and other mechanical properties can be calculated.

The analysis of the general axisymmetrical problem, even for a plastic-rigid body, involves substantial mathematical difficulties. This is an inducement to search for various possibilities of simplification in the formulation of problems.

58.2. *Saint Venant-von Mises relations*

For a plastic-rigid body it is necessary to begin with the Saint Venant-von Mises relations (13.12). In our axisymmetric case these take the form

$$\frac{\xi_r}{H}=\frac{\sigma_r-\sigma}{2\tau_s}, \qquad \frac{\xi_\varphi}{H}=\frac{\sigma_\varphi-\sigma}{2\tau_s}, \qquad \frac{\xi_z}{H}=\frac{\sigma_z-\sigma}{2\tau_s}, \qquad \frac{\eta_{rz}}{H}=\frac{\tau_{rz}}{\tau_s}, \tag{58.3}$$

where the intensity of shear strain-rate is

$$H=\sqrt{\tfrac{2}{3}}\sqrt{(\xi_r-\xi_\varphi)^2+(\xi_\varphi-\xi_z)^2+(\xi_z-\xi_r)^2+\tfrac{3}{2}\eta_z^2}\,. \tag{54.4}$$

The stress components given by equations (58.3) satisfy the von Mises yield criterion

$$(\sigma_r-\sigma_\varphi)^2+(\sigma_\varphi-\sigma_z)^2+(\sigma_z-\sigma_r)^2+6\tau_{rz}^2=6\tau_s^2\,. \tag{58.5}$$

The strain-rate components satisfy the incompressibility equation

$$\frac{\partial u}{\partial r}+\frac{u}{r}+\frac{\partial w}{\partial r}=0\,. \tag{58.6}$$

These relations, together with the differential equations of equilibrium, constitute a system of six equations in six unknown functions $\sigma_r$, $\sigma_\varphi$, $\sigma_z$, $\tau_{rz}$, $u$, $w$. In the general case this system is elliptic [50].

It should be noted that there are only three equations (58.1), (58.5) for the determination of the four stress components $\sigma_r$, $\sigma_\varphi$, $\sigma_z$, $\tau_{rz}$. In contrast with the plane strain and plane stress cases the axisymmetric problem is not locally statically determinate; for this reason a separate analysis of the stress and velocity field is not feasible.

A number of exact particular solutions have been obtained by semi-inverse methods. In this regard we may note the problem of flow of a plastic mass in a circular cone (Sokolovskii [44]), the problem of the compression of a cylinder subject to forces distributed on its ends in a prescribed way (Hill [54]), and the problem of extrusion from a contracting cylindrical container ([54]). In addition there are approximate solutions for a number of important practical problems, based on various supplementary hypotheses.

58.3. *Condition of total plasticity*

One of the methods which to some extent overcomes the difficulties, and which is extensively used in engineering practice, is due to Hencky [94].

Hencky put forward the hypothesis that in axisymmetric stress a regime of so-called *total plasticity* is achieved, in which *two of the principal stresses are equal.*

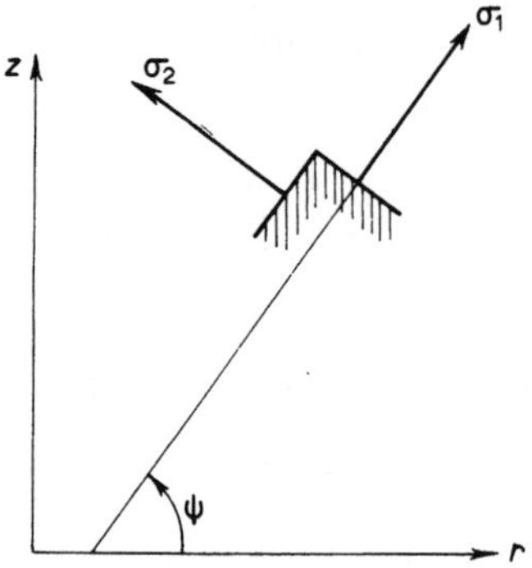


Fig. 179.

The stress $\sigma_\varphi$ is obviously principal. Let $\sigma_1, \sigma_2$ $(\sigma_1 \geqslant \sigma_2)$ denote the principal stresses in the $r, z$-plane (fig. 179), and $\psi$ the angle between the first principal direction and the $r$-axis. Then

$$\sigma_r, \sigma_z = p \pm q \cos 2\psi\,, \qquad \tau_{rz} = q \sin 2\psi\,, \tag{58.7}$$

where we have put

$$p = \tfrac{1}{2}(\sigma_1 + \sigma_2)\,, \qquad q = \tfrac{1}{2}(\sigma_1 - \sigma_2)\,.$$

The condition $\sigma_1 = \sigma_2$ leads to a very special case (cf. §59, regime A). Therefore we assume that in the regime of total plasticity $\sigma_\varphi = \sigma_1$ (or $\sigma_2$). Imposing on this state a hydrostatic pressure $\sigma = -\sigma_1$, we arrive at the stress state

$$0\,, \qquad 0\,, \qquad \sigma_2 - \sigma_1\,,$$

which corresponds to a *uniaxial tension or compression.* Thus, for total plasticity the stress state lies on one of the edges of the Tresca-Saint Venant yield prism (fig. 14); it is easy to see that $\tau_{max} = \frac{1}{2}|\sigma_1 - \sigma_2|$ here, and $T = (2/\sqrt{3})\,\tau_{max}$.

The condition of total plasticity greatly simplifies the system of equations for the stresses and leads to a locally *statically determinate problem for the stress state.*

In fact we now have four equations for the four stress components: the differential equations of equilibrium (58.1), the yield criterion (58.5) and the requirement $\sigma_1 = \sigma_\varphi$. This system of equations for the stresses is of hyperbolic type; its characteristics coincide with the slip lines. A detailed analysis of this system will be given in the next section (regime B). With the aid of methods similar to those used in the case of plane strain it is now possible to consider a wide range of particular problems.

It does not seem possible to carry out an analysis of the velocity field on the basis of the von Mises equations, since the system of equations turns out to be overdetermined. However, this difficulty can be overcome by passing to the Tresca-Saint Venant yield criterion and the associated flow law (see below, §59). Even so it is not possible to construct a solution of the axisymmetric problem in the general case on the basis of the total plasticity condition alone, although this condition can be useful in individual particular problems (cf. below, §§60, 61). Thus, for a solid body $\sigma_r = \sigma_\varphi$ on the axis of symmetry $Oz$, and sometimes this relation is approximately true throughout the whole body.

It is evident that in a number of cases the total plasticity condition leads to an acceptable approximation for the limit load.

A.Yu. Ishlinskii [113] examined the problem of indentation of a rigid sphere in a plastic medium, under the total plasticity hypothesis. This problem is particularly interesting because of its connection with Brinel's well-known method of testing the hardness of materials.

The total plasticity condition is widely used in engineering calculations of the working of metals under pressure (forging, stamping, pressing, cf. [10, 51]).

## §59. Equations of axisymmetric strain with Tresca-Saint Venant yield criterion

59.1. *Basic relations*

The mathematical formulation of the problem is greatly simplified by using the Tresca-Saint Venant plasticity condition and the associated flow law. The advantages are especially great in problems where the principal directions are known.

The yield criterion now has the form

$$\tau_{\max} = \text{const.} = k \qquad (2k = \sigma_s)\,, \tag{59.1}$$

with $2\tau_{\max} = \max(|\sigma_1 - \sigma_2|,\ |\sigma_2 - \sigma_\varphi|,\ |\sigma_\varphi - \sigma_1|)$. In the space of principal stresses this condition defines the surface of a right hexahedral prism (fig. 14). By the associated flow law the direction of the strain-rate vector is along the normal to the yield surface; along the edges of the prism the flow is undetermined (cf. §16).

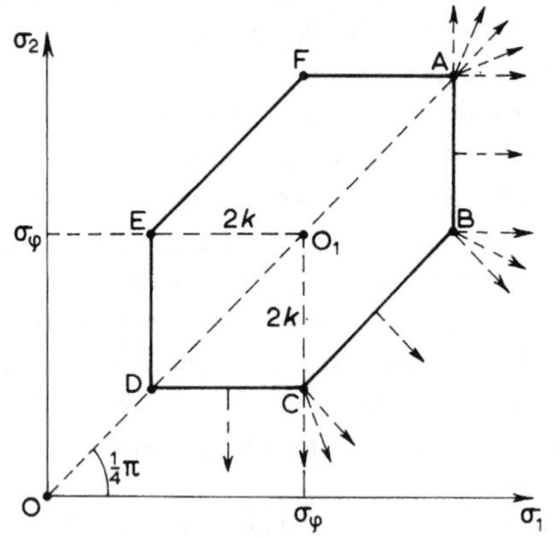


Fig. 180.

From the incompressibility condition we have

$$\xi_r + \xi_\varphi + \xi_z = 0\ . \tag{59.2}$$

The azimuthal stress $\sigma_\varphi$ is principal ($\sigma_\varphi = \sigma_3$). Consider the intercept of the Tresca-Saint Venant prism on the plane $\sigma_\varphi = \text{const}$. This will be the hexagon shown in fig. 180; the coordinates of its centre $O_1$ are $(\sigma_\varphi, \sigma_\varphi)$.

The yield criterion and the principal strain rates in different regimes are shown in the table, where $\lambda_1'$, $\lambda_2'$ are arbitrary, non-negative scalar functions (different for each regime).

We now turn to a detailed analysis of the flow in the different regimes.

59.2. *Regime* A

Here $\sigma_1 = \sigma_2$ and this, as we noted above, results in a very special stress state. In particular $q = 0$ and from (58.7) we have $\tau_{rz} = 0$, and $\sigma_r = \sigma_z = p$. Integrating the differential equations of equilibrium (58.1), we now obtain

$$\sigma_r = 2k \ln (c/r)\ , \qquad \sigma_\varphi = \sigma_r - 2k\ , \tag{59.3}$$

where $c > 0$ is an arbitrary constant. By the flow law $\xi_\varphi \leqslant 0$, and consequently $u \leqslant 0$. There are no tangential stresses in the $r$, $z$-plane, and therefore the projections of the velocities $u$, $w$ are continuous functions of $r$, $z$ (since a discontinuity in a tangential velocity component can only occur on a slip surface, where $\tau_{max} = k$).

Regime D is similar to regime A.

**Yield criterion and strain rates**

| Regime | Yield criterion | Principal strain rates | | |
|---|---|---|---|---|
| | | $\xi_1$ | $\xi_2$ | $\xi_\varphi$ |
| A | $\sigma_1 = \sigma_2 = \sigma_\varphi + 2k$ | $\lambda_1'$ | $\lambda_2'$ | $-\lambda_1' - \lambda_2'$ |
| D | $\sigma_1 = \sigma_2 = \sigma_\varphi - 2k$ | $-\lambda_1'$ | $-\lambda_2'$ | $\lambda_1' + \lambda_2'$ |
| AB | $\sigma_1 - \sigma_\varphi = 2k\ (\sigma_1 > \sigma_2 > \sigma_\varphi)$ | $\lambda_1'$ | 0 | $-\lambda_1'$ |
| CD | $\sigma_2 - \sigma_\varphi = -2k\ (\sigma_\varphi > \sigma_1 > \sigma_2)$ | 0 | $-\lambda_1'$ | $\lambda_1'$ |
| BC | $\sigma_1 - \sigma_2 = 2k\ (\sigma_1 > \sigma_\varphi > \sigma_2)$ | $\lambda_2'$ | $-\lambda_2'$ | 0 |
| B | $\sigma_2 = \sigma_\varphi = \sigma_1 - 2k$ | $\lambda_2' + \lambda_2'$ | $-\lambda_2'$ | $-\lambda_1'$ |
| C | $\sigma_1 = \sigma_\varphi = \sigma_2 + 2k$ | $\lambda_2'$ | $-\lambda_1' - \lambda_2'$ | $\lambda_1'$ |

59.3. *Regime* AB

In this case $\xi_\varphi \leqslant 0$, and hence $u \leqslant 0$. Also, we have

$$\xi_2 = 0\,. \tag{59.4}$$

Since the principal strain rates are

$$\xi_1, \xi_2 = \tfrac{1}{2}(\xi_r + \xi_z) \pm \tfrac{1}{2}\sqrt{(\xi_r - \xi_z)^2 + \eta_{rz}^2}\,,$$

the condition (59.4) leads to the differential equation

$$4\frac{\partial u}{\partial r}\frac{\partial w}{\partial z} = \left(\frac{\partial u}{\partial z} + \frac{\partial w}{\partial r}\right)^2. \tag{59.5}$$

Combining this with the incompressibility condition

$$\frac{\partial u}{\partial r} + \frac{u}{r} + \frac{\partial w}{\partial z} = 0\,, \tag{59.6}$$

we arrive at a system of two equations for the velocities $u$, $w$, with $u \leqslant 0$. If the velocity field is known, the stresses can be found from the equilibrium equations (58.1), the yield criterion

$$\sigma_1 - \sigma_\varphi = 2k \tag{59.7}$$

and the condition that the stresses and strain rates are coaxial. Thus, this case is locally *kinematically determinate.*

A similar flow takes place in the regime CD (but $u \geqslant 0$).

We return to the analysis of equations (59.5), (59.6) for the velocities. We suppose $u \neq 0$, since when $u = 0$ we have the elementary solution $w = \text{const.}$

The incompressibility condition (59.6) can be written in the form

$$\frac{\partial}{\partial r}(ru) + \frac{\partial}{\partial z}(rw) = 0 .$$

It is evident that we can introduce a velocity potential:

$$u = \frac{1}{r}\frac{\partial V}{\partial z}, \qquad w = -\frac{1}{r}\frac{\partial V}{\partial r}, \tag{59.8}$$

whereupon (59.5) leads to the differential equation

$$\left(\frac{\partial^2 V}{\partial r^2} - \frac{\partial^2 V}{\partial z^2} - \frac{1}{r}\frac{\partial V}{\partial r}\right)^2 + 4\frac{\partial^2 V}{\partial r \partial z}\left(\frac{\partial^2 V}{\partial r \partial z} - \frac{1}{r}\frac{\partial V}{\partial z}\right) = 0 . \tag{59.9}$$

To determine the type of this equation we shall ascertain whether it is possible to construct a solution near some line L if the values of the velocities $u$, $w$ are prescribed on the line (i.e. the derivatives $\partial V/\partial r$, $\partial V/\partial z$ are prescribed). Along the line L we have

$$\mathrm{d}\left(\frac{\partial V}{\partial r}\right) = \frac{\partial^2 V}{\partial r^2}\,\mathrm{d}r + \frac{\partial^2 V}{\partial r \partial z}\,\mathrm{d}z , \qquad \mathrm{d}\left(\frac{\partial V}{\partial z}\right) = \frac{\partial^2 V}{\partial r \partial z}\,\mathrm{d}r + \frac{\partial^2 V}{\partial z^2}\,\mathrm{d}z .$$

From these equations and (59.9) it is possible to calculate the second derivatives of the potential $V$ along L. Next, we write down the increments in the second derivatives along L and use (59.9); this immediately shows that third-order derivatives cannot be determined uniquely if the curve L is such that either of the following two conditions holds along it:

$$\frac{1}{2}\left(\frac{\partial^2 V}{\partial z^2} - \frac{\partial^2 V}{\partial r^2} - \frac{1}{r}\frac{\partial V}{\partial r}\right)\frac{\mathrm{d}z}{\mathrm{d}r} = \text{either } -\frac{\partial^2 V}{\partial r \partial z} + \frac{1}{r}\frac{\partial V}{\partial z}, \qquad \text{or } -\frac{\partial^2 V}{\partial r \partial z} .$$

With the aid of (59.8) and the well-known formulae

$$\begin{aligned} \xi_r, \xi_z &= \tfrac{1}{2}(\xi_1 + \xi_2) \pm \tfrac{1}{2}(\xi_1 - \xi_2)\cos 2\psi , \\ \eta_{rz} &= (\xi_1 - \xi_2)\sin 2\psi , \end{aligned} \tag{59.10}$$

where $\psi$ is the angle between the first principal direction of the strain rates

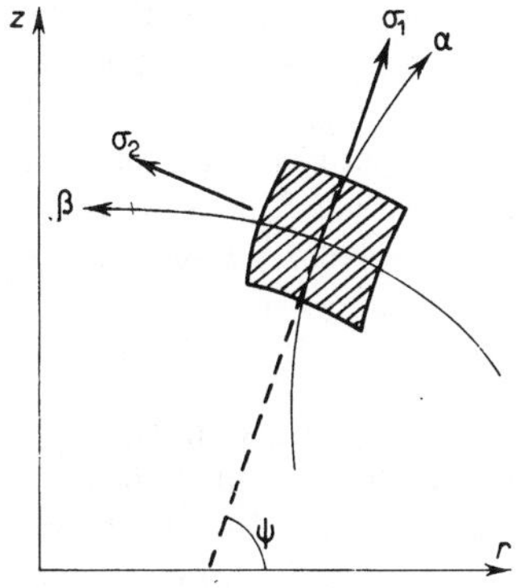


Fig. 181.

and the $r$-axis (fig. 181), we can write the above conditions in the form

$$\text{either} \quad \frac{dz}{dr} = \tan\psi\,, \qquad (\alpha\text{-lines})\,,$$
$$\text{or} \quad \frac{dz}{dr} = -\cot\psi\,, \qquad (\beta\text{-lines})\,. \tag{59.11}$$

Thus, equation (59.9) has two distinct families of characteristics; the latter are *orthogonal and coincide with the trajectories of the principal directions of the strain rates.* The relations along the characteristics have a rather complicated structure, and we shall omit their derivation here (cf. [177]). The solutions to various boundary value problems for the velocity potential $V$ can be obtained by the finite-difference method of Masso.

We now turn to the system of equations for the stresses, namely, the differential equations of equilibrium (58.1) and the yield criterion (59.7). The latter can also be written in the form $\sigma_\varphi = p + q - 2k$. Substituting this value of $\sigma_\varphi$, and the stresses $\sigma_r$, $\sigma_z$, $\tau_{rz}$ from (58.7), into the differential equations of equilibrium, we arrive at the system of equations

$$\frac{\partial p}{\partial r} + \frac{\partial q}{\partial r}\cos 2\psi + \frac{\partial q}{\partial z}\sin 2\psi =$$
$$= 2q\left(\sin 2\psi \frac{\partial \psi}{\partial r} - \cos 2\psi \frac{\partial \psi}{\partial z}\right) - \frac{1}{r}(2k - q + q\cos 2\psi)\,,$$
$$\frac{\partial p}{\partial z} + \frac{\partial q}{\partial r}\sin 2\psi - \frac{\partial q}{\partial z}\cos 2\psi = -2q\left(\cos 2\psi \frac{\partial \psi}{\partial r} + \sin 2\psi \frac{\partial \psi}{\partial z}\right) - \frac{1}{r} q \sin 2\psi\,. \tag{59.12}$$

Assuming the angle $\psi$ to be known from the solution of the kinematic problem, we have here a system of two equations for the unknown functions $p$, $q$. It is easy to show (for example, by the usual "determinantal" method) that the system (59.12) is hyperbolic, with the same characteristics (59.11). Thus, the equations for stresses and for velocities have the same characteristics – the trajectories of the principal stresses in the $r$, $z$-plane.

59.4. *Regime* BC

Here $\sigma_\varphi$ is the intermediate principal stress, so that $\xi_\varphi = 0$, i.e. $u = 0$. It follows from the incompressibility equation that $w$ depends only on $r : w = w(r)$. Furthermore, we find that $\xi_r = \xi_z = 0$, and $\eta_{rz} = w'(r)$. In consequence the coordinate directions $r$, $z$ coincide with the directions of the surfaces of maximum shear rate. On these surfaces the stresses $\sigma_r = \sigma_z$ and $\tau_{rz} = \pm k$ act. Integrating the second of the equilibrium equations, we obtain

$$\sigma_r = \sigma_z = \mp k\frac{z}{r} + R(r) ,$$

where $R(r)$ is an arbitrary function. It now follows from the first of the equilibrium equations that $\sigma_\varphi = (\mathrm{d}/\mathrm{d}r)\, rR(r)$.

This regime is connected with the trivial velocity field ($u = 0$, $w = w(r)$).

59.5. *Regime* B

corresponds to the state of "total plasticity", where $\sigma_\varphi$ is equal to one of the principal stresses; in the present case $\sigma_\varphi = \sigma_2$. We have already remarked in the preceding section that there are now four equations for the determination of the four unknown stress components, i.e. *the problem is statically determinate.*

The regime C is similar to B, but now the principal stress $\sigma_1$ is algebraically the largest, and $\sigma_\varphi = \sigma_1$. Since the systems of equations for the singular regimes B and C are analogous, it is sufficient to consider one of them.

For definiteness we shall concentrate on regime C. A corresponding system of equations for the stress state was investigated by Hencky. This system consists of the differential equations of equilibrium (58.1), the plasticity condition $\sigma_1 - \sigma_2 = 2k$, and the equality $\sigma_\varphi = \sigma_1$. The $\alpha, \beta$ slip lines (the trajectories $\tau_{\max}$) in the $r$, $z$-plane are inclined at angles $\mp\frac{1}{4}\pi$ to the principal directions (fig. 182). A normal stress $p = \frac{1}{2}(\sigma_1 + \sigma_2)$ acts on the slip surfaces; note that $p$ is different from the mean pressure $\sigma$.

It is evident that $\sigma_\varphi = p + k$ and that the angle of inclination of the slip sur-

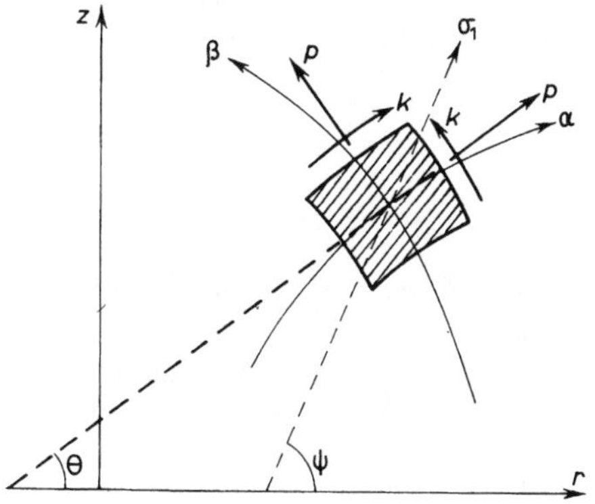


Fig. 182.

face is $\theta = \psi - \frac{1}{4}\pi$. Substituting these into (58.7), we find

$$\sigma_r, \sigma_z = p \mp k \sin 2\theta\ , \qquad \tau_{rz} = k \cos 2\theta\ . \tag{59.13}$$

Introducing the stresses into the equilibrium equations (58.1), we obtain a system of differential equations for the unknown functions $p, \theta$:

$$\frac{\partial p}{\partial r} - 2k\left(\cos 2\theta \frac{\partial \theta}{\partial r} + \sin 2\theta \frac{\partial \theta}{\partial z}\right) = \frac{k}{r}(1 + \sin 2\theta)\ ,$$
$$\frac{\partial p}{\partial z} - 2k\left(\sin 2\theta \frac{\partial \theta}{\partial r} - \cos 2\theta \frac{\partial \theta}{\partial z}\right) = -\frac{k}{r}\cos 2\theta\ . \tag{59.14}$$

We now show, using the "determinantal" method, that *this system of equations is hyperbolic.* Suppose the values of the functions $p$, $\theta$ are given along some line L in the $r$, $z$-plane. For the integral surface passing through L we have

$$\mathrm{d}p = \frac{\partial p}{\partial r}\,\mathrm{d}r + \frac{\partial p}{\partial z}\,\mathrm{d}z\ , \qquad \mathrm{d}\theta = \frac{\partial \theta}{\partial r}\,\mathrm{d}r + \frac{\partial \theta}{\partial z}\,\mathrm{d}z\ , \tag{59.15}$$

where the coefficients in equations (59.14), (59.15) are known on L; these equations constitute a system of four linear inhomogeneous algebraic equations for the partial derivatives.

If L is a characteristic, then these derivatives are indeterminate along it. Then the determinant of the algebraic system and the relevant numerators are zero.

Equating to zero the determinant of the system, we obtain the differential equations of the characteristic lines:

$$\frac{dz}{dr} = \tan\theta \qquad (\alpha\text{-lines}),$$

$$\frac{dz}{dr} = -\cot\theta \qquad (\beta\text{-lines}). \tag{59.16}$$

Thus, *there exist two families of orthogonal characteristic lines which coincide with the slip lines.*

If we set the numerators equal to zero (in Cramer's formulae), we obtain the *relations along the characteristic lines*:

$$d\left(\frac{p}{k} - 2\theta\right) - (\sin\theta + \cos\theta)\frac{ds_\alpha}{r} = 0 \qquad (\text{on } \alpha\text{-lines}),$$

$$d\left(\frac{p}{k} + 2\theta\right) + (\sin\theta + \cos\theta)\frac{ds_\beta}{r} = 0 \qquad (\text{on } \beta\text{-lines}), \tag{59.17}$$

where $ds_\alpha$, $ds_\beta$ are elements of length along the $\alpha$- and $\beta$-lines. These relations express the conditions of plastic equilibrium of an element of the slip grid (fig. 182).

Thus, the construction of a solution for a given regime reduces to the investigation of a series of boundary value problems (Cauchy, initial characteristic, etc.). The solution can be obtained by the finite-difference method of Masso (as in the case of plane strain, ch. 5). It is necessary to take into account possible discontinuities in the stress field.

We turn now to the determination of the velocities $u$, $w$; these can be found from the incompressibility equation (59.6) and the coaxial condition:

$$\sin 2\theta\left(\frac{\partial u}{\partial z} + \frac{\partial w}{\partial r}\right) + \cos 2\theta\left(\frac{\partial u}{\partial r} - \frac{\partial w}{\partial z}\right) = 0. \tag{59.18}$$

It is necessary here to satisfy the restrictions $\xi_2 \leqslant 0$, $\xi_\varphi \geqslant 0$; from the second inequality it follows that $u \geqslant 0$.

It is easy to verify that *the system of equations for the velocities is also hyperbolic*; its characteristics are the slip lines (59.16), as before. Along the characteristic lines the equations for the velocities take the form

$$\cos\theta\, du + \sin\theta\, dw + \frac{u}{2r} ds_\alpha = 0 \qquad \text{on } \alpha\text{-lines},$$

$$\sin\theta\, du - \cos\theta\, dw - \frac{u}{2r} ds_\beta = 0 \qquad \text{on } \beta\text{-lines}. \tag{59.19}$$

59.6. *Concluding remarks*

Analysis of the flows in the regimes DE, E, EF, F, FA leads to a repetition of the preceding results, with obvious changes.

In general, the boundaries of the zones which correspond to the various regimes are unknown a priori. This considerably complicates the solution. The construction of the solutions requires an accurate analysis of the distribution of the zones with the various regimes, as well as the fulfillment of all the imposed restrictions and compatibility conditions; a correct use of discontinuous solutions is of great value. In this regard a number of incorrect solutions which have been published can serve as a useful warning (cf. [150]).

The works [11, 132, 177] present more detailed analysis of certain questions in the general theory, as well as solutions of particular problems. Additional references to the literature will also be found in those works.

## §60. Stress in a thin plastic layer with tension (compression)

60.1. *General remarks*

A series of practical problems has stimulated the investigation of the stress state in a thin plastic layer fastened to rigid sides. We shall give some examples.

One of these is the soldered joint; it is well-known that a soldered joint is significantly stronger than a bar of the same cross-section and made of the same material. It is sometimes convenient to have welded joints in which the metal of the joint is considerably softer than that of the rigid portions. The stress state in thin layers has the feature that it can shed light on certain instances of rupture which would not be anticipated at first glance.

An important quantity in estimating the strength of a metal is its resistance to fracture. Measurement of this property on standard specimens is difficult for plastic metals; experiments are usually conducted at very low temperatures (to lower the plasticity). If, however, a composite specimen (fig. 183), made up of two solid cylindrical portions 1 (of diameter $2a$) joined by a thin ($h \ll a$) layer of softer metal 2, is subject to tension, then at some value of the load $P$ brittle (for small deformations) rupture of the layer takes place along some uneven "plane" n-n. Knowledge of the stress state in the layer enables the determination of the amount of resistance to fracture.

We note certain special features in the extension of these specimens. For steels the differences in the coefficients of elasticity are small, and in the sequel we shall assume them to be equal. Then for extension within the elastic limits the specimen is seen to be in a state of uniform uniaxial tension. When the yield limit of the material of the disk is attained, the latter immediately and completely goes over to the plastic state. As plastic deformations

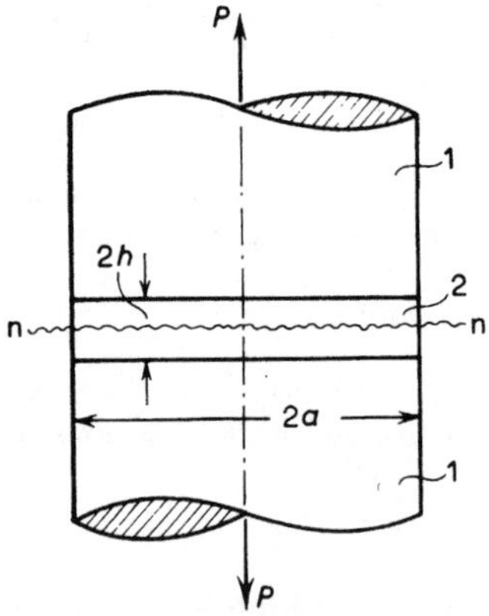


Fig. 183.

develop, the stress state of the disk diverges increasingly from uniform tension and assumes a complex spatial character, since the "rigid" portions of the specimen, which remain elastic, inhibit the deformation of the disk. On the surfaces of contact of the layer with the rigid surroundings tangential stresses develop. The greatest value of the latter is determined by the yield limit $\tau_s$.

The case of plane strain, when the tangential stress on the contact surface has everywhere reached the yield limit, has been studied by Prandtl (cf. §47). Prandtl's solution is relevant to the final stage of plastic flow.

In conditions of axial symmetry an analogous approximate solution can be obtained for the final stage, by modifying somewhat Siebel's result for compression of a cylinder. We shall also analyse below the process of development of stresses in a layer with increase in the load.

60.2. *Final stage of flow*

Basing ourselves on the stress pattern given by Prandtl's solution, we assume that the tangential stresses attain the yield limit $\tau_s$ on the contact surface, while in the greater part of the lamina the normal stresses are substantially larger than the tangential in magnitude (in other words, the stress state is close to hydrostatic), and are approximately constant through the thickness of the layer. Further, we note that at the centre ($r = 0$) $\sigma_r = \sigma_\varphi$, and we assume this relation to hold throughout the layer. We introduce non-dimensional coordinates $\rho = r/a$, $\zeta = z/a$, where $z$ is measured from the median plane of the disk, and non-dimensional stresses $\sigma_r \sim \sigma_r/\sigma_s$, $\sigma_\varphi \sim \sigma_\varphi/\sigma_s$, $\sigma_z \sim \sigma_z/\sigma_s$,

$\tau \sim \tau_{rz}/\tau_s$ where $\sigma_s = \sqrt{3}\tau_s$. Then in the case of tension the von Mises plasticity condition takes the form

$$\sigma_z - \sigma_r = \sqrt{1-\tau^2}\,. \tag{60.1}$$

Integrating the first of the equilibrium equations (58.1) with respect to $z$, and taking into account the fact that $\tau_{rz} = \pm\,\tau_s$ when $z = \pm\,h$, we obtain

$$\frac{d\sigma_r}{d\rho} + \frac{1}{\sqrt{3}\kappa} = 0\,, \qquad \left(\kappa = \frac{h}{a} \ll 1\right) \tag{60.2}$$

On the contour $\rho = 1$ we have $\sigma_r = 0$; the solution of (60.2) satisfying this condition has the form

$$\sigma_r = \frac{1}{\sqrt{3}\kappa}(1-\rho)\,. \tag{60.3}$$

The component $\sigma_z$ is determined from the yield criterion (60.1); since the right-hand side of this condition is zero on the contact surfaces, and since the layer is thin, it can be assumed that

$$\sigma_z \approx \sigma_r \tag{60.4}$$

approximately, everywhere in the layer.

This solution breaks down near the contour $\rho = 1$ and near the axis (as does Prandtl's solution). Since $\kappa \ll 1$, high normal stresses, significantly in excess of the yield limit, arise in a thin layer.

The mean stress across a transverse section of the layer is

$$\bar{p} = \frac{1}{\pi a^2}\int_0^{2\pi a}\int_0 \sigma_z r dr\, d\varphi = \frac{\sigma_s}{3\sqrt{3}\kappa}\,. \tag{60.5}$$

60.3. *Development of the stress state*

The layer joins sufficiently rigid regions which have approximately the same modulus of elasticity as the layer, but a substantially higher yield limit. For moderate force $P$ the whole unit experiences the elastic deformation of uniaxial tension:

$$\sigma_r = \sigma_\varphi = \tau_{rz} = 0\,, \qquad \sigma_z = \frac{P}{\pi a^2} \equiv p\,.$$

When the load $p = \sigma_s$, plastic flow immediately begins in the whole of the layer; this flow, however, is restrained by the rigid portions, and as a consequence tangential stresses arise on the contact surfaces. We shall assume that the contact surfaces remain plane; due to the "softness" of the layer this is an acceptable approximation.

In terms of the above dimensionless variables the differential equations of equilibrium and the von Mises yield criterion have the form

$$\frac{\partial \sigma_r}{\partial \rho} + \frac{\sigma_r - \sigma_\varphi}{\rho} + \frac{1}{\sqrt{3}} \frac{\partial \tau}{\partial \zeta} = 0 \,, \tag{60.6}$$

$$\frac{\partial \tau}{\partial \rho} + \frac{\tau}{\rho} + \sqrt{3} \frac{\partial \sigma_z}{\partial \zeta} = 0 \,, \tag{60.7}$$

$$(\sigma_r - \sigma_\varphi)^2 + (\sigma_\varphi - \sigma_z)^2 + (\sigma_z - \sigma_r)^2 + 2\tau^2 = 2 \,. \tag{60.8}$$

The boundary conditions at $\rho = 1$ are satisfied in the sense of Saint Venant, i.e.

$$\int_{-\kappa}^{\kappa} \tau \mathrm{d}\zeta = 0 \,, \qquad \int_{0}^{\kappa} \sigma_r \mathrm{d}\zeta = 0 \,. \tag{60.9}$$

We assume further that the incompressibility condition holds both in the elastic and in the plastic state; this does not materially affect the results, but greatly simplifies the solution.

Because of the oddness of the tangential stress $\tau$ and of the smallness of $\zeta$ we look for a solution of the form

$$\tau = R(\rho)\, \zeta/\kappa \,. \tag{60.10}$$

It is evident that the first condition of (60.9) is satisfied. The incompressibility condition has the form

$$\frac{\partial u_r}{\partial \rho} + \frac{u_r}{\rho} + \frac{\partial u_z}{\partial \zeta} = 0 \,. \tag{60.11}$$

In this problem the elastic and plastic strains are of the same order, and in general it is necessary to proceed from the equations of plastic flow theory. This, however, involves considerable mathematical difficulties. Taking into account the monotonic character of the loading, we shall start from the equations of deformation theory. Then from (14.5) we have

$$\frac{\partial u_r/\partial \rho - u_r/\rho}{\sigma_r - \sigma_\varphi} = \frac{u_r/\rho - \partial u_z/\partial \zeta}{\sigma_\varphi - \sigma_z} = \frac{\partial u_z/\partial \zeta - \partial u_r/\partial \rho}{\sigma_z - \sigma_r} = \tfrac{1}{2}\sqrt{3}\, \frac{\partial u_r/\partial \zeta + \partial u_z/\partial \rho}{\tau} \,. \tag{60.12}$$

Since the layer is thin, we shall in what follows construct an approximate solution to the problem by finding the stresses on the contact surfaces $\zeta = \pm\kappa$ and then continuing these by some method into the interior of the layer.

The sections $\zeta = 0$, $\zeta = \kappa$, $\zeta = -\kappa$ remain plane. Since the layer is thin, it is natural to use the hypothesis of plane sections $u_z = u_z(\zeta)$. It then follows from the incompressibility equation that the displacement is

$$u_r = C_1(\zeta)/\rho - \tfrac{1}{2}u_z'(\zeta)\,\rho\,,$$

where $C_1(\zeta)$ is an arbitrary function, and the prime denotes differentiation with respect to $\zeta$.

When $\rho = 0$ $u_r = 0$, and hence $C_1(\zeta) = 0$. The relations (60.12) now take the form

$$\frac{0}{\sigma_r - \sigma_\varphi} = \frac{-\frac{3}{2}u_z'}{\sigma_\varphi - \sigma_z} = \frac{\frac{3}{2}u_z'}{\sigma_z - \sigma_r} = \tfrac{1}{4}\sqrt{3}\,\frac{-\rho u_z''}{\tau}\,. \tag{60.13}$$

From this we conclude that

$$\sigma_r = \sigma_\varphi$$

everywhere in the layer.

We consider next equations (60.13) for $\zeta = \kappa$. The plasticity condition (60.8) with $\zeta = \kappa$ gives

$$\sigma_z - \sigma_r = \pm\sqrt{1 - R^2}\,. \tag{60.14}$$

The + sign corresponds to tension. From (60.13) we now obtain

$$\frac{\sqrt{3}u_z'}{\sqrt{1 - R^2}} = -\frac{\rho u_z''}{2R}\,.$$

Introducing the arbitrary parameter

$$\frac{1}{2\sqrt{3}}\left(\frac{u_z''}{u_z'}\right)_{\zeta=\kappa} = C\,,$$

we derive the law of expansion of the contact tangential stresses:

$$R = \pm\frac{C\rho}{\sqrt{1 + C^2\rho^2}}\,. \tag{60.15}$$

The + sign corresponds to extension of the layer. We now find from the differentual equation of equilibrium (60.6) with $\zeta = \kappa$:

$$\sigma_r = \frac{1}{\sqrt{3}\kappa} \int_\rho^1 R \mathrm{d}\rho + \beta \qquad (\beta = \text{const.}) . \tag{60.16}$$

Since the stress $\sigma_z$ is large in magnitude and even with respect to $\zeta$, we assume that $\sigma_z$ is independent of $\zeta$. Then $\sigma_z$ in the layer is determined from (60.14).

The stress $\sigma_r = \sigma_\varphi$ in the layer is given by the yield criterion

$$\sigma_r = \sigma_z \mp \sqrt{1-\tau^2} . \tag{60.17}$$

The constant $\beta$ is found by satisfying the second boundary condition (60.9). Substituting for this constant in (60.16) and (60.14), we obtain

$$\sigma_z = \frac{1}{\sqrt{3}\kappa} \left( \frac{1}{C_1} - \frac{\sqrt{1 + C^2\rho^2}}{C} \right) + \frac{1}{2} \left( \frac{1}{C_1} \arcsin C_1 - \frac{C_1}{C} \right) + \frac{1}{\sqrt{1 + C^2\rho^2}} , \tag{60.18}$$

where we have put

$$C_1 = \frac{C}{\sqrt{1 + C^2}}$$

The stresses $\sigma_z$ are statically equivalent to the force $P$, i.e.

$$2 \int_0^1 \sigma_z \rho \, \mathrm{d}\rho = p \qquad (p = P/\pi a^2 \sigma_s \geqslant 1) .$$

This equation defines the relation between the constant $C$ and the mean stress $p$:

$$p = \frac{1}{\sqrt{3}\kappa} \left[ \frac{1}{C_1} - \frac{2}{3} \left( \frac{1}{C_1^3} - \frac{1}{C^3} \right) \right] + \frac{2}{C} \left( \frac{1}{C_1} - \frac{1}{C} \right) + \frac{1}{2} \left( \frac{1}{C_1} \arcsin C_1 - \frac{C_1}{C} \right) . \tag{60.19}$$

When $C = 0$ (which corresponds to the initiation of plastic flow), we have $R = 0$, $\sigma_z = 1$, $p = 1$. For small values of $C$ the distribution of the contact tangential stress $R$ is nearly linear, and as $C$ increases this distribution deviates increasingly from a linear law (fig. 184). As $C \to \infty$ the tangential stress tends to the yield limit $\tau_s$ which corresponds to the final plastic state (dotted line,

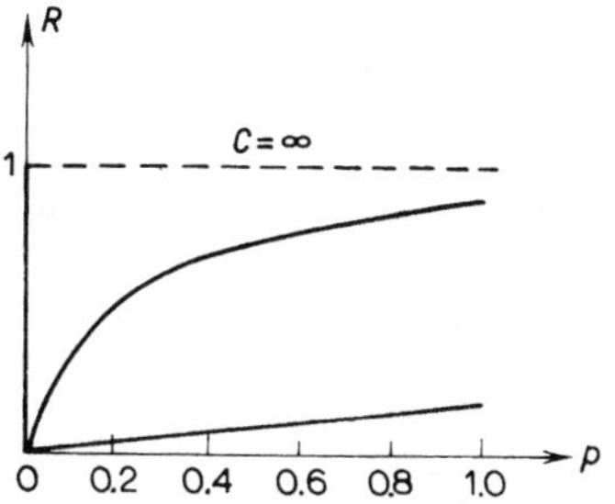


Fig. 184.

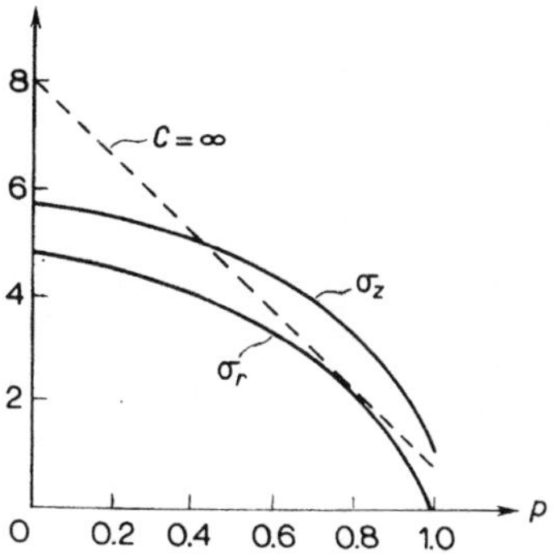


Fig. 185.

$R = 1$). The mean stress is then

$$p = p'_* = \tfrac{1}{4}\pi + \frac{1}{3\sqrt{3}\kappa} \,. \tag{60.20}$$

For thin layers there is little difference between the formulae (60.5) and (60.20). The value $p_*$ is the limit mean stress which can be applied to the layer. When $p_*$ is attained developed plastic flow sets in.

The normal stresses $\sigma_z$ and $\sigma_r = \sigma_\varphi$ are not uniformly distributed across a section; their graphs for $\zeta = 0$ with some value of $p$ are shown in fig. 185; the graph of $\sigma_z$ for the final plastic state is shown by the dotted line. As the load increases, the middle portion of the layer experiences almost "hydrostatic" tension. The normal stress can attain significant values (can sometimes exceed the value of the yield limit $\sigma_s$ for uniaxial tension). In such cases brittle fracture can occur in a soft layer.

Fig. 186 shows the variation of the maximum stress $\sigma_{z,\max}$ (at $r = 0$) with the mean axial stress $p$, for a series of values of the thickness parameter $\kappa$.

## §61. Stress distribution in the neck of a tension specimen

The question of the stress distribution in a neck formed under tension is complicated and has not been fully solved. Since it is important to know the magnitudes of the stresses at the instant preceding rupture, approximate solutions have been constructed which are based on various assumptions stimulated by experimental data. We consider one of these solutions, put forward by Davidenkov and Spiridonova [105].

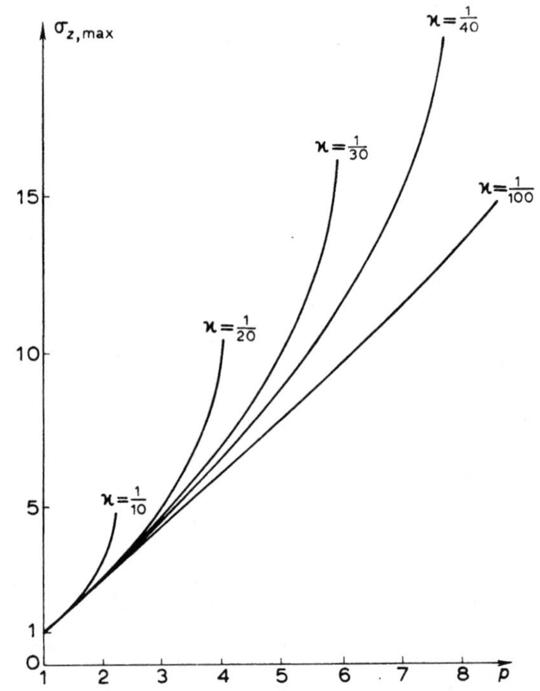


Fig. 186.

When the neck appears the stress distribution ceases to be uniaxial and uniform. The difficulty of the analysis is compounded by the fact that the shape of the neck is unknown. The approximate solution utilizes the experimentally observed fact that in the minimum section of the neck the natural strains in the radial and tangential directions are equal and uniformly distributed. Hence is follows that on the section $z = 0$

$$\xi_r = \xi_\varphi = \text{const.}$$

at a given instant of time.

Since the elastic deformations in the neck are negligibly small compared with the plastic deformations, the incompressibility equation gives $\xi_z = -2\xi_r = \text{const.}$, and from the Saint Venant-von Mises relations it follows that

$$\sigma_r = \sigma_\varphi \tag{61.1}$$

in the section $z = 0$. Further, we have from the symmetry condition that $\tau_{rz} = 0$ when $z = 0$. In this section the differential equations of equilibrium (58.1) take the form

$$\frac{d\sigma_r}{dr} + \left(\frac{\partial \tau_{rz}}{\partial z}\right)_{z=0} = 0\,, \qquad \frac{\partial \sigma_z}{\partial z} = 0\,, \tag{61.2}$$

and the yield criterion is

$$\sigma_z - \sigma_r = \sigma_s\,. \tag{61.3}$$

We take a meridional plane and consider in it the trajectories of the principal stresses $\sigma_3$, $\sigma_1$ (fig. 187) close to the plane $z = 0$. The angle $\omega$ of inclination of the tangent to the trajectory of the stress $\sigma_3$ is small, and formulae (58.7), with indices 1, 2 replaced by 1, 3 respectively, take the simple form

$$\sigma_z \approx \sigma_3\,, \qquad \sigma_r \approx \sigma_1\,, \qquad \tau_{rz} \approx (\sigma_3 - \sigma_1)\,\omega\,.$$

In consequence we have near the plane $z = 0$

$$\sigma_3 - \sigma_1 \approx \sigma_s\,, \qquad \tau_{rz} \approx \sigma_s \omega \tag{61.4}$$

and

$$\left(\frac{\partial \tau_{rz}}{\partial z}\right)_{z=0} = \sigma_s \left(\frac{\partial \omega}{\partial z}\right)_{z=0} = \frac{\sigma_s}{\rho}\,, \tag{61.5}$$

where $\rho$ is the radius of curvature of the trajectory of the principal stress for $z = 0$. The contour of the neck is one of these trajectories; let $\rho = R$ for the contour. From the differential equation (61.2) we obtain

$$\frac{\sigma_r}{\sigma_s} = \int_r^a \frac{dr}{\rho}\,,$$

since $\sigma_r = 0$ when $r = a$.

When $r = 0$, $\rho = \infty$ and when $r = a$, $\rho = R$; on the basis of observations we assume that

$$\rho = R\,a/r\,.$$

Then

$$\frac{\sigma_r}{\sigma_s} = \frac{a^2 - r^2}{2aR}\,, \qquad \frac{\sigma_z}{\sigma_s} = 1 + \frac{a^2 - r^2}{2aR}\,. \tag{61.6}$$

This stress distribution in the neck is shown on the left-hand side of fig. 187.

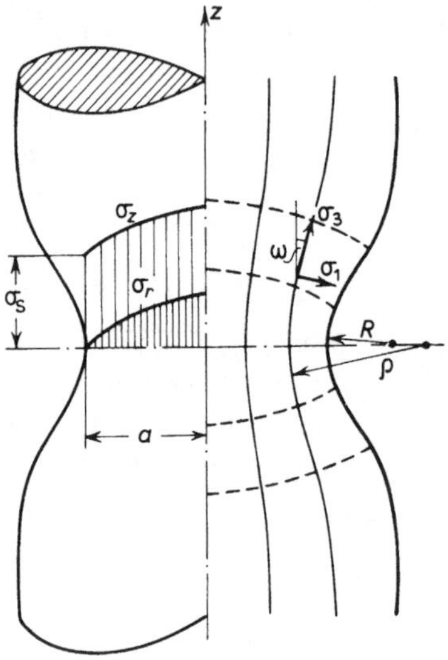


Fig. 187.

Fig. 188.

To calculate the stresses it is necessary to have experimental measurements of the quantities $a$, $R$.

The maximum stresses arise in the central portion of the neck and for this reason rupture begins at the centre. Fig. 188 shows an X-ray photograph (taken from Nadai's book [25]) of the neck of a specimen directly before rupture; it supports the above remark.

## §62. Plastic bending of circular plates

We consider the plastic bending of a circular plate (fig. 189) under an axisymmetric load $p = p(r)$, where $r$ is the radius vector; the thickness $2h$ of the plate is constant. The $z$-axis of a cylindrical coördinate system $r$, $\varphi$, $z$ is directed downwards. We shall proceed from a plastic-rigid model for the material. Then the plate remains undeformed right until the limit load is reached (characterizing the *load-bearing capacity* of the plate).

### 62.1. *Basic propositions*

In the classical theory of bending of elastic plates the basic propositions

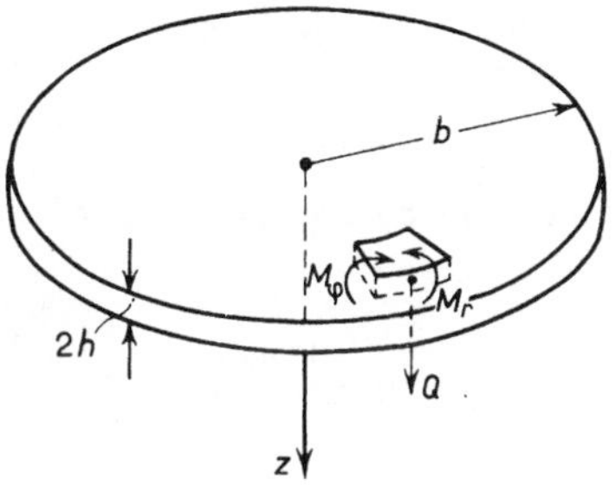


Fig. 189.

have a geometric character, and therefore remain valid for plastic bending as well. This means that we retain Kirchhoff's assumptions:

1. the median plane is not extended;
2. straight lines perpendicular to the median plane before deformation become straight lines perpendicular to the median surface after deformation.

The stress components $\sigma_z$, $\tau_{rz}$ can be ignored by comparison with the components $\sigma_r$, $\sigma_\varphi$; the tangential stresses $\tau_{r\varphi}$, $\tau_{\varphi z}$ are zero, by symmetry. In the plate a "plane stress state" is realized. In the sections $r$ = const., $\varphi$ = const. respectively, bending moments $M_r, M_\varphi$ act:

$$M_r = \int \sigma_r z \mathrm{d}z \,, \qquad M_\varphi = \int \sigma_\varphi z \mathrm{d}z \,,$$

where the integration is over the thickness of the plate $(-h, +h)$. The shearing force in the section $r$ = const. is

$$Q = \int \tau_{rz} \mathrm{d}z$$

and is related to the bending moments by the well-known differential equations of equilibrium (cf. for example, [47]):

$$\frac{\mathrm{d}M_r}{\mathrm{d}r} + \frac{M_r - M_\varphi}{r} = Q \,. \tag{62.1}$$

The tangential stresses on the circle $r$ = const. balance the external load, therefore

$$Q = -\frac{1}{r} \int_a^r pr \,\mathrm{d}r \,,$$

where $a$ is the radius of the hole in an annular plate ($a = 0$ for a solid plate).

Denote by $w = w(r)$ the rate of bending of the plate; the strain-rate components are determined by the well-known relations

$$\xi_r = z\kappa_r\,, \qquad \xi_\varphi = z\kappa_\varphi\,, \tag{62.2}$$

where we have introduced the coefficients of rate of curvature of the median surface of the plate:

$$\kappa_r = -\frac{\mathrm{d}^2 w}{\mathrm{d}r^2}\,, \qquad \kappa_\varphi = -\frac{1}{r}\frac{\mathrm{d}w}{\mathrm{d}r}\,.$$

It is evident that *the ratio of the strain rates* $\xi_r/\xi_\varphi$ *is constant along the normal* to the plane of the plate.

62.2. *Bending of a plate with von Mises yield criterion*

Here the strain rates and the stress components are connected by the Saint Venant-von Mises relations (§ 13) – the elastic components being neglected:

$$\xi_r = \tfrac{1}{3}\lambda'(2\sigma_r - \sigma_\varphi)\,, \qquad \xi_\varphi = \tfrac{1}{3}\lambda'(2\sigma_\varphi - \sigma_r)\,, \tag{62.3}$$

and the non-zero stress components $\sigma_r$, $\sigma_\varphi$ satisfy the yield criterion

$$\sigma_r^2 - \sigma_r\sigma_\varphi + \sigma_\varphi^2 = \sigma_s^2\,. \tag{62.4}$$

As we know (§ 16) the relations (62.3) can be written in the form

$$\xi_r = \tfrac{1}{3}\lambda'\frac{\partial f}{\partial \sigma_r}\,, \qquad \xi_\varphi = \tfrac{1}{3}\lambda'\frac{\partial f}{\partial \sigma_\varphi}\,,$$

where $f$ represents the expression on the left-hand side of the yield condition (62.4). In other words the strain-rate vector is normal to the yield curve (fig. 190a, dotted arrow).

Denoting the ratio $\xi_r/\xi_\varphi$ by $\eta$, we easily find from (62.3) that $\sigma_\varphi$ is proportional to $\sigma_r$ along the normal. But then it follows from the yield criterion that $\sigma_r = \pm f_1(\eta)\,\sigma_s$, $\sigma_\varphi = \pm f_2(\eta)\,\sigma_s$ where $f_1, f_2$ are some functions of $\eta$.

Thus, the stresses $\sigma_r$, $\sigma_\varphi$ are constant along the normal for positive $z$ and take the opposite sign (with respect to the bending of the plate) for negative $z$; the stresses $\sigma_r$, $\sigma_\varphi$ are discontinuous on the median plane (analogous to the situation for the bending of a beam) and are depicted on the yield ellipse (fig. 190a) by opposite points. As a consequence of this the bending moments are (here $\sigma_r$, $\sigma_\varphi$ are the values of the stresses for $z > 0$)

$$M_r = \sigma_r h^2\,, \qquad M_\varphi = \sigma_\varphi h^2 \tag{62.5}$$

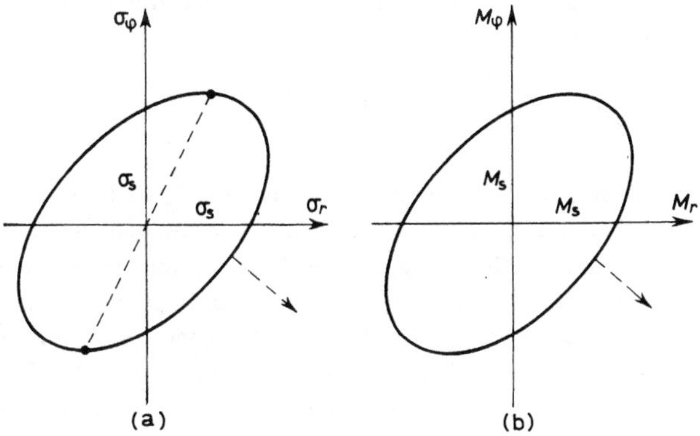


Fig. 190.

and, because of (62.4), they satisfy

$$M_r^2 - M_r M_\varphi + M_\varphi^2 = M_s^2 \,, \tag{62.6}$$

where $M_s = \sigma_s h^2$ is the maximum value of the bending moment. This equation is a special case of the finite relation between forces and moments in plastic shells, which was stated by Il'yushin [12], and also represents an ellipse in the plane $M_r$, $M_\varphi$ (fig. 190b).

Replacing the stresses $\sigma_r$, $\sigma_\varphi$ in (62.3) by the moments $M_r$, $M_\varphi$, and the strain rates by the curvatures $\kappa_r$, $\kappa_\varphi$, we easily obtain the relation

$$\kappa_r = \lambda^* \frac{\partial F}{\partial M_r}, \qquad \kappa_\varphi = \lambda^* \frac{\partial F}{\partial M_\varphi},$$

where $F$ represents the expression on the left-hand side of (62.6), and $\lambda^*$ is a scalar multiplier. Thus, the vector rate of curvature is normal to the limiting curve in the $M_r$, $M_\varphi$-plane (fig. 190b). Consequently, the associated flow law remains valid for a relationship between generalized quantities – the moments $M_r$, $M_\varphi$ and the rates of curvature $\kappa_r$, $\kappa_\varphi$.

With the aid of (62.6) we eliminate the moment $M_\varphi$ from the equilibrium equation (62.1) and obtain a non-linear differential equation for $M_r$:

$$\frac{dM_r}{dr} + \frac{1}{r}\left(\tfrac{1}{2}M_r \mp \sqrt{M_s^2 - \tfrac{3}{4}M_r^2}\right) = Q \,. \tag{62.7}$$

The solution of equation (62.7) with appropriate boundary conditions determines the limit load.

From (62.3) and (62.2) we obtain the differential equation for the rate of bending of the lamina:

$$r(2M_\varphi - M_r)\frac{d^2w}{dr^2} - (2M_r - M_\varphi)\frac{dw}{dr} = 0\,. \tag{62.8}$$

This is easily integrated if the bending moments $M_r$, $M_\varphi$ are known.

Analogous to the occurrence of plastic joints in the bending of a beam, it is possible for a jointed circle to appear in laminas. Along it the rate of bending is continuous, the derivative $dw/dr$ is discontinuous, and, consequently, the rate of curvature $\kappa_r = -d^2w/dr^2$ is not bounded. Since the moments $M_r$, $M_\varphi$ are everywhere bounded, it follows from (62.8) that $M_r = 2M_\varphi$ on the jointed circle. It is also necessary to assume that some (annular) region of the lamina can remain rigid.

We consider now the question of *boundary conditions*; these have the form

1. on a free boundary $M_r = 0$;
2. on a supported boundary $M_r = 0$; $w = 0$;
3. along a fixed boundary $w = 0$, $dw/dr = 0$.

The last condition means that $\kappa_\varphi = 0$, i.e. $M_r - 2M_\varphi = 0$; consequently

$$M_r = 2M_\varphi = \pm\frac{2}{\sqrt{3}}M_s\,.$$

62.3. *Bending of a lamina with Tresca-Saint Venant yield criterion*

The solution of equation (62.7) involves considerable difficulties. The problem is substantially simplified if we replace the von Mises ellipse by a hexagon, corresponding to the Tresca-Saint Venant yield criterion (fig. 191). The plate is then divided into annular zones, in each of which the yield criterion is linear and integration is easily effected.

The error involved in doing this is not too large; it can be reduced if we replace the inscribed hexagon by a hexagon which lies halfway between the inscribed and circumscribed hexagons; for this purpose it is sufficient to replace $\sigma_s$ by the value $\sigma_s' \approx 1.08\,\sigma_s$.

The equations for the rate of bending $w$ are easily formulated with the aid of the associated flow law, suitably applied to the plane stress distribution in §52. Here $\xi_1$, $\xi_2$ become $\xi_r$, $\xi_\varphi$.

On the circle which separates the annular regions of the different solutions, the bending moment $M_r$ and the shearing force $Q$ must be continuous – this follows from the equilibrium condition; the bending moment $M_\varphi$ can be discontinuous.

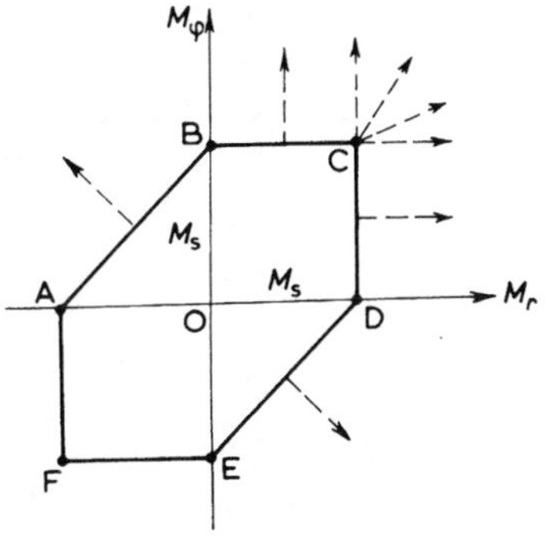


Fig. 191.

The rate of bending $w$ must be continuous, but, in general, the strain rates $\xi_r$, $\xi_\varphi$ can be discontinuous. The circle on which $\xi_\varphi$ is discontinuous (i.e. the tangent to the bending rate surface experiences a finite rotation) is called *jointed.*

On the jointed circle $M_r = \pm M_s$.

In fact the strain rate $\xi_\varphi$ is discontinuous on this circle, and $\xi_r$ is unbounded. By the associated flow law this situation is possible only for the vertical sides of the hexagon (fig. 191), for which $M_r = \pm M_s$.

In the plastic-rigid model we have to assume that part of the lamina (some annular zone) can remain undeformed and, in general, experiences a rigid vertical displacement.

The boundary conditions 1. and 2. (for free and supported boundaries) are obviously preserved. For the case of the fixed boundary $dw/dr = 0$ and $M_r = \pm M_s$.

62.4. *Examples*

We consider some particular problems on the basis of Tresca-Saint Venant criterion.

1. Supported lamina, uniform load $p$ distributed over a circle of radius $c$ (fig. 192). The stresses in the corresponding elastic problem are maximum at the centre ($r = 0$), and it is here that plastic deformations first arise. When $r = 0, M_r = M_\varphi = M_s$, and near the centre there will be one of the plastic regimes C, BC, CD. The regime CD contradicts the equation of equilibrium (on CD $M_r = M_s$ and $dM_r/dr = 0$); on the other hand, $M_\varphi < M_s$, $p > 0$, and from the differential equation (62.1) it follows that $dM_r/dr < 0$. In exactly the

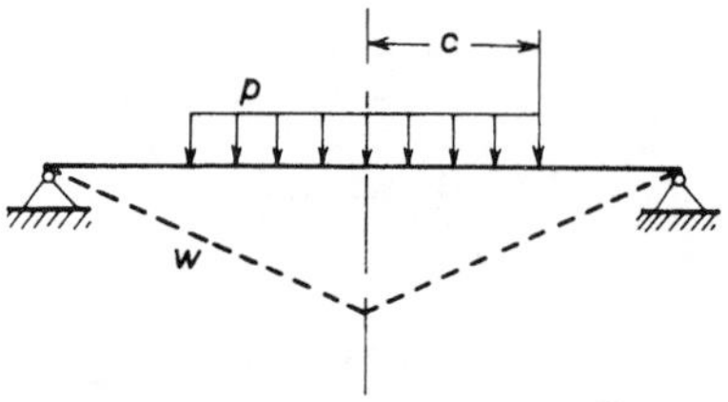


Fig. 192.

same way the regime C is impossible. Therefore the regime BC is realized, i.e. $M_\varphi = M_s$. From the differential equation (62.1) we then obtain ($p = 0$ when $r > c$)

$$M_r = \begin{cases} C_1/r - \frac{1}{6}pr^2 - M_s & \text{for} \quad r \leqslant c\,, \\ C_2/r - \frac{1}{6}pc^2 - M_s & \text{for} \quad r \geqslant c\,, \end{cases} \tag{62.9}$$

where $C_1$, $C_2$ are arbitrary constants. From the condition that $M_r$ be bounded at the centre it follows that $C_1 = 0$. From the condition that $M_r$ is continuous when $r = c$ we find that $C_2 = \frac{1}{3}pc^2$.

The bending moment $M_r$ decreases as $r$ increases, i.e. the representative point in fig. 191 moves from C to B. At the support, $M_r = 0$ and the regime B is achieved, while in the remainder of the lamina we have regime BC. By satisfying the condition $M_r = 0$ when $r = b$ we obtain the limiting load

$$p_* = \frac{6b}{c^2(3b-2c)} M_s\,.$$

The picture of plastic flow of a lamina in the limit state is determined in the following way. By the associated law we have $\kappa_r = 0$ or $\mathrm{d}^2w/\mathrm{d}r^2 = 0$ for the segment BC. Hence the boundary condition $w = 0$ when $r = b$ leads to the result

$$w = w_0(1 - r/b)\,,$$

where $w_0$, the value of the bending rate at the centre, remains undetermined. Thus, in the flow state the lamina assumes the form of the surface of a cone (fig. 192, dotted line).

If the load acts over the whole surface of the plate, then $c = b$; the limit load is now

$$p_* = 6M_s/b^2\,.$$

For this case numerical integration of the differential equation (62.7) has been performed on the basis of the von Mises plasticity condition, and leads to the result

$$p_* = 6.5\, M_s/b^2 .$$

2. Fixed plate under a uniform loading pressure (fig. 193). As in the previous example the regime BC is realized near the centre. The bending moment $M_r$ decreases with increasing $r$ and tends to zero at some value $r = \rho$. Further, the regime BA occurs and extends up to the contour $r = b$, where $M_r = -M_s$ i.e. the regime A occurs.

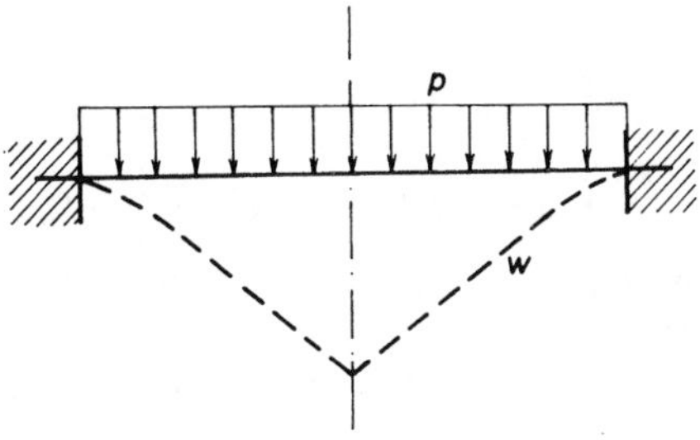


Fig. 193.

When $r \leqslant \rho$, (62.9) gives that

$$M_r = M_s - \tfrac{1}{6} p r^2 .$$

When $r = \rho, M_r = 0$ and

$$\rho^2 = 6M_s/p .$$

When $r > \rho$, $M_\varphi - M_r = M_s$ and we obtain from the differential equation of equilibrium:

$$M_r = M_s \ln (r/\rho) - \tfrac{1}{4} p(r^2 - \rho^2) ,$$

where we have used the condition $M_r|_{r=\rho} = 0$. Let $M_r = -M_s$ on the contour $r = b$, i.e. a joint is formed along the line of fixing. Hence we obtain the transcendental equation

$$5 + 2 \ln (b/\rho) = 3\, b^2/\rho^2 ,$$

which determines $\rho$ and, consequently, the limit load; from calculations we obtain $\rho \approx 0.73b$. Thus,

$$p_* = 11.3\, M_s/b^2 \,.$$

We now turn to the determination of the rate of bending. In the central region $r \leqslant \rho$ we have, as in the first example,

$$\frac{d^2w}{dr^2} = 0\,, \qquad w = w_0 + C_1 r\,,$$

where $w_0$, $C_1$ are arbitrary constants.

When $r > \rho$ (regime AB) the associated flow law gives $\kappa_r : \kappa_\varphi = -1 : 1$, i.e.

$$\frac{d^2w}{dr^2} + \frac{1}{r}\frac{dw}{dr} = 0\,.$$

On the contour $r = b$, $w = 0$; with this condition we obtain

$$w = C_2 \ln (r/b)\,,$$

where $C_2$ is an arbitrary constant. From this it is clear that the condition $dw/dr = 0$ on the contour is not fulfilled, and therefore a plastic joint is in fact formed on the contour.

The arbitrary constants $C_1$, $C_2$ are determined from the conditions that $w$ and $dw/dr$ are continuous for $r = \rho$. The bending rate $w_0$ at the centre remains undetermined. The nature of the bending of the lamina is shown by the dotted line in fig. 193.

62.5. *Concluding remarks*

The limiting state of a lamina under bending has been examined in numerous works; we mention here the works of Gvozdev [7], Prager [29], Hodge [55], Grivor'ev [101], Il'yushin [12] and other authors (see the surveys [70, 78]). The Tresca-Saint Venant yield criterion and the associated flow law have been widely used; here there is a direct connection with the generalized quantities – the moments and rates of curvature. This scheme has also been developed for analyzing the limit equilibrium of axisymmetric shells [55].

The elastic-plastic bending of circular plates has been studied by Sokolovskii [44] on the basis of deformation theory equations, and by Tekinalp [193] on the basis of flow theory.

The limit load for plates and shells can conveniently be found by energy methods, using the extremum properties of the limit state. This question will be considered in the next chapter, where one of the problems treated as an example is that of finding the limit load for a circular plate by energy methods (§66).

## PROBLEMS

1. Find the solution to the differential equation (62.7) for the case of an annular lamina, supported along its outer boundary and uniformly loaded along the inner boundary.

2. Using the Tresca-Saint Venant yield criterion find the limit load for a uniformly loaded annular plate, supported along its outer boundary.

# 8

# Extremum Principles and Energy Methods

## §63. Extremum principles

General theorems play an important role in the theory of plasticity, as they do in elasticity theory. Most important of all in this regard are theorems on the extremum properties of a solution, and uniqueness theorems. From a practical point of view less interest attaches to problems of existence of solutions, which are always more difficult, have received less attention, and are usually bound up with a number of restrictions dictated by the method of proof. Nevertheless these results are necessary for a rigorous justification of the validity of the equations.

Apart from their general value, these theorems open the way to a direct construction of solutions, by-passing the integration of the differential equations. In the non-linear problems which make up plasticity theory this possibility is extremely important.

Of great significance are *extremum principles in the theory of a plastic-rigid body.* In the preceding chapters we have said much concerning the difficulties arising from the non-uniqueness of the solution process in the plastic-rigid model. This compelled us to introduce the concepts of kinematically possible velocity fields and statically possible yield-stress distributions, and to formulate, without proof, a *selection criterion.* As we shall indicate below (§§64, 65), this criterion originates from extremum theorems.

When the limit load is attained in an ideal plastic-rigid body, free plastic flow occurs. The limit state can be interpreted as the state which precedes collapse. For this reason the limit state is sometimes called the *state of plastic collapse*, and the extremum theorems which characterize the limit load are called *theorems on plastic collapse*.

Extremum theorems for a plastic-rigid body lead to effective methods of obtaining the limiting load using successive approximations by upper and lower estimates (§65). In §66 we present a variety of examples illustrating the energy method.

In the *deformation theory* of plasticity extremum theorems are generalizations of the corresponding minimum theorems for an elastic body (namely, the theorem of minimum total energy of a system and Castigliano's theorem). These theorems are widely used as the basis for the approximate solution of various particular problems by direct methods (especially Ritz's method). The minimum theorems of deformation theory and some of their simplest applications are given in §§67, 68.

*Extremum principles for an elastic-plastic medium* which obeys the equations of flow theory (for perfect plasticity and for the presence of hardening) are formulated in the concluding section of this chapter (§69). In contrast with the preceding case these principles determine the extremum properties of *increments* (or rates) of displacements and *increments* in stresses corresponding to small increments in external forces or in given displacements. Naturally enough, these "local" properties, which are connected with the differential character of the equations of plastic flow theory, are more difficult to use in the effective construction of solutions, and their major interest is one of principle.

Related to these general theorems there are also *theorems of shakedown* of elastic-plastic structures subject to cyclic loads. Because of the special nature of these problems, we postpone this subject to a later chapter (chap. 9).

A rigorous account of the general theorems of the theory of elastic-plastic media, as well as a detailed bibliography, can be found in the excellent work of Koiter [68]. Numerous applications of energy theorems are presented in the books of Gvozdev [7], Il'yushin [12], Neal [26], Prager [29], Prager and Hodge [31], Rabotnov [33], Hodge [55] and other authors. Various additional results on this question can also be found in a number of review articles [68, 70, 72].

## §64. Extremum principles for a plastic-rigid body

64.1. *General remarks*

We have noted above that extremum principles are important for a plastic-rigid body (with yield plateau), and are widely applied in the construction of approximate solutions.

The conditions under which the plastic-rigid body model is useful have been discussed earlier; they essentially depend on the nature of the particular problem.

If the elastic strain-rates are neglected, we have the simpler relations of the Saint Venant-von Mises theory (§ 13)

$$\xi_{ij} = \lambda s_{ij} \qquad \text{or} \qquad \frac{\xi_{ij}}{H} = \frac{s_{ij}}{2\tau_s}. \tag{64.1}$$

In the following discussion we shall consider only small deformations of a plastic-rigid body, so that changes in the configuration of the body and in the positions of its points can be neglected. However, the results can often be applied to problems of steady plastic flow by focussing on the instantaneous state.

The limit state of a plastic-rigid body is determined by a finite combination of loads at the instant when plastic flow sets in. It is obvious that the loading path is not taken into account, nor are the initial stresses and strains. In this sense one can say that the limit load is independent of the loading path and the initial stresses.

The practical value of this conclusion is evidence by experimental data and by the complete solutions of certain elastic-plastic problems. This property becomes intelligible if we take into account the fact that for deformation which develops in a definite direction (cf. § 15) the stresses tend to some "steady" value independent of the strain path.

As the limit state is approached the deformations of the body usually increase rapidly in the directions in which the loads act. If these loads increase in proportion to some parameter when they are close to their limiting value, the deformations develop in a definite direction and the influence of the loading path diminishes continuously.

### 64.2. *Basic energy equation*

We consider a body which occupies the volume $V$ and is bounded by the surface $S = S_F + S_v$ (fig. 194). Suppose that on the portion $S_F$ of the surface the force $\mathbf{F}_n$ is given, where the components of the latter with respect to the axes $x_i$ ($i = 1, 2, 3$) are denoted by $X_{ni}$. On the portion $S_v$ of the surface we suppose that the velocity $v_0$ is prescribed; its components are denoted by $v_{0i}$. For simplicity we shall assume that body forces are absent. In the sequel we use tensor notation (cf. § 1).

Let $\sigma_{ij}$ be some stress field which satisfies the differential equations of equilibrium (cf. § 4) inside the body:

$$\partial \sigma_{ij} / \partial x_i = 0 \tag{64.2}$$

and which is consistent with the prescribed loads $X_{ni}$ on the boundary $S_F$ in

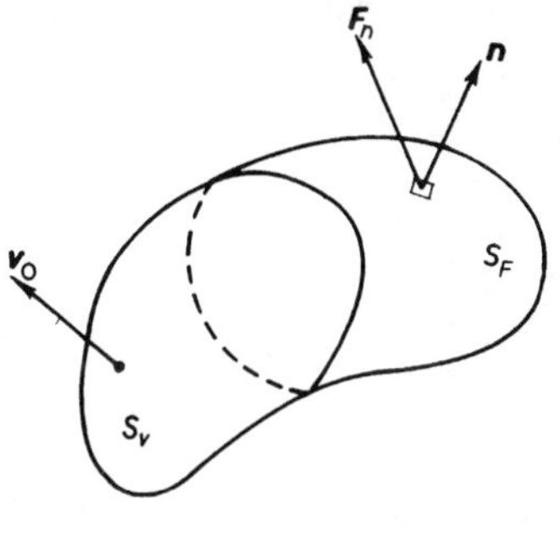


Fig. 194.

the sense of Cauchy's formula (1.2):

$$\sigma_{ij} n_j = X_{ni} \qquad \text{on} \qquad S_F\,, \tag{64.3}$$

where $n_j$ are the direction cosines of the normal $\mathbf{n}$.

On the other hand, let us introduce some continuous velocity field $v_i$ which satisfies the prescribed conditions on $S_v$, i.e.

$$v_i = v_{0i} \qquad \text{on} \qquad S_v\,. \tag{64.4}$$

Corresponding to this field we have the strain-rate components (§3)

$$\xi_{ij} = \frac{1}{2}\left(\frac{\partial v_i}{\partial x_j} + \frac{\partial v_j}{\partial x_i}\right). \tag{64.5}$$

These stress ($\sigma_{ij}$) and velocity ($v_i$) fields are in other respects arbitrary, and, in general, are not interrelated. We assume that these fields are continuous (later we shall remove this restriction). Either the configuration of the body is only slightly different from its initial position (in which case $V$ and $S$ are the volume and surface up to deformation) or else the configuration is characteristic of some yield state.

For a complete solid body the following basic equation holds:

$$\int \sigma_{ij}\xi_{ij}\,\mathrm{d}V = \int X_{ni} v_i\,\mathrm{d}S\,, \tag{64.6}$$

where the first integration extends over the whole volume of the body, and the second over the whole surface $S$.

To prove this we use Cauchy's formula (64.3) to write the surface integral in the form

$$\int X_{ni} v_i\,\mathrm{d}S = \int \sigma_{ij} v_i n_j\,\mathrm{d}S$$

and transform it to a volume integral by Gauss's formula

$$\int Q_j n_j \mathrm{d}S = \int \frac{\partial Q_j}{\partial x_j} \mathrm{d}V \,. \tag{64.7}$$

It is obvious that

$$\int \sigma_{ij} v_i n_j \mathrm{d}S = \int v_j \frac{\partial \sigma_{ij}}{\partial x_i} \mathrm{d}V + \int \sigma_{ij} \xi_{ij} \mathrm{d}V \,.$$

The first integral on the right-hand side is equal to zero because of the differential equation of equilibrium (64.2), and this establishes (64.6).

This equation needs to be generalized, first to the case of a body having rigid (non-deformable) regions, and second to the case of discontinuous stress and velocity fields.

The first generalization is almost obvious. If the body contains a deforming ($V_{\mathrm{D}}$) and a rigid ($V_{\mathrm{R}}$) region, separated by a surface $\Sigma$ on which the velocities and stresses are continuous, then, from (64.6), we have for each region

$$\int \sigma_{ij} \xi_{ij} \mathrm{d}V_{\mathrm{D}} = \int X_{ni} v_i \mathrm{d}S_{\mathrm{D}} + \int X_{ni} v_i \mathrm{d}\Sigma \,,$$

$$0 = \int X_{ni} v_i \mathrm{d}S_{\mathrm{R}} - \int X_{ni} v_i \mathrm{d}\Sigma \,.$$

The domain of integration (here and subsequently) is indicated by the form of the corresponding differential. Thus, the first integral on the left is taken over the domain $V_{\mathrm{D}}$, the first integral in the second line over the surface $S_R$ which bounds the domain $V_{\mathrm{R}}$, and so on. Adding the two equations (and noting that $S = S_{\mathrm{D}} + S_{\mathrm{R}}$) we arrive at our previous formula (64.6). Thus, *the basic energy equation (64.6) can be written with respect to the whole body (including rigid regions).*

64.3. *Generalization of the basic energy equation to discontinuous fields*

The preceding results are based on the assumption of continuous stress and velocity fields. However, simple examples (bending, torsion, the plane problem) testify to the fact that discontinuities in the stresses occur very frequently in the limit state. In the plastic-rigid body model velocity discontinuities are also inevitable. Finally, it is sometimes convenient to construct approximate solutions which are discontinuous. For these reasons we consider the generalization of the basic energy equation to the case of discontinuous fields.

Discontinuities in stresses. We begin by examining the case where the stresses are discontinuous on certain surfaces $S_k$ ($k = 1, 2, 3, \ldots$). The surfaces $S_k$ divide the body into a finite number of parts, in each of which the stresses vary continuously and, consequently, the equation obtained above is valid; here the respective surface integrals are performed over the surface of each individual region. Suppose that forces $X_{ni}^+$ act on one side of the surface $S_k$, and forces $X_{ni}^-$ on the other side.

From the condition of equilibrium for an element of any surface it follows that

$$X_{ni}^+ + X_{ni}^- = 0 \qquad (i = 1, 2, 3) .$$

Consequently, if we add all such equations written out for each part of the body, we find that all the integrals over the surface of discontinuity cancel, i.e. *the presence of stress discontinuities does not affect the structure of the basic energy equation.*

Discontinuities in velocities. We next consider discontinuities in the velocity field on certain surfaces $S_l$ ($l = 1, 2, 3, \ldots$).

First of all we note that the discontinuities can only occur in the velocity components which lie in the tangent plane to $S_l$ (tangential velocity components), since otherwise "cracks" appear in the body.

An exception to this is the case of a thin plate (or shell), when a steep thinning ("neck") or thickening ("bulge") can arise along certain lines. Such discontinuities have been discussed in ch. 6. We shall return to this case later; for the present we shall assume that the normal velocity component on $S_l$ is continuous.

At some point on the surface of discontinuity $S_l$ we construct a local coordinate system $x$, $y$, $z$, with the $z$-axis pointing along the normal to the surface

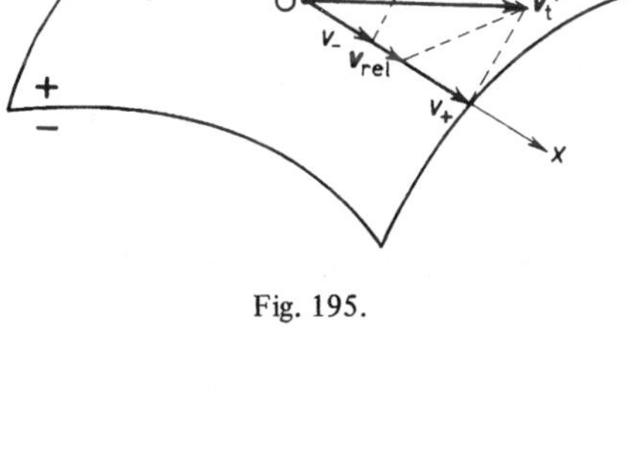

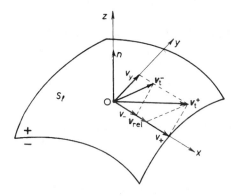

Fig. 195.

(fig. 195). We denote by $v_t^+$, $v_t^-$ the tangential velocity components on the positive and negative sides of the surface $S_l$ respectively. Let $v_{rel} = v_t^+ - v_t^-$ be the relative velocity vector; we take the $x$-axis in the direction of this vector. The surface of discontinuity must be regarded as the limiting position of a thin layer, through which the velocity changes continuously but rapidly (fig. 196a). Only the component $v_x$ experiences a rapid variation, while $v_y$, $v_z$ are almost constant through the layer. It is evident that the shear rate $\eta_{xz}$ is considerably greater than the other strain-rate components; as the thickness of the layer tends to zero, $\eta_{xz} \to \infty$, while the other strain-rate components remain bounded.

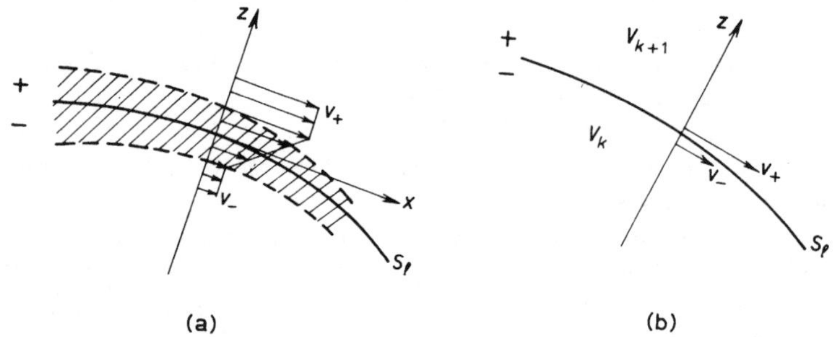


Fig. 196.

We denote by $\tau$ the tangential component of stress on the surface $S_l$ in the $x$-direction.

The surfaces of discontinuity $S_l$ divide the body into domains $V_1, V_2, \ldots$, in each of which the stresses and velocities have the necessary continuity properties, and therefore the equation obtained above can be applied to each of the domains. If we write down the equations for each domain of the body and then add them, we shall always find that the two integrals on either side of the surface of discontinuity (on the positive and negative sides, fig. 196b) coincide.

Consider an element of surface $dS_l$; let the region $V_k$ lie on the negative side of $dS_l$, and the region $V_{k+1}$ on the positive side. The corresponding rate of work of the stresses for the region $V_k$ is

$$(\sigma_n v_n + \tau_y v_y + \tau v_-)\, dS_l\,,$$

where $\tau_y$ and $v_y$ denote the components of tangential stress and tangential velocity in the $y$-direction.

The corresponding rate of work for the region $V_{k+1}$ is

$$-(\sigma_n v_n + \tau_y v_y + \tau v_+)\, dS_l\,.$$

Consequently, the sum of the rates of work of the stresses for the element is

$$-\tau[v]\ dS_l\,,$$

where $[v]$ denotes the jump in velocity $v_+ - v_- = |v_{\text{rel}}|$. Thus, the stresses acting on the surface of velocity discontinuity develop a rate of work

$$-\sum \int \tau[v]\ dS_l\,,$$

where the summation covers all surfaces of discontinuity $S_l$. This rate of work should be inserted in the basic energy equation, and then

$$\int X_{ni} v_i dS = \int \sigma_{ij} \xi_{ij} dV + \int \tau[v]\ dS_{\text{p}}\,. \tag{64.8}$$

For simplicity the summation sign has been omitted; the integration extends over all surfaces of discontinuity $S_{\text{p}} = S_1 + S_2 + \ldots$ .

The left-hand side of (64.8) represents the "power" of the surface forces, and the right-hand side the "dissipation".

*Remarks*

1. Equation (64.8) holds for any continuous medium in equilibrium. The velocities and stresses which occur in (64.8) are, in general, not inter-related.

2. Suppose the medium obeys the equations of Saint Venant-von Mises theory (64.1). We have already observed that in the neighbourhood of the surface of discontinuity the rate of shear $\eta_{xz} \to \infty$, and therefore it follows from equations (64.1) that

$$\begin{array}{lll} \sigma_x \to \sigma\,, & \sigma_y \to \sigma\,, & \sigma_z \to \sigma\,, \\ \tau_{xy} \to 0\,, & \tau_{yz} \to 0\,, & \tau_{xz} \to \tau_s\,, \end{array} \tag{64.9}$$

i.e. the surface of discontinuity is in essence a surface of maximum tangential stress (slip surface).

Since the stresses and strain rates are now connected by formulae (64.1), the shear takes place in the direction of the tangential stress; therefore

$$\tau[v] = \tau_s[v] > 0\,. \tag{64.10}$$

3. In the case of plane stress it is also possible to have a discontinuity in the *normal* velocity component on the line of discontinuity L (§53, fig. 165). The dissipation per unit length of the neck is given by formula (53.28). In

place of the second term on the right-hand side of the basic equation (64.8) it is necessary to write the quantity

$$\tau_s h \int v\sqrt{1+\sin^2\gamma}\ \mathrm{d}s_L\,,$$

where $\mathrm{d}s_L$ is an element of the line L (or the sum of such terms, if there are several lines of discontinuity). Here $v$ is the magnitude of the velocity discontinuity, and $\gamma$ is the inclination of the velocity $v$ to the line L; when $\gamma = \frac{1}{2}\pi$ there is no relative slip and only thinning takes place; when $\gamma = 0$ only slipping occurs.

64.4. *Minimum properties of an actual velocity field*

We apply the energy equation (64.8) to a plastic-rigid body. Let the quantities $\sigma_{ij}$, $\xi_{ij}$, $v_i$ be the actual solution of the problem, in which case the stresses and strain-rates are connected by the Saint Venant-von Mises relations and satisfy all the conditions of equilibrium and continuity. The energy equation (64.8) is obviously valid with respect to this actual solution.

Together with the actual state we consider another, *kinematically possible field* $v_i'$, satisfying the incompressibility condition and the prescribed boundary conditions on $S_v$. According to (64.5), the velocities $v_i'$ have associated with them strain rates $\xi_{ij}'$, and to these there corresponds a stress deviatoric $s_{ij}^*$, as we see from (64.1) with $\xi_{ij}' \neq 0$. In general the stress deviatoric will not satisfy the equilibrium equations. Further, Cauchy's formula (64.3) shows that the stresses $s_{ij}^*$ are associated with some surface forces $X_{ni}^*$ (which are known precisely apart from a hydrostatic pressure). Finally, let the kinematically possible field $v_i'$ be discontinuous on certain surfaces $S_l'$, $l = 1, 2, \ldots$ .

Thus we are led to a comparison of the actual velocity field $v_i$ with the kinematically possible field $v_i'$.

The basic energy equation (64.8) is applicable both to the actual stress distribution $\sigma_{ij}$ and to the kinematically possible velocity field $v_i'$, and can be rewritten in the form

$$\int \sigma_{ij}\xi_{ij}'\mathrm{d}V - \int X_{ni}v_i'\mathrm{d}S + \int \tau[v']\ \mathrm{d}S_p' = 0\,, \tag{64.11}$$

where $[v']$ denotes the jump $v_+' - v_-'$ on $S_p'$.

As we have seen (§16) the stresses $\sigma_{ij}$ and the strain rates $\xi_{ij}$, $\xi_{ij}'$ can be represented by vectors in 9-dimensional stress space. The yield criterion will be represented by a convex surface (hypersurface) – the yield surface (cf. fig. 23). By virtue of the Saint Venant-von Mises equation (64.1) the vectors $\sigma_{ij}$ and $\xi_{ij}$ will be parallel; the quantity $\sigma_{ij}\xi_{ij}$, being the scalar product of parallel vectors, will be equal to the product of their moduli, i.e.

$$\sigma_{ij}\xi_{ij} = s_{ij}\xi_{ij} = TH = \tau_s H\,.$$

The other expression $\sigma_{ij}\xi'_{ij}$ is in general the scalar product of non-parallel vectors, therefore

$$\sigma_{ij}\xi'_{ij} = s_{ij}\xi'_{ij} \leqslant TH' \; . \tag{64.12}$$

The equality sign will apply in the plastic zone when $\xi'_{ij} = c\xi_{ij}$, where $c$ is some scalar multiplier. But then it follows from (64.1) that $\overset{*}{s}_{ij} = s_{ij}$, i.e. the stress state $\overset{*}{\sigma}_{ij}$ corresponding to the kinematically possible field differs from the actual stress state $\sigma_{ij}$ by an amount equal to the hydrostatic pressure. Since $\sigma_{ij}$ satisfy the differential equations of equilibrium (64.2), it is easy to see that this additive pressure must be constant. If $S_v \neq 0$ (if the velocities are given somewhere on the surface) then $c = 1$ and the velocity fields $v_i$ and $v'_i$ coincide. If $S_F \neq 0$ (if the stresses are given somewhere on the surface) then the boundary conditions show that the additive pressure must be zero.

Passing now to the large quantity $\tau_s H'$, we obtain from (64.11) the inequality

$$\tau_s \int H' \mathrm{d}V - \int X_{ni} v'_i \mathrm{d}S + \int \tau [v'] \; \mathrm{d}S'_\mathrm{p} \geqslant 0 \; . \tag{64.13}$$

On the left-hand side there is here an unknown quantity – the actual stress $\tau$. But the maximum tangential stress $\tau_{\max}$ and the intensity $T$ are related by the inequality (1.20), from which it follows that $\tau_{\max} \leqslant \tau_s$, and hence that $|\tau| \leqslant \tau_s$.

If we replace $\tau[v']$ by $\tau_s |[v']|$ we only strengthen inequality (64.13).

For the actual velocity field the expression which is the analogue of (64.13) (i.e. without primes) vanishes; consequently,

$$\tau_s \int H \, \mathrm{d}V - \int X_{ni} v_i \mathrm{d}S_F + \tau_s \int [v] \; \mathrm{d}S_\mathrm{p} \leqslant$$
$$\leqslant \tau_s \int H' \; \mathrm{d}V - \int X_{ni} v'_i \mathrm{d}S_F + \tau_s \int |[v']| \; \mathrm{d}S'_\mathrm{p} \; . \tag{64.14}$$

The equality sign is achieved only when the kinematically possible field $v'_i$ coincides with the actual field $v_i$. We call the expression on the right-hand side the *total rate of work*.

Thus, *the total rate of work attains an absolute minimum for the actual velocity field.*

Note that because of (64.8) the left-hand side of the inequality is equal to $\int X_{ni} v_{0i} \mathrm{d}S_v$.

### 64.5. *Maximum properties of the actual stress distribution*

As before let $\sigma_{ij}$, $\xi_{ij}$, $v_i$ be the actual solution to the problem, where the stresses and strain-rates are connected by the Saint Venant-von Mises relation (64.1) and satisfy the equations of equilibrium and continuity.

We now introduce the concept of a *statically possible yield-stress distribution* $\sigma'_{ij}$. This is any stress state $\sigma'_{ij}$ which satisfies (i) the differential equations of equilibrium inside the body

$$\partial\sigma'_{ij}/\partial x_i = 0 , \tag{64.15}$$

(ii) satisfies prescribed boundary conditions on the portion $S_F$ of the surface

$$\sigma'_{ij}n_j = X_{ni} \qquad \text{on} \qquad S_F \tag{64.16}$$

(iii) does not exceed the yield limit, i.e.

$$T' \equiv (\tfrac{1}{2}s'_{ij}s'_{ij})^{\frac{1}{2}} \leqslant \tau_s . \tag{64.17}$$

The stress distribution $\sigma'_{ij}$ may be discontinuous.

We compare the actual stress distribution $\sigma_{ij}$ with the statically possible yield-stress distribution $\sigma'_{ij}$.

For the actual stresses we have the basic energy equation (64.8), where $v_i$ is the actual velocity field.

On the other hand, since the stress state $\sigma'_{ij}$ is in equilibrium, it follows from the basic energy equation that

$$\int X'_{ni}v_i \mathrm{d}S = \int \sigma'_{ij}\xi_{ij}\mathrm{d}V + \int \tau'[v]\ \mathrm{d}S_\mathrm{p} , \tag{64.18}$$

where $X'_{ni} = X_{ni}$ on $S_F$, and the forces $X'_{ni}$ on $S_v$ are determined by Cauchy's formula (64.16); $\tau'$ is the tangential component of the statically possible stress state $\sigma'_{ij}$ in the $x$-direction on the surface of discontinuity $S_\mathrm{p}$ of the actual velocities $v_i$.

Subtracting (64.8) from (64.18), and using (64.16), we obtain

$$\int (\sigma'_{ij}-\sigma_{ij})\ \xi_{ij}\mathrm{d}V = \int (X'_{ni}-X_{ni})\ v_{0i}\mathrm{d}S_v + \int (\tau_\mathrm{s}-\tau')\ [v]\ \mathrm{d}S_\mathrm{p} . \tag{64.19}$$

We now utilize an earlier geometrical representation. The vector $\xi_{ij}$ is normal to the yield surface $\Sigma$ (fig. 197); the vector $\sigma_{ij}$ is parallel to the vector $\xi_{ij}$ and reaches the yield surface. In general the vector of the statically possible stresses lies inside the yield surface. The difference vector $\sigma'_{ij}-\sigma_{ij}$ is shown by a dotted line in fig. 197. By virtue of the convexity of the yield surface the vectors $(\sigma'_{ij}-\sigma_{ij})$ and $\xi_{ij}$ generate an obtuse angle, and therefore their scalar product is negative:

$$(\sigma'_{ij}-\sigma_{ij})\ \xi_{ij} \leqslant 0 . \tag{64.20}$$

The equality sign here can only occur when the stresses $\sigma'_{ij}$ and $\sigma_{ij}$ differ by an amount equal to a uniform hydrostatic pressure (since the yield criterion is independent of the latter). This additive pressure is equal to zero if $S_F \neq 0$, i.e. if the load is prescribed anywhere on the surface.

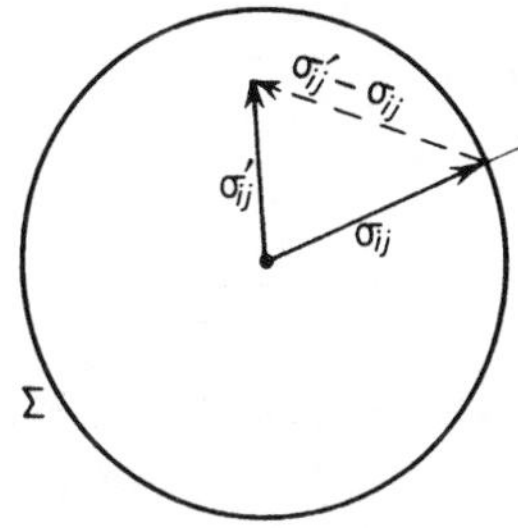


Fig. 197.

Thus it follows from (64.19) that its right-hand side is negative, so that

$$\int X_{ni} v_{0i} \mathrm{d}S_v \geqslant \int X'_{ni} v_{0i} \mathrm{d}S_v + \int (\tau_s - \tau') \; [v] \; \mathrm{d}S_\mathrm{p} \,. \tag{64.21}$$

On the right-hand side the value of the discontinuity in the actual field is unknown. Since $\tau_s \geqslant |\tau'|$ and $\tau_s[v] > 0$, the second term on the right-hand side is non-negative. Inequality (64.21) is strengthened, and we obtain

$$\int X_{ni} v_{0i} \mathrm{d}S_v \geqslant X'_{ni} v_{0i} \mathrm{d}S_v \,. \tag{64.22}$$

The right-hand side of this inequality can be calculated for a given field $\sigma'_{ij}$ since the velocities $v_{0i}$ are prescribed on $S_v$.

Thus, *the rate of work of the actual surface forces on prescribed velocities is greater than the rate of work developed by surface forces corresponding to any other statically possible yield-stress system.*

From the inequalities (64.14), (64.22) there emerges a two-sided estimate of the power of the actual surface forces for given velocities:

$$\tau_s \int H' \, \mathrm{d}V - \int X_{ni} v'_i \, \mathrm{d}S_F + \tau_s \int |[v']| \, \mathrm{d}S'_\mathrm{p} \geqslant \int X_{ni} v_{0i} \, \mathrm{d}S_v \geqslant \int X'_{ni} v_{0i} \, \mathrm{d}S_v \,. \tag{64.23}$$

The left-hand side of the inequality is evaluated by taking a kinematically possible velocity field, while the right-hand side is found by taking a statically possible yield-stress state.

In the case of plane stress it is possible to have a discontinuity in the normal velocity component, and in this case the last term on the left-hand side of the inequality is replaced by

$$\tau_s h \int v' \sqrt{1 + \sin^2 \gamma'} \, \mathrm{d}S_\mathrm{L} \,,$$

where $v'$ is the magnitude of the jump in the kinematically possible velocity, and $\gamma'$ is the angle of inclination of the velocity $v'$ to the line of discontinuity.

Fundamental results on extremum principles for a plastic-rigid body were obtained by Markov [134], Hill [54], Prager and Hodge [31], Koiter [68], and Feinberg [161].

## §65. Theorems on the limit load coefficient

The inequalities obtained in the preceding section open the way to the evaluation of limit loads by means of successive approximation by upper and lower bounds. However, the inequalities (64.14), (64.21) do not indicate directly a method of approximating the limit load except in the simplest cases (for example, if the velocity on $S_v$ is given to be constant in both magnitude and direction, then knowing the rate of work is equivalent to knowing the load on $S_v$ in this direction. This case occurs in the problem of indentation by a smooth flat die).

When several loads are present the limit state is represented by some surface ("yield surface"); for this surface the above inequalities allow, in principle, the construction of a two-sided estimate. Simple, but important, results can be obtained in the case of proportional loading.

65.1. *Proportional loading*

We consider the important case of surface forces which increase in proportion to a single parameter $m > 0$; in this case an estimate for the limit load is easy to find.

We shall dwell in more detail on the original assumptions. The loads on the portion $S_F$ of the surface grow in a definite ratio:

$$X_{ni} = mX^0_{ni} \qquad \text{on} \qquad S_F, \tag{65.1}$$

where $X^0_{ni}$ is some fixed distribution of loads on $S_F$. Moreover, we assume that on the portion $S_v$ of the surface the velocities are equal to zero:

$$v_{0i} = 0 \quad (\text{"stationary supports"}).$$

The limit state of the body is reached at some value of the parameter $m = m_*$. We shall call $m_*$ the *limit load coefficient.*

65.2. *Upper bound for the limit load*

The upper bound $m_*$ is obtained by considering inequality (64.16), which now takes the form (see remark at the end of §64.5)

$$\int X_{ni} v'_i \, \mathrm{d}S_F \leqslant \tau_s \int H' \, \mathrm{d}V + \tau_s \int |[v']| \, \mathrm{d}S'_p,$$

where the kinematically possible velocities $v_i'$ vanish on $S_v$. By virtue of (65.1) we have for the limit state

$$m_* \leqslant \tau_s \frac{\int H' \, dV + \int |[v']| \, dS_p'}{\int X_{ni}^0 v_i' \, dS_F} \equiv m_k . \tag{65.2}$$

It is assumed that the rate of work of the prescribed surface forces on the kinematically possible velocities which occur in the denominator of (65.2) is positive. The equality sign in (65.2) is attained only when the kinematically possible field $v_i'$ coincides with the actual field $v_i$ (with the stipulations indicated in the preceding section).

The dimensionless number on the right-hand side of the inequality is denoted by $m_k$, and we shall call it the *kinematic coefficient.* Thus,

$$m_* \leqslant m_k . \tag{65.3}$$

*The limit load coefficient $m_*$ cannot be greater than the kinematic coefficient $m_k$.*

It follows from (65.2) that *the kinematic coefficient $m_k$ is obtained by equating the rate of work of the loads on the kinematically possible velocities with the corresponding rate of work of deformation.*

*Remark.* In the case of plane stress the second term in the numerator must be replaced by the quantity.

$$h \int v' \sqrt{1 + \sin^2 \gamma'} \, ds_L .$$

65.3. *Lower bound of the limit load*

The lower bound of the limit load coefficient $m_*$ is found from the second extremum principle. We consider a statically possible yield-stress state $\sigma_{ij}'$ which satisfies somewhat different boundary conditions on $S_F$:

$$X_{ni}' = m_s X_{ni}^0 \qquad \text{on} \qquad S_F \tag{65.4}$$

(in place of (64.16) which in this case can be written in the form $X_{ni} = m_* X_{ni}^0$). Here $m_s$ is some value of the parameter $m$. The inequality obtained at the end of the preceding section cannot be applied directly to this case. In place of (64.19) we now obtain

$$\int (\sigma_{ij}' - \sigma_{ij}) \, \xi_{ij} dV = (m_s - m_*) \int X_{ni}^0 v_i \, dS_F + \int (\tau_s - \tau') \, [v] \, dS_p . \tag{65.5}$$

Equation (64.20) shows that, in general, the left-hand side is negative; consequently

$$m_* - m_s \geqslant \frac{\int (\tau_s - \tau')\,[v]\,\mathrm{d}S_p}{\int X_{ni}^0 v_i\,\mathrm{d}S_F}. \tag{65.6}$$

The equality sign is attained only if the stress systems $\sigma_{ij}$, $\sigma'_{ij}$ differ by a uniform pressure. Since the numerator is non-negative ($|\tau'| \leqslant \tau_s$, $\tau_s[v] > 0$), while the denominator is positive, *the limit load coefficient* $m_*$ *can not be less than the static coefficient* $m_s$:

$$m_s \leqslant m_* . \tag{65.7}$$

65.4. *Corollaries*

We consider a number of corollaries which emerge from the foregoing inequalities.

1. *The limit load coefficient* $m_*$ *is unique.*

One way of demonstrating this is to consider inequality (65.3), according to which the limit load coefficient $m_*$ attains an absolute minimum for the actual velocity field. The hypothesis that two limit load coefficients $m_{*1}$, $m_{*2}$ exist is consistent with the condition of the absolute minimum for $m_*$ only if they coincide.

2. *Addition of matter to the body cannot reduce the limit load.*

We shall clarify this statement by a simple example. Consider a circular tube (fig. 198a); we denote by

$$p_0 = 2\sigma_s \ln (b/a)$$

the limit pressure corresponding to an axisymmetrical stress field (§26).

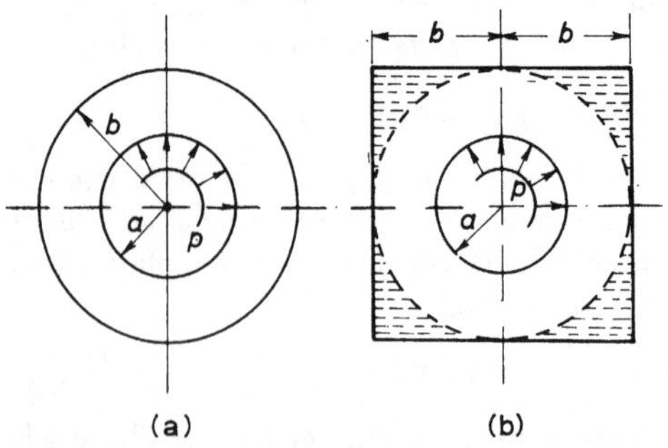


Fig. 198.

Now turn to the problem of finding the limit load $p_*$ for a thin square prism with a circular hole at the centre, subject to a uniform pressure; the outer surface of the prism is stress-free (fig. 198b).

To calculate the lower bound – the statically possible coefficient – we can take the following discontinuous stress distribution: we inscribe the tube shown in fig. 198a inside the prism; the stresses in the shaded corner regions will be assumed equal to zero. It is obvious that the equations of equilibrium and boundary conditions are satisfied, and that the yield criterion is nowhere exceeded. Evidently

$$p_* \geqslant p_0 \,.$$

This result is, of course, obvious, and permits a simple generalization. It is clear that attaching matter to the free boundary ("addition of matter to the body") does not reduce the limit load, since for the new body we can take a statically possible yield-stress state generated by zero stresses in the additional material and the stresses of the limit state in the initial body (analogous to fig. 198). But then the lower bound $m_*$ for the limit load remains the same.

In a similar manner we can show that

3. *Removal of material cannot increase the limit load.*

4. *Increase of the yield limit $\tau_s$ in some parts of the body cannot reduce the limit load* (since any statically possible yield-stress state for the original body will also be a statically possible yield-stress state for the new body).

5. *Of two kinematically possible solutions the more acceptable is the one which leads to the smaller limit load.*

This proposition was previously (§40) called the selection criterion.

6. *Of two statically possible solutions the more acceptable is the one which leads to the larger limit load.*

## 65.5. *Extension of limit load theorems to general yield criterion*

The theorems established above relate only to the von Mises yield criterion. At the same time we have repeatedly emphasized the significance of other yield criteria, in particular the Tresca-Saint Venant yield criterion. Theorems on limit loads can easily be proved for a general convex plasticity criterion $f(\sigma_{ij}) = \kappa$ with the associated flow law (§16).

*Theorem on the kinematic coefficient $m_k \geqslant m_*$.* For the determination of $m_k$ for the kinematically possible field $v_i'$ we have

$$m_k \int X_{ni}^0 v_i' \, dS_F = \int \sigma_{ij}^* \xi_{ij}' \, dV + \int \tau^* [v'] \, dS_p' \,.$$

Here $\sigma_{ij}^*$ is the stress tensor corresponding (according to the associated law) to the kinematically possible strain rates $\xi'_{ij}$, and $\tau^*$ is the corresponding tangential stress on the surface of discontinuity $S'_p$. The stresses $\sigma^*_{ij}$ lie on the yield surface $\Sigma$, but, in general, do not satisfy the equilibrium equations. On the other hand equation (64.11) gives

$$m_* \int X^0_{ni} v'_i \, \mathrm{d}S_F = \int \sigma_{ij} \xi'_{ij} \, \mathrm{d}V + \int \tau [v'] \, \mathrm{d}S'_p ,$$

where $\sigma_{ij}$, $\tau$ are the actual stresses. Subtracting, we obtain

$$(m_k - m_*) \int X^0_{ni} v'_i \, \mathrm{d}S_F = \int (\sigma^*_{ij} - \sigma_{ij}) \, \xi'_{ij} \, \mathrm{d}V + \int (\tau^* - \tau) \, [v'] \, \mathrm{d}S'_p . \tag{65.8}$$

Since the yield surface is convex (fig. 199a), $(\sigma^*_{ij} - \sigma_{ij}) \, \xi'_{ij} \geqslant 0$; this remains true for singular points of the surface. Furthermore, as we have seen earlier, the surface of discontinuity is a slip surface; the tangential stress $\tau^*$ is associated with the field $\xi'_{ij}$ and reaches its maximum value on $S'_p$; hence $(\tau^* - \tau) \, [v'] \geqslant 0$. Thus, the right-hand side of (65.8) is non-negative. The fact that the rate of work of the given loads is positive leads to the required assertion.

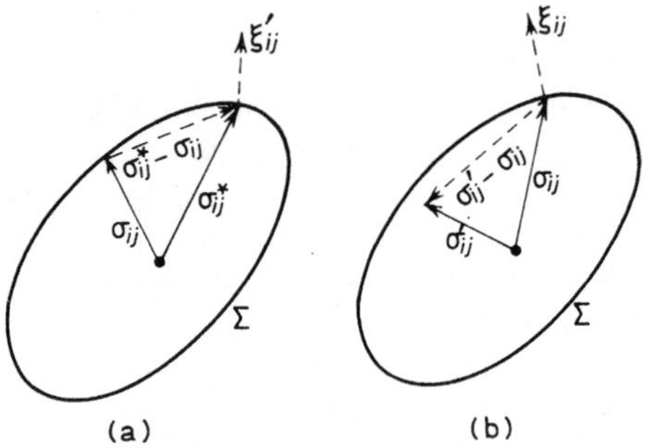


Fig. 199.

We consider, in particular, the widely used *Tresca-Saint Venant yield criterion*. Here the kinematically possible coefficient $m_k$ is determined by equating the rate of work of the given loads to the rate of work of plastic deformation:

$$m_k \int X^0_{ni} v'_i \, \mathrm{d}S_F = 2\tau_s \int \xi'_{max} \, \mathrm{d}V , \tag{65.9}$$

where $\xi_{max}$ is the absolute value of the numerically largest principal strain-rate (§ 16). If the kinematically possible velocity field $v'_i$ is discontinuous, it is necessary to include in the right-hand side of (65.9) the rate of work of plastic deformation dissipation in the discontinuities.

*Theorem on the static coefficient* $m_s \leqslant m_*$ follows in an obvious way from a variant of (65.5):

$$(m_s - m_*) \int X^0_{ni} v_i \, dS_F = \int (\sigma'_{ij} - \sigma_{ij}) \, \xi_{ij} \, dV + \int (\tau' - \tau) \, [v] \, dS_p \,, \tag{65.10}$$

where $\sigma_{ij}$, $\tau$ are the actual stresses, and $\sigma'_{ij}$, $\tau'$ are statically possible stresses, lying inside or on the yield surface. Since the yield surface is convex (fig. 199b), $(\sigma'_{ij} - \sigma_{ij}) \, \xi_{ij} \leqslant 0$, where this holds at singular points of $\Sigma$ also. Since $v_i$ is the actual velocity, then, as in the previous case, $(\tau - \tau') \, [v] \geqslant 0$. The second assertion follows at once from the fact that the left-hand side of (65.10) is negative.

### 65.6. *Limit load in contact problems*

It is sometimes necessary to find limit loads for a system of contiguous bodies. In cases when very simple conditions (for example, constancy of the tangential stress) apply on the contact surface, it is possible to use the preceding theorems. A number of problems of this type will be examined in the next section.

If Coulomb friction acts on the contact surface, it is not difficult to show [107] that the required limit load does not exceed the limit load for the same system of bodies with *soldered contact surfaces*, and is not less than the limit load for the same system with *perfectly smooth contact surfaces*.

### 65.7. *Concluding remarks*

Extremum properties of the limit load and the possibility of using them for approximating the limit load were first formulated (in terms of structural mechanics) by Gvozdev in 1936. The theorem on the lower bound (for continuous solid beams) was stated as long ago as 1914 by Kazinchi. A rigorous proof of the lower bound theorem was given by Feinberg. Subsequent works on theorems of plastic collapse include the investigations of Drucker, Feinberg, Prager (cf. [68, 70]), and Hill [54, 163].

## §66. Determination of the limit load by energy methods

### 66.1. *General remarks*

As we have already noted, the energy method permits the determination of an effective solution to problems of load-bearing capacity. This method is widely applied in various branches of the theory of limiting equilibrium – in the structural mechanics of systems of rods, in problems of the limit equilibrium of plates and shells, etc. With the aid of comparatively simple calculations it is often possible to construct upper and lower bounds which coincide, i.e. to find the exact value of the limit load. A simple example of this sort – the extension of a strip with a circular hole – was discussed in §40. Some other problems are presented below.

Usually it is a simple matter to establish rough upper and lower estimates with the aid of energy methods. It is considerably more difficult to obtain good estimates. It is even more difficult to develop methods of successive approximation for the upper and lower bounds; here the use of mathematical programming methods is promising, but requires the application of modern numerical techniques and the development of the corresponding algorithms [191].

66.2. *Trapezium of discontinuous stresses for plane strain*

To construct statically possible yield-stress distributions it is convenient to make use of discontinuous fields, compounded from regions of uniform stress. A simple field [1]) of this type is shown in fig. 200. The boundaries AC, BD are stress-free, a uniform normal stress $\sigma_1'$ acts on the boundary AB, and a uniform normal stress $\sigma_1''$ acts on the boundary CD. Each triangle contains a uniform stress distribution. In Δ AOB we denote the principal stresses by $\sigma_1'$, $\sigma_2'$, and in Δ COD by $\sigma_1''$, $\sigma_2''$; in the triangles AOC and BOD there is uniaxial tension $s$, parallel to the edges AC or BD respectively. The rays OA, OB, OC, OD are lines of stress discontinuity, along which the normal stress $\sigma_n$ and the tangential stress $\tau_n$ are continuous (cf. §39). Writing out these continuity conditions along OB with the aid of (35.1) (the $x$-axis is along the $\sigma_1'$ axis in

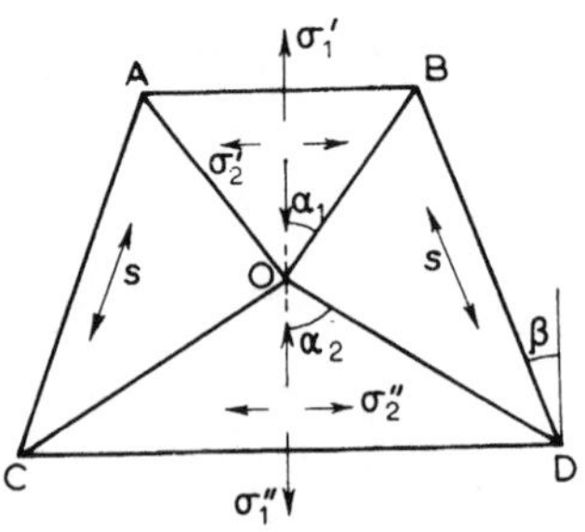


Fig. 200.

[1]) The conditions for intersection of lines of stress discontinuity have been studied in the works [31, 110, 180].

Δ AOB and in the direction of $s$ in Δ BOD), we obtain

$$\sigma_n = \sigma_1' \sin^2 \alpha_1 + \sigma_2' \cos^2 \alpha_1 = s \sin^2(\beta + \alpha_1)\,,$$

$$\tau_n = (\sigma_1' - \sigma_2') \sin \alpha_1 \cos \alpha_1 = s \sin(\beta + \alpha_1) \cos(\beta + \alpha_1)\,.$$

Hence we find

$$\sigma_1' = s\,\frac{\sin(\beta + \alpha_1)\cos\beta}{\sin\alpha_1}\,, \qquad \sigma_2' = s\,\frac{\sin(\beta + \alpha_1)\sin\beta}{\cos\alpha_1}\,. \tag{66.1}$$

Next we write out the continuity conditions for $\sigma_n$, $\tau_n$ on the line of discontinuity OD:

$$\sigma_1'' \sin^2 \alpha_2 + \sigma_2'' \cos^2 \alpha_2 = s \sin^2(\alpha_2 - \beta)\,,$$

$$(\sigma_1'' - \sigma_2'') \sin \alpha_2 \cos \alpha_2 = s \sin(\alpha_2 - \beta) \cos(\alpha_2 - \beta)\,,$$

from which

$$\sigma_1'' = s\,\frac{\sin(\alpha_2 - \beta)\cos\beta}{\sin\alpha_2}\,, \qquad \sigma_2'' = s\,\frac{\sin(\alpha_2 - \beta)\sin\beta}{\cos\alpha_2}\,. \tag{66.2}$$

The stress states in these triangles must not violate the von Mises yield criterion. We assume that $s$ is a tensile stress, whereupon, in keeping with formulae (66.1), (66.2), it can be assumed that $\sigma_1' \geqslant \sigma_2' \geqslant 0$, $\sigma_1'' \geqslant \sigma_2'' \geqslant 0$; since $\sigma_1''$ must be tensile, then $\alpha_2 > \beta$.

In the case of *plane strain* the yield criteria are not violated provided

$$\text{in} \quad \Delta\,\text{AOB} \qquad \sigma_1' - \sigma_2 \leqslant 2k\,, \tag{66.3}$$

$$\text{in} \quad \Delta\,\text{BOD} \qquad s \leqslant 2k\,, \tag{66.4}$$

$$\text{in} \quad \Delta\,\text{COD} \qquad \sigma_1'' - \sigma_2'' \leqslant 2k\,, \tag{66.5}$$

where $k$ is the yield limit for shear (§31).

The stresses $\sigma_1'$, $\sigma_1''$ are proportional to $s$; since we are constructing statically possible fields, leading to a lower bound, it is expedient to take the largest value $s = 2k$. Substituting now the stress components (66.1), (66.2) in the inequalities (66.3), (66.5), and effecting some simple transformations, we find

$$\text{in} \quad \Delta\,\text{AOB} \qquad \tan 2\alpha_1 \geqslant \cot\beta\,,$$

$$\text{in} \quad \Delta\,\text{COD} \qquad \sin 2\beta / \sin 2\alpha_2 \geqslant 0\,.$$

The second condition is obviously fulfilled. From the first it follows that

$$\alpha_1 \geqslant \tfrac{1}{4}\pi - \tfrac{1}{2}\beta \,. \tag{66.6}$$

The angle $\alpha_1$ is conveniently chosen so that the stress $\sigma_1'$ is largest. Equation (66.1) for $\sigma_1'$ can then be written in the form

$$\sigma_1' = s\cos^2\beta(1 + \tan\beta\cot\alpha_1) \,. \tag{66.7}$$

It is clear that we have to choose the least value of $\alpha_1$, i.e.

$$\alpha_1 = \tfrac{1}{2}\pi - \tfrac{1}{2}\beta \,. \tag{66.8}$$

66.3. *Bounds for limit load in extension of strip with circular notches*

The problem of the extension of a strip weakened by circular notches (in conditions of plane strain – fig. 107) was considered in §41, where the upper ("kinematic") bound of the tensile force was found:

$$P_{\mathrm{k}} = 4kh(1 + (a/h))\ln(1 + (h/a)) \,.$$

We shall concentrate on the special case $h = a$, where the width of the strip is $4a$ (fig. 201). Then

$$P_{\mathrm{k}} = 5.55\, ak \,.$$

It should be understood that it is not necessary to calculate the upper bound for the load in constructing the velocity and slip fields. It is sufficient

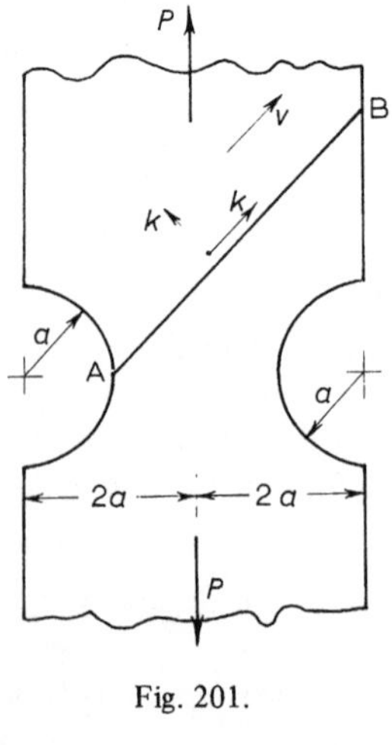


Fig. 201.

to choose any kinematically possible velocity field. For example, in the present problem we can choose the discontinuous field shown in fig. 201. The lower part of the strip is stationary, and the upper slides along the line of discontinuity AB like a rigid body. The line AB traverses a slip line, and consequently it is inclined at an angle $\frac{1}{4}\pi$ to the tensile force. The tangential stress along AB is equal to the yield limit $k$; from the equilibrium condition it follows that the normal stress has the same value. The line AB must be taken such that it has the smallest length. It is easy to see that the corresponding bound $P_{\mathrm{k}}'$ is somewhat worse than $P_{\mathrm{k}}$, namely,

$$P_{\mathrm{k}}' = 6\,ak\ .$$

It is not difficult to obtain a crude lower bound, by inscribing in the strip a smooth band of width $2a$ with uniaxial stress $2k$; then $P_{\mathrm{s}}^0 = 4ak$. As Prager has shown, a substantially better estimate can be obtained by inscribing in the strip the trapezium discussed in the preceding paragraph. Here the angle $\beta$ (fig. 202) must be chosen so that the tensile force is greatest. In the shaded

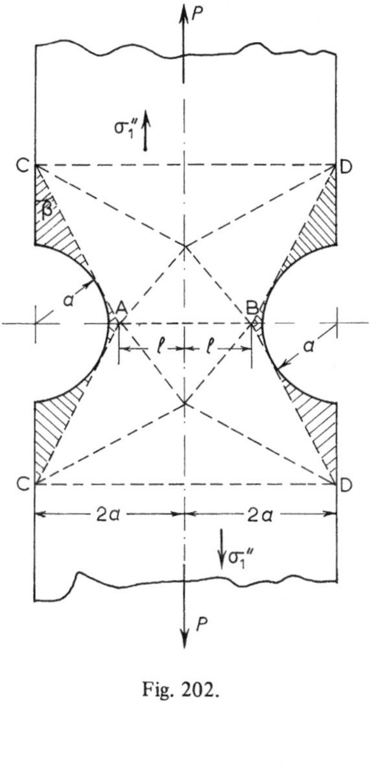


Fig. 202.

zones the stresses are zero, and in the regions above the line CD there is uniaxial tension $\sigma_1''$. It is clear from the diagram that $l = a(2-1/\cos\beta)$, $\beta < \frac{1}{3}\pi$. The corresponding tensile force is

$$P_s = 2l\sigma_1' = 4ak(2\cos\beta - 1)\,\frac{\sin\left(\frac{1}{4}\pi + \frac{1}{2}\beta\right)}{\sin\left(\frac{1}{4}\pi - \frac{1}{2}\beta\right)}. \tag{66.9}$$

To find the maximum $P^s$ we equate to zero its derivative with respect to $\beta$ and find that $\beta = 26°14'$. Then

$$P_s = 5.12\, ak\,.$$

For the limit load we can take the mean value $P_* = 5.33\, ak$; this involves an error of ± 4%.

66.4. *Discontinuous stress trapezium for plane stress*

In the case of plane stress, formulae (66.1), (66.2) are retained, but the yield criteria will be different. From the von Mises criterion (§52) we now have

$$\text{in} \quad \triangle\,\text{AOB} \qquad \sigma_1'^2 - \sigma_1'\sigma_2' + \sigma_2'^2 \leqslant \sigma_s^2\,, \tag{66.10}$$

$$\text{in} \quad \triangle\,\text{BOD} \qquad s \leqslant \sigma_s\,, \tag{66.11}$$

$$\text{in} \quad \triangle\,\text{COD} \qquad \sigma_1''^2 - \sigma_1''\sigma_2'' + \sigma_2''^2 \leqslant \sigma_s^2\,. \tag{66.12}$$

Substituting for the stresses from (66.1), (66.2) in the inequalities (66.10), (66.12), we obtain

$$\sin^2(\beta + \alpha_1)\left[\frac{\cos 2\beta}{\sin^2\alpha_1} + \frac{\sin^2\beta}{\cos^2\alpha_1}\right] \leqslant \left(\frac{\sigma_s}{s}\right)^2, \tag{66.13}$$

$$\sin^2(\alpha_2 - \beta)\left[\frac{\cos 2\beta}{\sin^2\alpha_2} + \frac{\sin^2\beta}{\cos^2\alpha_2}\right] \leqslant \left(\frac{\sigma_s}{s}\right)^2. \tag{66.14}$$

The maximum value of $\sigma_1'$ which is admissible by the von Mises yield criterion corresponds to the point $\omega = \frac{1}{6}\pi$ on the ellipse (fig. 157); here

$$\sigma_1' = \frac{2}{\sqrt{3}}\,\sigma_s\,, \qquad \sigma_2' = \frac{1}{\sqrt{3}}\,\sigma_s\,.$$

Substituting the stresses as given by (66.1) we easily find

$$\cot \beta \cot \alpha_1 = 2 ; \qquad \frac{\sin (\beta + \alpha_1) \cos \beta}{\sin \alpha_1} = \frac{2}{\sqrt{3}} \frac{\sigma_s}{s},$$

from which it follows that

$$\cos^2 \beta = 2 \left( 1 - \frac{1}{\sqrt{3}} \frac{\sigma_s}{s} \right) .$$

Using inequality (66.11) we obtain

$$\cos^2 \beta \leqslant 2 \left( 1 - \frac{1}{\sqrt{3}} \right), \qquad \text{i.e. } \beta \geqslant 23^\circ .$$

When $\beta = 23^\circ$ inequality (66.14) is satisfied for $\alpha_2 \leqslant 61^\circ 15'$; if the equality sign is attained here, then $\alpha_1 = 49^\circ 40'$.

66.5. *Lower bound for limit load in extension of strip with angular notches.* We consider the problem of the extension of a strip with angular notches (fig. 175), under conditions of plane stress. The upper bound for the limit load was found in §56. We now evaluate the lower bound, using the discontinuous stress trapezium. To do this we inscribe two trapezia in the strip (shown by dotted lines in fig. 203); adjoining them above and below there are

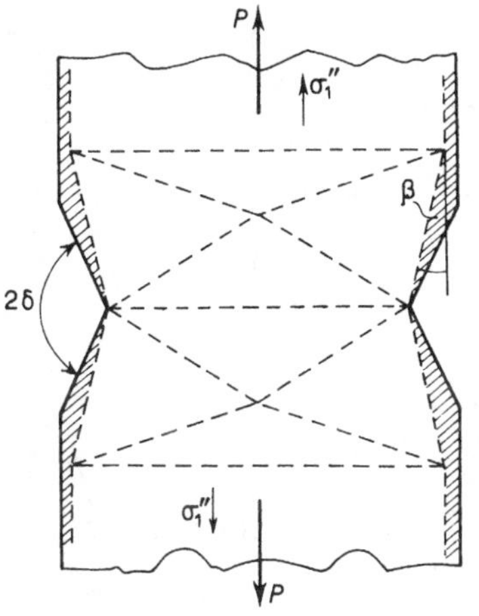


Fig. 203.

fields of uniform uniaxial stress $\sigma_1''$. When $\delta \leqslant \frac{1}{2}\pi-\beta \leqslant 67°$ the trapezia lie inside the strip. The load $P_s$ is then constant and equal to

$$P_s = \frac{4}{\sqrt{3}} h\sigma_s . \tag{66.15}$$

For angles $0 \leqslant \delta \leqslant 67°$ this value coincides with the upper bound (56.8) which was obtained by Hill for $\delta < 70°32'$. Consequently, in the interval $0 \leqslant \delta \leqslant 67°$ formula (66.15) gives the exact value of the limit load.

For angles $\delta \geqslant 67°$, we assume that the sides of the trapezia coincide with the sides of the notches; then in (66.1) and (66.2) we have $\beta = \frac{1}{2}\pi-\delta$ and $s = \sigma_s$. It is necessary to select a value of the angle $\alpha_1$ which is consistent with inequality (66.13) and which leads to the maximum possible-value of $\sigma_1'$. The results of the calculations are given in the table:

| $\delta$ | 70° | 75° | 80° | 85° | 90° |
|---|---|---|---|---|---|
| $P_s/P_*^0$ | 1.152 | 1.132 | 1.103 | 1.058 | 1.00 |

Here $P_*^0 = 2h\sigma_s$ denotes the limit load for a smooth strip of width $2h$. The upper and lower bounds for the strength coefficient $P_*/P_*^0$ are shown in fig. 204.

A similar construction can be effected for the extension of a strip with

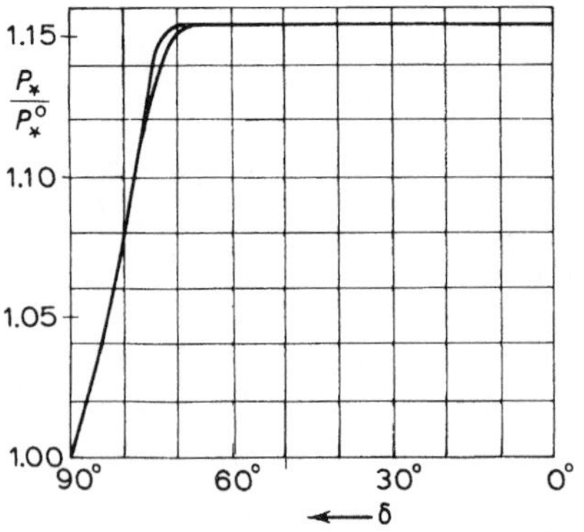


Fig. 204.

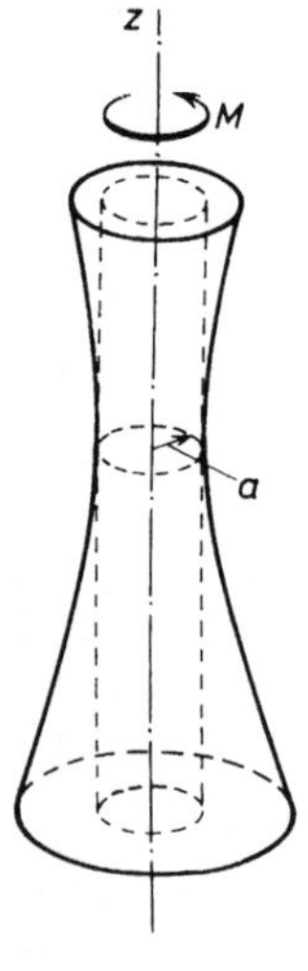


Fig. 205.

circular notches (fig. 173), for bending of strips weakened by notches (fig. 176), and in other problems.

66.7. *Torsion of a circular rod of variable diameter*

We consider the question of the limit value of the moment for twisting of a circular rod of variable diameter (fig. 205). Introduce a cylindrical coordinate system $r$, $\varphi$, $z$, with the $z$-axis along the axis of the rod. As in the case of elastic torsion it may be assumed that cross-sections of the rod remain plane, but that the radii are distorted. Consequently, the velocity components are

$$v_r = v_z = 0 ; \qquad v_\varphi = v_\varphi(r, z) \equiv v .$$

The strain-rate components are

$$\xi_r = \xi_\varphi = \xi_z = \eta_{rz} = 0 ,$$

$$\eta_{r\varphi} = r \frac{\partial}{\partial r}\left(\frac{v}{r}\right) , \qquad \eta_{\varphi z} = r \frac{\partial}{\partial z}\left(\frac{v}{r}\right) .$$

It follows from the Saint Venant-von Mises equations (13.11) that

$$\sigma_r = \sigma_\varphi = \sigma_z = \tau_{rz} = 0 , \qquad \eta_{r\varphi} = 2\lambda' \tau_{r\varphi} , \qquad \eta_{\varphi z} = 2\lambda' \tau_{\varphi z} . \tag{66.16}$$

The non-zero stress components $\tau_{r\varphi}$, $\tau_{\varphi z}$ satisfy the differential equation of equilibrium

$$\frac{\partial \tau_{r\varphi}}{\partial r} + \frac{\partial \tau_{\varphi z}}{\partial z} + \frac{2\tau_{r\varphi}}{r} = 0 \tag{66.17}$$

and the yield criterion

$$\tau_{r\varphi}^2 + \tau_{\varphi z}^2 = k^2 . \tag{66.18}$$

The corresponding stress field in the plastic region was studied by Sokolovskii [44]. The yield criterion will be satisfied if we put

$$\tau_{r\varphi} = k \sin\theta , \qquad \tau_{\varphi z} = k \cos\theta ,$$

where $\theta$ is the unknown angle of inclination of the tangential stress vector to the $z$-axis. Eliminating the multiplier $\lambda'$ from the above Saint Venant-von Mises equations, we find

$$\frac{\partial}{\partial r}\left(\frac{v}{r}\right) \cos\theta - \frac{\partial}{\partial z}\left(\frac{v}{r}\right) \sin\theta = 0 . \tag{66.19}$$

The characteristics of this equation coincide with the slip lines. The differential equation (66.19) has the obvious solution

$$v = Cr, \tag{66.20}$$

where $C$ is an arbitrary constant. This solution corresponds to rigid-body rotation of the shaft or of part of it.

We introduce the plane $z$ = const. of velocity discontinuity. Above and below this plane the solution (66.20) holds with different values of the arbitrary constant. On the plane of discontinuity we have $\eta_{r\varphi} \to 0, \eta_{\varphi z} \to \infty$.

It now follows from (66.16) that

$$\tau_{r\varphi} = 0\,, \qquad \tau_{\varphi z} = \text{const.} = k\,. \tag{66.21}$$

This solution corresponds to the torque

$$M = 2\pi \int_0^a kr^2 \mathrm{d}r = \tfrac{2}{3}\pi a^3 k\,, \tag{66.22}$$

where $a$ is the radius of the section.

The solution we have constructed corresponds to a kinematically possible velocity field (with a "cut" in the plane $z$ = const.), and therefore $M$ is the upper bound of the limit load. It is natural to suppose that $a$ is the radius of the smallest cross-section of the shaft.

On the other hand it is easy to construct a statically possible stress field, which does not violate the yield criterion. To do this it is sufficient to inscribe a circular rod (shown by dotted lines in fig. 205) of constant radius $a$ in the shaft, with the limit field (66.21), and to assume that the stresses are zero for $r > a$. Then by the static theorem for limit loads, $M$ will also be the lower bound. Thus, the complete solution has been found and $M$ is the exact value of the limit moment for torsion of a shaft with variable diameter.

### 66.8. *Shear and compression of a thin layer*

In ch. 5 (§47) we examined Prandtl's problem of compression of a thin plastic layer between rough, rigid plates. It was shown that the flow in the layer is considerably affected by the presence of a shearing force $2Q$ (fig. 206). We shall now derive a statically possible solution to this problem. In the absence of the shear, the upper and lower bounds of the compressive force for a thin plastic layer were found by Shield.

If there is no shear ($Q = 0$), then maximum tangential stresses $\tau_{xy} = \pm k$ develop on the contact planes $y = \pm h$, where $k$ is the yield limit for shear. In the presence of a shearing force $2Q$ the tangential stresses on the segments

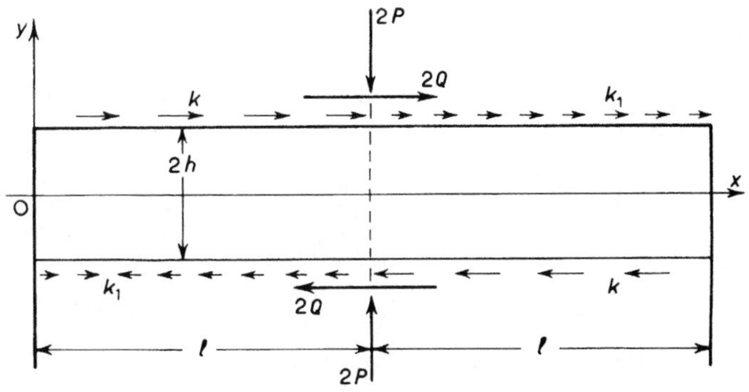


Fig. 206.

$y = h$, $x < l$ and $y = -h$, $x > l$ are of magnitude $k$ as before. On the remaining segment the tangential stresses "unload", and their magnitude is less than $k$; we assume that they are equal to $k_1$, $|k_1| \leqslant k$. It is an easy matter to construct the following solutions to the equilibrium equations (31.9) and the yield criterion (31.8) which satisfy prescribed boundary conditions for $\tau_{xy}$ on the lines $y = \pm h$:

$$\tau_{xy}/k = \tfrac{1}{2}(1+\kappa) + \tfrac{1}{2}(1-\kappa)\, y/h\,, \qquad \sigma_y/k = -C - \tfrac{1}{2}(1-\kappa)\, x/h\,,$$
$$\sigma_x/k = \sigma_y/k + 2\sqrt{1-(\tau_{xy}/k)^2}\,, \qquad x < l\,, \tag{66.23}$$

where $\kappa = k_1/k$, $|\kappa| \leqslant 1$, and $C$ is an arbitrary constant. When $\kappa = -1$ Prandtl's well-known formulae follow from (66.23); the case $\kappa = 1$ corresponds to the problem of pure shear of the layer ($\sigma_x = \sigma_y = 0$, $\tau_{xy} = k$).

The constants $C$ and $\kappa$ must be determined from the condition of static equivalence. First we satisfy, in the Saint Venant sense, the condition that there are no normal stresses on the edge of the layer, namely,

$$\int_{-h}^{h} (\sigma_x)_{x=0}\, \mathrm{d}y = 0\,.$$

From this we find

$$C = \frac{1}{1-\kappa}\left(\tfrac{1}{2}\pi - \kappa\sqrt{1-\kappa^2} - \arcsin \kappa\right).$$

Next, the condition that the normal stresses $\sigma_y$ are equivalent to a compression force $2P$ gives

$$C = p - \tfrac{1}{4}(1-\kappa)\, l/h \qquad (p \equiv P/kl)\,.$$

Finally, the condition that the contact tangential stresses are equivalent to a shearing force yields

$$1 + \kappa = 2q \qquad (q \equiv Q/kl \leqslant 1)\,.$$

Eliminating $C$ and $\kappa$ from these equations, we obtain the condition of limiting equilibrium:

$$(1-q)\,[2p-(1-q)\,l/h] = \tfrac{1}{2}\pi + 2(1-2q)\sqrt{q(1-q)} - \arcsin(2q-1)\,. \qquad (66.24)$$

When $q = 0$ Prandtl's formula (47.3) follows from this. It is also easy to see that $\mathrm{d}p/\mathrm{d}q \to -\infty$ when $q \to 1$.

Fig. 207 shows the limiting curves for the values $l/h = 10$ and $l/h = 20$. It is

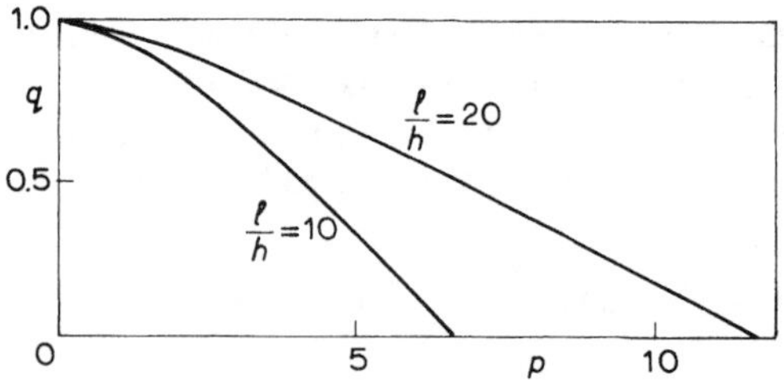


Fig. 207.

evident that adding a shearing force to the compressive load significantly lowers the load-bearing capacity of the layer. It can be shown that our solution gives a lower bound for the compressive force, i.e. $P > P'$ (for fixed $Q$).

66.9. *Compression of a cylinder between rough plates*

We consider the compression of a circular cylinder (height $2h$, diameter $2a$) between rough, parallel plates (fig. 208). The tangential stresses $\tau_{rz}$ attain the yield limit $\tau_s = \sigma_s\sqrt{3}$ on the contact planes. We introduce the dimensionless coordinates $\rho = r/a$, $\zeta = z/a$, $\kappa = h/a$.

*Upper bound.* We assume that the kinematically possible radial velocity $u'$ is

$$u' = A\rho(1-\beta\,\zeta^2/\kappa^2)\,,$$

where $A$ is a constant, and $0 \leqslant \beta \leqslant 1$ is a parameter characterizing the degree

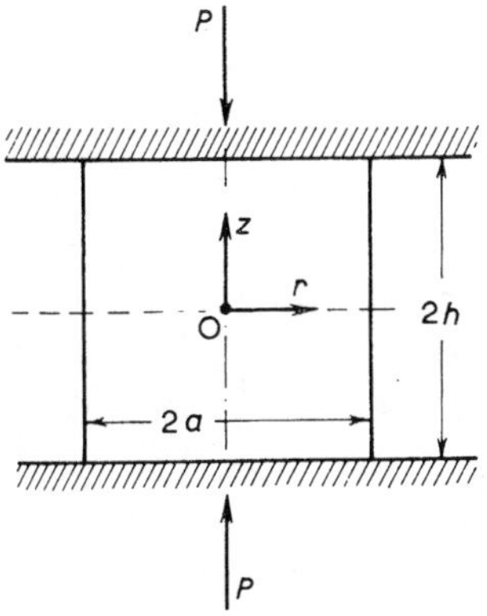


Fig. 208.

of "barrel-formation". When $\beta = 0$ there is no "barrel-formation". The axial velocity component $w'$ is found from the incompressibility equation (58.6):

$$w' = -2A(\zeta - \beta\zeta^3/3\kappa^2)\,.$$

Let $c$ be the velocity of motion of the plate; then $w' = -c$ when $\zeta = \kappa$ from which we find $A = 3c/2\kappa(3-\beta)$.

In accordance with (64.14) the upper bound of the compressive force is found from the relation

$$P_* c \leqslant 2\pi\tau_s \int_0^\kappa \int_0^1 H'\rho\, d\rho\, d\zeta + 2\pi\tau_s \int_0^1 (u')_{\zeta=\kappa}\, \rho\, d\rho\,. \quad (66.25)$$

The intensity of the shear strain-rates $H'$ can be calculated from the velocity field; by symmetry we can consider half the cylinder $\zeta \geqslant 0$. Performing the calculations, we obtain

$$\frac{p_*}{\sigma_s} \leqslant \frac{2}{3} + \frac{1}{\sqrt{3}\kappa}\,\frac{1-\beta}{3-\beta} + \frac{2}{3-\beta} \int_0^1 \frac{(1-\beta\rho^2)^2 + \frac{1}{3}\beta^2\rho^2/\kappa^2}{[(1-\beta\rho^2)^2 + \frac{1}{3}\beta^2\rho^2/\kappa^2]^{\frac{1}{2}} + (1-\beta\rho^2)}\, d\rho \equiv \frac{p'}{\sigma_s}\,.$$

Here we have introduced the mean pressure $p_* = P_*/\pi a^2$. The parameter $\beta$ is chosen such that $p'$ is a minimum for given $\kappa$. Fig. 209 shows the results of calculations [185] for cylinders of various heights. The dotted line corres-

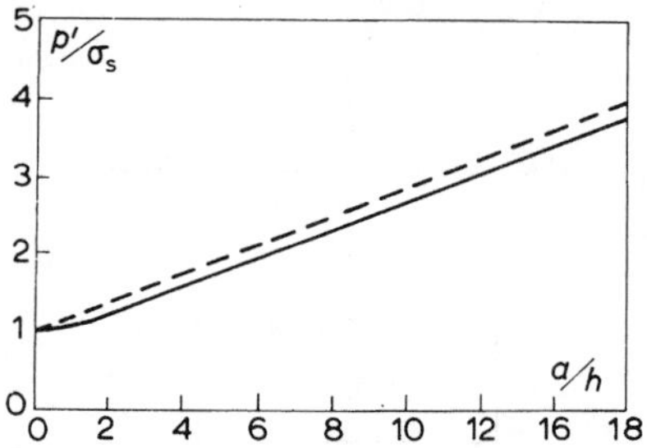


Fig. 209.

ponds to the elementary solution for $\beta = 0$; for which

$$\frac{p'}{\sigma_s} = 1 + \frac{1}{3\sqrt{3\kappa}} .$$

The *lower bound* is much more difficult to construct, since in this problem the statically possible yield-stress distribution satisfies a complex system of equations and boundary conditions. The construction of a discontinuous stress field is given in the above-mentioned work by Kobayashi and Thomsen [185].

66.10. *Bending of plates*

The energy method is very convenient for finding the upper bound of the limit load in the bending of plates of different shapes. To do this it is necessary to prescribe the kinematically possible form of the rate of bending of the plate.

By way of illustration we consider first *axisymmetric plates* (cf. §62). In this case the kinematically possible limit load coefficient is found from (65.2), which gives

$$m_k = \tau_s \int_a^b \int_{-h}^{h} H' r \, dr \, dz \Big/ \int_a^b p(r) w' r \, dr ,$$

where $w'$ is the kinematically possible velocity of bending. Using (62.2) and the incompressibility condition we can easily show that the intensity of possible shear strain-rates is

$$H' = 2|z| \left[ \left( \frac{d^2 w'}{dr^2} \right)^2 + \frac{1}{r^2} \left( \frac{dw'}{dr} \right)^2 + \frac{1}{r} \frac{dw'}{dr} \frac{d^2 w'}{dr^2} \right]^{\frac{1}{2}} \equiv 2|z| W .$$

Integrating with respect to $z$ we obtain

$$m_k = \frac{2}{\sqrt{3}} M_s \int_a^b W r dr \Big/ \int_a^b p(r)\, w' r dr \qquad (M_s = \sigma_s h^2)\,. \quad (66.26)$$

*Example.* A solid plate ($a = 0$), supported at its edge $r = b$, is bent by a uniform load $p$ = const. When $r = b$, $w' = 0$; assuming that the bending surface is smooth, we have that $dw'/dr = 0$ when $r = 0$. These conditions are satisfied, for example, by the function $w' = b^2 - r^2$. Then from (66.26) we obtain $m_k = 8M_s/pb^2$, and consequently $p_* < 8M_s/b^2$ (we recall that in the exact solution $p_* \approx 6.5M_s/b^2$).

If for $w'$ we take the form of bending of an analogous elastic plate at $\nu = \frac{1}{2}$, i.e.

$$w' = w_0' \left[ 1 - \frac{14}{11} \left( \frac{r}{b} \right)^2 + \frac{3}{11} \left( \frac{r}{b} \right)^4 \right],$$

where $w_0'$ is an arbitrary multiplier, we can substantially improve the estimate:

$$p_* < 6.7\, M_s/b^2\,.$$

Returning to the general case we observe that it is easy to find [36] the upper bound of the limit load for *polygonal plates supported at the perimeter and bent by a concentrated force P* (fig. 210). It may be assumed that in the limit state the median surface of such a plate has the form of the surface of a pyramid with vertex O at the point where the force is applied. Along the edges yield joints develop, while the triangular regions of the plate

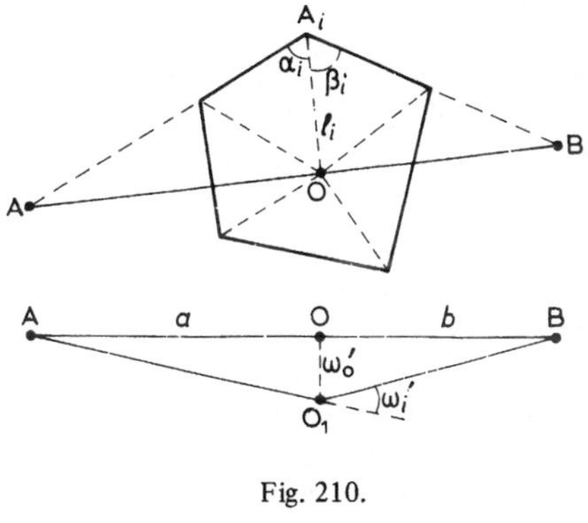


Fig. 210.

between them remain rigid. We denote by $w_0'$ the velocity of bending under the force, by $\omega_i'$ the corresponding rate of change of the dihedral angle along the $i$-th joint line, and by $l_i$ the length of the latter. According to (65.2) the kinematically possible load $P_\text{k}$ is determined from the rate of work equation

$$P_\text{k} w_0' = M_\text{s} \sum_i \omega_i' l_i .$$

The limit load $P_* \leqslant P_\text{k}$.

We calculate the angular velocity $\omega_i'$; to do this we construct the plane perpendicular to the line $\text{A}_i\text{O}$. In unit time the deflection of the point O increases by a small amount $w_0' \cdot 1$, and the angle $\text{AO}_1\text{B}$ by a small angle $\omega_i' \cdot 1$. From the diagram it is easy to see that

$$\omega_i' \approx w_0' \left(\frac{1}{a} + \frac{1}{b}\right) = \frac{w_0'}{l_i} (\cot \alpha_i + \cot \beta_i) ,$$

where $\alpha_i$, $\beta_i$ are the angles of the $i$-th joint line with the adjacent sides of the plate. Thus,

$$P_\text{k} = M_\text{s} \sum_i (\cot \alpha_i + \cot \beta_i) . \tag{66.27}$$

For a *right polygonal plate* loaded at the centre we have from (66.27)

$$P_\text{k} = 2nM_\text{s} \tan (\pi/n) ,$$

where $n$ is the number of sides of the polygon. For example, for a square plate ($n = 4$) we have $P_\text{k} = 8M_\text{s}$.

## §67. Minimum principles in deformation theory

In elasticity theory energy methods based on the principle of minimum potential energy and Castigliano's principle are of great importance. In this section we establish analogous theorems in the deformation theory of plasticity.

### 67.1. *Work of external forces (generalization of Clapeyron's theorem)*

Let the body occupy a volume $V$, bounded by a surface $S$. On a portion of the surface $S_F$ let external forces $\mathbf{F}_n$ with components $X_{ni}$ be prescribed, while on the other part $S_u$ the displacements are prescribed; let the displacements $u_i$ correspond to the state of equilibrium of the body. For simplicity we assume that body forces are absent.

The stress components $\sigma_{ij}$ satisfy the differential equations of equilibrium (64.2) and the boundary conditions on $S_F$ (64.3).

As the analogue of (64.6) we have the equation

$$\int X_{ni}u_i \, dS = \int \sigma_{ij}\epsilon_{ij} \, dV , \tag{67.1}$$

where, in general, the stress field $\sigma_{ij}$ and the displacement field $u_i$ need not be inter-related.

The proof proceeds just as for (64.6).

Let the stresses and strains be in accord with the equations of deformation theory (§ 14), i.e.

$$\epsilon_{ij} = \psi(\sigma_{ij} - \sigma\delta_{ij}) + k\sigma\delta_{ij} , \tag{67.2}$$

or, inversely,

$$\sigma_{ij} = \frac{1}{\psi}(\epsilon_{ij} - \tfrac{1}{3}\epsilon\delta_{ij}) + \frac{1}{3k}\epsilon\delta_{ij} . \tag{67.3}$$

It is easy to see that

$$\sigma_{ij}\epsilon_{ij} = 2(U + \psi T^2) . \tag{67.4}$$

Since $\epsilon = 3k\sigma$ the elastic energy of volumetric compression is

$$U = \tfrac{3}{2}k\sigma^2 = \tfrac{1}{2}\sigma\epsilon . \tag{67.5}$$

By virtue of the relation $\Gamma = 2\psi T$ we have $\psi T^2 = \frac{1}{2}T\Gamma$; consequently

$$\sigma_{ij}\epsilon_{ij} = \sigma\epsilon + T\Gamma , \tag{67.6}$$

and the work of the external forces on the corresponding displacements is

$$\bar{A} \equiv \int X_{ni}u_i \, dS = \int (\sigma\epsilon + T\Gamma) \, dV . \tag{67.7}$$

It is clear that $\sigma\epsilon$ is twice the elastic energy of volumetric compression (fig. 211a).

Consider the curve $T = g(\Gamma)\Gamma$ (fig. 211b); the work of shape-deformation

$$A_\varphi = \int T d\Gamma$$

is shown by the shaded area. Further, $dA_\varphi = T d\Gamma$; let the work of shape-deformation be a homogeneous function of $\Gamma$ of degree $m$; then $T\Gamma = mA_\varphi$ and

$$\bar{A} = \int (2U + mA_\varphi) \, dV .$$

*Clapeyron's theorem.* Consider an elastic medium obeying Hooke's law,

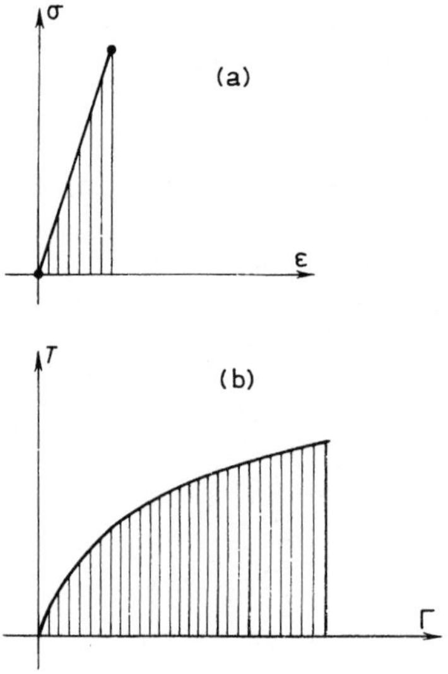


Fig. 211.

for which $A_\varphi = \frac{1}{2}G\Gamma^2$, $m = 2$; consequently,

$$\bar{A} = 2\int \Pi \mathrm{d}V\,,$$

i.e. *the work of the external forces on the corresponding displacements is equal to twice the elastic energy of the body.* This theorem is useful in calculating the elastic potential.

For *developed plastic deformations* we can neglect the elastic strains. Then for the yield state $T = \tau_s$ and

$$\bar{A} = \tau_s \int \Gamma \mathrm{d}V\,,$$

i.e. *the work of external forces on the corresponding displacements is equal to the work of plastic shape-deformation.*

67.2. *Principle of minimum total energy*

We consider the minimum properties of an actual distribution of displacements.

We focus attention on points of a body which is in equilibrium under the action of given forces and displacements, and take these points to have infinitesimal continuous displacements $\delta u_i$, compatible with the boundary conditions (*kinematically possible displacements*); it is assumed that there is no unloading here (more precisely, we shall investigate the minimum principle for the corresponding non-linear elastic medium). In accordance with the origin of the possible displacements the sum of the works of all the external and internal forces on the possible displacements near the equilibrium state is zero, i.e.

$$\int \sigma_{ij}\delta\epsilon_{ij}\,\mathrm{d}V - \int X_{ni}\delta u_i\,\mathrm{d}S_F = 0\,. \tag{67.8}$$

We note that this equation can be obtained by formal means – by transforming the surface integral and using the equilibrium equations (analogous to the derivation of equations (64.6)).

In the deformation theory of plasticity (cf. §14) we have

$$\sigma_{ij}\delta\epsilon_{ij} = \delta\Pi\,,$$

where $\Pi$ is the potential of the work of deformation. Since the external forces do not vary, the work of the external forces is

$$\delta\bar{A} = \delta\int X_{ni}u_i\,\mathrm{d}S_F$$

and equation (67.8) can be reduced to the form

$$\delta\left(\int \Pi\,\mathrm{d}V - \bar{A}\right) = 0\,. \tag{67.9}$$

The quantity inside the brackets is called the *total energy*; we denote it by $\mathscr{E}$, and then

$$\delta\mathscr{E} = 0\,. \tag{67.10}$$

*The actual form of the equilibrium of a body differs from all other possible forms by the fact that it renders the total energy a minimum* (cf. below).

The variational equation (67.10) replaces the boundary conditions and the differential equations of equilibrium in displacements (20.2); it generalizes the Lamé equations in elasticity theory (§20).

We have that

$$\delta\Pi = \frac{\partial\Pi}{\partial\epsilon_{ij}}\delta\epsilon_{ij}\,, \qquad \delta\epsilon_{ij} = \frac{1}{2}\left(\frac{\partial}{\partial x_j}\delta u_i + \frac{\partial}{\partial x_i}\delta u_j\right).$$

Using Gauss's formula we find

$$\int \frac{\partial\Pi}{\partial\epsilon_{11}}\delta\epsilon_{11}\mathrm{d}V = \int \frac{\partial\Pi}{\partial\epsilon_{11}}\delta u_1 \cos(n,1)\,\mathrm{d}S - \int \frac{\partial}{\partial x_1}\left(\frac{\partial\Pi}{\partial\epsilon_{11}}\right)\delta u_1\mathrm{d}V$$

and so on. Substituting these expressions into (67.9), we obtain

$$-\int\left[\frac{\partial}{\partial x_i}\left(\frac{\partial\Pi}{\partial\epsilon_{ij}}\right)\right]\delta u_{ij}\mathrm{d}V + \int\left[\frac{\partial\Pi}{\partial\epsilon_{ij}}\cos(n,i) - X_{ni}\right]\delta u_j\mathrm{d}S = 0\,. \tag{67.11}$$

Since the displacements are prescribed on $S_u$, we have that $\delta u_i = 0$ on $S_u$; inside the body and on the surface $S_F$ the variations $\delta u_i$ are arbitrary, and the differential equations of equilibrium in displacements (20.2) and the corresponding boundary conditions on $S_F$ follow from (67.11).

We now consider separately the various states of a medium – elastic, yield and hardening.

*Hooke's elastic medium* is characterized by the fact that

$$\Pi = U + \tfrac{1}{2} G\, \Gamma^2 \qquad (U = \epsilon^2/6k).$$

With the condition that the external forces are independent of the displacements, *the total energy of an elastic medium attains its minimum value.*

This can easily be shown by evaluating the second variation of the potential energy; since $\Pi$ is here a homogeneous positive quadratic form in the strain components, its second variation will be the same as the quadratic form of the *variation* in the strain components $\delta \epsilon_{ij}$, multiplied by 2; consequently

$$\delta^2 \Pi = \delta^2 U + \delta^2 \tfrac{1}{2} G\, \Gamma^2 > 0\,,$$

where

$$\delta^2 U = \frac{1}{3k} (\delta \epsilon)^2 \geqslant 0\,,$$

$$\delta^2 \tfrac{1}{2} G\, \Gamma^2 = G\, 2\, \delta e_{ij}\, \delta e_{ij} \geqslant 0\,.$$

We shall denote by $\Gamma^2(\delta \epsilon_{ij})$ the non-negative quadratic form of the variation of the strains. Thus, $\delta^2 \Pi > 0$ and then $\delta^2 \mathcal{E} > 0$ also.

In the *yield state* the increment in the potential of the work of deformation is

$$\delta \Pi = \delta \left( \frac{\epsilon^2}{6k} + \tau_s \Gamma \right)\,,$$

and the basic variational equation takes the form

$$\delta \left[ \int \left( \frac{\epsilon^2}{6k} + \tau_s \Gamma \right) \mathrm{d}V - \bar{A} \right] = 0\,. \tag{67.12}$$

The previous condition of independence of the external forces of the displacements in this case only implies the necessity of the minimum for the actual equilibrium form.

Here we have

$$\delta^2 \Pi = \delta^2 U + \tau_s \delta^2 \Gamma\,,$$

where $\delta^2 U \geqslant 0$; on the other hand

$$\delta \Gamma = \frac{1}{2\Gamma} \delta (\Gamma^2)\,, \qquad \delta^2 \Gamma = \frac{1}{\Gamma} \{\Gamma^2(\delta \epsilon_{ij}) - \frac{1}{4\Gamma^2} [\delta(\Gamma^2)]^2\}\,.$$

The quantity inside the brace brackets is non-negative; in fact

$$\Gamma^2 = \tfrac{2}{3}(\gamma_1^2 + \gamma_2^2 + \gamma_3^2)\,,$$

and after some simple transformations we find

$$\{\ldots\} = \frac{2}{3}\,\frac{(\gamma_1\delta\gamma_2 - \gamma_2\delta\gamma_1)^2 + (\gamma_2\delta\gamma_3 - \gamma_3\delta\gamma_2)^2 + (\gamma_3\delta\gamma_1 - \gamma_1\delta\gamma_3)^2}{\gamma_1^2 + \gamma_2^2 + \gamma_3^2} \geqslant 0\,,$$

i.e. $\delta^2\Gamma \geqslant 0$. Since $\tau_s > 0$, $\Gamma > 0$, we have $\delta^2\Pi \geqslant 0$, as the expression inside the brace bracket can vanish even if $\delta\gamma_1, \delta\gamma_2, \delta\gamma_3$ are not zero.

It should be noted that if there are elastic zones in the body, $\delta^2\Pi > 0$ in them and so $\delta^2\mathcal{E} > 0$.

For a *hardening medium* the potential of deformation is expressed by formula (14.25), so that

$$\delta\left\{\int\left(U + \int g(\Gamma)\,\Gamma\mathrm{d}\Gamma\right)\mathrm{d}V - A\right\} = 0\,.$$

With the same condition, that the external forces are independent of the displacements, the energy of the system attains its minimum in the actual equilibrium state.

The second variation is

$$\delta^2\Pi = \delta^2 U + \delta^2\int T\mathrm{d}\Gamma\,.$$

Moreover,

$$\delta^2\int T\mathrm{d}\Gamma = \delta T\delta\Gamma + T\delta^2\Gamma\,.$$

Substituting for $\delta T$ from the equation $T = g(\Gamma)\Gamma$, we obtain

$$\delta^2\int g(\Gamma)\,\Gamma\mathrm{d}\Gamma = \frac{\mathrm{d}T}{\mathrm{d}\Gamma}(\delta\Gamma)^2 + T\delta^2\Gamma\,.$$

The second term on the right-hand side cannot be negative; we now suppose that the shear stress increases with increasing shear strain (fig. 212). This condition, characterizing the "stability" of the material, is obviously satisfied for all solid bodies (cf. §18). Then

$$\mathrm{d}T/\mathrm{d}\Gamma > 0\,, \qquad (67.13)$$

and $\delta^2\Pi$ is a positive definite quadratic form of the variations $\delta\epsilon_{ij}$. This is easily verified by analysing the condition that the quantities $\delta^2 U$, $\delta(\Gamma^2)$ and $\delta^2\Gamma$ should vanish simultaneously.

We consider the case of the *mixed boundary-value problem*. In the equilibrium state let the volume $V$ be divided into the parts $V_1$, $V_2$ by a surface $\Sigma$; in each of the two parts of the material let the deformation follow its own characteristic law. Take the respective expressions for the potential of the work of deformation to be $\Pi_1$ and $\Pi_2$. On the surface $\Sigma$ the two states pass continuously into each other, and the quantities $T$ and $\Gamma$ are constant (§21).

If the deformation state is varied, the surface $\Sigma$ in general moves to a neighbouring surface $\Sigma'$, which separates volumes $V_1'$, $V_2'$, close to the previous ones. The change in the surface of separation $\Sigma$ depends only on the increment in the quantity $\Gamma$; $\Gamma$ must have the same constant value on the surfaces $\Sigma$ and $\Sigma'$.

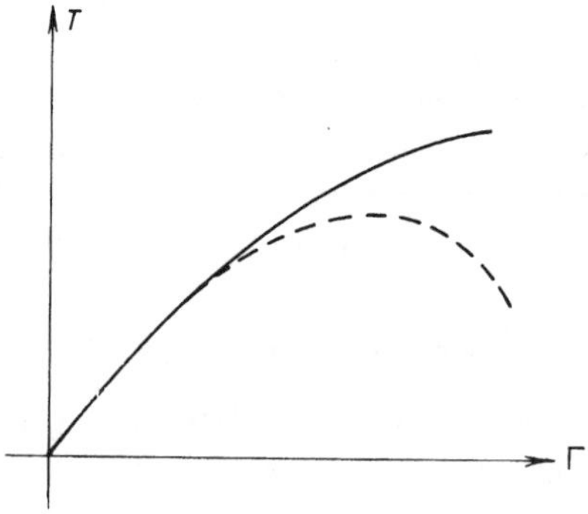


Fig. 212.

Consider the triple integral

$$I(\lambda) = \iiint_D F(x, y, z; \lambda) \,\mathrm{d}x\mathrm{d}y\mathrm{d}z ,$$

over the domain D bounded by some surface $\Sigma$, which varies with some parameter $\lambda$. The variation of this integral (cf. Goursat: Course of Mathematical Analysis, Vol. I. Part I, Supplement) is

$$\delta I(\lambda) = \iiint_D \delta F \,\mathrm{d}x\mathrm{d}y\mathrm{d}z + \iint_\Sigma F \delta n \mathrm{d}S ,$$

where $\delta n$ is the infinitesimal displacement of a point on the surface $\Sigma$ in the direction of the outward normal, corresponding to variation of $\lambda$.

In our case the potential of the work of deformation has the form

$$\Pi = \int \Pi_1 \mathrm{d}V_1 + \int \Pi_2 \mathrm{d}V_2 ,$$

and the preceding formula has to be applied twice – to the volume $V_1$ and to the volume $V_2$. Since $\Pi$ is continuous though the whole volume of the body, the respective integrals over the surface $\Sigma$, being equal in magnitude and opposite in sign, cancel. It follows that for the mixed problem, with continuity of displacements, stress components and strain components on the surface $\Sigma$, we obtain

$$\delta \left\{ \int \Pi_1 \mathrm{d}V_1 + \int \Pi_2 \mathrm{d}V_2 - \bar{A} \right\} = 0 , \tag{67.14}$$

i.e. *the actual equilibrium of a body, parts of which are in different states, has the same behaviour as the simple case, i.e. the minimum of the total energy is attained.*

67.3. *Principle of minimum additional work*

We have been considering the minimum properties of actual displacements. We now turn to an elucidation of the minimum properties of actual stress distributions (neglecting unloading, as before).

*Equation of statically possible stress variations.* We compare the actual stress state $\sigma_{ij}$ in a body under the action of given forces and displacements with all possible neighbouring stress states $\sigma_{ij} + \delta\sigma_{ij}$ which satisfy static equations inside the body

$$\frac{\partial}{\partial x_i}(\sigma_{ij} + \delta\sigma_{ij}) = 0 \tag{67.15}$$

and obey

$$(\sigma_{ij} + \delta\sigma_{ij})\, n_j = X_{ni} + \delta X_{ni} \tag{67.16}$$

on the portion of the surface $S_F$. These states are called *statically possible.*

It is clear that the variations in the stresses $\delta\sigma_{ij}$ and the variations in the external forces $\delta X_{ni}$ constitute a system in balance. Consequently, the work of these internal and external forces on each possible displacement must vanish. We take as possible displacements the actual displacements $u_i$; then

$$\int \epsilon_{ij}\delta\sigma_{ij}\mathrm{d}V = \int u_i\delta X_{ni}\mathrm{d}S\ . \tag{67.17}$$

Here we have to assume that $\delta X_{ni}$ are given on $S_F$; on $S_u$ these variations are found from the stress variations, in accordance with (67.16).

An important, narrower class of stress variations is that which occurs when the work of the variations in the external forces on the actual displacements is zero:

$$\int u_i\delta X_{ni}\mathrm{d}S = 0\ . \tag{67.18}$$

With this condition we have

$$\int \epsilon_{ij}\delta\sigma_{ij}\mathrm{d}V = 0\ . \tag{67.19}$$

Equation (67.18) is satisfied, for example, if the external forces are *prescribed* on the whole surface, for then $\delta X_{ni} = 0$. Alternatively, we can prescribe only some components of the external forces, if the displacements corresponding to the others are zero,

The derivation of (67.19) did not at all involve the mechanical properties of the medium; only its continuity was used. The actual stress state implies deformations for which the Saint Venant compatibility conditions are satisfied. It can be shown that these compatibility conditions follow from (67.19). Consequently the variational equation (67.19) is the energy formulation of the condition of continuity of deformation (for a proof see Leibenzon [20]).

*Concept of additional work.* We consider in more detail the expression for the work of the stress variations on the actual displacements; substituting for the strain components from Hencky's formulae (67.2) and rearranging, we obtain

$$\epsilon_{ij}\delta\sigma_{ij} = \psi\delta T^2 + \delta U .$$

Since the body we are examining satisfies one of the conditions

$$\psi = \text{const.} = \frac{1}{2G} \text{ (Hooke's elastic medium)} ,$$

$$T = \text{const.} = \tau_s \text{ (yield state)} ,$$

$$\psi = \tfrac{1}{2}\bar{g}(T) \text{ (hardening state)} ,$$

it follows that the right-hand side will be the total differential of some stress function $R$:

$$\epsilon_{ij}\delta\sigma_{ij} = \delta R . \tag{67.20}$$

For Hooke's elastic medium

$$R = U + \frac{1}{2G}T^2 = \Pi ,$$

for the hardening state

$$R = U + \int \bar{g}(T)\, T\mathrm{d}T , \tag{67.21}$$

for the yield state

$$R = U .$$

We shall call $R$ the density of additional work, or, simply, the *additional work.* To clarify this concept, we discard for the time being the term $\delta U$, representing the change in volume, which operates in all cases according to the

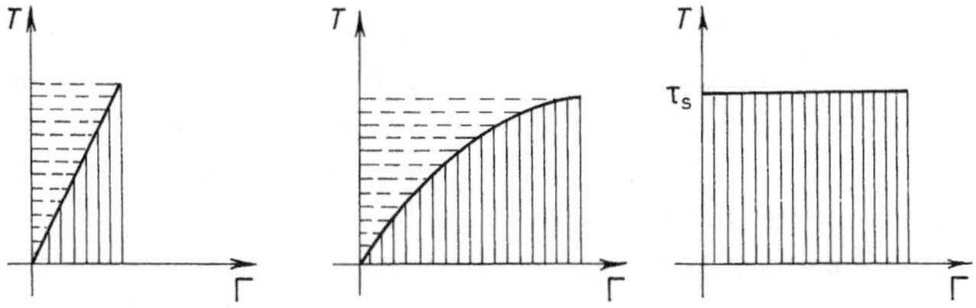


Fig. 213.

same law; we then have, by virtue of the relation $\Gamma = 2\psi T$, that [1]):

$$R_\varphi = \int \Gamma \mathrm{d}T \,,$$

$$\delta R_\varphi = \Gamma \delta T \,.$$

We consider the different cases of the curve $T = g(\Gamma)\Gamma$ (fig. 213). The work of deformation is represented by the area hatched with vertical lines, while the additional work is the area hatched with horizontal lines.

In the case of Hooke's elastic medium the areas are equal in magnitude, $R_\varphi = A_\varphi$, and it is not possible to distinguish the concepts of work of deformation and additional work; in the other cases this can be done. It is obvious that for a given relation $T = g(\Gamma)\Gamma$ the additional work $R_\varphi$ will be a definite function of the work of deformation $A_\varphi$.

*Generalization of Castigliano's formulae.* Since

$$\delta R = \frac{\partial R}{\partial \sigma_{ij}} \delta \sigma_{ij} \,,$$

then, if we compare this relationship with (67.20), we obtain the formulae

$$\epsilon_{ij} = \frac{\partial R}{\partial \sigma_{ij}} \,, \tag{67.22}$$

which replace the well-known Castigliano's formulae in the present case of non-linear relations between stresses and strains. It is evident that Hencky's equations (67.2) can be expressed in the form (67.22).

*Reciprocity principle.* A necessary consequence of (67.22) is the fulfillment of the 15 conditions

$$\frac{\partial \epsilon_{ij}}{\partial \sigma_{kl}} = \frac{\partial \epsilon_{kl}}{\partial \sigma_{ij}} \,,$$

[1]) We recall that the work of deformation of shape is $A_\varphi = \int T \mathrm{d}\Gamma$.

which should be regarded as a natural generalization of the reciprocity theorem to the case of non-linear dependence of stresses and strains [133].

*Generalization of Castigliano's theorem.* We shall extend Castigliano's theorem to the case of non-linear relationships between stresses and strains. Let $P_k$ ($k = 1, 2, 3, \ldots$) be concentrated forces applied to the body; $\alpha_k, \beta_k, \gamma_k$ the corresponding direction cosines of the vectors of these forces; $u_{1k}, u_{2k}, u_{3k}$ the components of the displacements of the points at which the forces are applied. Then proceeding from the general variational equation (67.17) and assuming that one of the forces $P_k$ receives an infinitesimal increment $\delta P_k$, while the supports are stationary, we find

$$\delta \widetilde{R} = (u_{1k}\alpha_k + u_{2k}\beta_k + u_{3k}\gamma_k)\, \delta P_k\,, \tag{67.23}$$

where $\widetilde{R}$ denotes the additional work of the whole body:

$$\widetilde{R} = \int R \mathrm{d}V\,.$$

The expression in brackets in (67.23) is the displacement of the point of application of the force along the line of action of this force. Thus,

$$\partial \widetilde{R} / \partial P_k = \Delta_k\,, \tag{67.24}$$

i.e. *the partial derivative of the additional work with respect to the magnitude of any of the applied forces $P_k$ equals the displacement of the point of application of this force in the direction in which the latter acts.* This result remains valid in respect of generalized forces and displacements.

In fact, if the forces applied to the body are proportional to some quantity $Q$, then it is clear that

$$\frac{\partial \widetilde{R}}{\partial Q} = \sum (u_{1k}\alpha_k + u_{2k}\beta_k + u_{3k}\gamma_k) \equiv q\,. \tag{67.25}$$

The quantities $Q$ and $q$ are generalized forces and displacements.

In the case of Hooke's elastic medium $\widetilde{R} = \widetilde{\Pi}$ and formula (67.25) reduces to Castigliano's theorem.

Suppose the stress state depends on $m$ superfluous unknowns $X_1, X_2, \ldots, X_m$. We have encountered mechanical systems of this type in calculations of beams, systems of rods, etc. In this case the condition of minimum additional work $\widetilde{R}$ leads to a system of $m$ equations

$$\partial \widetilde{R}/\partial X_1 = 0\,, \quad \ldots, \quad \partial \widetilde{R}/\partial X_m = 0\,. \tag{67.26}$$

*Example.* The generalized Castigliano theorem is suitable for calculating very simple bodies – rod lattices, beams, frameworks. We consider as an example a lattice made up of three identical rods (length $l$, sectional area $F$, fig. 35). The material under tension follows the law

$$\epsilon_x = \frac{B_1}{3\sqrt{3}} \sigma_x^3 ,$$

where $B_1$ is a constant. Comparing this with Hencky's formula (67.2), we find

$$k = 0 , \qquad T = \frac{\sigma_x}{\sqrt{3}}, \qquad \bar{g}(T) = B_1 T , \qquad R = \frac{B_1}{9\sqrt{3}} \sigma_x^3 .$$

We shall suppose the vertical rod to be superfluous – let $X$ be the force in it. Then we easily find

$$S_1 = S_3 P - X , \qquad S_2 = X .$$

The additional work of the lattice is

$$\widetilde{R} = lF \left[ \frac{2B_1}{9\sqrt{3}} \left( \frac{P-X}{F} \right)^3 + \frac{B_1}{9\sqrt{3}} \left( \frac{X}{F} \right)^3 \right] .$$

Constructing the equation $\partial \widetilde{R} / \partial X = 0$ we find that $\sqrt{2}(P-X) = \pm X$. Minimum additional work of the whole body is attained for the actual equilibrium state; it is easy to see that the plus sign corresponds to a minimum of $\widetilde{R}$. Consequently,

$$X = \frac{2P}{2+\sqrt{2}} .$$

*Principle of minimum additional work.* The variational equation (67.19) for the medium under consideration takes the form

$$\delta \widetilde{R} = 0 . \tag{67.27}$$

We examine now the particular states of the medium – elastic, yield and hardening.

*Hooke's elastic medium.* Here $R = \Pi$ and the variational equation (67.27) assumes the form

$$\delta \int \left( U + \frac{1}{2G} T^2 \right) \mathrm{d}V = 0 , \tag{67.28}$$

which is known as *Castigliano's rule.* In contrast with statically possible stress states corresponding to the *same external load,* the actual stress state in a body renders the value of the elastic potential energy of the body a *minimum.*

The fact that a minimum of potential energy is attained is easily verified by examining the sign of the second variation

$$\delta^2 \Pi = \delta^2 U + \delta^2 \left( \frac{1}{2G} \right) > 0 ,$$

where

$$\delta^2 U = 3k\,(\delta\sigma)^2 \geqslant 0\,,$$

$$\delta^2 T^2 = -\delta s_{ij}\delta s_{ij} \geqslant 0\,.$$

We denote by $2T^2(\delta\sigma_{ij})$ the non-negative quadratic form of the variation of the stresses.

In the *yield state* $T$ = const. and the variational equation (67.27) takes the form

$$\delta \int U \mathrm{d}V = 0\,, \tag{67.29}$$

i.e. the actual stress state differs from all neighbouring statically possible stress states in yield by the fact that only the former achieves an extremum value for the elastic potential energy of volumetric change of the body.

We assume that the material is incompressible; then $U = 0$ and we deduce the following conclusion: the actual displacements of points of an incompressible body in yield state are such that infinitesimal variations in the stresses lying within the yield regime do not produce any additional work on these displacements.

For the *hardening state* the variational equation (67.27) takes the form

$$\delta \int \left[ U + \int \bar{g}(T)\, T \mathrm{d}T \right] \mathrm{d}V = 0\,, \tag{67.30}$$

i.e. the stresses corresponding to the actual equilibrium state are such that the additional work $\tilde{R}$ of the whole body attains a *minimum value* with respect to all neighbouring values which are consistent with the equilibrium equations.

We show that a minimum of $\tilde{R}$ is attained. The second variation is

$$\delta^2 R = \delta^2 U + \delta^2 \int \Gamma \mathrm{d}T\,.$$

We know that $\delta^2 U \geqslant 0$. Furthermore,

$$\delta^2 \int \Gamma \mathrm{d}T = \delta T \delta \Gamma + \Gamma \delta^2 T\,,$$

where, as in the case considered on page 360, $\delta^2 T \geqslant 0$ and

$$\delta^2 R = \delta^2 U + \frac{\mathrm{d}\Gamma}{\mathrm{d}T}(\delta T)^2 + \Gamma\delta^2 T\,.$$

It is easy to see that if

$$\mathrm{d}\Gamma/\mathrm{d}T > 0\,, \tag{67.31}$$

then $\delta^2 R > 0$. The condition (67.31), just as (67.13), is obviously always satisfied for real materials.

If parts of the body $V_1$, $V_2$, . . . are in different states, then the increments in additional work are of the form $\delta R_1, \delta R_2, \ldots$ respectively. Since $R$

changes continuously in passing through the surfaces $\Sigma_1, \Sigma_2, \ldots$ which separate the different regions, and since on each surface $\Sigma_1, \Sigma_2, \ldots$ the intensity of tangential stresses is constant, it is easy to see that for the mixed problem

$$\delta\left\{\int R_1 \mathrm{d}V_1 + \int R_2 \mathrm{d}V_2 + \ldots\right\} = 0\,,$$

just as before.

Note that the distribution of the surfaces of separation $\Sigma_1, \Sigma_2, \ldots$ corresponds to a minimum of additional work of the whole body.

A generalization of (67.24) can be established in a similar manner for the case when concentrated forces are applied to the body:

$$\int \frac{\partial R_1}{\partial P_k}\mathrm{d}V_1 + \int \frac{\partial R_2}{\partial P_k}\mathrm{d}V_2 + \ldots = \Delta_k\,.$$

67.4. *Concluding remarks*

The above energy theorems of deformation theory are presented in [118]; corresponding equations for non-uniformly heated bodies are given in [16]. The case of a finite number of generalized coordinates, which is important in structural mechanics, was studied by Lur'ye [133]. In the article by Phillips [190] minimum principles have been generalized to the case of large plastic deformations. In Hill's work [165] it is shown that an absolute minimum of total energy and additional energy is attained for the actual state.

## §68. Ritz's method. Example: elastic-plastic torsion

### 68.1. *Ritz's method*

The variational equations considered above open up the possibility of constructing approximate solutions by direct methods. At first glance the most natural approach appears to be the direct application of Ritz's method in its usual form.

We consider for definiteness the variational equation (67.10) which describes the minimum properties of the displacements $u_i$ $(i = 1, 2, 3)$. Let $u_{i0}$ be three functions which satisfy given conditions on the parts $S_u$ of the surface, and let $u_{is}$ $(s = 1, 2, \ldots, n)$ be a sequence of coordinate functions which satisfy zero conditions on $S_u$. We shall seek an approximate solution to the minimum total energy problem in the form

$$u_i = u_{i0} + \sum_{s=1}^{n} c_{is} u_{is}\,, \tag{68.1}$$

where the $c_{is}$ are Ritz coefficients. If now we calculate the expression for the total energy, it will be a function of the coefficients $c_{ik}$. The latter are determined from the condition that the total energy is a minimum:

$$\frac{\partial}{\partial c_{ik}} \mathcal{E} = 0 . \tag{68.2}$$

For an elastic body the total energy will be a quadratic form of the coefficients $c_{ik}$; in this case the conditions (68.2) generate a system of linear inhomogeneous algebraic equations for the $c_{ik}$.

For plastic deformation, however, the total energy will not be a quadratic form of the $c_{ik}$, and conditions (68.2) lead to a non-linear system of equations for the determination of the Ritz coefficients. Even when $n$ is comparatively small the construction and solution of this system involves substantial computational difficulties.

As a consequence, extensive use has been made of a one-term approximation (with zero conditions on $S_u$):

$$u_i = c_{i1} u_{i1} ,$$

where $u_{i1}$ is usually taken to be the solution of the corresponding linear (elastic) problem. This particular method is used for solving approximately various engineering problems. It should be noted, however, that these solutions could involve very substantial errors.

As more approximating functions are taken the difficulty of constructing Ritz's system increases sharply. Even if the non-linear system has been derived somehow, it still remains to solve it, and this in its turn is very difficult and requires the use of various numerical methods.

Ritz's method can also be formulated for the problem of minimum additional work $\breve{R}$.

68.2. *Modification of Ritz's method*

The difficulty of applying Ritz's method directly has led to a number of attempts to modify it. We present below one such modification [123] which can be used for finding the minimum in several other problems as well. This method overcomes the difficulty arising from the non-quadratic nature of the functionals, and enables the direct construction of the solution to required accuracy.

We consider as an example the application of this method to finding the minimum of additional work. Let $\sigma_{ij0}$ be a particular solution of the equilibrium equations (64.2), which satisfies the given boundary conditions on the portion $S_F$ of the surface, and let $\sigma_{ijs}$, $s = 1, 2, \ldots, n$ be a set of particular solutions of (64.2) with zero boundary conditions on $S_F$. We construct a solution to the problem of minimum additional work

$$\int \left[ U + \int \bar{g}(T)\, T \mathrm{d}T \right] \mathrm{d}V = \min. \tag{68.3}$$

by successive approximations of the form

$$\sigma_{ij}^{(r)} = \sigma_{ij0} + \sum_{s=1}^{n} c_s^{(r)} \sigma_{ijs}, \tag{68.4}$$

where $c_s^{(r)}$ are coefficients to be determined.

In the zeroth approximation $\sigma_{ij}^{(0)}$ we put $\bar{g}(T) = 1/G_0$, where $G_0$ is the shear modulus (or some quantity representing the inclination of the straight approximating strain curve in the initial segment). The zeroth approximation corresponds to an elastic body and is determined from the condition for the minimum of the quadratic functional

$$\int \left( U + \frac{1}{2G_0} T^2 \right) \mathrm{d}V = \min.$$

The coefficients of the zeroth approximation $c_s^{(0)}$ are found from a system of linear inhomogeneous algebraic equations. Having found the stresses and calculated the intensity $T^{(0)} = (\frac{1}{2} s_{ij}^{(0)} s_{ij}^{(0)})^{\frac{1}{2}}$ we put $G_1 = g(T^{(0)}/G_0)$ and determine the next ("first") approximation from the condition for a minimum of the quadratic functional:

$$\int \left( U + \frac{1}{2G_1} T^2 \right) \mathrm{d}V = \min.$$

Here the "current modulus" $G_1$ is a known function of coordinates. With the above method of choosing $G_1$, the intensity of tangential stresses $T$, which relates to the intensity of shear strain $\Gamma$ by some linear law $T = G_0 \Gamma$, is "restored" in the next approximation to the strain curve $T = g(\Gamma)\Gamma$ (fig. 214). It is apparent that $G_1 = G_0 T_*^{(0)}/T^{(0)}$..

This procedure is repeated until the required accuracy is obtained. For the $r$-th approximation we obtain

$$\int \left( U + \frac{1}{2G_r} T^2 \right) \mathrm{d}V = \min.$$

where

$$G_r = G_{r-1} \frac{T_*^{(r-1)}}{T^{(r-1)}}, \tag{68.5}$$

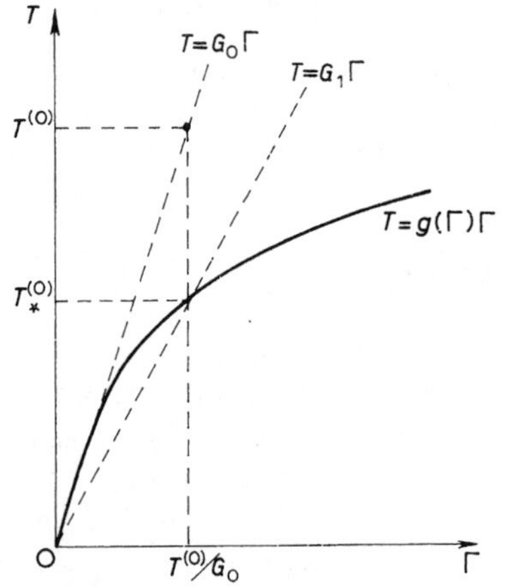


Fig. 214.

and

$$T_*^{(r-1)} = g\left(\frac{T^{(r-1)}}{G_{r-1}}\right)T^{(r-1)}.$$

The presence of a variable modulus $G_r$ in the $r$-th approximation does not greatly complicate the quadratures, since the $r$-th approximation has the same form as for the elastic body. In each approximation the coefficients $c_s^{(r)}$ are determined from a linear system of algebraic equations.

In the representation (68.4) it is expedient to retain the number of terms which ensures the required accuracy of the solution to the elastic problem. Of course with fixed $n$, calculation of higher approximations is not very meaningful. The quadratures can conveniently be found numerically. To calculate the current modulus $G_r$ we can proceed directly from an experimental deformation curve "$T$–$\Gamma$". The fact that the solution retains the same form in each approximation (only the coefficients $c_s^{(r)}$ change) considerably simplifies the calculations, and makes the results less awkward compared with other methods of successive approximation.

We can apply an analogous method to the problem of minimum total energy (67.10). In this case we seek a solution by successive approximation in the form

$$u_i^{(r)} = u_{i0} + \sum_{s=1}^{n} c_{is}^{(r)} u_{is},$$

where $u_{i0}$ satisfy given conditions on $S_u$, $u_{is}$ vanish on $S_u$, and $c_{is}^{(r)}$ are arbitrary constants. In the zeroth approximation we obtain $g(\Gamma) = \text{const.} = G_0$, and in the $r$-th approximation $g(\Gamma) = g(\Gamma^{(r-1)})$.

Other variations of this approximation method are possible (cf. the review [66]), and also an analogous modification of Galerkin's method.

In practice we can use this method to solve those elastic-plastic problems for which a solution by Ritz's method can be found in the elastic state.

68.3. *Example: elastic-plastic torsion of a rod with square cross-section* (length of side = $2a$)

Suppose the relation between $T$ and $\Gamma$ is characterized by linear hardening (fig. 215):

$$T = \begin{cases} G_0\Gamma & \text{for} \quad \Gamma \leqslant 0.0025\,, \\ (19.4 + 236\Gamma)\ \text{kg/mm}^2 & \text{for} \quad \Gamma \geqslant 0.0025\,. \end{cases} \tag{68.6}$$

The shear modulus is $G_0 = 7.85 \times 10^3$ kg/mm$^2$. This relation describes the behaviour of nickel steel.

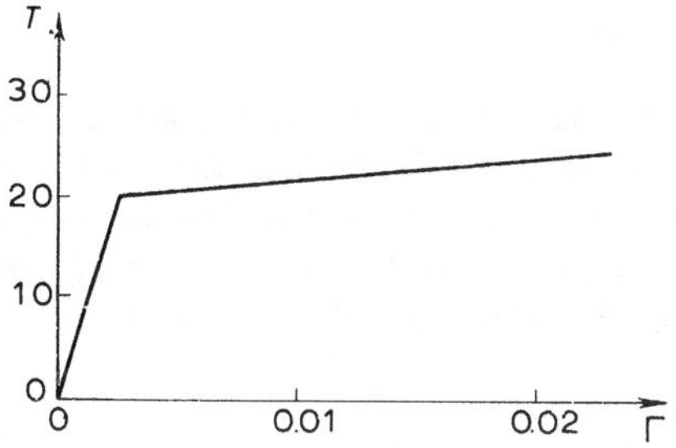


Fig. 215.

In §30 we derived the differential equation of torsion of a hardening rod. The variational equation for the stress function $F$ (using the notation of §30) can be obtained from the general variational equation (67.17). The work of the variations in surface forces on the lateral surface and the fixed base $z = 0$ is zero; at the free end $z = l$ we have $u_x = -\omega yl$, $u_y = \omega xl$; thus

$$\int u_i \delta X_i \mathrm{d}S = \omega l \iint \left(-y \frac{\partial}{\partial y}\,\delta F - x \frac{\partial}{\partial x}\,\delta F\right) \mathrm{d}x\mathrm{d}y$$

$$= -\omega l \iint \left[\frac{\partial}{\partial x}(x\delta F) + \frac{\partial}{\partial y}(y\delta F)\right] \mathrm{d}x\mathrm{d}y + 2\omega l \iint \delta F \mathrm{d}x\mathrm{d}y\,.$$

The first integral on the right-hand side can be transformed into an integral along the contour of the section, and equals zero since $\delta F = 0$ on the contour. The variational equation (67.17) then takes the form

$$\delta \int_0^1 \int_0^1 \left[ \int_0^T \bar{g}(\lambda)\, \lambda \mathrm{d}\lambda - 2a\omega F \right] \mathrm{d}\xi \mathrm{d}\eta = 0 \,, \tag{68.7}$$

where we have introduced the dimensionless coordinates $\xi = x/a$, $\eta = y/a$. On the contour of the section $F = 0$.

For an elastic rod $\bar{g}(\lambda) = \text{const.} = 1/G_0$ and the problem is linear.

We write the variational equation (68.7) in the form

$$\delta \int_0^1 \int_0^1 \left[ \frac{1}{G_r} \frac{T^2}{2} - 2a\omega F \right] \mathrm{d}\xi \mathrm{d}\eta = 0 \tag{68.8}$$

and look for a solution of the form

$$F^{(r)} = c_1^{(r)} F_1 + c_2^{(r)} F_2 \,,$$

where $c_1^{(r)}$, $c_2^{(r)}$ are arbitrary constants, and

$$F_1 = (\xi^2 - 1)(\eta^2 - 1) \,, \qquad F_2 = F_1(\xi^2 + \eta^2) \,. \tag{68.9}$$

The solution of the elastic problem in this approximation leads to the following results: the torque is $M = 0.1404 G_0 \omega (2a)^4$, and is less than the exact value by only 0.15%; the maximum tangential stress in the middle of a side is $\tau_{\max} = 1.40 G_0 a\omega$, compared with the exact value $1.35 G_0 a\omega$. Here the quadratures are very easy to perform, and the coefficients are

$$c_1^{(0)} = \frac{5}{8} \frac{259}{277} G_0 a\omega \,,$$

$$c_2^{(0)} = \frac{5}{16} \frac{105}{277} G_0 a\omega \,.$$

For plastic torsion the stress distribution is smoother than in the elastic case, and therefore we can reasonably assume that in general the approximation in the form (68.9) is not worse than for the elastic rod.

The current modulus $G_r$ can be calculated from (68.5), and the intensity $T_*^{(r-1)}$ determined from (68.6). Calculations have been carried out for the case $a\omega = 0.015$, with the integrals computed numerically by Gauss's method.

In the zeroth approximation ($r = 0$), the coefficients $c_1^{(0)}$, $c_2^{(0)}$ differ only in the sixth decimal place from the exact values mentioned above. In order to check the stability of the results, 10 approximations have been calculated; the values of the coefficients are given in the table.

**Coefficients $c_1^{(r)}$, $c_2^{(r)}$**

| $r$ | $c_1^{(r)} \cdot 1.02 \cdot 10^{-3}$ | $c_2^{(r)} \cdot 1.02 \cdot 10^{-3}$ | $r$ | $c_1^{(r)} \cdot 1.02 \cdot 10^{-3}$ | $c_2^{(r)} \cdot 1.02 \cdot 10^{-3}$ |
|---|---|---|---|---|---|
| 0 | 0.070125 | 0.014217 | 5 | 0.017551 | –0.003740 |
| 1 | 0.018023 | –0.003382 | 6 | 0.017543 | –0.003889 |
| 2 | 0.017872 | –0.003765 | 7 | 0.017461 | –0.003757 |
| 3 | 0.017728 | –0.003802 | 8 | 0.017469 | –0.003663 |
| 4 | 0.017554 | –0.003641 | 9 | 0.017400 | –0.003693 |

Once the values of the constants have been found (in practice we can stop after three or four approximations), the stress components and the intensity $T$ can be evaluated. Fig. 216 shows the graph of the tangential stress in the section $y = 0$; the divergences from linear law are clearly visible. The boundaries of the plastic zones when $T = \tau_s = 19.6$ kg/mm$^2$, are shown shaded in fig. 217. Details of the calculations can be found in the author's work [66].

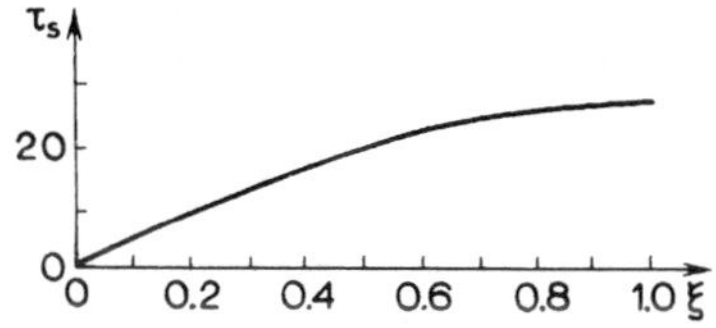


Fig. 216.

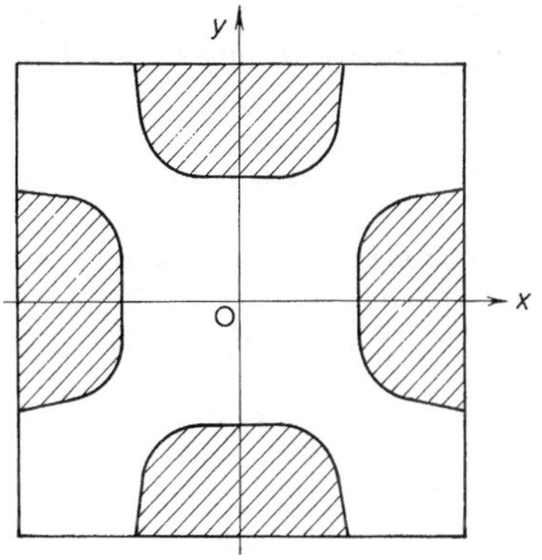


Fig. 217.

## *§69. Extremum principles in plastic flow theory

Under the action of prescribed surface loads $\mathbf{F}_n$ on $S_F$, and displacements $\mathbf{u}$ on $S_u$, a body experiences a distribution of stresses and strains $\sigma_{ij}$, $\epsilon_{ij}$ which, as before, we shall call actual and which we shall assume to be known.

Let the surface loads be given an increment $\mathrm{d}\mathbf{F}_n$ on $S_F$, and the displacements an increment $\mathrm{d}\mathbf{u}$ on $S_u$; to these increments there correspond increments in the actual stresses and strains.

In plastic flow theory it is possible to establish extremum properties of *actual increments* of strain (stress) with respect to virtual increments.

69.1. *Minimum properties of actual strain increments*

Let $\mathrm{d}u_i'$ be any continuous increments in displacements assuming prescribed values on the surface $S_u$. To these kinematically possible displacements there correspond, by equations (3.8), increments of the strain components $\mathrm{d}\epsilon_{ij}'$ and, by (13.7), certain increments in the stress components $\mathrm{d}\sigma_{ij}'$, which, in general, do not satisfy the equilibrium equations.

Using the fact that the actual increments $\mathrm{d}\sigma_{ij}$ satisfy the equilibrium equations, we can easily obtain by the usual methods the equation [1])

$$\int \mathrm{d}\sigma_{ij}\mathrm{d}\epsilon_{ij}'\mathrm{d}V = \int \mathrm{d}X_{ni}\mathrm{d}u_i'\mathrm{d}S\ .$$

Here the fields of stress-increments $\mathrm{d}\sigma_{ij}$ and displacement-increments $\mathrm{d}u_i'$ are not in general inter-related. If $\mathrm{d}u_i$ is the increment in actual displacement, then

$$\int \mathrm{d}\sigma_{ij}\mathrm{d}\epsilon_{ij}\mathrm{d}V = \int \mathrm{d}X_{ni}\mathrm{d}u_i\mathrm{d}S\ . \tag{69.1}$$

Subtracting the latter equation from the preceding one we obtain

$$\int \mathrm{d}\sigma_{ij}(\mathrm{d}\epsilon_{ij}'-\mathrm{d}\epsilon_{ij})\,\mathrm{d}V = \int \mathrm{d}X_{ni}(\mathrm{d}u_i'-\mathrm{d}u_i)\,\mathrm{d}S_F\ . \tag{69.2}$$

It is easy to verify the identity

$$2\mathrm{d}\sigma_{ij}(\mathrm{d}\epsilon_{ij}'-\mathrm{d}\epsilon_{ij}) \equiv$$
$$\equiv (\mathrm{d}\sigma_{ij}'\mathrm{d}\epsilon_{ij}'-\mathrm{d}\sigma_{ij}\mathrm{d}\epsilon_{ij})-[\mathrm{d}\epsilon_{ij}'(\mathrm{d}\sigma_{ij}'-\mathrm{d}\sigma_{ij}) + \mathrm{d}\sigma_{ij}(\mathrm{d}\epsilon_{ij}-\mathrm{d}\epsilon_{ij}')]\ . \tag{69.3}$$

We consider now the expression inside the square brackets, using (13.7)

[1]) The continuity of increments in stress and strain components is assumed in what follows; this restriction can be removed, but we shall not do so here.

for a yield state and (13.14) for a hardening material; we obtain respectively

$$\Big[\ldots\Big] = \frac{1}{2G}(\mathrm{d}s'_{ij}-\mathrm{d}s_{ij})(\mathrm{d}s'_{ij}-\mathrm{d}s_{ij}) + 3k(\mathrm{d}\sigma'-\mathrm{d}\sigma)^2 +$$

$$+\begin{cases} [(\kappa'\mathrm{d}\lambda'-\kappa\mathrm{d}\lambda)\,\mathrm{d}T'^2+\kappa\mathrm{d}\lambda(\mathrm{d}T^2-\mathrm{d}T'^2)]\,, \\ [(\kappa'\mathrm{d}T'^2-\kappa\mathrm{d}T^2)\,\mathrm{d}T'^2+\kappa\mathrm{d}T^2(\mathrm{d}T^2-\mathrm{d}T'^2)]\;F(T)\,. \end{cases}$$

The multipliers $\kappa$, $\kappa'$ have the following meaning: the multiplier $\kappa$, which relates to actual increments, is equal to unity if loading takes place, while $\kappa = 0$ for unloading and neutral changes; the multiplier $\kappa'$ assumes analogous values in respect of virtual increments, which give rise to "loading" and "unloading" in accordance with the equations of plastic flow theory.

In the above equations the first two terms on the right are positive; they can only vanish if the equations $\mathrm{d}s'_{ij} = \mathrm{d}s_{ij}$, $\mathrm{d}\sigma' = \mathrm{d}\sigma$ are satisfied simultaneously. We now show that the quantities inside the following (square) brackets are non-negative.

If unloading takes place, $\kappa = 0$, $\kappa' = 0$ and the quantity concerned is zero. If loading occurs ($\kappa = 1$, $\kappa' = 1$) the right-hand bracket (for a material with yield plateau) gives zero, since $\mathrm{d}T^2 = 0$, $\mathrm{d}T'^2 = 0$; in the second case also this quantity is non-negative, being equal to $(\mathrm{d}T'^2-\mathrm{d}T^2)^2 \geqslant 0$.

If $\kappa = 1$, $\kappa' = 0$ the first bracket gives $-2\mathrm{d}\lambda\mathrm{d}T'^2 \geqslant 0$, while the second is $(\mathrm{d}T^2)^2-2\mathrm{d}T^2\mathrm{d}T'^2 \geqslant 0$, since $\mathrm{d}T'^2 \leqslant 0$, $\mathrm{d}\lambda \geqslant 0$. Finally when $\kappa = 0$, $\kappa' = 1$ the first bracket is zero and the second is non-negative $(\mathrm{d}T'^2)^2 \geqslant 0$. Thus

$$\tfrac{1}{2}(\mathrm{d}\sigma'_{ij}\mathrm{d}\epsilon'_{ij}-\mathrm{d}\sigma_{ij}\mathrm{d}\epsilon_{ij}) > \mathrm{d}\sigma_{ij}(\mathrm{d}\epsilon'_{ij}-\mathrm{d}\epsilon_{ij}) \tag{69.4}$$

whenever the virtual increments differ from the actual increments. Thus, omitting the case when they coincide, we find from (69.2)

$$\tfrac{1}{2}\int \mathrm{d}\sigma_{ij}\mathrm{d}\epsilon_{ij}\mathrm{d}V-\int \mathrm{d}X_{ni}\mathrm{d}u_i\mathrm{d}S_F < \tfrac{1}{2}\int \mathrm{d}\sigma'_{ij}\mathrm{d}\epsilon'_{ij}\mathrm{d}V-\int \mathrm{d}X_{ni}\mathrm{d}u'_i\mathrm{d}S_F\,. \tag{69.5}$$

We call the expression (functional) on the right-hand side of the inequality the *energy increment* $\mathcal{E}(\mathrm{d}u'_i)$.

*The actual increments in displacements* $\mathrm{d}u_i$ *render the energy increment* $\mathcal{E}(\mathrm{d}u'_i)$ *an absolute minimum with respect to all kinematically possible increments.*

### 69.2. *Maximum properties of actual stress increments*

We now compare with the actual stress increments $\mathrm{d}\sigma_{ij}$ statically possible increments $\mathrm{d}\sigma'_{ij}$ (which satisfy the equilibrium equations inside the body and on the portion $S_F$ of the surface).

Let $de'_{ij}$ be the increments in the strain components (in respect of plastic flow theory) for the statically possible increments $d\sigma'_{ij}$ under consideration. It is evident that in general $de'_{ij}$ will not satisfy the conditions of continuity for strain. Since the increments $d\sigma'_{ij}, d\sigma_{ij}$ balance, the same method as before leads easily to the equation

$$\int (d\sigma'_{ij} - d\sigma_{ij})\, d\epsilon_{ij} dV = \int (dX'_{ni} - dX_{ni})\, du_i dS_u\,. \tag{69.6}$$

Consider the identity

$$2d\epsilon_{ij}(d\sigma'_{ij} - d\sigma_{ij}) \equiv (d\sigma'_{ij} d\epsilon'_{ij} - d\sigma_{ij} d\epsilon_{ij}) - [d\sigma'_{ij}(d\epsilon'_{ij} - d\epsilon_{ij}) + d\epsilon_{ij}(d\sigma_{ij} - d\sigma'_{ij})]\,.$$

It can be shown by arguments similar to those used above that the quantity inside the square brackets is non-negative; it will be zero if $d\sigma'_{ij} = d\sigma_{ij}$. If therefore we exclude the latter case we obtain the inequality

$$-\tfrac{1}{2}\int d\sigma_{ij} d\epsilon_{ij} dV + \int dX_{ni} du_i dS_u > -\tfrac{1}{2}\int d\sigma'_{ij} d\epsilon'_{ij} dV + \int dX'_{ni} du_i dS_u\,, \tag{69.7}$$

the right-hand side of which we denote by $\mathcal{E}(d\sigma'_{ij})$.

*The actual stress increments effect an absolute maximum for the energy increment $\mathcal{E}(d\sigma'_{ij})$ with respect to all statically possible stress increments.*

Extremum principles for plastic flow theory were formulated by Hodge and Prager, and Greenberg [181]. Various generalizations are given in the works of Hill [54], Yamamoto [175] and others (cf. review in [68]).

## PROBLEMS

1. Derive the equations of the minimum principles for total energy and additional work (§67) when body forces are present.

2. When the temperature $\theta$ of the body is non-uniform the equations of deformation theory have the form

$$\epsilon_{ij} = (k\sigma + \alpha\theta)\,\delta_{ij} + \psi s_{ij}\,,$$

where $\alpha$ is the coefficient of linear expansion; equation (12.2) remains valid.

Derive the equations of the minimum principles for total energy and additional work for this case.

Show that in the case of non-uniform heating the problem of determining the displacements $u_i$ reduces to the "usual" isothermal problem by adding a fictitious body force $-(\alpha/k)$ grad $\theta$ to the actual body forces $\rho X_i$, and a fictitious normal tension $(\alpha/k)\,\theta$ (on the portion $S_F$ of the surface) to the actual surface loads $X_{ni}$.

3. A lamina has identical circular holes (of diameter $d$) distributed in staggered rows (interval = $l$). Tension is applied uniformly in the $x$, $y$ directions. Calculate the statically possible load $p_s$ (the statically possible field consists of square regions of "hydrostatic" tension and rectangular regions of uniaxial tension).

*Answer:* $p_s = (1-d/l)\,\sigma_s$ .

4. An infinite lamina weakened by a series of holes of diameter $d$, located along the $x$-axis at intervals $l$, is stretched in the $y$-direction. Indicate the simplest lower bound for the limit load.

*Answer:* $p_s = (1-d/l)\,\sigma_s$ ($p$ is the mean stress) .

5. In the same problem find the upper bound, assuming that a neck develops along the weakened section (along the $x$-axis).

*Answer:* $p_k = \frac{2}{\sqrt{3}}\left(1-\frac{d}{l}\right)\sigma_s$ .

6. In the same problem, find the upper bound, using the solution of §56 (cf. fig. 173).

7. Use (66.27) to find the upper bound of the limit load for a circular joint-supported plate loaded at the centre.

*Answer:* $P_k = 2\pi M_s$ .

8. A circular joint-supported plate of radius $b$ is loaded eccentrically (at a distance $a$ from the centre) with a concentrated force. Find the upper bound assuming that the bending shape is the surface of a cone with vertex at the point where the load is applied.

*Answer:* $P_k = 2\pi M_s\, b/\sqrt{b^2-a^2}$ .

9. Find the additional work for a bending beam with a step-like relation between stress and strain (cf. problem 3, ch. 3).

*Answer:* $$R = \int_0^l \frac{|M|^{1+1/\mu}}{(1+1/\mu)\,D}\,dx .$$

10. With the same conditions derive the variational equation for the bending of the beam.

*Answer:* $$\delta \int_0^l \left[\frac{D^\mu}{1+\mu}\left|\frac{d^2v}{dx^2}\right|^{1+\mu} - q(x)\,v\right] dx = 0 .$$

# 9

# Theory of Shakedown

## §70. Behaviour of elastic-plastic bodies under variable loads

70.1. *Variable loads*

In the plastic flow problems considered hitherto it has been understood that the loading is simple. In practice, however, machines and structures are often subject to the action of varying loads and temperatures. If a body is deformed elastically, then in the presence of variable loads its strength is determined by the fatigue properties of the material; fracture occurs after a large number of cycles. But if the body experiences elastic-plastic deformation, a load less than limiting can cause the attainment of a critical state with a comparatively small number of cycles. In this situation it is necessary to distinguish two cases.

1. Fracture occurs as the result of plastic deformations alternating in sign (for example, plastic compression succeeds plastic extension, and so on). This is called *alternating plasticity* (*plastic* or *few-cycled fatigue*).

2. Plastic deformations do not change sign, but grow with each cycle (*progressive deformation, progressive fracture*). This leads to an inadmissible accumulation of plastic deformations.

70.2. *Alternating plasticity*

As an example of the development of alternating plasticity we consider the elastic-plastic state of a hollow sphere under the action of an internal pressure (§25), where the latter varies according to the scheme $0 \to p \to 0 \to p \ldots$ . With the first loading $0 \to p$ a zone of plastic deformation ($a \leqslant r \leqslant c$) develops in the sphere. After the unloading $p \to 0$ the residual stresses are described by formulae (25.12). The graph of the residual stress $\sigma_\varphi^0$ is shown on the left-hand side of fig. 41 and in fig. 218a. It is assumed that the residual stresses are not large enough to induce secondary plastic deformation; by (25.13) this will be so if $\tilde{p} \leqslant 2\tilde{p}_0$, and then

$$p \leqslant \tfrac{4}{3}(1-a^3/b^3)\,\sigma_s \equiv p_1\,. \tag{70.1}$$

With this condition the interval in which the intensity of tangential stresses (according to the elastic solution) varies does not exceed twice the yield limit, $2\tau_s$, anywhere in the sphere.

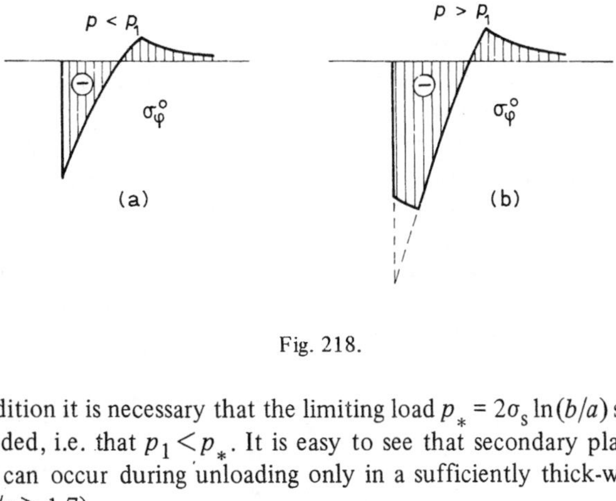


Fig. 218.

In addition it is necessary that the limiting load $p_* = 2\sigma_s \ln(b/a)$ should not be exceeded, i.e. that $p_1 < p_*$. It is easy to see that secondary plastic deformations can occur during unloading only in a sufficiently thick-walled shell (when $b/a > 1.7$).

If condition (70.1) is satisfied, renewed loading will induce only elastic deformations, at the expense of the development in the sphere of a field of residual stresses of favourable ("inverse") sign. It is as if the sphere were strengthened by comparison with its first loading. As we have remarked earlier, this effect is called *hardening,* or *autofrettage,* of the structure.

In recent years a different term has been increasingly in use – the term "*shakedown*", which was introduced by Prager. We say that the structure shakes down to the loading cycle, due to the development of a favourable field of residual stresses. The inequality (70.1) can be regarded as the shakedown condition for the sphere; it defines the *region of shakedown* (the region of admissible load variations).

If $p > p_1$, then unloading in some zone adjacent to the cavity (fig. 218b) results in plastic deformation of inverse sign with respect to the plastic deformation under loading. If the sphere is now loaded again under the same pressure $p > p_1$, this zone experiences plastic deformation of the original sign. After a moderate number of such cycles the zone undergoes fracture due to "plastic fatigue" (a well-known example is the rapid breaking of a wire with alternating plastic bending). Thus, safety considerations require that the loads should be confined to the shakedown region.

These ideas can be extended to bodies of arbitrary shape, and this leads to a *sufficient condition* for the development of alternating plasticity; alternating plasticity occurs if the interval of variation of the intensity of tangential stresses exceeds, anywhere in the corresponding perfectly-elastic body, twice the yield limit $2\tau_s$ (according to the von Mises criterion).

70.3. *Progressive deformation*

To illustrate how this type of plastic deformation can arise, we consider the simple model shown in fig. 219. A circular rod 1 and a tube 2 which surrounds it are joined by a rigid plate 3. A constant force $2P$ is applied to the latter. Let the cross-sectional areas of the rod and the tube be the same, and equal to $F$. The temperature of the rod is constant (zero, say), while the temperature of the tube varies periodically between $0^\circ$ and $\theta^\circ$ ($0 \to \theta \to 0 \to \to \theta \to \ldots$). The modulus of elasticity is assumed invariant, and the yield limit is $\sigma_{s0}$ at $0^\circ$, and $\sigma_{s\theta}$ at $\theta^\circ$; $\alpha$ is the coefficient of linear expansion. We introduce the notation $p = P/F$, $q = \frac{1}{2}E\alpha\theta$.

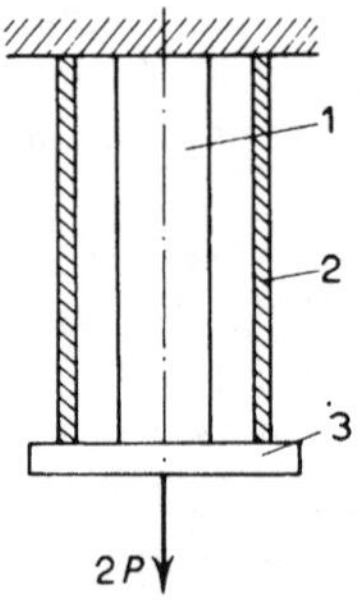


Fig. 219.

From the equilibrium condition we have

$$\sigma_1 + \sigma_2 = 2p \,, \tag{70.2}$$

where $\sigma_1, \sigma_2$ are the stresses in the rod and tube respectively. The state of the model depends on the relationship between the magnitudes of $p$, $q$ and the yield limits $\sigma_{s0}$ and $\sigma_{s\theta}$. We shall not consider all variations, but merely discuss a few of the possible states.

*Elastic state.* During heating the stresses in the rod and the tube are $\sigma_1 = p + q$, $\sigma_2 = p - q$. In order that plastic deformations be absent, it is necessary that

$$p + q < \sigma_{s0} \,, \qquad p - q > -\sigma_{s\theta} \,.$$

*Shakedown.* The rod always remains elastic, but the tube experiences plastic deformation when heated. At temperature $\theta$ the stress in the tube is $\sigma_2' = -\sigma_{s\theta}$; the stress in the rod must not reach the yield limit, i.e.

$$\sigma_1' = 2p + \sigma_{s\theta} < \sigma_{s0} \,. \tag{70.3}$$

After cooling we have $\sigma_2'' = -\sigma_{s\theta} + q$, $\sigma_1'' = 2p + \sigma_{s\theta} - q$. It is necessary for shakedown that these stresses should not exceed the elastic limit $\sigma_{s\theta}$, i.e.

$$-\sigma_{s\theta} + q < \sigma_{s0} \,, \qquad |2p + \sigma_{s\theta} - q| < \sigma_{s0} \,. \tag{70.4}$$

The inequalities (70.3), (70.4) characterize the conditions of shakedown.

*Progressive deformation.* Suppose that in each cycle the rod experiences plastic deformation when heated, and the tube when cooled. In this case it is easy to see that the rod flows with continuing thermal expansion of the tube (i.e. the rod "acquires" plastic deformation). During cooling the tube also flows, with stress $\sigma_{s0}$, but retains a constant length (because of the constancy of length of the rod). This pattern is repeated in each cycle, and general plastic extension of the system develops.

During heating the stress is $\sigma_{s0}$ in the rod, and $2p - \sigma_{s0}$ in the tube, where

$$|2p - \sigma_{s0}| < \sigma_{s\theta} \,. \tag{70.5}$$

During cooling the stress is $\sigma_{s0}$ in the tube, and $2p - \sigma_{s0}$ in the rod. With *elastic* unloading (cooling), the stress in the tube is $2p - \sigma_{s0} + q$, and must not be less than $\sigma_{s0}$; consequently

$$q > 2\sigma_{s0} - 2p \,. \tag{70.6}$$

After each cycle the general strain increases by an amount

$$\epsilon = \alpha\theta + \frac{4p}{E} - \frac{2\sigma_{s0}}{E}$$

and can reach inadmissible values as the number of cycles increases.

70.4. *The influence of hardening, the Bauschinger effect and creep*

In real bodies the conditions of shakedown depend on the increase in the elastic limit with plastic deformation (hardening), and on its decrease with loading in the reverse direction (Bauschinger effect). These effects can be computed, although the analysis becomes very complicated. In a similar fashion it is possible to take into account changes in mechanical characteristics during the temperature cycles. If the cycle lasts a sufficiently long time, shakedown depends significantly on creep, which can in large measure change the field of residual stresses; in a number of cases this contracts the region of shakedown.

## §71. Shakedown theorems for elastic-plastic bodies

The examples discussed in the preceding section show that clarification of shakedown conditions requires an analysis of the elastic-plastic equilibrium of a body. This analysis, however, can only be effected in very simple problems.

There exist a number of theorems regarding shakedown which eliminate this difficulty, by enabling upper and lower bounds for the shakedown region to be found. An analysis of the elastic-plastic state is then no longer necessary; we only require the detailed application of the solution of an appropriate elastic problem. This is, of course, incomparably simpler.

71.1. *Statical shakedown theorem (Melan's theorem)*

We consider an ideal elastic-plastic body experiencing the action of a system of loads, which vary slowly in time, between given limits. This condition enables dynamical effects to be neglected. We denote by $\sigma_{ij}^*$, $\epsilon_{ij}^*$ the instantaneous values of the stresses and strains in the corresponding *perfectly elastic body* (for the instantaneous values of the loads, i.e. at some point of the loading programme), and denote by $\sigma_{ij}$, $\epsilon_{ij}$ the instantaneous values of the stresses and strains in the actual *elastic-plastic state*. Let $\sigma_{ij}^0$, $\epsilon_{ij}^0$ be the residual stresses and strains in the body, given by the differences

$$\sigma_{ij}^0 = \sigma_{ij} - \sigma_{ij}^* , \tag{71.1}$$

$$\epsilon_{ij}^0 = \epsilon_{ij} - \epsilon_{ij}^* , \tag{71.2}$$

and let $\epsilon_{ij}^{0e}$ be the elastic strains corresponding to the residual stresses. Because the loads are variable, all the above stresses and strains are slowly-changing functions of time. Note also that the strains $\epsilon_{ij}^*$, $\epsilon_{ij}$ are *kinematically possible,* i.e. they satisfy the compatibility conditions, and the corresponding displacements satisfy prescribed kinematic boundary conditions.

The actual strains $\epsilon_{ij}$ are made up of elastic and plastic components:

$$\epsilon_{ij} = \epsilon_{ij}^e + \epsilon_{ij}^p . \tag{71.3}$$

Consequently:

$$\epsilon_{ij}^0 = \epsilon_{ij}^p + \epsilon_{ij}^{0e} , \tag{71.4}$$

$$\epsilon_{ij} = \epsilon_{ij}^{0e} + \epsilon_{ij}^* + \epsilon_{ij}^p . \tag{71.5}$$

We now suppose that we have found some field of fictitious residual stresses $\bar{\sigma}_{ij}$, *independent of time.* This field $\bar{\sigma}_{ij}$ can be taken to be any non-trivial solution of the homogeneous equilibrium equations (64.2) which satisfies *zero* boundary conditions on the portion $S_F$ of the body's surface. We denote by $\bar{\epsilon}_{ij}$ the strain components corresponding to the fictitious stresses $\bar{\sigma}_{ij}$ *according to Hooke's law.* Note that, in general, the $\bar{\epsilon}_{ij}$ are not kinematically possible strains.

We shall call the stress field

$$\bar{\sigma}_{ij} + \sigma_{ij}^* \equiv \sigma_{ij}^s$$

*safe* if no arbitrary load-variation in the prescribed limits causes the yield limit to be reached, i.e. if ($f$ is the yield function, cf. §16)

$$f(\sigma_{ij}^s) < K . \tag{71.6}$$

The stress field

$$\bar{\sigma}_{ij} + \sigma_{ij}^* \equiv \sigma_{ij}^a$$

will be called *admissible* if the stress distribution can reach the yield surface, i.e. if

$$f(\sigma_{ij}^a) \leqslant K . \tag{71.7}$$

Melan's theorem. *Shakedown occurs if it is possible to find a field of fictitious residual stresses* $\bar{\sigma}_{ij}$, *independent of time, such that for any variations of loads within the prescribed limits the sum of this field with the stress field* $\sigma^*_{ij}$ *in a perfectly elastic body is safe* (sufficient condition).

*Shakedown cannot occur if there does not exist any time-independent field of residual stresses* $\bar{\sigma}_{ij}$ *such that the sum* $\bar{\sigma}_{ij} + \sigma^*_{ij}$ *is admissible* (necessary condition).

The necessary condition is obvious: if there is no distribution of residual stresses for which $f(\sigma^{\mathrm{a}}_{ij}) \leqslant K$, then by definition shakedown cannot occur.

Suppose now that a suitable field of residual stresses $\sigma_{ij}$ exists. We show that shakedown then occurs.

Consider the fictitious elastic energy $\tilde{\Pi}$ of the stress difference $\sigma^0_{ij}-\bar{\sigma}_{ij}$:

$$\tilde{\Pi} = \tfrac{1}{2}\int (\sigma^0_{ij}-\bar{\sigma}_{ij})\,(\epsilon^{0\mathrm{e}}_{ij}-\bar{\epsilon}_{ij})\,\mathrm{d}V\,.$$

The stress differences $\sigma^0_{ij}-\bar{\sigma}_{ij}$ are connected with the strain differences $\epsilon^{0\mathrm{e}}_{ij}-\bar{\epsilon}_{ij}$ by the linear homogeneous relations of Hooke's law, and therefore the derivative of the energy $\tilde{\Pi}$ with respect to time is:

$$\frac{\mathrm{d}\tilde{\Pi}}{\mathrm{d}t} = \int (\sigma^0_{ij}-\bar{\sigma}_{ij})\,\frac{\mathrm{d}}{\mathrm{d}t}\,(\epsilon^{0\mathrm{e}}_{ij}-\bar{\epsilon}_{ij})\,\mathrm{d}V\,.$$

But the stresses $\bar{\sigma}_{ij}$ and the strains $\bar{\epsilon}_{ij}$ are by definition independent of time, consequently

$$\frac{\mathrm{d}\tilde{\Pi}}{\mathrm{d}t} = \int (\sigma^0_{ij}-\bar{\sigma}_{ij})\,\xi^{0\mathrm{e}}_{ij}\,\mathrm{d}V\,.$$

From (71.5) we have:

$$\xi^{0\mathrm{e}}_{ij} = \xi_{ij}-\xi^{\mathrm{p}}_{ij}-\xi^*_{ij}\,. \tag{71.8}$$

Thus

$$\frac{\mathrm{d}\tilde{\Pi}}{\mathrm{d}t} = \int (\sigma^0_{ij}-\bar{\sigma}_{ij})\,(\xi_{ij}-\xi^{\mathrm{p}}_{ij}-\xi^*_{ij})\,\mathrm{d}V\,.$$

We note now that the stress differences $\sigma^0_{ij}-\bar{\sigma}_{ij}$ satisfy the equilibrium conditions with zero external forces, and the strain rates $\xi_{ij}-\xi^*_{ij}$ are kinematically possible. The rate of work of the internal forces is equal to the rate of work of the corresponding external forces. But the latter are zero on $S_F$, while $v_i-v^*_i = 0$ on $S_u$; hence:

$$\int (\sigma^0_{ij}-\bar{\sigma}_{ij})\,(\xi_{ij}-\xi^*_{ij})\,\mathrm{d}V = 0\,.$$

This result can also be established by formal transformation of the volume integral into a surface integral (cf. §64). Thus

$$\frac{\mathrm{d}\tilde{\Pi}}{\mathrm{d}t} = -\int (\sigma_{ij}^0 - \bar{\sigma}_{ij})\, \xi_{ij}^{\mathrm{p}} \mathrm{d}V \,.$$

With the aid of (71.1) this equation can be written in a different form:

$$\frac{\mathrm{d}\tilde{\Pi}}{\mathrm{d}t} = -\int (\sigma_{ij} - \sigma_{ij}^{\mathrm{s}})\, \xi_{ij}^{\mathrm{p}} \mathrm{d}V \,.$$

Since the vector of the plastic strain rate $\xi_{ij}^{\mathrm{p}}$ is in the direction of the normal to the convex yield surface $\Sigma$, the vector $\sigma_{ij}$ reaches the yield surface, while the vector $\sigma_{ij}^{\mathrm{s}}$, being safe, lies inside $\Sigma$ (cf. fig. 199b, with $\sigma_{ij}^{\mathrm{s}}$ in place of $\sigma'_{ij}$), therefore we have a local maximum principle:

$$(\sigma_{ij} - \sigma_{ij}^{\mathrm{s}})\, \xi_{ij}^{\mathrm{p}} > 0 \,. \tag{71.9}$$

Thus $\mathrm{d}\tilde{\Pi}/\mathrm{d}t < 0$ as long as $\xi_{ij}^{\mathrm{p}} \neq 0$. Since the elastic energy $\tilde{\Pi}$ is non-negative, a time will be reached when plastic flow ceases (i.e. $\xi_{ij}^{\mathrm{p}} = 0$, $\mathrm{d}\tilde{\Pi}/\mathrm{d}t = 0$). The residual stresses will no longer change with time, and the body will experience only elastic deformations as the loads are varied.

In practice, structures shake down to some field of residual stresses which depends on the loading programme.

The field of residual stresses $\bar{\sigma}_{ij}$ is expediently chosen such that the region of admissible load variation is greatest.

In this sense the application of Melan's theorem leads to a *lower bound* for the limits of load variation. The actual realization of this scheme in concrete problems is difficult, especially in cases where the loads depend on several parameters. In general the determination of the optimal field of residual stresses $\bar{\sigma}_{ij}$ which gives the maximum extent to the region of shakedown constitutes a problem of mathematical programming. In lattices of rods and framework structures the safety conditions are, as a rule, linear inequalities; in such cases we can use well-known methods of linear programming. Note also that to determine the admissible loads it is only necessary to consider loads below limiting.

A simple method of constructing an approximate solution based on Melan's theorem is illustrated in the next section.

The statical shakedown theorem for the general case was proved by Melan in 1938.

### 71.2. *Kinematical shakedown theorem (Koiter's theorem)*

Let the displacements be zero on a portion $S_u$ of the body's surface, and let loads, which vary slowly between prescribed limits, act on the remaining portion $S_F$.

Take some arbitrary field of plastic strain rates $\xi^{\mathrm{p}}_{ij0} = \xi^{\mathrm{p}}_{ij0}(t)$. We shall call this field *admissible*, if the plastic strain increments over some time interval $\tau$

$$\Delta\epsilon^{\mathrm{p}}_{ij0} = \int_0^\tau \xi^{\mathrm{p}}_{ij0}\mathrm{d}t$$

generate a kinematically possible field (i.e. $\Delta\epsilon^{\mathrm{p}}_{ij0}$ satisfy the compatibility conditions, while the corresponding displacement field satisfies zero conditions on $S_u$). Corresponding to the field $\xi^{\mathrm{p}}_{ij0}$ there is a stress field $\sigma_{ij0}$ (according to the associated law), and a unique field of "accompanying" residual stress rates $\dot{\sigma}^0_{ij0}$, which can be determined in the following way. From (71.3) and (71.2) we have

$$\xi^0_{ij} = \xi^{\mathrm{e}}_{ij} - \xi^*_{ij} + \xi^{\mathrm{p}}_{ij} .$$

If we here replace the components $\xi^{\mathrm{p}}_{ij}$ by the components $\xi^{\mathrm{p}}_{ij0}$, we obtain

$$\xi^0_{ij0} = \xi^{\mathrm{e}}_{ij} - \xi^*_{ij} + \xi^{\mathrm{p}}_{ij0} .$$

The strain rates $\xi^{\mathrm{e}}_{ij}$ and $\xi^*_{ij}$ are connected with the stress rates $\sigma_{ij}$ and $\dot{\sigma}^*_{ij}$ by Hooke's law, and consequently the differences $\xi^{\mathrm{e}}_{ij} - \xi^*_{ij}$ and $\dot{\sigma}_{ij} - \dot{\sigma}^*_{ij} = \dot{\sigma}^0_{ij0}$ are related by the same law. Thus

$$\xi^0_{ij0} = c_{ijhk}\dot{\sigma}^0_{hk0} + \xi^{\mathrm{p}}_{ij0} , \tag{71.10}$$

where $c_{ijhk}$ are elastic constants. By considering the body's equilibrium with zero loads on $S_F$, zero displacements on $S_u$, and with the inhomogeneous linear relations (71.10), we find a unique distribution of accompanying residual stress rates $\dot{\sigma}^0_{ij0}$, residual strain rates $\xi^0_{ij0}$ and residual velocities $v_{i0}$. Here the $\xi^{\mathrm{p}}_{ij0}$ will play the role of given additional ("imposed") strains. The first terms on the right-hand side of (71.10) are the elastic strain rates $\xi^{\mathrm{e}}_{ij0}$ induced by the residual stress rates $\dot{\sigma}^0_{ij0}$.

The displacement increments over the interval $\tau$ are

$$\Delta u_{i0} = \int_0^\tau v_{i0}\mathrm{d}t .$$

By definition the plastic strain increments $\Delta\epsilon^{\mathrm{p}}_{ij0}$ over the time $\tau$ are kinematically possible, therefore the accompanying elastic strain-increments $\Delta\epsilon^{\mathrm{e}}_{ij0}$ are also kinematically possible. The residual stresses $\sigma^0_{ij0}$ at the end of the cycle $t = \tau$ are returned to their initial values at $t = 0$, i.e.

$$\sigma^0_{ij0}|_{t=0} = \sigma^0_{ij0}|_{t=\tau} . \tag{71.11}$$

Then

$$\int_0^\tau \xi^{e}_{ij0}\,dt = 0\,. \tag{71.12}$$

Koiter's theorem. *Shakedown does not occur if it is possible to find an admissible cycle of plastic strain rates* $\xi^{p}_{ij0}$ *and some programme of load variations between prescribed limits for which*

$$\int_0^\tau dt \int X_{ni} v_{i0}\, dS_F > \int_0^\tau dt \int \dot{A}(\xi^{p}_{ij0})\, dV\,, \tag{71.13}$$

where $\dot{A}(\xi^{p}_{ij0}) = \sigma_{ij0}\xi^{p}_{ij0}$ is the rate of work of the plastic strain on the admissible rates $\xi^{p}_{ij0}$.

Conversely. *Shakedown occurs if for all admissible cycles of plastic strain rates and arbitrary loads (between prescribed limits) it is possible to find a number* $\kappa > 1$ *such that*

$$\kappa \int_0^\tau dt \int X_{ni} v_{i0}\, dS_F \leqslant \int_0^\tau dt \int \dot{A}(\xi^{p}_{ij0})\, dV\,. \tag{71.14}$$

The first part of the theorem is proved by contradiction. Let there exist an admissible cycle for which inequality (71.13) holds, and at the same time let shakedown occur. Then by Melan's theorem there exists a time-independent field of residual stresses $\bar{\sigma}_{ij}$; the sum of this field with the elastic field $\sigma^{*}_{ij}$ generates an admissible stress field $\sigma^{a}_{ij}$. By the principle of virtual work we have

$$\int X_{ni} v_{i0}\, dS_F = \int \sigma^{a}_{ij}\xi^{0}_{ij0}\, dV\,. \tag{71.15}$$

Using the definition of $\sigma^{a}_{ij}$ and formula (71.10), we easily obtain

$$\int \sigma^{a}_{ij}\xi^{0}_{ij0}\, dV = \int \sigma^{*}_{ij} c_{ijhk}\dot{\sigma}^{0}_{hk0}\, dV + \int \bar{\sigma}_{ij} c_{ijkk}\dot{\sigma}^{0}_{hk0}\, dV + \int \sigma^{a}_{ij}\xi^{p}_{ij0}\, dV\,.$$

Since by Hooke's law $\epsilon^{*}_{hk} = c_{ijhk}\sigma^{*}_{ij}$, it follows that

$$\int \sigma^{*}_{ij} c_{ijhk}\dot{\sigma}^{0}_{hk0}\, dV = \int \dot{\sigma}^{0}_{hk0}\epsilon^{*}_{hk}\, dV = 0\,,$$

as the stresses $\sigma^{0}_{hk0}$ correspond to zero external forces. We integrate equation (71.15) with respect to time over $(0, \tau)$. Since $\bar{\sigma}_{ij}$ is time-independent, (71.12) now leads to

$$\int_0^\tau dt \int \bar{\sigma}_{ij} c_{ijhk}\dot{\sigma}^{0}_{hk0}\, dV = \int \bar{\sigma}_{ij}\, dV \int \xi^{e}_{ij0}\, dt = 0\,.$$

Thus

$$\int_0^T \mathrm{d}t \int X_{ni} v_{i0} \mathrm{d}S_F = \int_0^T \mathrm{d}t \int \sigma_{ij}^{\mathrm{a}} \xi_{ij0}^{\mathrm{p}} \mathrm{d}V,$$

which contradict the original inequality (71.13), since $(\sigma_{ij0}-\sigma_{ij}^{\mathrm{a}})\,\xi_{ij0}^{\mathrm{p}} \geqslant 0$.

The proof of the second part of Koiter's theorem is much more complicated and we shall not give it here (cf. [68]).

If we choose an admissible cycle of plastic strain rates and write (71.13) with the equality sign, we can use Koiter's theorem to find an *upper bound* for shakedown. The application of Koiter's theorem involves greater difficulties than the application of Melan's theorem (except in the simplest systems, rod lattices and frameworks, where linear programming methods can be used). A useful inverse method was proposed by V.I. Rozenblum; we also refer the reader to a series of works by D.A. Hochfield [100].

71.3. *Shakedown for non-uniformly heated bodies*

A matter of great practical interest is the case of simultaneous action of loads and a temperature field $\theta = \theta(x_1, x_2, x_3, t)$ varying between prescribed limits.

Melan's theorem can easily be generalized to non-uniformly heated bodies. The formulation of the theorem is as before, but by $\sigma_{ij}^*$ we now have to understand a field of *thermoelastic stresses in a perfectly elastic body* [149].

Koiter's theorem can also be extended to non-uniformly heated bodies [100, 149], though the formulation is slightly different. On the left-hand side of the inequality (71.13) it is necessary to include the term

$$3\alpha \int_0^T \mathrm{d}t \int \theta \dot{\sigma}_0^0 \mathrm{d}V,$$

where $\alpha$ is the thermal coefficient of linear expansion, $\dot{\sigma}_0^0$ is the rate of mean pressure for the residual stress field.

71.4. *Remark on the connection between shakedown theorems and limit load theorems*

Koiter has drawn attention to the fact that the limit load theorems (§65) are a consequence of the shakedown theorems if it is assumed that the prescribed limits of load variation coincide.

## §72. Approximate method of solution. Example

As we have already observed, if we wish to find the region of shakedown from Melan's theorem, we have to consider admissible fields of residual stresses and at the same time to set out the solution to the corresponding elastic problem with loads varying arbitrarily between prescribed limits. We encounter substantial difficulties in carrying through this scheme (especially in cases where there are several independent loads). Shakedown analysis for simple lattices and frameworks with one – or two – parameter load systems is usually effected by geometrical methods of constructing the region of admissible distributions; in more complicated cases we can use the methods of linear programming.

For bodies of arbitrary shape with one – or two – parameter load systems a convenient approximate method of finding the shakedown region was suggested by V.I. Rozenblum [148]. We now describe this method.

### 72.1. *Approximate method of solution*

Basing ourselves on Melan's theorem, we consider the case where the *loads are proportional to a single parameter p*; then the solution to the corresponding elastic problem has the form

$$\sigma_{ij}^{*} = p\sigma_{ij}',$$

where $\sigma_{ij}'$ are functions of coordinates only. Next we choose some field of residual stresses

$$\bar{\sigma}_{ij} = \lambda\bar{\sigma}_{ij}',$$

where $\bar{\sigma}_{ij}'$ are functions of coordinates only, and $\lambda$ is an undetermined multiplier. According to Melan's theorem it is necessary to construct a field

$$\sigma_{ij} = p\sigma_{ij}' + \lambda\bar{\sigma}_{ij}', \tag{72.1}$$

which is safe, i.e.

$$f(\sigma_{ij}) < K. \tag{72.2}$$

It is required to find the optimum value of the multiplier $\lambda$ for which the range of admissible variations of the coefficient $p$ is greatest.

Consider the plane of the variables $p$, $\lambda$. To every point of the plane there corresponds some stress distribution (72.1). In this plane the yield criterion

$$f(p\sigma_{ij}' + \lambda\bar{\sigma}_{ij}') = K \tag{72.3}$$

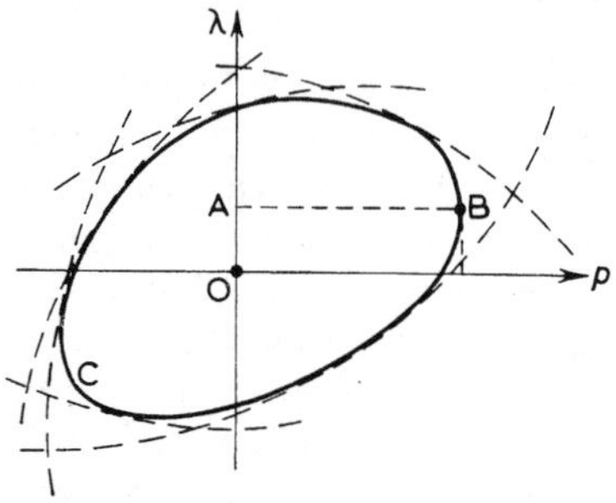


Fig. 220.

defines a three-parameter family of curves (dotted lines in fig. 220), which do not pass through the origin of coordinates $p = 0$, $\lambda = 0$ (since $K > 0$), and which therefore marks out some region of admissible values. The boundary of this region C is generated by the envelope of the family (72.3) or by individual curves of the family closest to the origin. When the boundary C has been constructed, it is easy to determine the admissible interval of variation of the load parameter $p$. Suppose $p > 0$; then the greatest possible deviation must be less than the largest abscissa B. The optimum value of $\lambda$ is equal to the segment OA.

A good approximation can be achieved with successful choice of the residual stress field.

This method can be extended without difficulty to two-parameter load systems, when

$$\sigma_{ij}^* = p\sigma_{ij}' + q\sigma_{ij}'' , \tag{72.4}$$

and in place of (72.3) we have

$$f(p\sigma_{ij}' + q\sigma_{ij}'' + \lambda\bar{\sigma}_{ij}') = K . \tag{72.5}$$

We specify a series of values $q = q_1, q_2, q_3, \ldots$ and, as before, construct the curves of admissible distribution $C_1, C_2, C_3, \ldots$ (fig. 221). On each curve we then mark the greatest value of the parameter $p$ ($p = p_1, p_2, p_3, \ldots$). In the parameter plane $p, q$ we draw a curve through these points (fig. 222); this curve bounds the region of admissible loads.

72.2. *Example. Combined torsion and tension of a rod*

We shall find the shakedown region for a circular rod of radius $a$, with a tensile force $P$ and a twisting moment $M$. In the cylindrical coordinate system

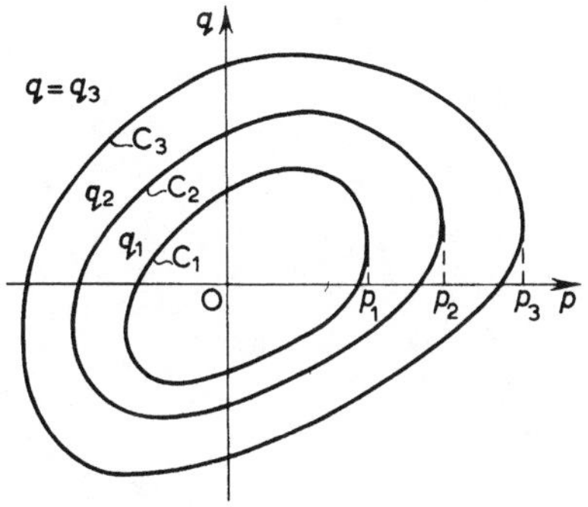


Fig. 221.

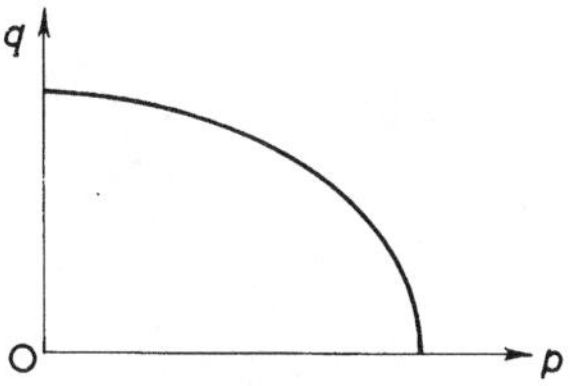


Fig. 222.

$r, \varphi, z$ the stress components $\sigma_z, \tau_{\varphi z}$ are different from zero. We introduce the dimensionless quantities

$$\rho = r/a\,, \qquad \tau = \tau_{\varphi z}/\tau_s\,, \qquad \sigma = \sigma_z/\sigma_s\,, \qquad p = \frac{P}{\pi a^2 \sigma_s}\,, \qquad q = \frac{2M}{\pi a^3 \tau_s}\,.$$

The elastic solution can be written in the form

$$\sigma^* = p\,, \qquad \tau^* = q\rho\,.$$

We choose the following residual stress distribution

$$\bar{\sigma} = 0\,, \qquad \bar{\tau} = \lambda(1 + c\rho)\,.$$

The constant $c$ is determined from the condition that the moment of the stress $\bar{\tau}$ is equal to zero; we easily find that $c = -\frac{4}{3}$.

The total field

$$\sigma = p\,, \qquad \tau = q\rho + \lambda(1 - \tfrac{4}{3}\rho) \tag{72.6}$$

must be safe, i.e. (by the von Mises yield criterion)

$$\sigma^2 + \tau^2 < 1\,. \tag{72.7}$$

We consider first the case of pure torsion ($p = 0$); then from the yield criterion we find

$$q\rho + \lambda(1 - \tfrac{4}{3}\rho) = \pm 1\,. \tag{72.8}$$

In the $q$, $\lambda$-plane (fig. 223) this equation defines two pencils of straight lines with centres A ($q = \frac{3}{4}$, $\lambda = 1$) and A′($q = -\frac{4}{3}$, $\lambda = -1$). Since $0 \leqslant \rho \leqslant 1$, these pencils bound a parallelogram of admissible values ABA′B′; its sides are obtained from (72.8) with $\rho = 0$ and $\rho = 1$.

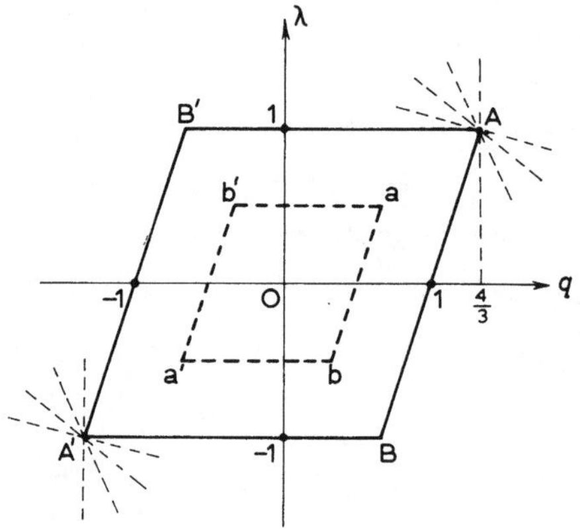


Fig. 223.

It is worth noting one or two details. First of all we observe that the first plastic deformations appear in the outer layer of the rod when $q = \pm 1$. If the twisting moment does not change sign (i.e. $0 \leqslant q$), then the greatest admissible interval of variation of $q$ is characterized by the abscissa of the point A ($q = \frac{4}{3}$). This value corresponds to the limiting twisting moment. If the moment changes sign, the greatest interval of variation of $q$ is determined by the length of the horizontal segment between the sides AB and A′B′; obviously $q_{max} - q_{min} = 2$ for any $\lambda$.

We now pass to the analysis of the general case when the force $p \neq 0$ varies arbitrarily between limits $(-p, +p)$. Then from the yield criterion we obtain

$$q\rho + \lambda(1 - \tfrac{4}{3}\rho) = \pm\sqrt{1-p^2}\,. \tag{72.9}$$

This equation defines the parallelogram aba′b′ in the $q$, $\lambda$-plane (fig. 223), similar to the parallelogram ABA′B′, and lying inside it. It cuts off segments of length $\sqrt{1-p^2}$ on the axes $q$, $\lambda$. Thus, when an axial force acts the interval of admissible variations of the twisting moment is reduced. If the twisting moment does not change sign, then $q_{max} - q_{min} = \frac{4}{3}\sqrt{1-p^2}$; otherwise $q_{max} - q_{min} = 2\sqrt{1-p^2}$.

## PROBLEMS

1. Find the shakedown region for a circular tube under the action of an internal pressure and a longitudinal force (by the approximate method, § 72).

2. Find the shakedown region for a circular rod twisted by a moment $M$ and experiencing a pressure $P$ at the end surfaces; there is no axial force.

# 10

# Stability of Elastic-Plastic Equilibrium

## §73. Criteria of stability

### 73.1. *Remarks on the stability of mechanical systems*

The stability of the equilibrium of a mechanical system depends on the system's parameters. For some mechanical systems a number of these parameters involve the forces which act. A similar situation occurs in *elastic* systems; for certain values of the loads, thin rods, plates and shells under pressure become unstable and buckle.

We assume that the reader is familiar with the basic ideas of stability theory, and merely concentrate on a number of details.

In rigid-body mechanics the question of stability of equilibrium is solved by examining the motion of a system close to the original equilibrium configuration (*dynamic criterion*). If small disturbances generate a motion which diverges from the neighbourhood of the equilibrium state, then the latter is unstable; if oscillations about the equilibrium state occur, then we have stability (stability *in the small*).

Stability depends on the magnitude of the disturbances. In the case of substantial permissible deviations we speak of stability *in the large*, and the concept of degree of stability is connected with the magnitude of the deviations. In the sequel we shall consider only small disturbances.

If the system is conservative, its oscillations need not be considered; the well-known Lagrange-Dirichlet rule provides a sufficient condition for stability: in a stable equilibrium state the potential energy of the system is a minimum (*energy criterion*).

73.2. *Stability of elastic systems*

The general dynamical criterion obviously remains valid for elastic systems; but its application is associated with great mathematical difficulties, because the motion of such systems is described by a system of partial differential equations. For this reason the analysis of stability of elastic bodies is usually effected with the aid of different, simpler but less general, criteria. We shall now briefly discuss these.

*The static stability criterion* consists in the following. We consider an equilibrium state infinitesimally close to the original (basic, "trivial") equilibrium state. For a certain value of the load it is possible to have a second equilibrium configuration, at the same time as the basic one. In other words, different equilibrium configurations can exist for the same load (*bifurcation point, branching of equilibrium configurations*). This state can be regarded as the transition from stability to instability. The smallest load for which different equilibrium configurations are possible is called *critical*. For the case of a compressed rod this situation is shown schematically by the solid line in fig. 224, the load $P$ is represented by the ordinate axis, and the abscissa is the flexure $\Delta$. When $P > P_{cr}$ a rectilinear equilibrium configuration is possible for a straight rod. When $P > P_{cr}$ ($P_{cr}$ is the bifurcation point) the flexured equilibrium is stable.

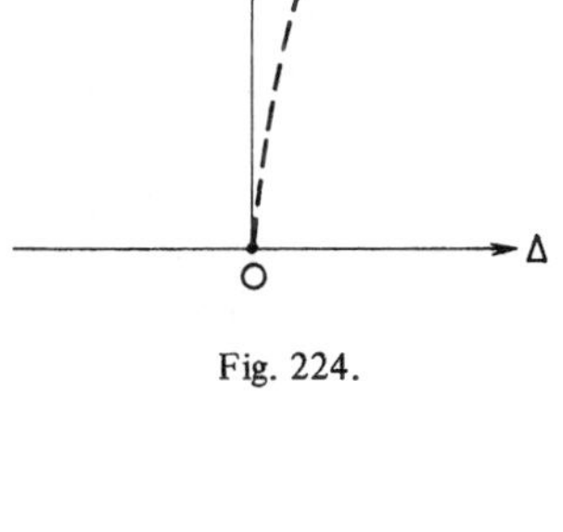

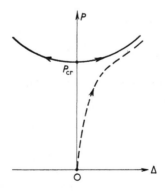

Fig. 224.

For conservative systems the static and dynamic criteria lead to the same values of the critical load. In its mathematical formulation the static criterion leads to the classical problem of eigenvalues for linear differential equations. The first to use the static method was Euler, who investigated the stability of a compressed elastic rod.

*Energy criterion of stability.* If the elastic system under consideration is conservative, a sufficient condition for its stability is that the potential energy is a minimum. If $\Pi$ is the elastic potential, $A$ the potential of external forces, and $\mathcal{E} = \Pi - A$ the total energy of the system, then the second variation of the energy must be positive definite for stability:

$$\delta^2 \mathcal{E} > 0 \, .$$

The critical value of the load is the least value for which this inequality ceases to hold, i.e. for which

$$\delta^2 \mathcal{E} = 0 \, . \tag{73.1}$$

The energy criterion can also be formulated in a somewhat different way, suggested by S.P. Timoshenko. In a position of stable equilibrium the energy $\mathcal{E} = \Pi - A$ is a minimum, and consequently, for every small deviation from the equilibrium position the increment in the total energy $\delta \mathcal{E} > 0$, or $\delta \Pi > \delta A$. If the equilibrium ceases to be stable for some value of the load, then $\delta \mathcal{E} = 0$, i.e.

$$\delta \Pi = \delta A \, . \tag{73.2}$$

Let $p$ be the load parameter, i.e. $\delta A = p \overline{\delta A}$; then

$$p = \delta \Pi / \overline{\delta A} \, .$$

The minimum value of the load which satisfies this relation for non-zero deviations of the system from the basic equilibrium position is the critical load

$$p_{\mathrm{cr}} = \min \, (\delta \Pi / \overline{\delta A}) \, . \tag{73.3}$$

In mathematical terms we have here the problem of the minimum of a quadratic functional.

*For conservative systems the static and energy criteria are equivalent.* The differential equations of stability obtained by using the static method are Euler's differential equations for the variational problem, to which the energy criterion reduces.

For an elastic system it is essential that the external forces should have a potential; otherwise the static and energy criteria can lead to erroneous conclusions (cf. [195]).

If the system is non-conservative then, in general, only a dynamic criterion is valid. The simplest example of such a system is the elastic rod compressed by a force acting along the tangent to the axis of the rod.

In the sequel we shall assume that the external forces have a potential, i.e., that the work of the external forces is independent of the path traversed.

The magnitude of the critical load can be approached by tracing the deformation of a system having *initial deviations.* For example, it is possible to consider the compression of a rod with initial twisting (or with an additional transverse load). The dotted line in fig. 224 shows the growth of flexure with increase in compressive force. As the latter approaches the critical value $P_{cr}$ the flexure increases sharply.

In this conception a stability analysis is by definition eliminated, but the method has its deficiencies. The behaviour of the system depends on the initial deviations which are a priori unknown. Furthermore, the mathematical problem turns out to be more complicated than the use of the static or energy criterion.

### 73.3. *Stability for elastic-plastic deformations*

Experiments show that the classical solutions of elastic stability problems are often very inadequate. One of the principal reasons for the discrepancy lies in the inelastic properties of real materials, which cause a sharp decrease in the resistance to buckling.

In practical treatments of the problem (in large measure loss of stability in modern structures occurs beyond the elastic limit), wide use has been made of empirical formulae obtained through experimental studies of the stability of compressive rods. More recently a number of theoretical methods have been developed to analyze the stability of structures operating beyond the elastic limit.

A system which experiences elastic-plastic deformations is not conservative. For this reason the investigation of stability beyond the elastic limit must in general be based on an analysis of motion near the basic equilibrium state, when certain disturbances are applied to the system. This analysis is extremely difficult, for two reasons: first, we do not have available reliable equations of plasticity for cyclic deformations; secondly, even when the simplest equations are used (for example, the equations of flow theory or deformation theory) the mathematical difficulties are enormous. In many cases (though not always) it is possible to obtain the required answer by investigating the elastic-plastic deformation of a system having *initial deviations.* This analysis, however, leads to non-linear problems and is also associated with great mathematical difficulties. Usually, one proceeds from some static criterion, and looks for a load for which different neighbouring equilibrium configurations are possible under some or other additional conditions. These criteria, which we consider in the next section, do not have a reliable theoretical foundation; their value is illustrated by analyzing the behaviour of very simple models of elastic-plastic bodies and is confirmed by experimental data.

## §74. Stability of a compressed rod. Reduced-modulus and tangent-modulus loads

### 74.1. *Stability of an elastic rod*

The stability of a compressed elastic rod was investigated by Euler, in about 1757. We briefly present the solution to this problem on the basis of the static criterion, where for simplicity we consider a rod of constant and symmetric section (fig. 225); the axes $x$, $y$ are principal central axes.

Suppose that for some value of the compressive force $P$, buckling of the rod takes place in the plane of least rigidity $Ox$; let $u = u(x)$ denote the displacement of the axis of the rod under buckling (fig. 226).

According to the static stability criterion buckling occurs in the "neutral" equilibrium state and is realized at the same value of the axial force, i.e. for buckling the increment in the axial force is zero:

$$\delta P = 0 \, . \tag{74.1}$$

During buckling the rod is twisted and experiences additional infinitesimal

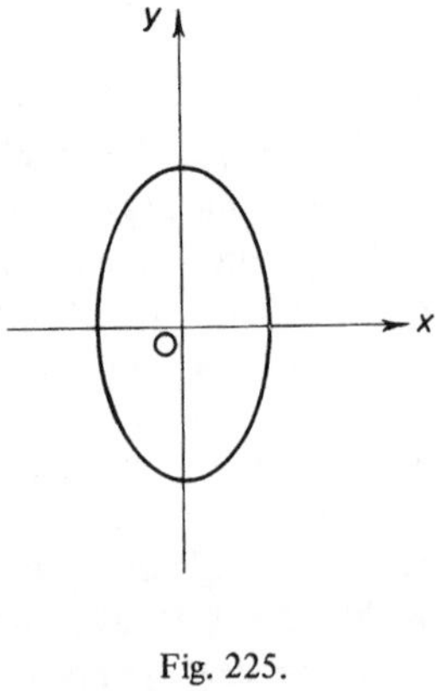


Fig. 225.

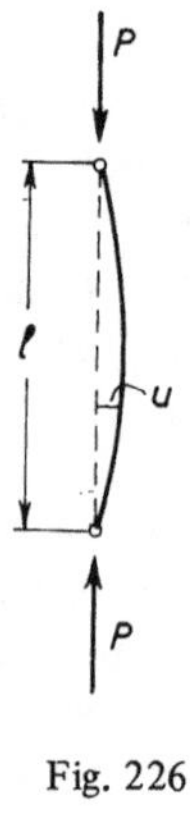


Fig. 226.

strains $\delta\epsilon_z$; since the rod is thin, these additional strains satisfy the hypothesis of plane sections, i.e.

$$\delta\epsilon_z = \epsilon_0 - qx\,, \tag{74.2}$$

where $\epsilon_0$ is an infinitesimal additional axial extension, and $q$ is an infinitesimal change in curvature. From Hooke's law the corresponding additional stresses will be

$$\delta\sigma_z = E\delta\epsilon_z\,. \tag{74.3}$$

It is obvious that $\delta P = EF\epsilon_0$, where $F$ is the cross-sectional area of the rod; in accordance with (74.1) it is necessary to put $\epsilon_0 = 0$.

The infinitesimal bending moment $Pu$ which arises with buckling is balanced by the moment of the internal forces

$$-\iint \delta\sigma_z x \mathrm{d}x \mathrm{d}y = EJq\,,$$

where $J$ is the moment of inertia of the section with respect to the $y$-axis. Since $q = \mathrm{d}^2u/\mathrm{d}z^2$, we now have the differential equation

$$\frac{\mathrm{d}^2u}{\mathrm{d}z^2} + \frac{P}{EJ}u = 0\,. \tag{74.4}$$

For a rod supported at its ends the boundary conditions are: $u = 0$ for $z = 0$ and $z = l$. The corresponding critical load (Euler force) is

$$P = \pi^2 EJ/l^2\,. \tag{74.5}$$

If we introduce a *flexure parameter*

$$\lambda = Fl^2/J = (l/\rho)^2\,,$$

where $\rho$ is the radius of inertia, and a *loading parameter*

$$\mu = P/EF\,,$$

equal to the ratio of the critical load to Young's modulus, formula (74.5) takes the form

$$\lambda\mu = \pi^2\,. \tag{74.6}$$

Consequently, the stability boundary in the $\lambda$, $\mu$-plane will be a hyperbola (fig. 227a).

Note that the results are not changed if we assume $\delta P$ to be different from zero.

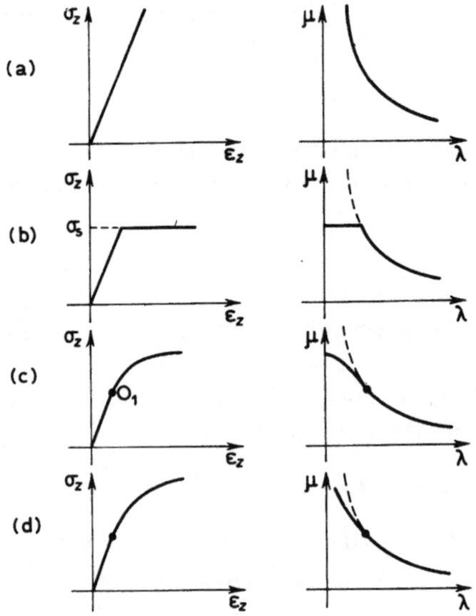

Fig. 227.

### 74.2. *Stability of a compressed rod beyond the elastic limit; tangent-modulus load*

If the rod is compressed beyond the elastic limit (point C in fig. 228), the above analysis is unsuitable.

In 1889 Engesser proposed a simple method of calculating plastic properties. We assume that the curve OC is a curve of *nonlinear-elastic* deformation,

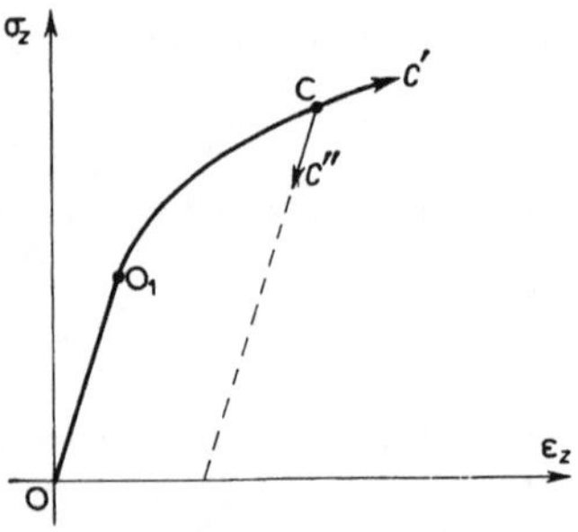

Fig. 228.

i.e. we consider the stability of some *nonlinear-elastic rod.* Then the relation between stress increments $\delta\sigma_z$ and strain increments $\delta\epsilon_z$ will be

$$\delta\sigma_z = E'\delta\epsilon_z\,, \tag{74.7}$$

where $E' = (\mathrm{d}\sigma_z/\mathrm{d}\epsilon_z)_c$ is a local modulus of elasticity, sometimes called the *tangent modulus.* It is evident that in this case the critical load $P'$ (which we shall call the *tangent-modulus load*) is determined by formula (74.5) with Young's modulus replaced by the modulus $E'$, i.e.

$$P' = \pi^2 E'J/l^2\,. \tag{74.8}$$

The stability boundary depends on the form of the deformation curve; some typical cases are shown in fig. 227. For materials with explicit yield plateau, $E' = 0$, i.e. stability breaks down on transition to the yield limit (fig. 227b); this result is confirmed by experiments. The stability boundary for gradual transition to the yield plateau is shown in fig. 227c. Finally, for a hardening material the stability boundary deviates laterally from the hyperbola which corresponds to the Euler force, and then rises (fig. 227d). Up to the elastic limit $O_1$ Euler's solution (74.6) is valid; subsequently it is shown by the dotted line.

In general this scheme gives a qualitatively accurate picture.

### 74.3. *Stability of a compressed rod beyond the elastic limit; reduced modulus load.*

Shortly after the publication of Engesser's work, Yasinskii noticed that in the buckling of real materials a part $F_1$ of the section (fig. 229) experiences additional compression, so that (74.7) is valid here, while the other part $F_2$ of the section experiences unloading, which proceeds according to Hooke's law, so that Engesser's suggestion cannot be assumed valid.

Subsequently Engesser and von Karman gave a solution to the problem of the stability of a compressive rod beyond the elastic limit which took into account Yasinskii's objection. We shall reproduce this solution.

The infinitesimal increments in the stress $\delta\sigma_z$ are

$$\delta\sigma_z = \begin{cases} E'\delta\epsilon_z & \text{in the loading region} \quad F_1\ (\delta\epsilon_z < 0)\,, \\ E\delta\epsilon_z & \text{in the unloading region} \quad F_2\ (\delta\epsilon_z > 0)\,. \end{cases}$$

As we have noted earlier (§24), the hypothesis of plane sections is geometrical in character and is not connected with the properties of the material; therefore the additional strains $\delta\epsilon_z$ are represented by the preceding formula (74.2). Along the line n–n, $\delta\epsilon_z = 0$; we shall call this line the *separation line*

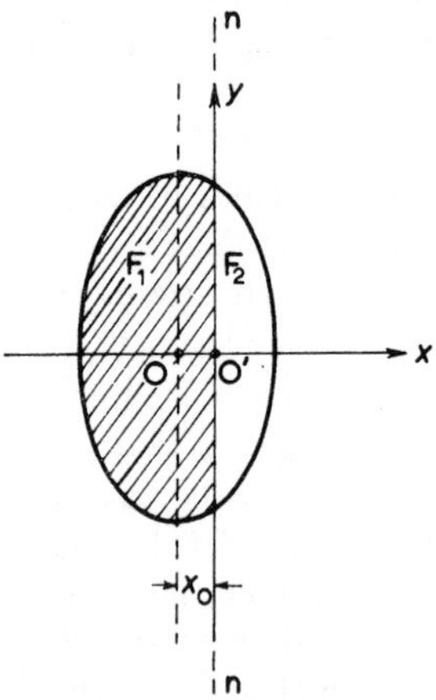


Fig. 229.

and shall henceforth measure $x$ from it. Then

$$\delta\epsilon_z = -qx\,. \tag{74.9}$$

Since it is being assumed that *buckling occurs at the same value of the axial force,* we have

$$\delta P = \iint \delta\sigma_z \mathrm{d}x\mathrm{d}y = 0\,,$$

and hence

$$S_1 E' - S_2 E = 0\,, \tag{74.10}$$

where $S_1$, $S_2$ are the statical moments of the areas $F_1$, $F_2$ with respect to the separation line. This equation determines the position of the separation line (i.e. $x_0$), which obviously depends on the shape of the section and on the ratio $E'/E$. It is clear that the loading region $F_1$ is always larger than the unloading region $F_2$.

It is now possible to calculate the moment of the internal forces with respect to the neutral line:

$$-\iint \delta\sigma_z x \mathrm{d}x\mathrm{d}y = E_k J q\,, \tag{74.11}$$

where we have put

$$E_k = \frac{E'J_1 + EJ_2}{J}\,.$$

Here $J_1$, $J_2$ are respectively the moments of inertia of the areas $F_1$, $F_2$, with respect to the separation line n–n. The quantity $E_k$ is called the *reduced-modulus* (or the *Engesser-von Karman modulus*).

This distribution of additional stresses in the section is shown in fig. 230. Comparing the moments of the internal forces for the distribution $E'$n$E$ with those for the distribution $E'$nA, $E$nB, we conclude that

$$E' \leqslant E_k \leqslant E\,. \tag{74.12}$$

Thus we have the previous equation for bending, but with the reduced modulus $E_k$ in place of $E$; consequently the critical load is now

$$P_k = \pi^2 E_k J/l^2 \qquad \text{or} \qquad \lambda\mu = \pi^2 E_k/E\,. \tag{74.13}$$

The reduced modulus depends on the shape of the section. However, calculations show that for steel the influence of the shape of the section on the value of the reduced modulus $E_k$ is not great.

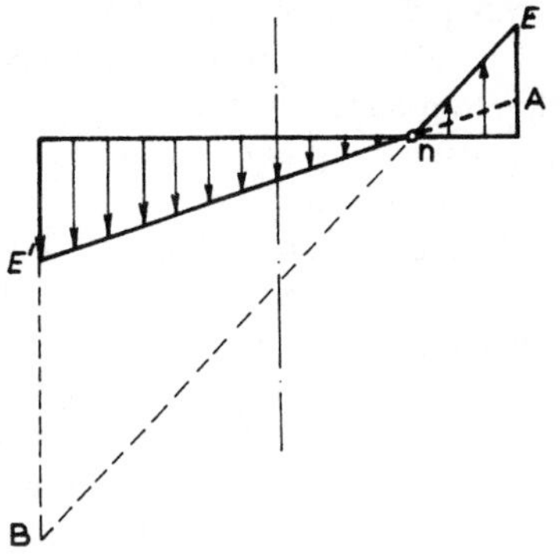


Fig. 230.

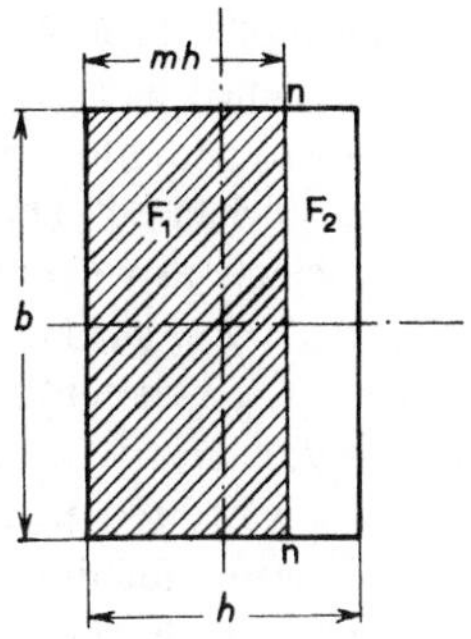


Fig. 231.

*Example.* Consider a rod of rectangular section (fig. 231); denoting by $mh$ ($m < 1$) the height of the loading zone $F_1$, we have

$$S_1 = \tfrac{1}{2}m^2h^2\,, \qquad S_2 = \tfrac{1}{2}(1-m)^2h^2$$

and equation (74.10), which determines $m$, takes the form

$$m^2E' - (1-m)^2E = 0\,,$$

hence

$$m = (1 + \sqrt{E'/E})^{-1}\,.$$

Further,

$$J_1 = \tfrac{1}{3}bm^3h^3\,, \qquad J_2 = \tfrac{1}{3}b(1-m)^3h^3\,, \qquad J = \tfrac{1}{12}bh^3\,.$$

The reduced modulus is

$$E_{\mathrm{k}} = \frac{4EE'}{(\sqrt{E}+\sqrt{E'})^2}\,.$$

In general the theory is satisfactorily confirmed by the experiments of von Karman and others. It is, however, necessary to bear in mind two facts. First, the tangent modulus $E'$ cannot in practice be determined very accurately because of the inevitable scatter of points and the rapid variation of the slope of the tangent to the deformation curve. Secondly, doubts have been expressed concerning the accuracy of the experimental data, because of the substantial influence of the end conditions and the loading characteristics in test machines.

### 74.4. *Shanley's model. Value of the tangent-modulus load*

The Engesser-von Karman solution is based on the use of the static stability criterion in the form which applies to problems of stability of elastic systems. It is assumed that the rod remains straight up to the moment of stability break-down, with the transition from straight to distorted state being achieved at an unchanged value of the compressive force, i.e. $\delta P = 0$. For a long time no objections were raised to the validity of this approach to solve problems of stability of a compressive rod beyond the elastic limit.

In a series of experimental investigations on compressed rods made of aluminium alloy, which were carried out recently in connection with aircraft construction, it was discovered that the critical load was usually somewhat closer to the tangent-modulus load $P'$ than to the reduced-modulus load $P_k$. Tests have shown that bending appears even before the reduced-modulus load $P_k$ is reached, and is not initially accompanied by unloading of the material. These facts acquired a new significance in the researches of Shanley [171] and a series of subsequent works by other authors.

If we abandon the restriction $\delta P = 0$ and seek the smallest load for which distortion is possible in conditions of increase of the compressive force ($\delta P > 0$), then this load turns out to be the tangent-modulus load $P'$. This was demonstrated by Shanley for the special case of an idealized column, consisting of two rigid rods of length $\tfrac{1}{2}l$ each, joined by an elastic-plastic joint (fig. 232); the two deformable layers of the latter (area $\tfrac{1}{2}F$ each, $h \ll l$) experience extension or compression. In the original state the system is rectilinear; fig. 232 shows the distorted state ($u$ is the "flexure"). Let $\epsilon_1, \epsilon_2$ denote the

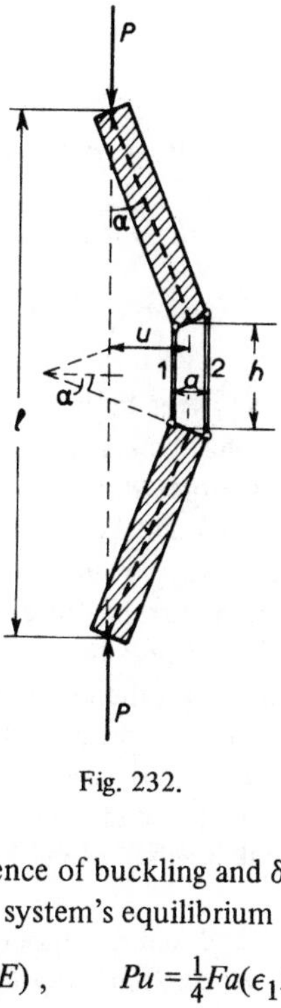


Fig. 232.

additional strains in the presence of buckling and $\delta P$ the load-increment. Substituting the condition of the system's equilibrium

$$\delta P = \tfrac{1}{2}F(\epsilon_1 E' - \epsilon_2 E)\,, \qquad Pu = \tfrac{1}{4}Fa(\epsilon_1 E' + \epsilon_2 E)$$

and using the relation $u = \frac{1}{2}\alpha l = (l/4a)(\epsilon_1 + \epsilon_2)$ we obtain, after elimination of $\epsilon_1$ and $\epsilon_2$,

$$u\left[\frac{4a}{l} - \frac{2}{a}\frac{P}{F}\left(\frac{1}{E'} + \frac{1}{E}\right)\right] = \frac{\delta P}{F}\left(\frac{1}{E'} - \frac{1}{E}\right).$$

A distorted ($u \neq 0$) state turns out to be possible:

1. for elastic layers ($E' = E$) with Euler load

$$P_{\mathrm{e}} = EFa^2/l\,,$$

2. beyond the elastic limit, according to the tangent-modulus scheme ($E = E'$), with load

$$P' = E'Fa^2/l\,,$$

3. beyond the elastic limit, according to the reduced-modulus scheme

$(\delta P = 0)$, with load

$$P_{\mathrm{k}} = P' \frac{2E}{E+E'} .$$

It is easy to show that $\epsilon_2$ is proportional to $(P-P')$; but $\epsilon_2 > 0$ and for $P < P'$ we shall have $u = 0$. If $P' \leqslant P \leqslant P_{\mathrm{k}}$, we have bifurcation of equilibrium, and in addition to the null solution, a further solution is possible,

$$u = \tfrac{1}{2}a \frac{P-\overline{P}}{P_{\mathrm{k}}-P} \frac{P_{\mathrm{k}}-P'}{P'} , \qquad (\delta P = P-\overline{P}) ,$$

where $\overline{P}$ is the value of the load at the bifurcation point; this relation is shown in fig. 233. The tangent-modulus load is the smallest load for which buckling is possible; as the compressive force $P$ increases flexure grows (curve 1) and tends to infinity as the reduced-modulus load $P_{\mathrm{k}}$ is approached. If we prevent displacements of the rod up to some load $\overline{P} > P'$ and then release it, the rod is bent according to the dotted curve 2. Buckling can occur for any load in the interval $(P', P_{\mathrm{k}})$. For a force $P > P'$ bending is accompanied by unloading and the presence of residual strains; therefore Shanley's analysis does not signify reversion to the original, rough scheme of Engesser. But at the instant of bifurcation unloading is absent. Thus, insignificant disturbances can lead to bending with the tangent-modulus load. These deviations, however, are not initially unsafe and only with further increase of the load do they create a dangerous situation.

The tangent-modulus $P'$ and reduced-modulus $P_{\mathrm{k}}$ loads are sometimes called the *lower* and *upper critical loads* respectively. They bound the region in which buckling takes place.

The lower bound $P'$ is safer and is easier to determine; it therefore has greater practical value.

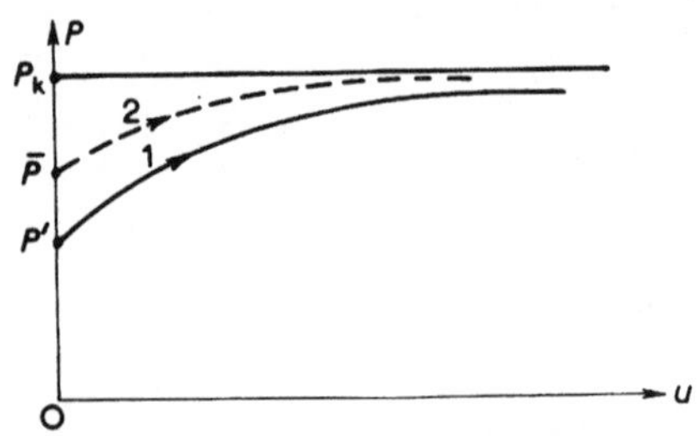


Fig. 233.

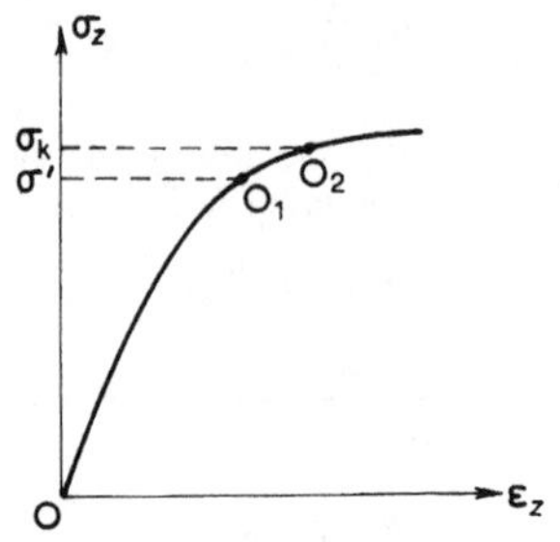


Fig. 234.

It should be noted that the difference between the loads $P'$ and $P_k$ is often not great. On the pressure diagram (fig. 234) the point $O_1$ corresponds to the tangent-modulus load, and the point $O_2$ to the reduced-modulus load. The stresses $\sigma'$ and $\sigma_k$ are often close in value, which is explained by the decrease in the tangent modulus $E'$ with transition along the deformation curve.

An analysis of buckling conditions for real rods, carried out by approximation or numerical methods by Yu.N. Rabotnov [145], Pflüger [189] and other authors, has confirmed Shanley's conclusions. V.D. Klyushnikov [126] examined the motion of an idealized model (fig. 232) on the assumption that the whole mass of the system was concentrated in the middle of the rod, and came to the same conclusion for the initiation of buckling with a tangent-modulus load. Cf. also the book by Ya.G. Panovko and I.I. Gubanova [27a].

### 74.5. *Lower critical load*

The preceding discussion has explained the significance of the tangent-modulus load for a compressive rod. The tangent-modulus load is called the *lower critical load.*

Matters are considerably worse as regards justification of the stability criterion beyond the elastic limit for more complex systems – strips, plates, shells. Usually one of the following methods is used.

The first method, analogously to the case of a perfectly elastic body, considers bifurcation of equilibrium for *fixed external forces.* Regions of unloading develop immediately with buckling. In application to the compressive rod this scheme yields the reduced modulus load. We shall assume that this method gives *upper critical load.*

The second method is based on Shanley's analysis. We seek bifurcation of equilibrium under the condition of continued loading (at the instant of bifurcation there is no unloading). Recently V.D. Klyushnikov [127] investigated the disturbed motion of an idealized plate (two-dimensional analogue of the model shown in fig. 232). The analysis shows that the second method leads to the lower critical load if the starting point is the equations of plastic flow theory.

Below we use the second method; we assume that it leads to the *lower critical load.* This concept, of course, does not have any reliable justification. Furthermore, it is by no means clear that it is always possible to realize "continued loading" in the above sense at the expense of an increment to the load parameter in problems of shell stability. Nevertheless, the relative simplicity and the more "safe" character of the lower critical load, as well as the significant error in determining the tangent modulus $E'$, which lessens the accuracy of the solution by both criteria, necessitates that the second method be preferred.

## §75. Stability of a strip bent by couples

### 75.1. *Basic equations*

We consider the problem of lateral buckling of a strip which is bent by the action of couples beyond the elastic limit (fig. 235a). The ends of the strip are fixed with joints. The section of the strip has the form of an elongated rectangle (fig. 235b). The material obeys the equations of plastic flow theory, where in the plastic zones $|y| > \xi$, shown shaded in fig. 235a, the von Mises yield criterion is fulfilled. When $|y| < \xi$ we have the elastic core; it is easy to see (cf. §24) that

$$\xi/h \equiv \zeta = \sqrt{3(1-\mu)}\,, \qquad \mu = |M|/M_*\,, \tag{75.1}$$

where $M_* = \sigma_s bh^2$ is the limit bending moment. Up to the onset of buckling, only the stress $\sigma_z$ is non-zero, with $|\sigma_z| = \sigma_s$ for $|y| \geqslant \xi$. For sufficiently large values of the bending moment lateral buckling occurs (in the $x$-direction), and is accompanied by twisting.

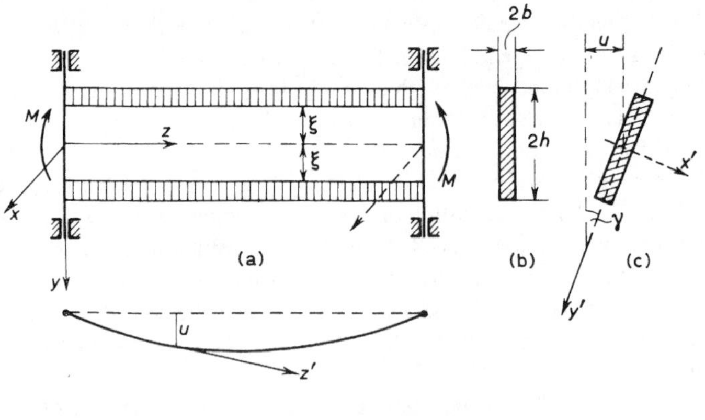


Fig. 235.

Denote by $x'$, $y'$, $z'$ the triad of axes for an arbitrary cross-section of the strip after buckling; the axes $x'$, $y'$ are directed along the principal axes at the centre of the cross-section (fig. 235c). Denote by $\gamma$ the angle of rotation of the section about the $z$-axis (angle of torsion); then the twisting per unit length is $d\gamma/dz$. Further, let $u$ be the lateral flexure with buckling, so that the corresponding curvature of the axis of the strip is $d^2u/dz^2$.

The moment-vector at the end $z = l$ of the strip is shown in fig. 235a (dotted line); its projections on the axes $y'$, $z'$ correspondingly are

$$L_{y'} = M\gamma\,, \qquad L_{z'} = -M\, du/dz\,. \tag{75.2}$$

With infinitesimal buckling the strip experiences additional deformations. As in the elastic case, these deformations are made up of the bending of the strip and its torsion. The stress components $\sigma_x$, $\tau_{xz}$, $\tau_{xy}$ can be neglected, since the lateral surfaces of the strip are stress-free and its thickness is small; the stress $\sigma_y$ is also negligible, since there is no mutual pressure of elements with bending and torsion. As a consequence, the only additional stresses which arise under buckling are $\delta\sigma_z$, $\delta\tau_{yz}$.

75.2. *Lower stability bound*

We shall seek the lower critical load, so that there is no unloading at the moment of bifurcation. In the plastic zones $|y| \geqslant \xi$ we therefore have $\delta\sigma_z = 0$. According to the flow theory equations (13.7),

$$\delta\epsilon_z = \frac{1}{E}\delta\sigma_z \qquad \text{for} \qquad |y| \leqslant \xi\,; \qquad \delta\gamma_{yz} = \frac{1}{G}\delta\tau_{yz}\,. \tag{75.3}$$

From the second of these relations we have that the increments in tangential stress and shear are connected by Hooke's law, and hence the twisting moment is proportional to the torsion, i.e.

$$L_{z'} = C_0 \frac{d\gamma}{dz}\,, \tag{75.4}$$

where $C_0$ is the rigidity for torsion of an *elastic* strip.

Further, since $\delta\sigma_z = 0$ in the plastic zones, the rigidity for lateral bending is determined by the rigidity of the elastic core $|y| < \xi$. The axial strain increments $\delta\epsilon_z$ obey the hypothesis of plane sections, and therefore the bending moment $L_{y'}$ during buckling is

$$L_{y'} = B\frac{d^2u}{dz^2}\,, \qquad B = \tfrac{4}{3}E\xi b^3 = B_0\zeta\,. \tag{75.5}$$

Here $B_0 = \frac{4}{3}Ehb^3$ denotes the bending rigidity of an *elastic* strip.

Combining the above formulae we obtain

$$-M\frac{du}{dz} = C_0\frac{d\gamma}{dz}\,, \qquad M\gamma = B\frac{d^2u}{dz^2}\,. \tag{75.6}$$

From the conditions at the jointed ends we have

$$u = 0\,, \qquad \gamma = 0 \qquad \text{for} \qquad z = 0 \qquad \text{and} \qquad z = l\,. \tag{75.7}$$

Integrating the first of equations (75.6) we find that $C_0\gamma = -Mu$; then

$$\frac{d^2u}{dz^2} + \frac{M^2}{BC_0}u = 0\,.$$

The solution of this equation satisfying zero conditions at $z = 0$ has the form

$$u = A \sin \frac{Mz}{\sqrt{BC_0}}\,.$$

where $A$ is arbitrary constant. From the second boundary condition we have

$$A \sin \frac{Ml}{\sqrt{BC_0}} = 0\,.$$

Consequently, in addition to the original (trivial) equilibrium state ($A = 0$), it is possible to have buckling with

$$\frac{Ml}{\sqrt{BC_0}} = \pi, 2\pi, \ldots$$

The critical moment is

$$M_{cr} = \frac{\pi}{l}\sqrt{BC_0}\,.$$

Introducing the flexibility parameter $\lambda = M_* l/\sqrt{B_0C_0}$, we can write the last expression in the form

$$\lambda\mu = \pi\sqrt{\zeta}\,, \tag{75.8}$$

where

$$\zeta = \begin{cases} 1 & \text{for} \quad \mu \leqslant \frac{2}{3}\,, \\ \sqrt{3(1-\mu)} & \text{for} \quad \frac{2}{3} \leqslant \mu \leqslant 1\,. \end{cases}$$

Fig. 236 shows the stability boundary in the $\lambda$, $\mu$-plane. The dotted line corresponds to a perfectly elastic strip ($\zeta = 1$). The point D, for which $\mu = \frac{2}{3}$, corresponds to the appearance of the first plastic deformation. In the presence of plastic deformations the critical load is sharply decreased.

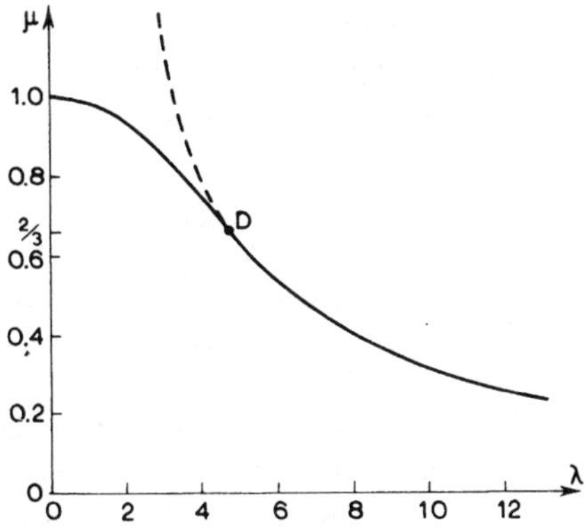


Fig. 236.

75.3. *Concluding remarks*

The general theory of stability of a plane shape bent beyond the elastic limit and its applications to various special problems were presented in the first edition of this book, in which the upper critical loads were also determined. Calculations showed that the upper and lower bounds were close to each other. The influence of hardening has been considered in [119]. Experiments by Neal [188] confirm well the theoretical values of the critical loads.

The stability of thin-walled rods with elastic-plastic deformations has been examined in [122] on the basis of flow theory with the "continued loading" scheme (lower critical load).

## §76. Stability of compressed plates

The buckling of plates and shells has been studied by Il'yushin and other authors (cf. [12, 39, 75]) on the basis of deformation theory and the classical representation of loss of stability with invariant external forces. Here buckling is accompanied by the presence of unloading regions, which substantially complicates the analysis. If flow theory and the same criterion are used much of the difficulty remains.

Such problems can be solved considerably more easily if one begins with flow theory and looks for the *lower critical load* corresponding to buckling under continued loading. We have already observed that the interpretation of this condition can give rise to difficulties here, nevertheless we shall use it, bearing in mind that we are looking for the safer stability boundary. We briefly consider the question of buckling of plates from this point of view.

76.1. *Basic equations* [1])

Let the plate (of thickness $h$) be deformed in its own plane $x, y$ beyond the elastic limit. For simplicity we suppose that up to the onset of buckling a uniform stress distribution exists in the plate, namely, compression in two perpendicular directions:

$$\sigma_x = -p\,, \qquad \sigma_y = -q\,, \qquad \tau_{xy} = 0\,.$$

With buckling the stresses in the plate experience infinitesimal increments

$$\delta\sigma_x, \delta\sigma_y, \delta\tau_{xy}$$

(the residual stress components are discarded as second-order for the plate). From the flow theory equations (13.14) we obtain for the general case of a hardening medium the corresponding infinitesimal strain-increment components:

$$\delta\epsilon_x = \frac{1}{E}(\delta\sigma_x - \nu\delta\sigma_y) - \tfrac{1}{3}F(T)\,\delta T(2p-q)\,,$$

$$\delta\epsilon_y = \frac{1}{E}(\delta\sigma_y - \nu\delta\sigma_x) - \tfrac{1}{3}F(T)\,\delta T(2q-p)\,, \tag{76.1}$$

$$\delta\gamma_{xy} = \frac{1}{G}\delta\tau_{xy}\,,$$

where

$$T^2 = \tfrac{1}{3}(p^2 - pq + q^2)\,,$$

$$-\delta T = \frac{1}{6T}\,[(2p-q)\,\delta\sigma_x + (2q-p)\,\delta\sigma_y]\,,$$

$$F(T) = \frac{1}{2T^2}\,\frac{\mathrm{d}\Phi}{\mathrm{d}T}\,.$$

Here $\Phi(T)$ is the work of plastic deformation; this function is characteristic for the given material, and is independent of the form of the stress distribution.

We consider simple tension; let the tension curve have the equation

[1]) It is assumed that the reader is familiar with the basic theory of flexure and stability of elastic plates (cf. [49]).

$\epsilon_x = f(\sigma_x)$, then the increment to the work of plastic deformation is

$$dA_p = \sigma_x d\epsilon_x^p = \sigma_x f'(\sigma_x)\, d\sigma_x - \frac{1}{E}\sigma_x d\sigma_x .$$

The derivative $f'(\sigma_x)$ is the inverse of the local (tangential) modulus $E'$, and therefore

$$\frac{1}{\sigma_x}\frac{dA_p}{d\sigma_x} = \frac{1}{E'} - \frac{1}{E} = \frac{1}{3T}\frac{d\Phi}{dT},$$

since $A_p = \Phi$; here $E'$ is the tangent modulus for the tensile stress $\sigma_x$, corresponding to a given intensity $T$.

Substituting in formulae (76.1) the values of $T$, $\delta T$ and $F(T)$, we obtain

$$E\delta\epsilon_x = A_{xx}\delta\sigma_x + A_{xy}\delta\sigma_y ,$$

$$E\delta\epsilon_y = A_{xy}\delta\sigma_x + A_{yy}\delta\sigma_y , \qquad (76.2)$$

$$G\delta\gamma_{xy} = \delta\tau_{xy} ,$$

where we have introduced the notation

$$A_{xx} = 1 + \theta(2p-q)^2 ,$$

$$A_{xy} = -\nu + \theta(2p-q)(2q-p) , \qquad (76.3)$$

$$A_{yy} = 1 + \theta(2q-p)^2 ,$$

with

$$\theta = \frac{1}{12}\frac{1}{T^2}\left(\frac{E}{E'} - 1\right) .$$

For an elastic body $\theta = 0$, while for transition to the yield plateau $\theta \to \infty$. Solving equations (76.2) with respect to the stress increments, we find

$$\delta\sigma_x = \frac{E}{\omega}(A_{yy}\delta\epsilon_x - A_{xy}\delta\epsilon_y) ,$$

$$\delta\sigma_y = \frac{E}{\omega}(-A_{xy}\delta\epsilon_x + A_{xx}\delta\epsilon_y) , \qquad (76.4)$$

$$\delta\tau_{xy} = G\delta\gamma_{xy} ,$$

where we have put

$$\omega = (1-\nu^2)\,\omega_1\,, \qquad \omega_1 = 1 + \frac{5-4\nu}{1-\nu^2}\left(\rho^2 - \frac{8-10\nu}{5-4\nu}pq + q^2\right)\theta\,.$$

By hypothesis bifurcation takes place in the absence of an unloading region, and therefore formulae (76.4) are valid in the whole plate.

According to Kirchhoff's theory for flexure of plates, the strain increments in the presence of buckling will be linear functions of the distance from median surface:

$$\begin{aligned}
\delta\epsilon_x &= e_1 - z\kappa_1\,,\\
\delta\epsilon_y &= e_2 - z\kappa_2\,,\\
\delta\gamma_{xy} &= 2(e_{12} - z\kappa_{12})\,,
\end{aligned} \tag{76.5}$$

where $e_1$, $e_2$, $2e_{12}$ are the infinitesimal strain increments of the median surface, and $\kappa_1, \kappa_2, \kappa_{12}$ are the infinitesimal changes in its curvature and torsion. As is well-known

$$\kappa_1 = \frac{\partial^2 w}{\partial x^2}\,, \qquad \kappa_2 = \frac{\partial^2 w}{\partial y^2}\,, \qquad \kappa_{12} = \frac{\partial^2 w}{\partial x \partial y}\,, \tag{76.6}$$

where $w = w(x, y)$ is the bending of the plate under buckling.

Evidently the stress increments also change linearly through the plate's thickness. It is then easy to calculate the increments in the bending moments, $\delta M_x$, $\delta M_y$, and in the twisting moment $\delta M_{xy}$:

$$\begin{aligned}
\delta M_x &= \int \delta\sigma_x z\mathrm{d}z = -\frac{D}{\omega_1}(A_{yy}\kappa_1 - A_{xy}\kappa_2)\,,\\
\delta M_y &= \int \delta\sigma_y z\mathrm{d}z = -\frac{D}{\omega_1}(-A_{xy}\kappa_1 + A_{xx}\kappa_2)\,,\\
\delta M_{xy} &= \int \delta\tau_{xy} z\mathrm{d}z = -D(1-\nu)\,\kappa_{12}\,,
\end{aligned} \tag{76.7}$$

where integration is performed between the limits $-\frac{1}{2}h$, $\frac{1}{2}h$, and $D = Eh^3/12(1-\nu^2)$ is the rigidity of the *elastic* plate.

If we project on the $z$-axis the forces which act on an element of the plate in the buckling state, we obtain the well-known equilibrium equation

$$\frac{\partial^2 \delta M_x}{\partial x^2} + 2\frac{\partial^2 \delta M_{xy}}{\partial x \partial y} + \frac{\partial^2 \delta M_y}{\partial y^2} - ph\kappa_1 - qh\kappa_2 = 0\,. \tag{76.8}$$

Hence with the aid of (76.7), (76.6) we obtain the differential equation of the plate's buckling (assuming the thickness $h$ of the plate is constant and the original stress state is homegeneous):

$$A_{yy}\frac{\partial^4 w}{\partial x^4}+2(\omega_1-\nu\omega_1-A_{xy})\frac{\partial^4 w}{\partial x^2\partial y^2}+A_{xx}\frac{\partial^4 w}{\partial y^4}+$$

$$+\frac{h\omega_1}{D}\left(p\frac{\partial^2 w}{\partial x^2}+q\frac{\partial^2 w}{\partial y^2}\right)=0\,. \tag{76.9}$$

An analogous differential equation was given by Pearson [143].

For elastic plates $\theta = 0$, and the equation of buckling takes the form

$$\Delta\Delta w+\frac{h}{D}\left(p\frac{\partial^2 w}{\partial x^2}+q\frac{\partial^2 w}{\partial y^2}\right)=0\,. \tag{76.10}$$

Equation (76.9) has a superficial resemblance to the stability equation for an elastic anisotropic plate.

76.2. *Boundary conditions*

If the edge of the plate is *rigidly attached*, then

$$w=0\,,\qquad \frac{\partial w}{\partial n}=0\,,$$

along the edge, where $n$ is the direction of the normal to the contour.

If the edge is *supported*, then

$$w=0\,,\qquad \delta M_{\rm n}=0\,,$$

along it, with $\delta M_{\rm n}$ the increment to the bending moment on the contour.

Finally, if the edge of the plate (for example, $x$ = const. = $a$) is *free*, then the bending moment $\delta M_x$ and the transverse force [1]) are zero along it,

$$\delta M_x=0\,,\qquad N_x+\frac{\partial\delta M_{xy}}{\partial y}=0\,. \tag{76.11}$$

It should be emphasized that the boundary conditions (like the differential equation itself) are homogeneous.

[1]) $N_x$, $N_y$ are the shearing forces:

$$N_x=\partial\delta M_x/\partial x+\partial\delta M_{xy}/\partial y\,,\qquad N_y=\partial\delta M_{xy}/\partial x+\partial\delta M_y/\partial y\,.$$

76.3. *Energy method*

We now see that the determination of the critical load reduces to the resolution of an eigenvalue problem. It is possible to seek directly for non-trivial solutions of the differential equation (76.9). In many cases, however, it is more convenient to proceed from the energy equation, which can be derived by going from the differential equation to the corresponding variational formulation.

It is not difficult (following the method of Timoshenko [49] and Bijlaard [39]) to derive the energy equation directly by equating the work of bending deformation with the work of external forces on the displacements from buckling.

76.4. *Buckling of a compressed rectangular strip*

We consider the problem of buckling of a thin rectangular plate compressed in one direction (fig. 237). Here $q = 0$, and since the plate is elongated ($b \gg a$), it can be assumed that the cylindrical form of loss of stability is realized, i.e. $w = w(x)$. The edges $x = 0, x = a$ are supported, i.e. $w = 0$, $\mathrm{d}^2w/\mathrm{d}x^2 = 0$ along them.

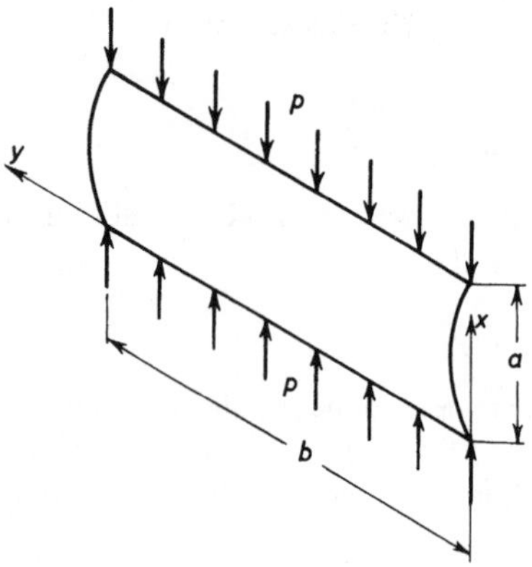


Fig. 237.

The differential equation of buckling (76.9) takes the form

$$A_{yy} \frac{\mathrm{d}^4 w}{\mathrm{d}x^4} + \frac{p\omega_1 h}{D} \frac{\mathrm{d}^2 w}{\mathrm{d}x^2} = 0 .$$

It is easy to see that the critical pressure is

$$p = p_0 \frac{A_{yy}}{\omega_1},$$

where $p_0 = \pi^2 D/a^2 h$ is the critical pressure for an *elastic* strip. The multiplier is

$$\frac{A_{yy}}{\omega_1} = \frac{1 + \frac{1}{4}(E/E'-1)}{1 + \{(5-4\nu)/(1-\nu^2)\}\, \frac{1}{4}(E/E'-1)} \leqslant 1 .$$

As the tangent modulus decreases ($E' \to 0$) the critical load decreases monotonically to the value $\{(1-\nu^2)/(5-4\nu)\} \cdot p_0$, attained on the yield plateau. Thus, in contrast with the case of a compressed rod, there is not a complete loss of stability on the yield plateau here.

76.5. *Stability of a supported rectangular plate compressed in one direction* (fig. 238)

In this case $q = 0$ and the boundary conditions have the form

$$\text{for} \quad x = 0 \quad \text{and} \quad x = a \qquad w = 0 , \quad \frac{\partial^2 w}{\partial x^2} = 0 ,$$

$$\text{for} \quad y = 0 \quad \text{and} \quad y = b \qquad w = 0 , \quad \frac{\partial^2 w}{\partial y^2} = 0 .$$

We seek a solution to the differential equation of buckling (76.9) in the

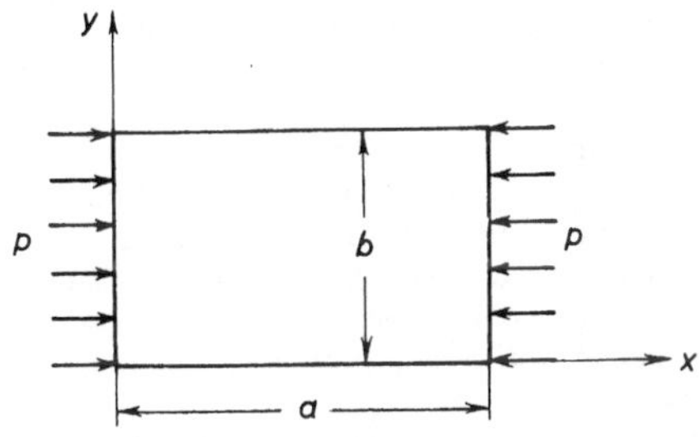


Fig. 238.

usual form:

$$w = c_{mn} \sin \frac{m\pi x}{a} \sin \frac{n\pi y}{b}, \qquad (76.12)$$

where $c_{mn}$ is an arbitrary constant, $m, n$ are integers; it is easy to see that the boundary conditions are satisfied. Substituting (76.12) into equation (76.9) we obtain

$$c_{mn} \left\{ \pi^2 \left[ A_{yy} \left(\frac{m}{a}\right)^4 + 2(\omega_1 - \nu\omega_1 - A_{xy}) \left(\frac{mn}{ab}\right)^2 + A_{xx} \left(\frac{n}{b}\right)^4 \right] - \frac{\omega_1 h p}{D} \left(\frac{m}{a}\right)^2 \right\} = 0 .$$

Since we are looking for a non-trivial solution, we must equate to zero the expression inside the brace brackets; then

$$p = \frac{\pi^2 D}{\omega_1 h} \left[ A_{yy} \left(\frac{m}{a}\right)^2 + 2(\omega_1 - \nu\omega_1 - A_{xy}) \left(\frac{n}{b}\right)^2 + A_{xx} \left(\frac{n}{b}\right)^4 \left(\frac{a}{m}\right)^2 \right].$$

It is evident that $\omega_1 - \nu\omega_1 - A_{xy} > 0$, so that all the terms are positive, and to find the smallest $p$ it is necessary to put $n = 1$ (i.e. one half-wave in the $y$-direction). The choice of the number $m$ must depend on the ratio $a/b$ and the coefficients of the equation in such a way that $p$ is a minimum.

76.6. *Concluding remarks*

As we have already noted, it is more convenient when considering more complex problems to proceed from the energy formulation and seek the critical load by Ritz's method.

Qualitatively the picture of the stability boundary for plates obtained by this method is correct, but the theoretical values of the critical loads sometimes differ considerably from experimental data. It should be noted that test data are found to be in better agreement with the results of Bijlaard [39] and Stowell, in which the equations of deformation theory are used and where it is assumed that loss of stability is accompanied by plastic deformation alone.

This fact is possibly explained by the influence of deviations from the ideal shape. It is known that even for elastic plates and shells classical stability theory leads to results which deviate from observed data. With plastic deformation the influence on the critical load of finite displacements, geometric deviations, the nature of the material and boundary conditions is strongly enhanced. To obtain more satisfactory quantitative results, however, we require a difficult analysis of the deformation of plates in the presence of initial disturbances.

The literature on the stability of plates and shells beyond the elastic limit is enormous. We merely refer to the book by Volmir and Timoshenko, and to some review articles [75, 76], in which references can be found.

## PROBLEMS

1. Calculate the reduced modulus $E_k$ for a circular section.

2. Derive the differential equation of buckling (for the lower critical load).

$$\frac{d^2\gamma}{dt^2} + \lambda^2\mu^2 \frac{(1-t)^2}{\sqrt{3[1-\mu(1-t)]}}\,\gamma = 0$$

for a cantilever strip bent by a force $P$ applied at the centre of gravity of the end section $t = 1$ (where $z = lt$, $\mu = Pl/M_*$).

3. Derive the differential equation of buckling (for the lower critical load).

$$\frac{d^2\gamma}{dt^2} + \lambda^2\mu^2 \frac{(1-t)^4}{\zeta}\,\gamma = 0$$

for a cantilever strip bent by a force $Q/l$ uniformly distributed along the axis (here $\mu = Ql/2M_*$, $Q$ is the total force).

4. Show that when the condition of incompressibility holds ($\nu = \frac{1}{2}$), the differential equation of buckling for a plate compressed in the $x$-direction has the form

$$\frac{A_{yy}}{\omega_1}\frac{\partial^2 w}{\partial x^4} + 2\frac{\partial^4 w}{\partial x^2 \partial y^2} + \frac{\partial^4 w}{\partial y^4} + \frac{ph}{D}\frac{\partial^2 w}{\partial x^2} = 0 .$$

# 11

# Dynamical Problems

## §77. Propagation of elastic-plastic waves in a rod

### 77.1. *General remarks*

The action of impulsively applied loads is not distributed instantaneously through a body, but is transferred from particle to particle in a wave-like manner.

Until quite recently only the motion of elastic waves has been investigated, i.e. the propagation of disturbances in elastic media. The dynamical theory of elasticity has important applications in seismology and technology.

At the end of the second world war great interest developed in problems of propagation of disturbances through elastic-plastic media. The reason for this is the following. Every moderately intensive impactive loading is accompanied by plastic deformation. Questions of the strength of different machines and structures which experience impacts (or which are liable to the effects of explosions) can only be investigated with a clear understanding of the mode of propagation of elastic-plastic deformations prior to fracture. Moreover, real media (for example, in seismology) are not completely elastic, and it is necessary to take into account plastic properties. Finally, dynamical problems can be of great significance in the analysis of high-speed technological processes in metal machining.

The first problems in the propagation of compressive (or expansive) elastic-plastic waves in a rod were considered by Rakhmatulin [146], von Karman and Duwez [116] and Taylor [192]. Various generalizations of this problem were investigated by Rakhmatulin [34], Shapiro [168], Sokolovskii and other authors. References can be found in the books [18, 28, 34] and articles [63–65].

77.2. *Basic propositions*

We consider the problem of waves propagating in a thin prismatic rod, whose axis coincides with the $x$-axis. We shall base ourselves on the following fundamental assumptions:

1. The cross-section of the rod remains plane and normal to the $x$-axis during deformation.
2. Deformations are small, and therefore we can neglect changes in the dimensions of the rod.
3. Inertia forces corresponding to the motion of particles of the rod in transverse directions (as a result of contraction or expansion of a section) can be neglected.
4. The influence of strain rates on the relationship between the stresses $\sigma_x$ and strains $\epsilon_x$ can be neglected.

Since in this section we are considering a one-dimensional problem, we shall replace $\sigma_x, \epsilon_x, u_x, v_x$ by $\sigma, \epsilon, u, v$ respectively.

The relative extension $\epsilon$ and stress $\sigma$ are distributed uniformly over a section. Up to the elastic limit the material obeys Hooke's law:

$$\sigma = E\epsilon \qquad \text{for} \qquad |\epsilon| \leqslant \epsilon_0 , \tag{77.1}$$

where $E$ is Young's modulus, and $\sigma_0$, $\epsilon_0$ correspond to the elastic limit (fig. 239a). We shall assume that the compression curve is similar to the tension curve. Unloading proceeds along a straight line. In keeping with experimental data we suppose that the loading branch ($\sigma > 0$) is concave downwards, with the angle of inclination of the tangent a decreasing function of strain:

$$0 < \mathrm{d}\sigma/\mathrm{d}\epsilon < E \qquad \text{for} \qquad |\epsilon| > \epsilon_0 .$$

In dynamical problems the strain rates are large, and their influence on the deformation curve can be substantial. Therefore the assumed relationship $\sigma = \sigma(\epsilon)$ between stress and strain is only a first approximation and relates to some average strain rate in the given interval. To calculate the influence of the strain rate it is necessary to consider an elastic-visco-plastic model of the medium (cf. ch. 12).

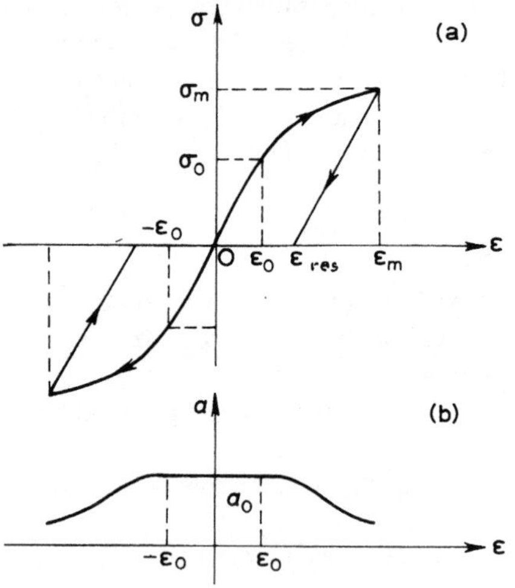


Fig. 239.

77.3. *Equations of motion*

From the differential equations of motion (4.1) for a continuous medium, we have

$$\frac{\partial \sigma}{\partial x} = \rho \frac{\partial^2 u}{\partial t^2}.$$

Because the deformations are small, $\rho = \text{const}$. Since

$$\frac{\partial \sigma}{\partial x} = \frac{d\sigma}{d\epsilon} \frac{\partial \epsilon}{\partial x}, \qquad \frac{\partial \epsilon}{\partial x} = \frac{\partial^2 u}{\partial x^2},$$

we obtain

$$\frac{\partial^2 u}{\partial t^2} = a^2 \frac{\partial^2 u}{\partial x^2}, \tag{77.3}$$

where the quantity

$$a = \sqrt{\frac{1}{\rho} \frac{d\sigma}{d\epsilon}}$$

is called the *local velocity of propagation of disturbances* ("local speed of sound").

In the elastic region the propagation speed is constant:

$$a = \sqrt{E/\rho} \equiv a_0 .$$

In the plastic region the propagation speed decreases with increasing deformation (fig. 239b).

The second-order equation (77.3) can be conveniently replaced by a system of two first-order equations:

$$\frac{\partial v}{\partial t} = a^2 \frac{\partial \epsilon}{\partial x}, \qquad \frac{\partial \epsilon}{\partial t} = \frac{\partial v}{\partial x} \tag{77.4}$$

for the functions $\epsilon$, $v$, where $v$ is the speed of the particles. It is easy enough to show that this system is of hyperbolic type (cf. appendix).

Suppose the functions $v$, $\epsilon$ are prescribed along some line L. Combining the relations

$$\mathrm{d}\epsilon = \frac{\partial \epsilon}{\partial t}\,\mathrm{d}t + \frac{\partial \epsilon}{\partial x}\,\mathrm{d}x ,$$

$$\mathrm{d}v = \frac{\partial v}{\partial t}\,\mathrm{d}t + \frac{\partial v}{\partial x}\,\mathrm{d}x ,$$

in the usual way, we obtain along L a system of four linear algebraic equations with respect to the first derivatives $\partial\epsilon/\partial t$, $\partial\epsilon/\partial x$, $\partial v/\partial t$, $\partial v/\partial x$. We then find

$$\frac{\partial \epsilon}{\partial t} = \frac{\Delta_1}{\Delta}, \ldots, \qquad \frac{\partial v}{\partial x} = \frac{\Delta_4}{\Delta},$$

where $\Delta$ is the determinant of the system, and $\Delta_1, \ldots, \Delta_4$ are the appropriate co-factors. It is easy to see that

$$\Delta = \mathrm{d}x^2 - a^2\mathrm{d}t^2 .$$

If L is a characteristic curve, then the derivatives are indeterminate along it, i.e. $\Delta = 0$, $\Delta_1 = \Delta_2 = \Delta_3 = \Delta_4 = 0$. Consequently

$$\mathrm{d}x \pm a\,\mathrm{d}t = 0 .$$

The condition that the co-factors are zero now leads to the relationship

$$a\,\mathrm{d}\epsilon \pm \mathrm{d}v = 0 .$$

Introduce the function

$$\varphi(\epsilon) = \int_0^\epsilon a(\epsilon)\, d\epsilon \,. \tag{77.5}$$

Then

$$d[v \pm \varphi(\epsilon)] = 0 \,.$$

Thus, the system of differential equations under consideration has two distinct real families of characteristics:

$$\begin{aligned} &dx - a\, dt = 0 \,, \\ &v - \varphi(\epsilon) = \text{const.} = \xi \,, \end{aligned} \tag{77.6}$$

$$\begin{aligned} &dx + a\, dt = 0 \,; \\ &v + \varphi(\epsilon) = \text{const.} = \eta \,, \end{aligned} \tag{77.7}$$

i.e. they are of hyperbolic type. The quantity $a$ is the speed of propagation of disturbances.

A solution of the system (77.4) which has continuous first derivatives in some domain $x$, $t$, will be called a *wave*. The relations (77.6) refer to a *positive (forward) wave* (propagating in the positive $x$-direction), and relations (77.7) to a *negative (backward) wave.*

The point of separation of the two waves, which is displaced with time along the rod, is called a *front.* In the $x$, $t$-plane the front is represented by some line.

The front is said to have a *weak discontinuity*, if the quantities $\epsilon$, $v$ are continuous and their first derivatives are discontinuous.

The front has a *strong discontinuity* if the functions $\epsilon$, $v$ are themselves discontinuous. These waves are called *discontinuous*, or *shock*, waves.

In each of (77.6), (77.7), the first relation defines the law of propagation of disturbances, while the second connects the speeds and the deformations of particles on the characteristics.

### 77.4. *Impactive loading of a semi-infinite rod*

Consider a semi-infinite rod $x \geqslant 0$ which is in a state of rest at time $t = 0$. Let the end $x = 0$ of the rod experience certain disturbances for $t \geqslant 0$. These disturbances may be of various types. For example, at the end $x = 0$ we can prescribe the velocity

$$v = v(t)$$

or the stress

$$\sigma = \sigma(t)\,.$$

If a body of mass $m$ impinges on the end of the rod with initial velocity $v_0$, then

$$m\,\partial v/\partial t = \sigma S \qquad \text{when} \qquad x = 0\,,$$

where $S$ is the cross-sectional area, and where $v = v_0$ at $t = 0$.

The case of impactive loading which is of greatest interest is that when the end $x = 0$ of the rod suddenly experiences a finite stress $\sigma_*$, or equivalently, some strain $\epsilon_*$.

We shall restrict ourselves to a consideration of the following basic problem: at time $t = 0$ the end of the rod experiences a tension $\sigma_*$ which is maintained constant (fig. 240) during some time-interval $0 \geqslant t > t_1$. At time $t = t_1$ the load is completely removed. Thus, the initial and boundary conditions take the form

$$\text{when} \quad t = 0 \quad u(x, t) = 0 \qquad (x \geqslant 0)\,,$$

$$\frac{\partial u}{\partial t} = 0 \qquad (x > 0)\,,$$

$$\text{when} \quad x = 0 \quad \sigma = \text{const.} = \sigma_* \qquad (0 \leqslant t < t_1)\,,$$

$$\sigma = 0 \qquad (t \geqslant t_1)\,.$$

77.5. *Propagation of elastic waves*

When $\sigma_* \leqslant \sigma_0$ the deformations in the rod are elastic, the speed of motion of the wave is constant ($a = a_0$), and equation (77.3) becomes the classical

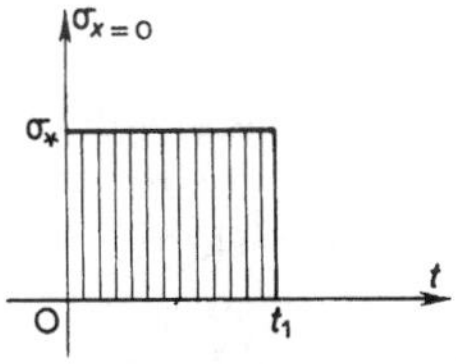


Fig, 240.

wave equation. Its solution, in the form given by d'Alembert, is

$$u = f(x-a_0 t) + \psi(x + a_0 t) ,$$

where $f$, $\psi$ are arbitrary functions determined by the initial and boundary conditions. In the $x$, $t$-plane the characteristics

$$x-a_0 t = \text{const.} , \qquad x + a_0 t = \text{const.}$$

represent families of straight lines.

For impact along the rod a positive wave is set into motion:

$$u = f(x-a_0 t) .$$

But when $x = 0$, $\partial u/\partial x = f'(-a_0 t) = \epsilon_*$, where $\epsilon_* = \sigma_*/E$. Hence it follows that $f(\zeta) = \epsilon_* \zeta + C$, i.e.

$$u = \epsilon_*(x-a_0 t) + C .$$

Since the displacement is continuous on the wave front, we have that $u = 0$ when $x = at$, i.e. $C = 0$. The solution then has the form

$$u = \epsilon_*(x-a_0 t) .$$

The picture in the $x$, $t$-plane (fig. 241a) is as follows. An undisturbed region lies below the front $x = a_0 t$. On the front the strain $\epsilon$, stress $\sigma$ and velocity $v$ undergo discontinuities. At a fixed instant of time $t'$ the distribution of strains, velocities $v$ and displacements $u$ are shown in fig. 241b. Before the passage of the wave front the particles of the rod are at rest, and after the wave has passed they acquire a constant speed $-\epsilon_* a$ (opposite in direction to the motion of the wave). The displacement increases linearly with distance from the front.

At the instant $t = t_1$ the end of the rod is unloaded (*impactive unloading*); an unloading wave front propagates to the right with velocity $a_0$, behind it a state of rest, $\epsilon = 0$, $\sigma = 0$, $v = 0$, $u = \text{const.} = u_1$.

Fig. 241c shows the strain distribution in the rod at time $t'' > t_1$. A picture of the time-variation in the distribution at a fixed point $x = x'$ is illustrated in fig. 241d.

## 77.6. *Transformation of the equations, simple waves*

We turn now to the propagation of elastic-plastic waves. When $\sigma_* > \sigma_0$ the wave velocity is variable, with small velocities corresponding to large deformations; the wave "diffuses" with the passage of time.

The system (77.4) of nonlinear differential equations is reducible and can be transformed analogously to the equations of the plane problem (cf. §33).

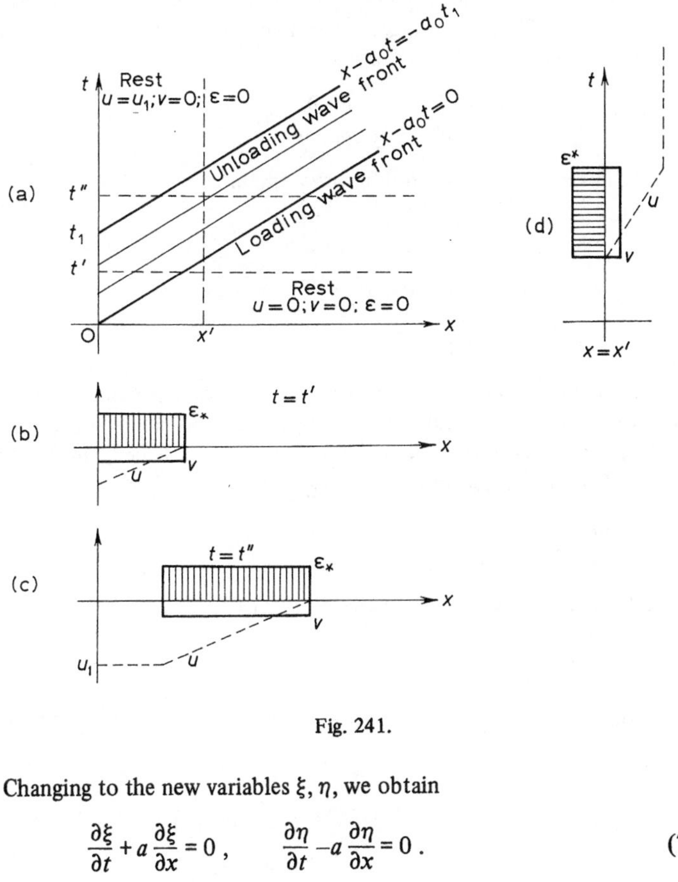


Fig. 241.

Changing to the new variables $\xi, \eta$, we obtain

$$\frac{\partial \xi}{\partial t} + a\frac{\partial \xi}{\partial x} = 0\,, \qquad \frac{\partial \eta}{\partial t} - a\frac{\partial \eta}{\partial x} = 0\,. \tag{77.8}$$

This system can be linearized by an inversion of the variables; if the Jacobian

$$\Delta(\xi, \eta) = \frac{D(\xi, \eta)}{D(x,t)} = \frac{\partial \xi}{\partial x}\,\frac{\partial \eta}{\partial t} - \frac{\partial \xi}{\partial t}\,\frac{\partial \eta}{\partial x} \neq 0$$

is non-zero, we easily obtain the canonical system

$$\frac{\partial x}{\partial \eta} - a\frac{\partial t}{\partial \eta} = 0\,, \qquad \frac{\partial x}{\partial \xi} + a\frac{\partial t}{\partial \xi} = 0\,. \tag{77.9}$$

This canonical system is not equivalent to the original equations (77.8), since in the process of inverting the variables we lose those solutions for which $\Delta(\xi, \eta) = 0$. These solutions play an important role and can be found directly. With the aid of (77.8) we obtain

$$\Delta(\xi, \eta) = 2a \frac{\partial \xi}{\partial x} \frac{\partial \eta}{\partial x} = -\frac{2}{a} \frac{\partial \xi}{\partial t} \frac{\partial \eta}{\partial t} = 0 .$$

Hence it follows that the lost solutions have the form

1. $\xi = \text{const.} = \xi_0 , \qquad \eta = \text{const.} = \eta_0 ,$

2. $\eta = \text{const.} = \eta_0 ,$

3. $\xi = \text{const.} = \xi_0 .$

In the *first case* it follows from (77.8) that $v = \text{const.}$, $\epsilon = \text{const.}$, i.e. there is a state of constant strain and velocity (in particular, a state of rest).

In the *second case* one of the equations (77.8) is satisfied, since $\eta = \eta_0$, while the other, after substitution of $\xi = \eta_0 - 2\varphi(\epsilon)$, takes the form

$$\frac{\partial \epsilon}{\partial t} + a \frac{\partial \epsilon}{\partial x} = 0 .$$

The differential equations of the characteristics

$$\frac{\mathrm{d}t}{1} = \frac{\mathrm{d}x}{a} = \frac{\mathrm{d}\epsilon}{0}$$

have the obvious integrals

$$\epsilon = C_1 , \qquad x - at = C_2 ,$$

where $C_1$, $C_2$ are arbitrary constants. The solution of the original system of equations is:

$$\eta = \varphi(\epsilon) + v = \eta_0 , \qquad x - at = \Phi(\epsilon) ,$$

where $\Phi$ is an arbitrary function. Hence the characteristics are straight lines; along each of the characteristics the strain, stress and particle velocity are constant.

An important special case is the central wave, for which the straight characteristics originate from some centre O (fig. 242).

The *third case* ($\xi = \text{const.}$) is analogous to the second, except that here the wave moves in the reverse direction.

These solutions are called *simple waves* and are represented in the $\xi, \eta$-plane by points (for regions of constant values) or by segments of straight

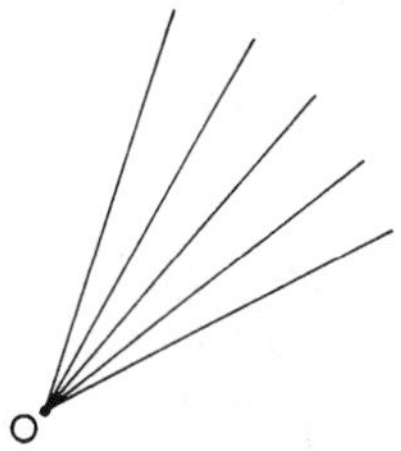

Fig. 242.

lines. As in the case of plane strain (§33) it is easy to establish an important theorem: adjacent to a region of constant values (in particular, to a region of rest), there is always a simple wave.

77.7. *Propagation of an elastic-plastic loading wave*

When the end of the rod is given a blow, a simple extension wave $v + \varphi(\epsilon) = \eta_0$ begins to propagate; ahead of the wave there is a state of rest, hence $\eta_0 = 0$ and

$$v + \varphi(\epsilon) = 0 .$$

Different deformations will be propagated with different speeds: elastic deformations with the maximum speed $a_0$, deformations beyond the elastic limit with the least speed. At the wave front $x - a_0 t = 0$ (fig. 243a), the strain experiences a jump from zero to $\epsilon_0$, and the particle velocity from zero to $v = -\varphi(\epsilon_0) = -a_0\epsilon_0$. Immediately after this a central expansion wave passes:

$$v + \varphi(\epsilon) = 0 , \qquad x - at = 0 .$$

It is characterized in the $x$, $t$-plane by a pencil of straight characteristics emanating from the coordinate origin, and is sometimes called a Riemann wave. The speed of propagation of strain is constant along each ray; this speed obviously decreases for the characteristic with large slope. When the maximum strain $\epsilon_*$ is attained, the state is fixed and further from the end of the rod a constant disturbance $\epsilon_*$ is propagated with the minimum speed $a(\epsilon_*)$.

If we look at the situation at a fixed point $x = x'$ (fig. 243c), we have a rest state for $t < x'/a_0$; the wave front arrives at the instant $t = x'/a_0$; the strain $\epsilon$, the stress $\sigma$ and the velocity $v$ experience the jumps described above. Next, in the interval $x'/a_0 < t < x'/a(\epsilon_*)$, plastic deformation appears and gradually increases; at the moment $t = x'/a(\epsilon_*)$ this deformation reaches its

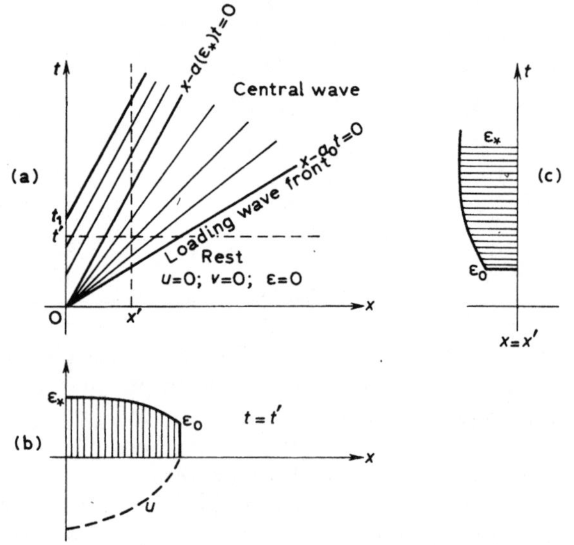


Fig. 243.

maximum value and does not change subsequently. The distributions of strain and displacement along the rod at some instant $t'$ are shown in fig. 243b.

In conclusion two remarks should be made. First, the deformation picture we have described is valid only up to the moment that unloading sets in; secondly, we have been considering here the propagation of expansion waves – for a compressive impact the signs of the strain $\epsilon$, the velocity $v$ and the displacement $u$ are reversed.

### 77.8. *Unloading wave*

At the instant $t = t_1$ a new wave begins to propagate in the positive $x$-direction: the *unloading wave.*

In an unloading region the *differences* in stresses and strains are related by Hooke's law

$$\sigma - \sigma_m = E(\epsilon - \epsilon_m)\,, \tag{77.10}$$

where $\sigma_m$, $\epsilon_m$ are the values of stress and strain attained in a given section of the rod at the instant when unloading begins (fig. 239a); these values are unknown functions of $x$. Substituting (77.10) in the equations of motion

(77.2), we obtain

$$\frac{\partial^2 u}{\partial t^2} = a_0^2 \frac{\partial^2 u}{\partial x^2} + \psi(x) ,$$

where

$$\psi(x) = \frac{1}{\rho} \frac{\partial}{\partial x} (\sigma_{\mathrm{m}} - E\epsilon_{\mathrm{m}})$$

is known.

It is necessary to construct a solution of this equation which is consistent with the solution of the differential equation of loading (77.3); the separation boundary between the solutions is a priori unknown. We shall not go into a discussion of the boundary value problem (cf. [34]), since this is unnecessary for the simple case of impactive unloading which we are considering.

We shall show that the *unloading wave moves with the velocity $a_0$ of the elastic wave.* We consider the kinematic and dynamic compatibility conditions at the unloading wave front. With passage of the front the stress and strain experience discontinuities

$$\sigma_+ - \sigma_- = [\sigma] ,$$

$$\epsilon_+ - \epsilon_- = [\epsilon] ,$$

where $\sigma_+$, $\epsilon_+$, $\sigma_-$, $\epsilon_-$ are the values of $\sigma$ and $\epsilon$ in front of and behind the front (fig. 244), which is moving with some velocity $g$. By Hooke's law

$$[\sigma] = E[\epsilon] . \tag{77.11}$$

At the wave front the condition of continuity of displacement must be satisfied (*kinematic compatibility condition*). Consider an element of the rod

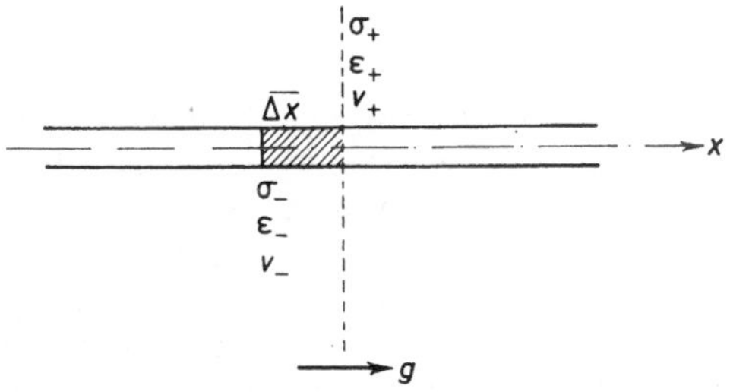


Fig. 244.

of length $\Delta x$ (before the unloading wave passes); after passage of the wave the length of this element is

$$\overline{\Delta x} = \Delta x - (\epsilon_+ - \epsilon_-)\, \Delta x \,.$$

On the other hand ($\Delta x/g$ is the time taken by the wave to travel a distance $\Delta x$)

$$\overline{\Delta x} = \Delta x + (v_+ - v_-)\, \Delta x/g \,.$$

Thus, the discontinuities in strain and velocity are connected by the relation

$$-[v] = g[\epsilon] \,. \qquad (77.12)$$

Next, the change in momentum of the element $\Delta x$ on passage of the unloading wave must obey a dynamical law (*dynamic compatibility condition*): by the conservation of momentum theorem

$$-\rho \Delta x [v] = [\sigma]\, \Delta x/g \,. \qquad (77.13)$$

Eliminating the discontinuities $[v]$, $[\sigma]$ with the aid of (77.11), (77.12), we find that the unloading wave is propagated with the velocity of the elastic wave

$$g = \sqrt{(E/\rho)} = a_0 \,.$$

Note that the general theory of kinematic and dynamic compatibility conditions and their application to the propagation of waves in elastic and plastic media is treated in the book of Thomas [50].

77.9. *Residual strain*

The residual strain can be calculated on the basis of the relation (77.11); since the load has been completely removed, we have $\sigma_- = 0$. But then the residual strain is

$$\epsilon_- = \epsilon_+ - \sigma_+/E \,.$$

The unloading waves, which move with the maximum speed $a_0$, overtake the simple waves; since the slowest of these corresponds to the largest plastic deformations, so as the unloading wave propagates along the rod, the magnitude of the residual strains beyond the front decreases.

The picture in the $x$, $t$-plane has the following form (fig. 245). The distance between the loading and unloading fronts, which move with constant speed $a_0$, is constant (fig. 245a). With the passage of time the simple waves are increasingly "diffused", and therefore the plastic strains in the interval be-

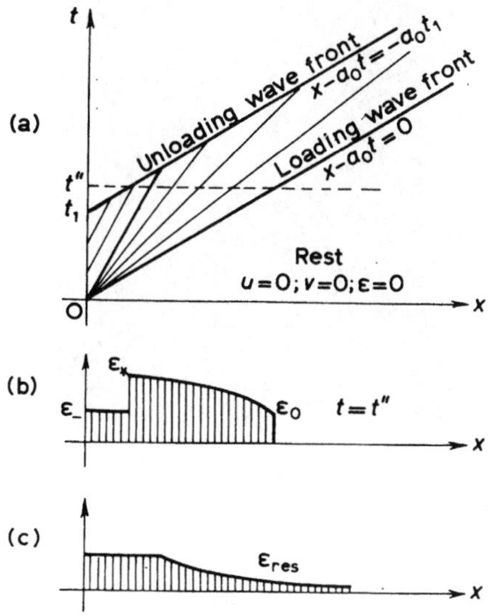


Fig. 245.

tween the fronts are continuously decreasing. Thus, the residual strain tends to zero with distance from the end of the rod (fig. 245c).

The results of the theory are satisfactorily confirmed by experiments [18, 34, 63–65]. Note that the assumption of small deformations is not essential. Calculations of deformations in a rod under impact agree well with observations (fig. 246).

## §78. Dynamical problems for a plastic-rigid body. Some energy theorems

### 78.1. *Dynamical problems for a plastic-rigid body*

If the influence of strain rates can be neglected, the relation between stress and strain can tentatively be represented by the $\sigma$-$\epsilon$ curve shown in fig. 247a; the initial linear segment is elastic. When plastic deformations are large it is possible to neglect elastic strains, i.e. to use the plastic-rigid model, as in statical problems. In this case we have the curve shown in fig. 247b. If hardening is insignificant, we can use directly the perfectly plastic-rigid model (fig. 247c).

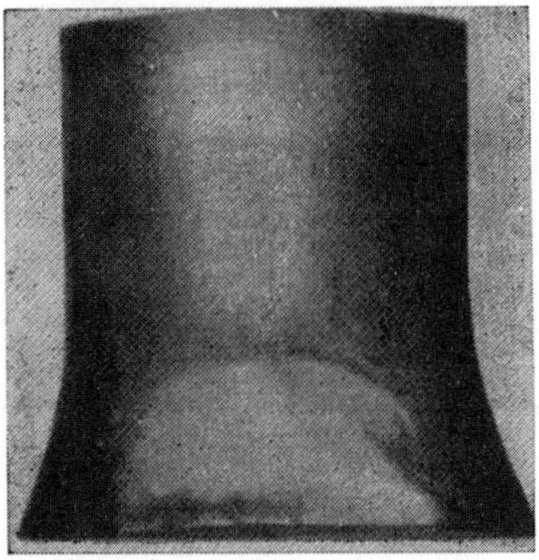

Fig. 246.

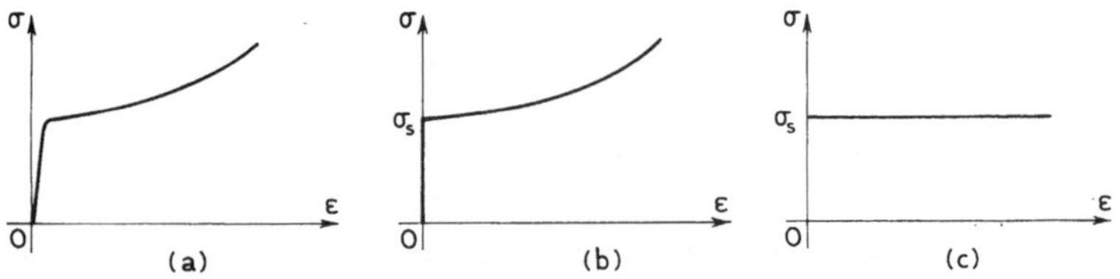


Fig. 247.

Neglect of elastic strains essentially simplifies the solution and allows simple results to be obtained in a series of dynamical problems.

The plastic-rigid model is convenient if the plastic work greatly (by an order of magnitude, say) exceeds the elastic energy.

This condition follows from the solutions of a number of elastic-plastic dynamical problems. It should be understood that a strict estimate of the plastic work and elastic energy cannot in practice be obtained from the original data of a problem. However, the particular features of a problem usually permit us to judge the possibility of neglecting the elastic deformations. For example, if it is necessary to determine large plastic changes in shape resulting from an impact, we can, as a rule, disregard the elastic strains. An example of a different type is the problem of a strong explosion (spherical) in an elastic-plastic medium; although the plastic work can here greatly exceed the elastic energy, elastic strains cannot be neglected if we need to know the intensity of the elastic waves developed in the explosion.

The plastic-rigid model has been widely used in problems of the action of an impulsive load on beams, circular plates, shells – these are of great practical interest. Thus, a beam remains rigid so long as the bending moment $M$ does not reach the limit value $M_*$. Subsequently, the beam develops plastic joints (or plastic zones) in which $|M| = \text{const.} = M_*$ (fig. 248). The motion of the rigid and plastic segments of the beam can be examined with the aid of conditions on their boundaries; this enables determination of the residual strains in the beam and the deformation time. The relative simplicity of the equations of motion for a plastic-rigid beam permit not only small, but also large, flexure to be determined.

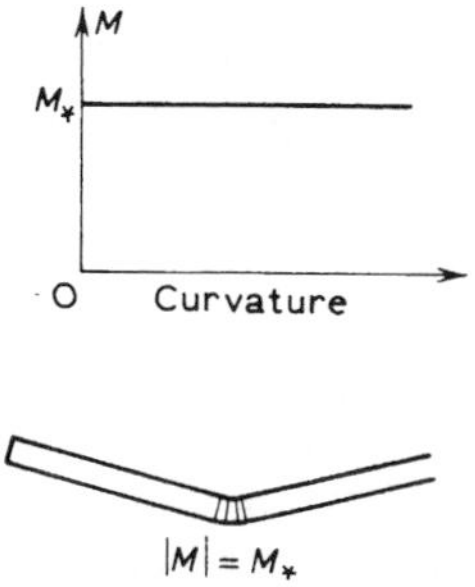


Fig. 248.

It is necessary, however, to take into account that, in general, the plastic-rigid model in dynamical problems leads only to qualitatively good agreement with experimental data for displacements; quantitative differences can be substantial.

The first use of the plastic-rigid model in dynamical problems was made by Gvozdev (1943) and Taylor (1946). An intensive development in this direction was begun some time later after the work of Lee and Symonds (1952), Prager and Hopkins (1954), and other authors, on the dynamics of plastic-rigid beams and plates (cf. [65, 99, 152, 169]).

78.2. *Some energy theorems*

In the dynamics of a plastic-rigid body hardly any sufficiently general theorems have been found which would enable effective estimates of the solutions to be obtained (as do the limit load theorems, ch. 8). Recently Martin [135] has established some simple theorems of a special character.

At the initial instant of time $t = 0$ suppose that velocities $v_i^0$ of the points of the given plastic-rigid body are known, and suppose that for $t > 0$ both the surface forces $X_{ni}$ and velocities $v_i$ are zero on the body's surface. It is assumed that the body's configuration does not change greatly under deformation; for simplicity body forces are disregarded.

With these assumptions the initial kinetic energy of the body is known; in the subsequent motion this energy is completely dissipated in the plastic deformation of the body, since external agencies do no work when $t > 0$. It is natural that the motion stops at some instant $t = \bar{t}$.

*Theorem* 1. *The deformation time t has the lower bound*

$$\bar{t} \geqslant \int \rho v_i^0 v_i' \mathrm{d}V \Big/ \int \sigma_{ij}^* \xi_{ij}' \mathrm{d}V \,, \tag{78.1}$$

where $\rho$ is the density of the body, $v_i'$ is a kinematically possible velocity field, *independent of time*, $\sigma_{ij}^*$ are the stresses determined from the strain rates $\xi_{ij}'$ in accordance with the associated law of plastic flow (§16),

$$\xi_{ij} = \lambda \frac{\partial f}{\partial \sigma_{ij}}, \qquad \lambda \geqslant 0 \,,$$

where $f(\sigma_{ij}) - K = 0$ is the equation of the yield surface.

In the denominator of the right-hand side of (78.1) we have the rate of work of the kinematically possible plastic strain; we denote it by $D'$.

*Proof.* From the general mechanical equation we have

$$\int (-\rho \dot{v}_i)\, v_i' \mathrm{d}V = \int \sigma_{ij} \xi_{ij}' \mathrm{d}V$$

($\cdot$ denotes time-differentiation), where $\sigma_{ij}$, $\dot{v}_i$ are the actual stresses and accelerations. By the local maximum principle $(\sigma_{ij}^* - \sigma_{ij})\, \xi_{ij}' \geqslant 0$ (cf. §§64, 65), and consequently

$$\int \sigma_{ij} \xi_{ij}' \mathrm{d}V \leqslant \int \sigma_{ij}^* \xi_{ij}' \mathrm{d}V \equiv D' \,.$$

Thus

$$\int (-\rho \dot{v}_i)\, v_i' \mathrm{d}V \leqslant D' \,.$$

Integrating this inequality with respect to time from $t = 0$ to $t = \bar{t}$, we find

$$\left[ -\int \rho v_i v_i' \mathrm{d}V \right]_0^{\bar{t}} \leqslant \bar{t} D' \,.$$

Substituting the initial values of $\bar{v}_i$ and putting $v_i|_{t=\bar{t}} = 0$, we arrive at inequality (78.1).

*Theorem* 2. *For surface displacements $\bar{u}_i$ over the deformation time the* *upper bound*

$$\int X'_{ni}\bar{u}_i \mathrm{d}S \leqslant \int \tfrac{1}{2}\rho v_i^{0^2} \mathrm{d}V \,, \tag{78.2}$$

*holds,* where $X'_{ni}$ are time-independent forces corresponding to the statically possible yield-stress state $\sigma'_{ij}$ (cf. §64.5), i.e. $f(\sigma'_{ij})-K \leqslant 0$.

*Proof.* The general mechanical principle gives

$$\int (-\rho \dot{v}_i)\, v_i \mathrm{d}V = \int \sigma_{ij}\xi_{ij}\mathrm{d}V \,.$$

By the local maximum principle

$$\int \sigma_{ij}\xi_{ij}\mathrm{d}V \geqslant \int \sigma'_{ij}\xi_{ij}\mathrm{d}V \,.$$

But from (64.6) we have

$$\int \sigma'_{ij}\xi_{ij}\mathrm{d}V = \int X'_{ni}v_i \mathrm{d}S \,.$$

Consequently

$$-\frac{\mathrm{d}}{\mathrm{d}t}\int \tfrac{1}{2}\rho\, v_i^2 \mathrm{d}V \geqslant \int X'_{ni}v_i \mathrm{d}S \,.$$

Integrating this inequality with respect to time from $t=0$ to $t=\bar{t}$, and using the fact that $u_i=0$, $v_i=v_i^0$ when $t=0$, and $u_i=\bar{u}_i$, $v_i=0$ when $t=\bar{t}$, we arrive at the required inequality.

These theorems can easily be formulated in terms of generalized forces and displacements, which is convenient for application to rods, plates and shells.

Comparison of the estimates from Martin's theorems with some exact solutions shows that the deformation time $\bar{t}$ is determined with moderate error; but the estimate of the displacements is rather high.

A weak point in these theorems is the assumption that the change in shape of the body is small; this is not in good agreement with the conditions for applicability of the plastic-rigid model in dynamical problems.

78.3. *Example*

Suppose that at the instant $t=0$ a freely supported beam (fig. 249a) experiences the action of an impulsive load which communicates to the whole section of the beam a constant initial velocity $v_0$. As an example of a time-independent, kinematically possible velocity field we take the triangular distribution shown in fig. 249b. Here there is a plastic joint at the centre of the beam, and each half rotates like a rigid body with angular velocity $\omega$ about the supports. Then $v' = \omega x$, and $D' = 2M_* \omega$. It is easy to see that

$$\int \rho v_i^0 v'_i \mathrm{d}V = 2\int m v_0 \omega x \mathrm{d}x = m v_0 l^2 \omega \,,$$

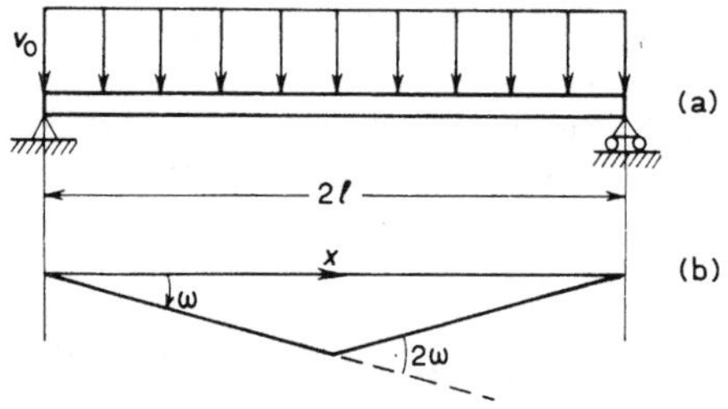


Fig. 249.

where $m$ is the mass of unit length of the beam. From the inequality (78.1) we obtain

$$\bar{t} \geqslant \tfrac{1}{2} m v_0 l^2 / M_* \,, \tag{78.3}$$

which coincides with the exact value found by Symonds. Calculation of the right-hand side of (78.2) gives

$$\int \tfrac{1}{2} \rho v_i^{0^2} \, \mathrm{d}V = 2 \int_0^l \tfrac{1}{2} m v_0^2 \mathrm{d}x = m l v_0^2 \,.$$

We take as a statically possible load a point force $P'$, applied at the centre of the beam and corresponding to its limiting state, i.e. $P' = 2M_*/l$. Then (78.2) gives the estimates for the flexure $\bar{u}$ at the centre of the beam:

$$\bar{u} \leqslant \tfrac{1}{2} m l v_0^2 / M_* \,. \tag{78.4}$$

This estimate is one and a half times greater than the exact value.

78.4. *Concluding remarks*

The foregoing theorems have been extended by Martin to bodies having hardening and viscosity.

In a recently published work Tamuzh [160] has shown that the actual accelerations minimize a certain functional. This principle can be used, in particular, for finding approximate solutions; cf. also [135].

## §79. Longitudinal impact of a plastic-rigid rod against a stationary barrier

Following Taylor [192], we consider the problem of the normal impact of a cylindrical rod (initial length $l_0$), moving with initial velocity $v_0$, against a stationary undeformable barrier.

Suppose the curve of uniaxial statical compression of the cylinder, $\sigma = \sigma(\epsilon)$, is known, where $\sigma$, $\epsilon$ are the stress and strain corresponding respectively to an initial area $F_0$ and initial length $l_0$ (fig. 247a). Neglecting elastic deformations we obtain the curve shown in fig. 247b; here plastic flow begins with the value $\sigma = \sigma_s$ ("yield limit").

Let the wave front of plastic deformation propagate from the motionless barrier with velocity $g$, and let the material behind it be left at rest; to the right of the front the material is rigid and moves like a solid body (fig. 250) with decreasing velocity $v$. At the front the stress and the cross-sectional area change discontinuously; ahead of the front we have $F_0$ and $\sigma_s$ (since the material ahead of the front is almost plastic), and beyond the front $F$ and $\sigma$.

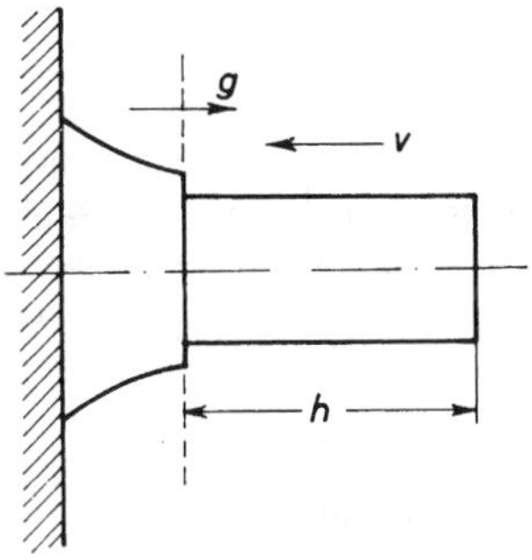


Fig. 250.

From the incompressibility equation it follows that

$$gF = (v + g)\, F_0 \,. \tag{79.1}$$

The strain beyond the wave front can be calculated in the following way: in unit time a column of the non-deforming part of length $v + g$ becomes a column of deforming material of length $g$. The relative compressive strain is then

$$\epsilon = -\frac{g-(v+g)}{v+g} = \frac{v}{v+g}\,. \tag{79.2}$$

Let $h$ be the length of the non-deforming part of the specimen at time $t$;

obviously

$$-\frac{dh}{dt} = v + g . \tag{79.3}$$

When the wave front passes through an element $dx = -dt(v+g)$ of the rod, the latter's velocity vanishes. By the conservation of momentum theorem

$$\rho F_0(v+g)\, dt \cdot v = F_0(\sigma - \sigma_s)\, dt ,$$

where $\rho$ is the density, assumed constant. Thus

$$\rho(v+g)\, v = \sigma - \sigma_s . \tag{79.4}$$

We construct the equation of motion for the non-deforming part of the rod; since the latter has variable length $h$, and, therefore, variable mass also, then in the right-hand side of the equation

$$\rho h F_0\, dv/dt = P$$

it is necessary to add the reaction of the separating mass to the applied forces. But the material of the rod ahead of the front is almost plastic, and therefore it can be assumed that the "total force" $P$ reaches the value $-\sigma_s F_0$. Thus, the equation of motion of the non-deforming part of the rod has the form

$$\rho h\, dv/dt = -\sigma_s . \tag{79.5}$$

From (79.2), (79.3) it follows that

$$dt = -\frac{\epsilon}{v}\, dh ,$$

and from (79.5) we obtain, after multiplying by $v$,

$$\tfrac{1}{2}\, d(\rho v^2) = \sigma_s \epsilon\, d(\ln h) .$$

But from (79.2), (79.4) we have

$$\rho v^2 = \epsilon(\sigma - \sigma_s) . \tag{79.6}$$

Consequently

$$d[\epsilon(\sigma - \sigma_s)] = 2\sigma_s \epsilon\, d(\ln h) .$$

Hence, for initial values

$$h = l_0 , \qquad v = v_0 , \qquad \epsilon = \epsilon_0 \qquad \text{when} \qquad t = 0$$

we obtain

$$\ln\frac{h}{l_0}=\frac{1}{2}\int_{\sigma_0}^{\sigma}\frac{\mathrm{d}[\epsilon(\sigma/\sigma_s-1)]}{\epsilon}\,. \tag{79.7}$$

The value of $\epsilon_0$ is determined from (79.6):

$$\rho v_0^2=\epsilon_0(\sigma_0-\sigma_s)\,,\qquad \sigma_0=\sigma(\epsilon_0)\,.$$

Since the curve $\sigma=\sigma(\epsilon)$ is known, equation (79.7) determines the residual stress $\epsilon$ as a function of the length $h$. From (79.6) we find the velocity $v$ as a function of $h$, and from (79.2) the velocity $g$ of the plastic wave front as a function of $h$. Finally, the dependence on time is given by equation (79.3). These calculations for the deformation of a rod under impact agree well with observations (fig. 246) and with the results of a more exact theory which takes into account the elasticity of the material.

## §80. Bending of a plastic-rigid beam under an impulsive load

### 80.1. *The plastic-rigid beam*

The beam remains rigid so long as the bending moment $M$ does not reach the limiting value $M_*$. At this point stationary or moving plastic joints can occur (fig. 251b), or, eventually, some plastic zones (fig. 251c). Between the

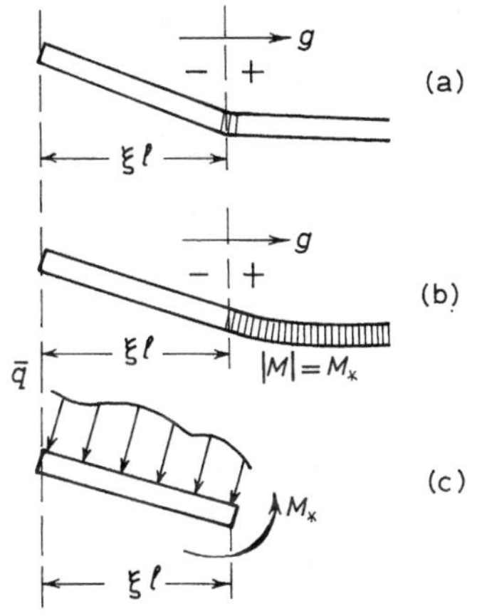


Fig. 251.

plastic joints and zones there will be rigid portions, for which $|M| < M_*$. In the process of motion the position of the joints and zones will, in general, change. Intensive dynamical loads frequently lead to large deformations of the beam. It is natural that an important role is played by the longitudinal forces which can arise when the supports are not displaced. We consider below simple problems in which axial forces are absent.

The plastic-rigid model leads to satisfactory results in dynamical problems if the plastic work is substantially (an order of magnitude, say) greater than the maximum elastic energy $M_*^2 l/2EJ$, where $J$ is the moment of inertia of a cross-section, and $l$ the length of the beam. This condition is usually realized in normal plastic flexure of beams.

80.2. *Kinematic compatibility conditions*

Let the $x$-axis be along the axis of the beam, and let $l$ be the length (or half length) of the beam. Introduce the following notation: $\theta = \partial u/\partial x$, $\omega = \dot{\theta}$, $\dot{\omega}$ are respectively the angle of inclination of the tangent to the axis, the angular velocity and the angular acceleration; $v = \dot{u}$, $\dot{v}$ are respectively the speed and acceleration of bending. Further, we denote by $\xi l$ the distance from the left-hand end of the beam to the plastic joint (fig. 251a) or to the plastic zone (fig. 251b), and by $g = l\dot{\xi}$ the velocity of motion of the joint (or of the expansion of the plastic zone). The value of quantities to the left of the joint (to the left of the boundary of the plastic zone) will be designated with the suffix $-$, and those to the right with the suffix $+$.

At the joint (boundary) $x = l\xi$ the kinematic compatibility must be satisfied. First of all we have the obvious condition of continuity of bending

$$u_- = u_+ \,. \tag{80.1}$$

Differentiating this expression with respect to time, we find

$$\frac{\partial u_-}{\partial t} + \frac{\partial u_-}{\partial x}\frac{\partial x}{\partial t} = \frac{\partial u_+}{\partial t} + \frac{\partial u_+}{\partial x}\frac{\mathrm{d}x}{\mathrm{d}t}$$

or

$$v_- - v_+ = -g(\theta_- - \theta_+) \,.$$

The right-hand side here is zero. For if the joint is stationary $g = 0$; while if the joint moves with finite velocity the adjacent section does not have time to be turned through a finite angle in the time it takes for the joint to traverse the section. Thus

$$v_- = v_+ \,. \tag{80.2}$$

Differentiating now with respect to time, we obtain

$$\dot{v}_- - \dot{v}_+ = -g(\omega_- - \omega_+) \,, \tag{80.3}$$

i.e. the discontinuity in acceleration of bending is proportional to the discontinuity in angular velocity.

### 80.3. *Equations of motion of a plastic-rigid beam*

The analysis of the motion of a plastic-rigid beam is based on simultaneous consideration of its rigid and plastic portions.

The equation of motion of particles in the plastic part is straight-forward. From the yield criterion we have that the bending moment is constant ($M = \text{const.} = M_*$), as a result of which the transverse force is zero and the equation of motion has the form

$$m\ddot{u} = \overline{q}(x, t) \,, \tag{80.4}$$

where $\overline{q}(x, t)$ is the distributed load.

The equation of motion of the rigid portion (fig. 251c) with stationary joint is also easy to write down. If the joint is non-stationary, the length of the rigid portion (and, consequently, its mass $m\xi l$) changes, and therefore it is in general necessary to take into account the reactive forces which arise as a result of the separation of particles from the body (or their attachment to the body). As we know these forces are proportional to the difference between the velocities of the particles and the body. In the problem of the motion of a plastic-rigid beam this difference is zero on the boundary because of (80.2); consequently additional reactive forces do not arise and we can usually base ourselves on the laws of mechanics in their customary formulation.

### 80.4. *Example: joint-supported beam under the action of a uniformly distributed impulsive load*

We consider the motion of a joint-supported beam (fig. 252a) which experiences the action of a uniformly distributed load $\overline{q} = Q/2l$ of explosive type; the load acts over an interval of time $(0, \tau)$, during which it is constant (fig. 253a) or decreases (fig. 253b) with the assumption that

$$\int_0^t Q\mathrm{d}t \geqslant tQ(t) \,. \tag{80.5}$$

We introduce the dimensionless loading parameter $q = ql/M_*$.

The beam will not be deformed so long as the maximum bending moment does not attain the limiting value $M_*$. It is easy to see that this takes place

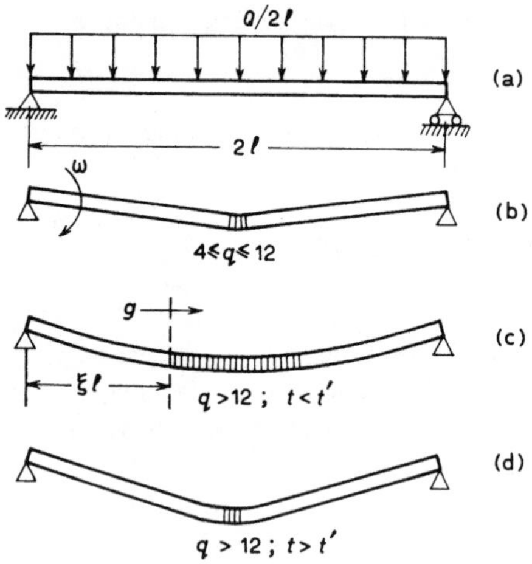


Fig. 252.

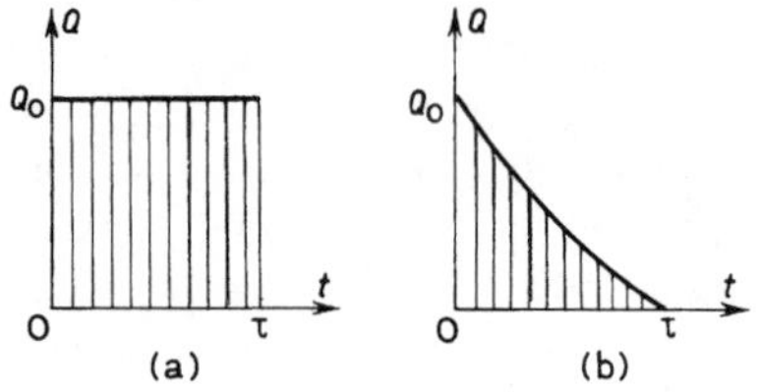


Fig. 253.

when $q = 4 \equiv q_1$ for the central section of the beam. Thus, for a *weak load* $q < 4$ the beam remains rigid.

*Motion of the beam for a moderate load* ($4 \leqslant q \leqslant 12$). For a load somewhat in excess of $q_1$ the rigid halves of the beam rotate about the supports (fig. 252b) with a stationary plastic joint in the middle. The equation of rotation of the left-hand half of the beam has the form

$$\frac{ml^3}{M_*} \dot{\omega} = \tfrac{3}{4}(q-4) . \tag{80.6}$$

Since the load reaches its maximum value instantaneously, the initial conditions are zero, i.e.

$$\text{when} \qquad t = 0 \qquad u = 0 , \qquad \omega = 0 . \tag{80.7}$$

Equation (80.6) retains its validity so long as the bending moment in the rigid half of the beam nowhere exceeds $M_*$. The bending moment is determined by the uniformly distributed load of intensity $Q/2l$, the inertial load $-m\dot{\omega}x$, and the limit moment $M_*$ in the joint. For small values of the angular acceleration the graph of the bending moment is shown in fig. 254 by the solid line, while for large accelerations it is represented by the dotted line. In the latter case the transverse force is negative near the joint, which will clearly occur when $ml\dot{\omega} > Q/2l$. It follows that equation (80.6) is valid when $ml\dot{\omega} < Q/2l$. Putting $ml\dot{\omega} = Q/2l$ in (80.6), we find the maximum load $q = 12 \equiv q_2$, below which motion takes place with a point-joint at the centre of the beam.

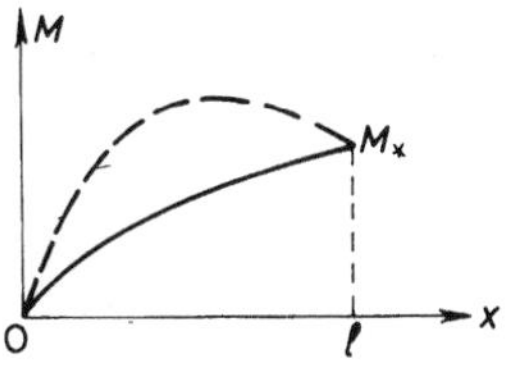


Fig. 254.

Integrating equation (80.6) we obtain

$$\frac{ml^3}{M_*}\omega = \tfrac{3}{4}I(t) - 3t \qquad \left(I(t) = \int_0^t q\,\mathrm{d}t\right). \tag{80.8}$$

Motion ceases at the instant $t = \bar{t}$, when the angular velocity vanishes; then

$$I(\bar{t}) - 4\bar{t} = 0\,. \tag{80.9}$$

Integrating next equation (80.8) and putting $t = \bar{t}$, we obtain the deformation graph of the beam (fig. 252b), determined by the angle $\bar{\theta} = \theta(\bar{t})$:

$$\frac{ml^3}{M_*}\bar{\theta} = \frac{3}{4}\int_0^{\bar{t}} I(t)\,\mathrm{d}t - \tfrac{3}{32}I^2(\bar{t})\,. \tag{80.10}$$

For a rectangular impulse (fig. 253a), we have

$$\frac{ml^3}{M_*}\bar{\theta} = \tfrac{3}{32}(q_0 - 4)\,q_0\tau^2 \qquad \left(q_0 = \frac{Q_0 l}{M_*} \leqslant 12\right). \tag{80.11}$$

The maximum flexure is $\bar{u} = l\bar{\theta}$.

*The motion of the beam for a strong force* ($q > 12$). In this case loading generates a plastic zone $|x| \geqslant \xi l$ (fig. 252c), in which the bending moment is constant. Under certain conditions (see below) this zone is shortened and contracts to a point – the plastic joint at the centre. For $t > t'$ motion proceeds with one stationary joint (fig. 252d).

Consider the motion for $t < t'$. It is described in the plastic zone by equation (80.4), where the load $\bar{q} = Q/2l$ is independent of $x$; thus the plastic segment moves down and does not deform (fig. 252c). We suppose first that the load $Q(t)$ increases, and find $\xi$. For an element on the boundary $x = \xi l$ we have

$$ml\xi\dot{\omega} - Q/2l = 0\,. \tag{80.12}$$

The rotation of the rigid portion is determined by the equation

$$\frac{ml^3\xi^3}{M_*}\dot{\omega} = \tfrac{3}{4}(q\xi^2 - 4)\,, \tag{80.13}$$

analogous to equation (80.6). From these equations it follows that

$$q\xi^2 - 12 = 0\,, \tag{80.14}$$

$$4\sqrt{3}\frac{ml^3}{M_*}\dot{\omega} = q^{\frac{3}{2}}\,. \tag{80.15}$$

These equations are valid for an increasing load $Q(t)$, when $g < 0$ and the left-hand part of the beam $x < l\xi$ remains straight. From (80.4) and (80.13) the jump in acceleration at the boundary is now

$$\dot{v}_- - \dot{v}_+ = \frac{M_*}{4ml^2}\left(q - \frac{12}{\xi^2}\right) = 0\,,$$

and vanishes because of (80.14). The jump in the angular velocities is also equal to zero.

Now suppose the load $Q(t)$ diminshes. Then

$$q\xi^2 - 12 \leqslant 0\,, \tag{80.16}$$

otherwise the compatibility condition (80.3) is violated. We construct the equation for conservation of moment of momentum for the half-beam $0 \leqslant x \leqslant l$, with respect to the left-hand support:

$$\int_0^t \tfrac{1}{4}Ql\,\mathrm{d}t - M_* t = \int_0^l mvx\,\mathrm{d}x\,. \tag{80.17}$$

When $x \leqslant \xi l$, $v = \omega x$, while when $x \geqslant \xi l$, (80.14) gives

$$m v = \int_0^t \frac{Q}{2l}\, \mathrm{d}t = \frac{M_*}{2l^2} I(t) .$$

Therefore equation (80.17) takes the form

$$\frac{1}{4}\xi^2 I(t) - t = \frac{m l^3 \xi^3}{3M_*}\, \omega . \tag{80.18}$$

Differentiating this equation with respect to time and using (80.13), we find that $\frac{1}{2}I(t) = ml^3\xi\omega/M_*$. It is then easy to see that

$$\xi^2 = 12t/I(t) , \tag{80.19}$$

$$\left(\frac{ml^3}{M_*}\right)^2 \omega^2 = \frac{I^3(t)}{48t} . \tag{80.20}$$

We clarify the conditions for the realization of this regime. Differentiating (80.9) we find that the condition (80.5), which can be written in the form $I(t) \geqslant tq$, gives

$$2\xi\dot{\xi} = \frac{12}{I}\left(1 - \frac{tq}{I}\right) \geqslant 0 ,$$

i.e. $g \geqslant 0$. Eliminating the quantity $I$ with the aid of (80.19) we obtain

$$l\xi(q\xi^2 - 12) = -24gt ,$$

i.e. the inequality (80.16) does in fact hold.

Taking the limit $t \to 0$ in (80.19), we arrive at the initial position of the boundary

$$q_0\xi_0^2 = 12 , \qquad \xi_0 = \xi(0) . \tag{80.21}$$

At the instant $t'$ we have $\xi = 1$, hence

$$I(t') - 12t' = 0 . \tag{80.22}$$

The angular velocity $\omega'$ at this instant can be found from (80.20):

$$\frac{ml^3}{M_*}\omega' = 6t' , \tag{80.23}$$

and the displacement $u'$ of the mid-point by integrating the equation for $u$:

$$\frac{ml^2}{M_*} u' = \tfrac{1}{2} \int_0^{t'} I(t)\,\mathrm{d}t\,. \tag{80.24}$$

*Motion for* $t > t'$. In this final stage of the motion the angular velocity of rotation of the "re-solidified" half of the beam is determined by the previous equation (80.6), but with the initial condition $\omega = \omega'$ when $t = t'$.

On finding the corresponding solution and using the relations (80.22), (80.23), we find

$$\frac{ml^2}{M_*} \omega = \tfrac{3}{4} I(t) - 3t \qquad (t \geqslant t')\,. \tag{80.25}$$

The motion ceases with $\omega = 0$ at the instant $\bar{t}$, determined by the formula

$$I(\bar{t}) - 4\bar{t} = 0\,. \tag{80.26}$$

Integrating (80.25) we find the angle turned through in this stage of the motion

$$\frac{ml^3}{M_*} \theta = \tfrac{3}{4} \int_{t'}^{t} I(t)\,\mathrm{d}t - \tfrac{3}{2}(t^2 - t'^2)\,.$$

We next determine the displacement $l\theta$ at the final moment $\bar{t}$, and add to it the previous displacement (80.24). Then the total displacement $\bar{u}$ of the mid-point is found to be:

$$\frac{ml^2}{M_*} \bar{u} = \tfrac{3}{4} \int_{t'}^{\bar{t}} I(t)\,\mathrm{d}t + \tfrac{1}{2} \int_0^{t'} I(t)\,\mathrm{d}t - \tfrac{3}{32} \left[I^2(\bar{t}) - \tfrac{1}{9} I^2(t')\right]\,. \tag{80.27}$$

In the course of the final stage of the motion a break can occur at the mid-point.

For the case of a rectangular impulse (fig. 253a)

$$I(t) = \begin{cases} q_0 t & \text{for} \quad 0 \leqslant t \leqslant \tau\,, \\ q_0 \tau & \text{for} \quad t \geqslant \tau\,, \end{cases}$$

and from (80.22), (80.26) it follows that

$$t' = \tfrac{1}{12} q_0 \tau\,, \qquad \bar{t} = \tfrac{1}{4} q_0 \tau\,.$$

By (80.27) the maximum residual flexure is

$$\frac{ml^2}{M_*}\bar{u} = \frac{q_0\tau^2}{12}(q_0-3) \qquad (q>12)\,. \tag{80.28}$$

In the present example (80.19) shows that in the time-interval $(0, \tau)$ the boundary is $\xi = \xi_0$, i.e. is motionless; it is displaced only when $t > \tau$.

It is not difficult to calculate the energy absorbed by the beam. For a moderate load ($q_0 \leqslant 12$), absorption takes place in the central joint. For a strong load ($q_0 > 12$) the energy is absorbed at the stationary boundary $\xi = \xi_0$, in the zone of continuous plastic strains $\xi_0 < \xi < 1$, and in the central joint $\xi = 1$.

80.5. *Concluding remarks*

Analysis of different cases of loading shows that the "shape" of the impulse has little influence on the bending which is basically determined by the maximum load $q_0$ and the total impulse

$$q_0\tau_0 = \int_0^\infty q\mathrm{d}t\,.$$

Other type of loads (concentrated, non-uniform, etc.) and clamped beams have been studied by similar methods in a series of works [152, 153]. Experimental data satisfactorily confirm the calculations of the plastic-rigid model for developing plastic deformations. The differences which are sometimes observed are usually connected with the effect of axial forces in the presence of large flexure, and with departure from perfect plasticity.

# PROBLEMS

1. A circular cylindrical rod $z \geqslant 0$ receives an impactive twist at the end $z = 0$. If the twisting moment is sufficiently large, an elastic-plastic deformation wave is propagated.

Derive the differential equation for the propagation of twisting strain on the assumption that the material obeys the elastic-plastic scheme with yield plateau, and that the section of the rod rotates as a whole.

*Answer:*

$$\frac{\partial^2\theta}{\partial t^2} = b_*^2\frac{\partial^2\theta}{\partial z^2}, \qquad b_* = b_0(c/a)^2, \qquad b_0 = \sqrt{G/\rho},$$

where $b_0$ is the velocity of propagation of the elastic wave, $c$ is the radius of the elastic core, and $\theta$ the angle of rotation of the section.

2. Find the propagation speed for the torsional wave of elastic-plastic strain in a hardening rod (§ 30.3).

3. Derive the differential equation of longitudinal vibrations of a rod of variable section ($S$ is the cross-sectional area).

*Answer:*

$$\frac{\partial\sigma}{\partial x}+\frac{S'(x)}{S(x)}\sigma=\rho\frac{\partial^2 u}{\partial t^2}.$$

4. For safe application of a tensile load to an elastic ($\sigma = E\epsilon$) rod, whose mass is negligible compared with the mass of the load, the dynamic strain $\epsilon^d$ can be twice as large the strain $\epsilon^s$ for static (slow) loading (cf. [47]). Find the ratio $\epsilon^d/\epsilon^s$ for a rod which obeys the deformation law $\sigma = B\epsilon^\beta$, where $B$, $\beta$ are constants ($0 < \beta \leqslant 1$). Show that $\epsilon^d/\epsilon^s \geqslant 2$ and tends to $e = 2.718\ldots$ as $\beta \to 0$.

# 12

# Composite Media. Visco-Plasticity

## §81. Composite media

### 81.1. *Influence of viscosity*

In the preceding chapters we have been considering plastic deformation which is independent of the time (*athermal plasticity*). By comparison with Hooke's law, the new equations of state describe more completely the mechanical properties of real bodies, and therefore lead to results which are of great importance in questions of the strength of machines and structures. Both deformation theory and plastic flow theory are relevant to a description of irreversible equilibrium processes in deformation.

Nevertheless it is not always possible to neglect the influence of viscosity (which is connected with the thermal motion of atoms); a temporal analysis of the deformation process then becomes necessary. In particular cases (for example, the deformation of steel in conditions of normal temperature) the influence of time is negligibly small, and the previous theories of plasticity can be used. In other cases this effect turns out to be substantial, and essentially changes the whole picture of the deformation. Thus, even hard steels in conditions of high temperature exhibit yield at small stresses and can accumulate large strains with the passage of time (presence of *creep*). Viscosity has frequently to be taken into account in rapid motions (for example, in connection with vibrations, shocks).

In modern technology an ever-increasing importance is being acquired by the use of the complex mechanical properties of high polymers, which include all kinds of resins and various artificial and natural fibrous materials. Those materials are characterized by the important role of the time; the deformation processes are here called *non-equilibrium.*

Study of the mechanical properties of complex media has led to the development of a new science – *rheology* (cf. [35, 41]). In this chapter we consider only *composite plastic media*; by this term we understand a "usual" plastic medium, complicated by the presence of viscosity alone. For the more complete equations of rheology the reader may refer to the books [35, 33, 62].

### 81.2. *Mechanical models*

The mechanical equations of state of composite media are usually illustrated by the aid of simple mechanical models. In this section we shall consider for simplicity a uniaxial stress distribution (extension of a rod); the corresponding stress will be denoted by $\sigma$, the relative extension by $\epsilon$, and the rate of relative extension by $\xi$.

An *elastic element*, obeying Hooke's law

$$\sigma = E\epsilon \,, \tag{81.1}$$

can be represented in the form of a spring (fig. 255a).

A *viscous element* which obeys Newton's law of viscosity

$$\sigma = \mu \, \mathrm{d}\epsilon/\mathrm{d}t \,, \tag{81.2}$$

where $\mu$ is the coefficient of viscosity, is represented by a model consisting of a piston moving in a cylinder containing viscous fluid (fig. 255b).

A *plastic-rigid body* with stresses below the elastic limit does not deform; flow develops only for stresses satisfying the yield criterion ($\sigma = \sigma_s$). This medium is represented in the form of a plane with cold friction (fig. 255c).

By combining those simple models we can introduce various composite media. For example, an *elastic-plastic medium* is characterized by a model in which elastic and plastic elements are joined in series (fig. 256). Other examples of composite media are presented below.

Note that an elastic (or viscous) element can follow a nonlinear law of elasticity (or viscosity). For example, the relation

$$\mathrm{d}\epsilon/\mathrm{d}t = B|\sigma|^{m-1}\sigma \,, \tag{81.3}$$

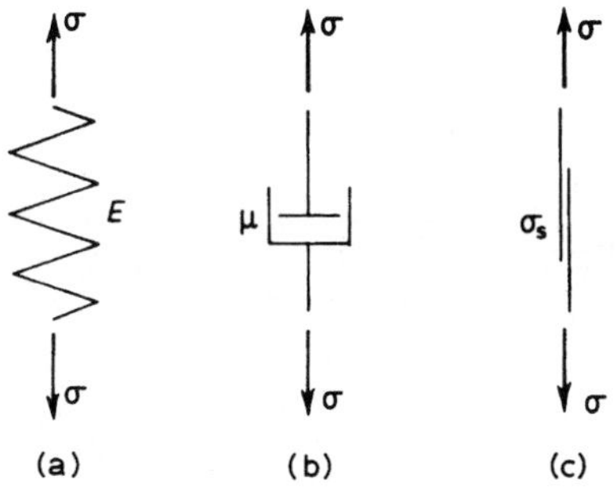


Fig. 255.

Fig. 256.

where $B$, $m$ are constants, corresponds to nonlinear viscous flow (creep) of metals.

81.3. *Visco-elasticity*

The joint of elastic and viscous elements leads to the so-called *visco-elastic medium.*

A parallel join (fig. 257a) of two elements – elastic and viscous – gives the *Voigt visco-elastic medium*

$$\sigma = E\epsilon + \mu \, \mathrm{d}\epsilon/\mathrm{d}t \, . \tag{81.4}$$

This equation is obtained if we use the fact that the total stress in the medium is compounded of a stress corresponding to elastic deformation, and a stress generated by viscous resistance. A visco-elastic medium in a state of rest ($\mathrm{d}\epsilon/\mathrm{d}t = 0$) behaves as if elastic. The stress in the medium increases with increasing strain-rate. If the medium undergoes a constant strain $\epsilon = \text{const.}$, a constant stress $\sigma = E\epsilon$ is present. If it is loaded with a constant stress

$$\sigma = \text{const.} = \sigma_0 \qquad \text{for} \qquad t \geqslant 0 \, ,$$

then from (81.4) we obtain

$$\epsilon = \frac{\sigma_0}{E}(1 - \mathrm{e}^{-Et/\mu}) \, ,$$

i.e. the strain increases gradually, and tends to the value $\sigma_0/E$.

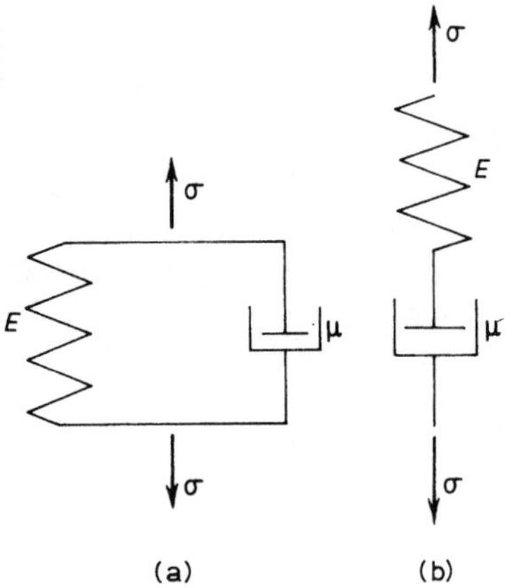


Fig. 257.

A visco-elastic medium was first studied in detail by Voigt in connection with the problem of attenuation of vibrations, and subsequently was investigated by many researchers [1, 52, 62].

For *connection in series* (fig. 257b), we have compounding of strain rates corresponding to one and the same stress. The deformation law for such a medium was first obtained by Maxwell, and has the form

$$\frac{d\epsilon}{dt} = \frac{1}{E}\frac{d\sigma}{dt} + \frac{\sigma}{\mu} . \tag{81.5}$$

We consider the properties of this medium. If a constant stress is communicated to the medium ($\sigma$ = const.), then it will be deformed with constant rate, i.e. it will flow like a viscous fluid. With rapid loading $\sigma = \sigma_0$, a strain $\sigma_0/E$ develops immediately in the medium (at the expense of the elastic term). If the stress is removed, the strain-rate also vanishes, but some constant strain remains in the medium.

Let the body experience a stress $\sigma_0$ at the instant $t = 0$; the corresponding initial elongation is $\epsilon_0 = \sigma_0/E$. We now fix the strain, putting $\epsilon = \text{const.} = \epsilon_0$ (for example, by clamping the ends of the rod). Then $d\epsilon/dt = 0$ and from (81.5) it follows that

$$\frac{1}{E}\frac{d\sigma}{dt} + \frac{\sigma}{\mu} = 0 .$$

Hence

$$\sigma = \sigma_0 e^{-t/t_0},$$

where $t_0 = E/\mu$ is called the *relaxation time.* Thus the stress decreases exponentially with time, and tends to zero (fig. 258). Maxwell's *relaxing medium* described qualitatively an important property of real materials – the weakening with time of the stress distribution under constant strain (so-called *stress-relaxation*). Maxwell's equation is often used for a qualitative description of the relaxation effect; but quantitative results are in poor agreement with observations.

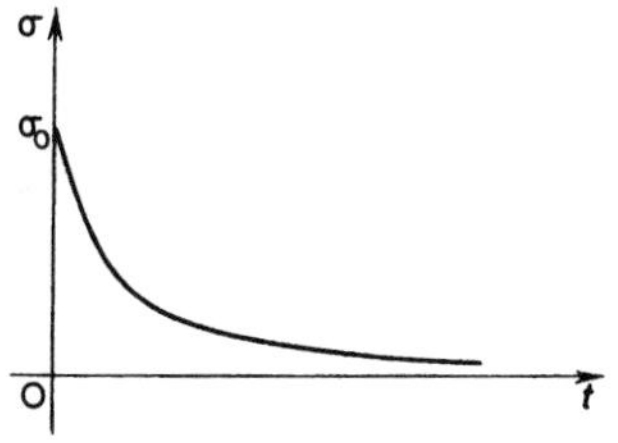


Fig. 258.

The above models contain two parameters: $E$, $\mu$.

A description of complex mechanical properties (for example, high polymers) requires *multi-element models,* characterized by a large number of elements. The model shown in fig. 259a consists of three parameters $E_1$, $E_2$, $\mu_1$.

An example of a model with four parameters $E_1$, $E_2$, $\mu_1$, $\mu_2$ is shown in fig. 259b.

Such media (with $n$ parameters or with continuously distributed parameters) are studied in the theory of linear visco-elasticity [1, 33, 62].

Nonlinear relaxing media play an important role in the theory of creep of models [2, 17, 33].

81.4. *Visco-plasticity*

The join of viscous and plastic elements leads to so-called *visco-plastic media.*

*Parallel connection* (fig. 260a) of two elements – viscous and plastic – gives a visco-plastic medium, first considered, apparently, by Shvedov (1900)

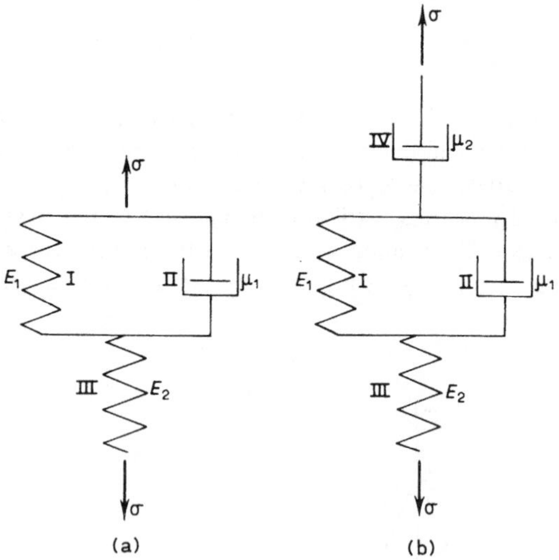


Fig. 259.

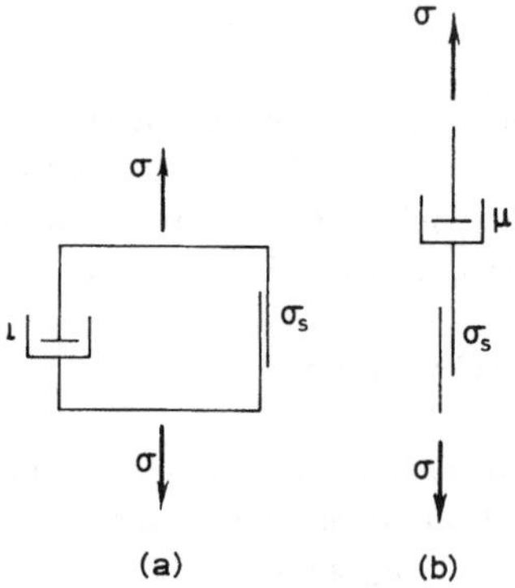


Fig. 260.

and Bingham (1922). Here the deformation law has the form

$$\sigma = \sigma_s + \mu \frac{d\epsilon}{dt} \qquad \text{for} \qquad \sigma \geqslant \sigma_s\,;$$

when $\sigma < \sigma_s$ the medium does not deform.

This model expresses the fact that for many substances noticeable flow appears only for a definite load; the rate of flow depends on the viscosity of the medium. Many real substances are characterized by visco-plastic properties – metals at sufficiently high temperatures, various heavy lubricating materials, paints, etc. The perfection of many technological processes (hot machining of metals, movements of various plastic masses in machines, conduits, etc.) requires the examination of the motion of visco-plastic materials; the hydrodynamic theory of lubrication for dense lubricating materials is also based on the equations of visco-plastic flow.

The equations of a visco-plastic medium are discussed in greater detail in the next section.

*Series connection* (fig. 260b) of two elements – viscous and plastic – leads to a *creep-plastic medium*, which is of great interest in problems of creep.

When $\sigma < \sigma_s$ this medium behaves like a viscous fluid, following Newton's law of viscosity (81.2) or a nonlinear flow law, for example, equation (81.3).

When $\sigma = \sigma_s$ the medium flows like a perfectly plastic body (cf. §83).

Connection of a large number of viscous and plastic elements leads to a complex visco-plastic medium.

81.5. *Elastic-visco-plasticity*

If elements of the three types are joined we obtain a much more complex medium. Thus, the model shown in fig. 261 is sometimes used in dynamical problems. To obtain the corresponding equation, it is necessary, as usual, to write laws of deformation for each element, and to construct conditions of equilibrium and continuity of strain.

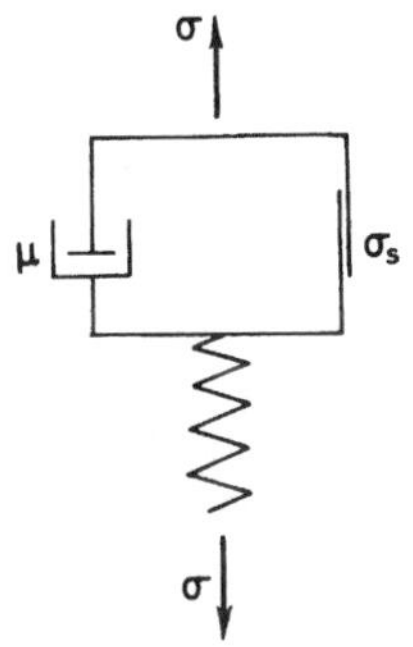


Fig. 261.

## §82. Visco-plastic medium

### 82.1. *Basic relations*

We consider in more detail the visco-plastic medium; by this we mean a medium, the model for which is described by *parallel* connection of viscous and plastic elements (fig. 257a). Visco-plastic media have been studied intensively in connection with a variety of practical applications.

The transition from equation (81.4) for a uniaxial stress state to the case of a complex stress state is achieved with the aid of the usual additional assumptions. First, we assume the *incompressibility condition*

$$\xi_{ij}\delta_{ij} = 0 . \tag{82.1}$$

Next, the stress components $\sigma_{ij}$ are combinations of the stress components $\sigma'_{ij}$, related to the plastic properties, and the stress components $\sigma''_{ij}$, generated by viscous resistance:

$$\sigma_{ij} = \sigma'_{ij} + \sigma''_{ij} . \tag{82.2}$$

The stress components $\sigma'_{ij}$ are determined by equations (13.12) of the von Mises plasticity theory, i.e.

$$s'_{ij} = \frac{2\tau_s}{H} \xi_{ij} , \tag{82.3}$$

where the von Mises yield criterion

$$s'_{ij}s'_{ij} = 2\tau_s^2 \tag{82.4}$$

is satisfied.

Let the viscous resistance obey Newton's linear viscosity law (5.5), i.e.

$$s''_{ij} = 2\mu'\xi_{ij} \qquad (3\mu' = \mu) . \tag{82.5}$$

Having constructed the stresses, we arrive at the relations of a visco-plastic medium

$$s_{ij} = 2\left(\frac{\tau_s}{H} + \mu'\right) \xi_{ij} . \tag{82.6}$$

From this we obtain the expression

$$T = \tau_s + \mu' H , \tag{82.7}$$

which is a generalization of equation (81.6). Note that the visco-plastic medium can be regarded as the limiting case of a nonlinear-viscous medium

$$s_{ij} = 2g(H)\,\xi_{ij}\,, \tag{82.8}$$

for which $T = g(H)\,H$. For a visco-plastic medium $g(H) = \mu + \tau_s/H$.

82.2. *Equations of visco-plastic flow*

The formulae (82.6), together with the incompressibility condition (82.1) and the three equations of motion, generate a system of ten equations for the ten unknown functions $s_{ij}$, $\sigma$, $v_i$.

Substituting the stress components from (82.6) into the differential equations of equilibrium, we obtain (together with the incompressibility equation (82.1)) a system of four equations for the mean pressure $\sigma$ and the components of velocity $v_i$; this system is of a complicated form and there is no point in writing it here.

In the case of *plane strain* $v_z = 0$, and therefore $\sigma_z = \sigma = \frac{1}{2}(\sigma_x + \sigma_y)$, $\tau_{xz} = \tau_{yz} = 0$. The differential equations of equilibrium can be satisfied by introducing the stress function $\Phi(x, y)$, namely

$$\sigma_x = 2\mu' \frac{\partial^2 \Phi}{\partial y^2}\,, \qquad \sigma_y = 2\mu' \frac{\partial^2 \Phi}{\partial x^2}\,, \qquad \tau_{xy} = -2\mu' \frac{\partial^2 \Phi}{\partial x \partial y}\,.$$

The incompressibility condition is satisfied by introducing the stream function $\Psi(x, y)$:

$$v_x = -\partial\Psi/\partial y\,, \qquad v_y = \partial\Psi/\partial x\,.$$

The functions $\Phi$ and $\Psi$ are determined from a system of two nonlinear, second order differential equations obtained from (82.7) and the relation

$$\sigma_x - \sigma_y = 4\left(\frac{\tau_s}{H} + \mu'\right)\xi_x\,, \tag{82.9}$$

which follows from (82.6).

Eliminating one of the functions, say $\Phi$, we can derive one nonlinear fourth order equation for the function $\Psi$. This equation was used by Il'yushin [112] and Ishlinskii [114] to analyze a series of problems of visco-plastic flow.

The equations of a visco-plastic medium are used in solving various problems of a technological type related to the processing of metals under pressure, to the flow of different plastic masses in tubes and slots, to the theory of lubrication by dense oils, etc. (cf. [80]). Equations (82.6), supplemented by elastic strains, are also used in problems of plastic dynamics when the influence of strain rates cannot be neglected. In hydrodynamic problems of visco-plastic flow much use has been made of boundary-layer theory.

As we have already remarked, a visco-plastic medium can be regarded as the limiting case of a nonlinear-viscous medium. This fact enables us to write down without difficulty variational equations for visco-plastic flow, namely, a principle of minimum total expansion (characterizing the minimal properties of the actual velocity field) and a principle of minimum additional expansion (characterizing the minimal properties of the actual stress distribution). A detailed analysis of the first variational problem is to be found in the recently published work by Molosov and Myasnikov [137].

### 82.3. *Steady flow in a tube*

We consider the problem of the flow of a visco-plastic mass in a thin circular tube. The motion is assumed to be slow, steady-state and axisymmetric; there is no rotation of the mass in the tube. Then in the cylindrical coordinate system $r, \varphi, z$ we have

$$v_r = 0\,, \qquad v_\varphi = 0\,.$$

Calculating the strain-rate components we find

$$\xi_r = \xi_\varphi = \eta_{r\varphi} = \eta_{\varphi z} = 0\,, \qquad \xi_z = \partial v_z/\partial z\,, \qquad \eta_{rz} = \partial v_z/\partial r\,.$$

But by the incompressibility condition $\xi_z = 0$, and hence $v_z = v_z(r)$. From (82.6) we obtain

$$\sigma_r = \sigma_\varphi = \sigma_z = \sigma\,; \qquad \tau_{r\varphi} = \tau_{\varphi z} = 0\,; \qquad \tau_{rz} = -\tau_s + \mu' \,\mathrm{d}v_z/\mathrm{d}r\,.$$

It is assumed here that $\mathrm{d}v_z/\mathrm{d}r \leqslant 0$, so that $\eta_{rz}/H = -1$.

Next, we have from the equilibrium equations (4.3)

$$\frac{\partial \sigma}{\partial r} = 0\,, \qquad \frac{\mathrm{d}\tau_{rz}}{\mathrm{d}r} + \frac{\tau_{rz}}{r} + \frac{\partial \sigma}{\partial z} = 0\,.$$

Hence it follows that $\sigma = \sigma(r)$, where the pressure gradient $\partial\sigma/\partial z = q$ is a constant. Substituting for $\tau_{rz}$ in the last equation and integrating, with the conditions that the visco-plastic mass adheres to the wall of the tube (i.e. $v_z = 0$ for $r = b$) and that the velocity $v_z$ is bounded, we obtain

$$v_z = \frac{q}{4\mu'}(b^2 - r^2) - \frac{\tau_s}{\mu'}(b - r)\,.$$

Since $|\tau_{rz}| \geqslant \tau_s$ inside the deformation zone, the solution is meaningful only if $\mathrm{d}v_z/\mathrm{d}r \leqslant 0$, i.e. when

$$r \geqslant 2\tau_s/q \equiv c\,.$$

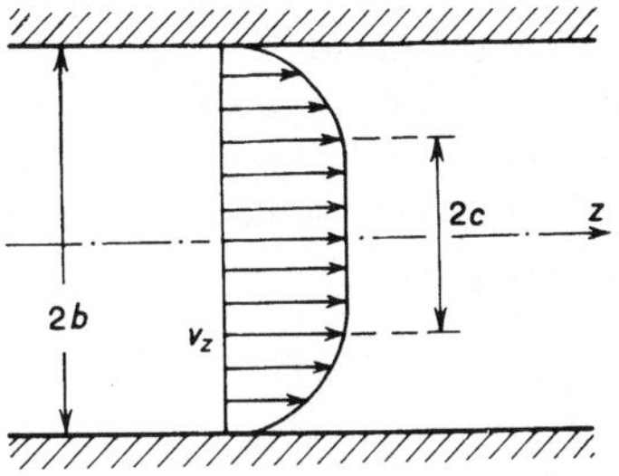


Fig. 262.

The remaining part of the mass does not deform and moves inside the tube like a rigid body. The velocity increases according to the parabolic law from a zero value at the wall to its maximum at $r = c$ (fig. 262). At the wall of the tube the magnitude of the tangential stress is $\frac{1}{2}bq$ and then falls off to the value $\tau_s$ at the boundary of the non-deforming core. Since $c \leqslant b$ this flow occurs only for a sufficiently large pressure gradient:

$$q \geqslant 2\tau_s/b \, .$$

The flux per unit time of the mass is

$$Q = 2\pi \int_0^b r v_z \mathrm{d}r = \frac{\pi}{\mu'} \left[\tfrac{1}{8}q\,(b^4 - c^4) - \tfrac{1}{3}\tau_s\,(b^3 - c^3)\right] \, .$$

In the case of a viscous fluid $\tau_s = 0$, hence $c = 0$, there is no rigid core, and the last equation reduces to the well-known formula of Poiseuille:

$$Q = \frac{\pi b^4}{8\mu'} q \, .$$

In the other limiting case – the plastic-rigid body – $\mu' = 0, c = b$, and slip occurs in a thin layer at the wall of the tube.

## §83. Creep-plastic medium

### 83.1. *Basic relations*

We now turn to a more detailed consideration of a creep-plastic medium, the model of which is generated by *series connection* of viscous and plastic elements (fig. 260b). This medium is of great interest in the theory of creep of metals where, incidentally, it is also frequently necessary to take into account elastic deformation and the influence of "hardening". Here we consider a simple version of the basic relations, which involve only nonlinear viscosity and perfect plasticity.

For the model shown in fig. 260b the strain rates are compounded, but the stress is the same in both elements.

Thus we have

$$\xi_{ij} = \xi'_{ij} + \xi''_{ij} . \tag{83.1}$$

The plastic strain rates are

$$\xi'_{ij} = \begin{cases} 0 & \text{when} \quad T < \tau_s , \\ \lambda' s_{ij} & \text{when} \quad T = \tau_s . \end{cases} \tag{83.2}$$

The viscous strain rates are

$$\xi''_{ij} = \tfrac{1}{2}\bar{g}(T)\, s_{ij} , \tag{83.3}$$

where $\bar{g}(T)$ is a known function. Note that

$$T = g(H'')\, H'' , \tag{83.4}$$

where $g(H'')\,\bar{g}(T) = 1$. These are the basic relations "according to the von Mises criterion". It is not difficult to formulate analogous relations corresponding to other criteria (such as the $\tau_{\max}$ criterion).

### 83.2. *Equations of creep-plastic flow*

If the stresses do not reach the yield limit, the body experiences only creep in accordance with relations such as (81.3), with the total rates of strain being now $\xi_{ij} = \xi''_{ij}$. For a sufficiently large load in the body "viscous" ($T < \tau_s$) and "plastic" ($T = \tau_s$) regions occur. In the former regions the strain-rate components $\xi_{ij}$ and the stress components are related by equations (83.3); in the latter by the formulae

$$\xi_{ij} = [\lambda' + \tfrac{1}{2}\bar{g}(\tau_s)]\; s_{ij} .$$

On the boundaries between different zones we must have suitable conditions of continuity of stresses and velocities. So long as the limit state is not reached the flow of the whole body is determined by the deformations of the "viscous core".

### 83.3. *Flow of a hollow sphere under pressure*

We consider the problem of steady flow of a hollow, creep-plastic sphere which experiences an internal pressure $p$ (fig. 41). We retain here the notation used in §25. Let $v = v(r)$ be the radial velocity. The incompressibility equation has the form

$$\frac{dv}{dr} + 2\frac{v}{r} = 0 ,$$

from which we find

$$v = C/r^2 ,$$

where $C$ is an arbitrary constant. Calculation of the strain rates and of the intensity of the shear strain-rates now gives

$$\xi_r = -\frac{2C}{r^3} , \qquad \xi_\varphi = \xi_\chi = \frac{C}{r^3} , \qquad H = 2\sqrt{3}\,\frac{C}{r^3} .$$

The intensity of the tangential stresses is $T = (\sigma_\varphi - \sigma_r)/\sqrt{3}$. The differential equation of equilibrium (25.1) now takes the form

$$\frac{d\sigma_r}{dr} = \frac{2\sqrt{3}}{r} T . \tag{83.5}$$

Let $T < \tau_s$. We take the power-law relation $T = \bar{B}H^{\bar{\mu}}$, where $\bar{B}$, $0 < \bar{\mu} \leqslant 1$ are respectively the coefficient and exponent of creep. Substituting for $T$ in equation (83.5) and integrating from $r = a$ (where $\sigma_r = -p$) to $r = b$ (where $\sigma_r = 0$), we obtain

$$\sigma_r = s\left[1 - \left(\frac{b}{r}\right)^{3\bar{\mu}}\right] , \qquad T = \frac{\sqrt{3}}{2}\bar{\mu} s \left(\frac{b}{r}\right)^{3\bar{\mu}} , \qquad s = \frac{p}{\beta^{3\bar{\mu}} - 1} . \tag{83.6}$$

It is evident that the intensity of tangential stresses decreases with increasing radius $r$. For sufficient pressure (namely, when $p > \{2\tau_s/\sqrt{(3\mu)}\} \times (1 - \beta^{-3\bar{\mu}})$), a plastic zone $a \leqslant r \leqslant c$ occurs in which $T = \tau_s$. In this zone the solution is determined by formulae (25.8); we rewrite then as

$$\sigma_r = 2\sigma_s \ln (r/a) - p , \qquad \sigma_\varphi = \sigma_r + \sigma_s . \tag{83.7}$$

In the outer region $c \leqslant r \leqslant b$ the solution can be obtained from formulae (83.6), with $p$ replaced by $-q = -(\sigma_r)_{r=c}$, and the radius $a$ by the radius $c$. On the boundary $r = c$ the quantities $\sigma_r$ and $T$ are continuous. It is easy to see that $q = 2\sigma_s \ln (c/a) - p$. From the condition $T = \tau_s$ at $r = c$ we can now find the equation which determines the radius $c$:

$$\frac{1}{3\bar{\mu}}\left[1 - \left(\frac{c}{b}\right)^{3\bar{\mu}}\right] + \ln\frac{c}{b} = \frac{p}{2\sigma_s} - \ln\frac{b}{a} . \tag{83.8}$$

The velocity $v(r)$ can be calculated with a suitable value of the arbitrary constant $C$.

When $c = b$ the plastic zone occupies the whole sphere; then the value of the limiting load $p_*$ is obtained from (83.8).

## PROBLEMS

1. Beginning with equation (81.4) construct the differential equation for longitudinal vibrations of a visco-elastic rod

$$\frac{\partial^2 u}{\partial t^2} - a^2 \frac{\partial^2 u}{\partial x^2} - b^2 \frac{\partial^3 u}{\partial x^2 \partial t} = 0\,,$$

where $u = u(x, t)$ is displacement along the axis of the rod, $a = \sqrt{E/\rho}$, $b = \sqrt{\mu/\rho}$.

2. Beginning with the equation of steady creep (81.3) and the usual hypothesis, derive the differential equation for the rate of bending of a beam

$$\frac{d^2 v}{dz^2} = \pm \frac{|M|^m}{D}\,,$$

where $M$ is the bending moment and $D$ the rigidity of the beam in the presence of creep.

3. Obtain the deformation law

$$E_1 \epsilon + \mu_1 \frac{d\epsilon}{dt} = \left(1 + \frac{E_1}{E_2}\right) \sigma + \frac{\mu_1}{E_2} \frac{d\sigma}{dt}$$

for the three-element medium shown in fig. 259a.

4. Obtain the deformation law

$$\mu_1 \frac{d^2\epsilon}{dt^2} + E_1 \frac{d\epsilon}{dt} = \frac{\mu_1}{E_2} \frac{d^2\sigma}{dt^2} + \left(1 + \frac{\mu_1}{\mu_2} + \frac{E_1}{E_2}\right) \frac{d\sigma}{dt} + \frac{E_1}{\mu_2}\, \sigma$$

for the four-element medium shown in fig. 259b.

5. Solve the problem of steady visco-plastic flow between parallel, rough plates in the case of plane strain. Show that the thickness of the rigid core is $2c = 2\tau_s/q$.

6. Derive the formula for the velocity $v(r)$ in the problem of flow of a hollow sphere (§83) when a plastic zone is present.

7. Solve the problem of the flow of a hollow creep-plastic tube under the action of an internal pressure.

# Appendix

## 1. The type of a system of partial differential equations

In plasticity, gas dynamics and other branches of mechanics one encounters systems of two quasi-linear, first order, partial differential equations for two functions $u$, $v$ of two independent variables $x$, $y$:

$$A_1 \frac{\partial u}{\partial x} + B_1 \frac{\partial u}{\partial y} + C_1 \frac{\partial v}{\partial x} + D_1 \frac{\partial v}{\partial y} = E_1 ,$$

$$A_2 \frac{\partial u}{\partial x} + B_2 \frac{\partial u}{\partial y} + C_2 \frac{\partial v}{\partial x} + D_2 \frac{\partial v}{\partial y} = E_2 , \tag{1}$$

where the coefficients $A_1, B_1, \ldots, E_2$ are given functions of $x, y, u, v$. The properties of the solutions of this system and the methods of determining them are essentially decided by the type of the system. We consider this question.

Let the values of $u$, $v$ be prescribed along some curve $x = x(s), y = y(s)$ in the $x, y$-plane:

$$u = u(s) , \qquad v = v(s) .$$

If we think in terms of the four-dimensional space $x$, $y$, $u$, $v$, then the equations $x = x(s)$, $y = y(s)$, $u = u(s)$, $v = v(s)$ represent some curve L in it; the solutions of the differential equations, $u = u(x, y)$, $v = v(x, y)$, generate some surface (integral surface).

A basic question is the possibility of making a definite integral surface pass through the given curve L (Cauchy's problem). The question is connected with the possibility of determining uniquely along L the derivatives of the unknown functions $u$, $v$ from the differential equations (1) themselves. In geometrical language, the unique determination along L of the first partial derivatives from the differential equations means the determination along L of the tangent plane to the integral surface.

Along L the differential equations (1) have known coefficients and can be used to determine the partial derivatives. Since $u$, $v$ are known along L, we obviously have the additional relations

$$\mathrm{d}u = \frac{\partial u}{\partial x}\mathrm{d}x + \frac{\partial u}{\partial y}\mathrm{d}y\,, \qquad \mathrm{d}v = \frac{\partial v}{\partial x}\mathrm{d}x + \frac{\partial v}{\partial y}\mathrm{d}y\,. \tag{2}$$

Equations (1), (2) generate along L a system of inhomogeneous, linear, algebraic equations for the first partial derivatives. These partial derivatives are determined in a non-unique way if the determinant $\Delta$ of the system and the relevant numerators $\Delta_1, \Delta_2, \Delta_3, \Delta_4$ in Cramer's formulae vanish along L. The first condition can be written

$$\Delta = \begin{vmatrix} A_1 & B_1 & C_1 & D_1 \\ A_2 & B_2 & C_2 & D_2 \\ \mathrm{d}x & \mathrm{d}y & 0 & 0 \\ 0 & 0 & \mathrm{d}x & \mathrm{d}y \end{vmatrix} = 0$$

or, in expanded form,

$$a\left(\frac{\mathrm{d}y}{\mathrm{d}x}\right)^2 - 2b\frac{\mathrm{d}y}{\mathrm{d}x} + c = 0\,, \tag{3}$$

where we have introduced the notation

$$a = C_1A_2 - A_1C_2\,,$$

$$2b = C_1B_2 - B_1C_2 + D_1A_2 - A_1D_2\,,$$

$$c = D_1B_2 - B_1D_2\,.$$

The differential equation (3) can be divided into two equations:

$$\frac{dy}{dx} = \frac{1}{a}(b \pm \sqrt{b^2 - ac}) . \tag{4}$$

If in some $x$, $y$-domain $b^2-ac>0$, then at each point of this domain we have two distinct *characteristic directions*; in this domain the system is of *hyperbolic type.*

If $b^2-ac=0$ in some $x$, $y$-domain, there is only one characteristic direction at each point of the domain; in this region the system (1) is of *parabolic type.*

Finally, if $b^2-ac<0$, no real characteristics exist in the given domain, and the system (1) is of *elliptic type.*

Since the coefficients of equation (1) are functions of $x$, $y$, $u$, $v$, the system (1) can in general be of different type in different regions.

The solution of a system of equations of hyperbolic type is closely connected with the characteristic curves determined by equations (4) and which cover the $x$, $y$-region with a curvilinear network.

Note that if the system is linear, i.e. if the coefficients of equation (1) are functions of $x$, $y$ only, the network of characteristic curves is independent of the solution. With nonlinear equations the characteristic curves depend on the required solution.

When the numerators are set equal to zero, $\Delta_1 = \Delta_2 = \Delta_3 = \Delta_4 = 0$, we obtain the relations between the unknown functions along the characteristics.

The theory of hyperbolic differential equations (1) is examined in the book by Courant and Friedrichs [19], ch. 2. Questions in the theory of hyperbolic differential equations with two independent variables are further considered in Courant and Hilbert, "Methods of mathematical physics", Vol. 2, ch. 5, in Smirnov, "A course of higher mathematics", Vol. 4, ch. 3, and in Rozhdestvenskii and Yanenko, "Systems of quasilinear equations".

## 2. Reducible equations

In many branches of mechanics of deformable media (plane strain and plane stress in plasticity, certain dynamical problems in plasticity theory, etc.), we encounter systems of *homogeneous* equations:

$$A_1 \frac{\partial u}{\partial x} + B_1 \frac{\partial u}{\partial y} + C_1 \frac{\partial v}{\partial x} + D_1 \frac{\partial v}{\partial y} = 0 , \tag{1a}$$

$$A_2 \frac{\partial u}{\partial x} + B_2 \frac{\partial u}{\partial y} + C_2 \frac{\partial v}{\partial x} + D_2 \frac{\partial v}{\partial y} = 0 , \tag{1b}$$

where the coefficients $A_1, B_1, \ldots, D_2$ are functions of $u$, $v$ only.

In this case the system (1) is called reducible, since a transformation of variables leads to a linear system. Let

$$x = x(u, v) , \qquad y = y(u, v) .$$

Differentiating we obtain

$$1 = \frac{\partial x}{\partial u}\frac{\partial u}{\partial x} + \frac{\partial x}{\partial v}\frac{\partial v}{\partial x} , \qquad 0 = \frac{\partial y}{\partial u}\frac{\partial u}{\partial x} + \frac{\partial y}{\partial v}\frac{\partial v}{\partial x} ,$$

$$0 = \frac{\partial x}{\partial u}\frac{\partial u}{\partial y} + \frac{\partial x}{\partial v}\frac{\partial v}{\partial y} ; \qquad 1 = \frac{\partial y}{\partial u}\frac{\partial u}{\partial y} + \frac{\partial y}{\partial v}\frac{\partial v}{\partial y} .$$

Hence we find the partial derivatives

$$\frac{\partial x}{\partial u} = \frac{1}{\Delta}\frac{\partial v}{\partial y} , \qquad \frac{\partial x}{\partial v} = -\frac{1}{\Delta}\frac{\partial u}{\partial y} ; \qquad \frac{\partial y}{\partial u} = -\frac{1}{\Delta}\frac{\partial v}{\partial x} , \qquad \frac{\partial y}{\partial v} = \frac{1}{\Delta}\frac{\partial u}{\partial x} ,$$

where $\Delta$ is the functional determinant

$$\Delta = \frac{D(u, v)}{D(x, y)} = \frac{\partial u}{\partial x}\frac{\partial v}{\partial y} - \frac{\partial u}{\partial y}\frac{\partial v}{\partial x} \neq 0 .$$

Substituting the partial derivatives in equation (1) and dividing by $\Delta$, we arrive at the *linear* system

$$A_1 \frac{\partial y}{\partial v} - B_1 \frac{\partial x}{\partial v} - C_1 \frac{\partial y}{\partial u} + D_1 \frac{\partial x}{\partial u} = 0 ,$$

$$A_2 \frac{\partial y}{\partial v} - B_2 \frac{\partial x}{\partial v} - C_2 \frac{\partial y}{\partial u} + D_2 \frac{\partial x}{\partial u} = 0 . \tag{2}$$

In general the system (2) is not equivalent to the system (1) since in the process of transformation certain solutions are lost which correspond to the vanishing of the determinant $\Delta$. These solutions, however, are particularly simple and can be obtained directly.

These simple solutions (simple waves, §77, simple stress states, §33, etc.) play an important role in applications.

A further treatment of reducible equations is given in the book by Courant and Friedrichs [19], and in the articles by Khristianovich [167] and Mikhlin [56].

# Bibliography[1]

## A. Books

[1] Alfrey, T., Mechanical behaviour of high polymers (Interscience, 1948).
[2] Arutyunyan, N.H., Some problems in the theory of creep (Pergamon, 1966).
[3] Bezukhov, N.I., Foundations of the theory of elasticity, plasticity and creep (Vys. Shkola, 1961) (in Russian).
[4] Birger, I.A., Circular plates and shells of revolution (Oborongiz, 1961) (in Russian).
[5] Boley, B.A. and Weiner, J.H., Theory of thermal stresses (Wiley, 1960).
[6] Bridgman, P.W., Studies in large plastic flow and fracture (McGraw Hill, 1952).
[6a] Volmir, A.S., Stability of elastic systems (Fizmatgiz, 1963) (in Russian).
[7] Gvozdev, A.A., Calculation of load-bearing capacity of structures by the method of limiting equilibrium (Stroyisdat, 1949) (in Russian).

[1]) The literature on plasticity theory is enormous; apart from the references cited in the text, this bibliography lists only a few works on the applications of the theory, and some survey articles. Additional references can be found in the books and articles mentioned, as well as those journals which are concerned with plasticity theory: Prikl. Matem. Mekhan. (PMM), Mekh. Tverd. Tela (MTT); Zh. Prikl. Mekhan. Tekh. Fizika (PMTF); Prikladnaya Mekhanika: Journal of Applied Mechanics (JAM); J. Mech. Phys. Solids (JMPS); Ing. Arch. Also see Mekhanika and Applied Mechanics Reviews.

[8] Golushkevich, S.S., The plane problem in the theory of limit equilibrium (Gostekhizdat, 1948) (in Russian).
[9] Goldenblat, I.I., Some questions in the mechanics of deformable bodies (G.I.T.T.L., 1955) (in Russian).
[10] Hoffman, O. and Sachs, G., Introduction to the theory of plasticity for engineers (McGraw Hill, 1953).
[11] Ivlev, D.D., Theory of perfect plasticity (Nauka, 1966) (in Russian).
[12] Il'yushin, A.A., Plasticity (Gostekhizdat, 1948) (in Russian).
[13] Il'yushin, A.A., Plasticity (Akad. Nauk USSR, 1963) (in Russian).
[14] Il'yushin, A.A. and Lenskii, V.S., Strength of materials (Fizmatgiz, 1959) (in Russian).
[15] Il'yushin, A.A. and Ogibalov, P.M., Elastic-plastic deformations of hollow cylinders (Moscow Univ., 1960) (in Russian).
[16] Kachanov, L.M., Mechanics of plastic media (Gostekhizdat, 1948) (in Russian).
[17] Kachanov, L.M., Theory of creep (Fizmatgiz, 1960) (in Russian).
[18] Kolsky, N., Stress waves in solids (Oxford, 1953).
[19] Courant, R. and Friedrichs, K., Supersonic flow and shock waves (Interscience, 1948).
[20] Leibenzon, L.S., Course in the theory of elasticity (Gostekhizdat, 1947) (in Russian).
[21] Love, A., A treatise on the mathematical theory of elasticity (Cambridge, 1927).
[22] Malmeister, A.K., Elasticity and anelasticity of concrete (Akad. Nauk. Latv. SSR, 1957) (in Russian).
[23] Mikhlin, S.G., Basic equations of the mathematical theory of plasticity (Akad. Nauk. USSR, 1934) (in Russian).
[24] Moskvitin, V.V., Plasticity with variable loads (Moscow Univ., 1965) (in Russian).
[25] Nadai, A., Theory of flow and fracture of solids, Vol. 1 (McGraw Hill, 1950).
[26] Neal, B.G., The plastic methods of structural analysis (Wiley, 1956).
[27] Novozhilov, V.V., Theory of elasticity (Sudpromgiz, 1958) (in Russian), see english translation.
[27a] Panovko, Ya.G. and Gubanova, I.I., Stability and vibrations of elastic systems (Nauka, 1967) (in Russian).
[28] Ponomarev, S.D. et al., Strength calculations in machine construction, Vols. I, II, III (Mashgiz, 1956) (in Russian).
[29] Prager, W., Introduction to plasticity (Addison Wesley, 1955).
[30] Prager, W., Introduction to mechanics of continua (Ginn, 1961).
[31] Prager, W. and Hodge, P.G., Theory of perfectly plastic solids (Wiley, 1951).
[32] Rabotnov, Yu.N., Strength of materials (Fizmatgiz, 1962) (in Russian).
[33] Rabotnov, Yu.N., Creep problems in structural members (North-Holland, 1969).
[34] Rakhmatulin, H.A. and Dem'yanov, Yu.A., Strength under intensive momentary loads (Fizmatgiz, 1961) (in Russian).
[35] Reiner, M., Deformation, strain and flow (Lewis, 1969).
[36] Rzhanitsin, A.R., Calculation of structures (Stroyzdat, 1961) (in Russian).
[37] Ruppeneit, K.V., Some problems in rock mechanics (Ugletekhizdat, 1954) (in Russian).
[38] Residual stresses in metals and metal construction, ed. W. Osgood (Reinhold, 1954)
[39] Theory of plasticity (I.L., 1948) (collected articles, in Russian).

[40] Advances in applied mechanics, eds. R. Mises and Th. von Karman, Vols. I, II, III.
[41] Rheology, ed. F. Eirich, Vol. I (Acad. Press, 1956).
[42] Sedov, L.I., Foundations of nonlinear mechanics of continua (Pergamon, 1966).
[43] Smirnov-Alyaev, G.A., Strength of materials with plastic deformations (Mashgiz, 1949) (in Russian).
[44] Sokolovskii, V.V., Theory of plasticity (Gostekhizdat, 1950) (in Russian).
[45] Sokolovskii, V.V., Statics of brittle media (Gostekhizdat, 1954) (in Russian),
[46] Tarnovskii, I.Y.A. et al., Theory of metal processing under pressure (Metallurgizdat, 1963) (in Russian).
[47] Timoshenko, S.P., Strength of materials (Van Nostrand, 1955).
[48] Timoshenko, S.P., Theory of elasticity (McGraw Hill, 1951).
[49] Timoshenko, S.P., Theory of elastic stability (McGraw Hill, 1961).
[50] Thomas, T.Y., Plastic flow and fracture in solids (Academic, 1961).
[51] Tomlenov, A.D., Theory of plastic deformation of metals (Mashgiz, 1951) (in Russian).
[52] Freudenthal, A.M. and Geiringer, H., Mathematical theories of the inelastic continuum, Handbuch der Physik, Vol. 6 (Springer, 1958).
[53] Fridman, Ya.B., Mechanical properties of metals (Oborongiz, 1952) (in Russian).
[54] Hill, R., Mathematical theory of plasticity (Oxford, 1950).
[55] Hodge, P.G., Plastic analysis of structures (McGraw Hill, 1959).
[56] Khristianovich, S.A., Mikhlin, S.G. and Devision, B.B., Some problems in the mechanics of continuous media (Akad. Nauk USSR, 1938) (in Russian).
[57] Ziegler, H.. Some extremum principles in irreversible thermodynamics with applications to continuum mechanics, in: Sneddon and Hill (eds.), Progress in Solid Mechanics, Vol. 4 (North-Holland, 1963).
[58] Nadai, A., Theory of flow and fracture in solids, Vol. II (McGraw Hill, 1963).
[59] Perzyna, P., Theory of visco-plasticity (Warsaw, 1966) (in Polish).
[60] Southwell, R., Relaxation methods (Oxford, 1946).
[61] Thomsen, E., Yang, C. and Kobayashi, S., Mechanics of plastic deformation in metal processing (MacMillan, 1965).

## B. Surveys

[62] Bland, D.R., Theory of linear viscoelasticity (Pergamon, 1960).
[63] Hopkins, H.G., Dynamic anelastic deformations of metals, Appl. Mech. Rev. 14 (1961) 417.
[64] Davies, R., Stress waves in solids, Appl. Mech. Rev. 6 (1953) 1.
[65] Zvolinskii, N.V., Malyshev, B.B. and Shapiro, G.S., Dynamics of plastic media, Proc. 2nd All-Union Congress on Mechanics, 1966 (in Russian).
[66] Kachanov, L.M., Variational methods in plasticity, Proc. 2nd All-Union Congress on Mechanics, 1966 (in Russian).
[67] Klyushnikov, V.D., Plasticity laws for materials with hardening, PMM 22 (1958) 1 (in Russian).
[68] Koiter, W., General theorems for elastic-plastic solids, in: Sneddon and Hill (eds.), Progress in Solid Mechanics, Vol. 1 (North-Holland, 1960).
[69] Naghdi, P.M., Stress-strain relations and thermoplasticity, in: Lee and Symonds (eds.), Plasticity (Pergamon, 1970).

[70] Olszak, W., Mruz, Z. and Perzyna, P., Recent trends in the development of plasticity (Pergamon, 1963).
[71] Olszak, W., Ryklewskii, Ya. and Ubranowskii, W., Plasticity under non-homogeneous conditions, Advan. Appl. Mech. 7 (1962).
[72] Prager, W., The theory of plasticity – a survey of recent achievements, Proc. Inst. Mech. Eng. 169 (1055) 41.
[73] Reitman, M.I. and Shapiro, G.S., Theory of optimal projection in structural mechanics, elasticity and plasticity (Itogi Nauki. Izd. Viniti, 1966) (in Russian).
[74] Hodge, P.G., Boundary value problems, in: Lee and Symonds (eds.), Plasticity (Pergamon, 1960).
[75] Horne, M.R., Stability of elastic-plastic structures, in: Sneddon and Hill (eds.), Progress in Solid Mechanics, Vol. 2 (North-Holland, 1961).
[76] Hoff, N., Longitudinal bending and stability, Res. Struct. Supp. 121 (1949).
[77] Goodier, J.N. and Hidge, P.G., Elasticity and plasticity (Wiley, 1958).
[78] Shapiro, G.S., The behaviour of plates and shells beyond the elastic limit, Proc. 2nd. All-Union Congress on Mechanics, 1966 (in Russian).
[79] Hundy, B., Plane plasticity, Metallurgia (1954).
[80] Perzyna, P., Fundamental problems in viscoplasticity, Adv. Appl. Mech. 9 (1966).
[81] Sawczuk, A. and Olszak, W., Problems of inelastic shells. Theory of plates and shells (Bratislava, 1966).
[82] Szczepinski, W., Load-bearing capacity of rods with notches, Mechan. teor. stosow. Warsaw, 1965 (in Polish).

## C. Articles

[83] Agamirzyan, L.S., Inzh. Zhurn. 1 (1961) 4; 2 (1962) 2.
[84] Arutyanyan, R.A. and Vakulenko, A.A., Izv. Akad. Nauk USSR, Mekh. 4 (1965).
[85] Batdorf, S.B. and Budiansky, B., NACA TN 1871 (1949).
[86] Birger, I.A., PMM, 15 (1951) 6.
[87] Birger, I.A., Izv. Akad. Nauk USSR, Mekh. 3 (1964).
[88] Budiansky, B., J. Appl. Mech. 26 (1959) 259.
[89] Vakulenko, A.A., Isled. uprug. plast. Leningrad Univ. No. 1, 1961; No. 2, 1962.
[90] Vakulenko, A.A. and Palley, I.Z., Isled. uprug. plast. Leningrad Univ. No. 5, 1966.
[91] Galin, L.A. PMM 10 (1946) 5–6.
[92] Galin, L.A., PMM 13 (1949) 3.
[93] Galin, L.A., PMM 8 (1944) 4.
[94] Hencky, H., Z. Angew. Math. Mekh. 3 (1923) 241.
[95] Hencky, H., Izv. Akad. Nauk USSR, otd. tekh. nauk. 2 (1937).
[96] Hencky, H., Z. Angew. Math. Mekh. 4 (1924) 323.
[97] Hencky, H., Proc. 1st Intern. Congr. Appl. Mech. (Delft) 1924.
[98] Hopkins, H.G. and Prager, W., JMPS 2 (1953) 1.
[99] Hopkins, H.G. and Prager, W., Z. Angew. Math. Phys. 5 (1954) 317.
[100] Hochfeld, D.A., Prikl. Mekh. 4 (1967).
[101] Grigoriev, A.S., PMM 16 (1952) 1.
[102] Green, A.P., JMPS 2 (1954) 197.
[103] Green, A.P., JMPS 3 (1954) 1.

[104] Green, A.P., Quart. J. Mech. Appl. Math. 6 (1953) 223.
[105] Davidenkov, N.N. and Spiridonova, N.I., Zavodsk. Lab. 11 (1945) 6.
[106] Drucker, D., Proc. Symp. Appl. Math. 8 (1958) 7.
[107] Drucker, D., JAM 21 (1954) 71.
[108] Drucker, D., Prager, W. and Greenberg, H., Quart. Appl. Math. 9 (1952) 381.
[109] Druyanov, B.A., PMTF 6 (1961).
[110] Ivanov, G.V., PMTF 4 (1960).
[111] Il'yushin, A.A., PMM 18 (1954) 3.
[112] Il'yushin, A.A., Uch. Zap. Mosk. Univ. 39 (1940).
[113] Ishlinskii, A.Yu., PMM 8 (1944) 3.
[114] Ishlinskii, A.Yu., Izv. Akad. Nauk USSR, otd. tekh. nauk. 3 (1945)
[115] Ishlinskii, A.Yu., Ukr. Math. Zh. 3 (1954).
[116] Von Karman, T. and Duwez, P.J., Appl. Phys. 21 (1950) 987.
[117] Kachanov, L.M., PMM 5 (1941) 3.
[118] Kachanov, L.M., PMM 6 (1942) 2–3.
[119] Kachanov, L.M., PMM 15 (1951) 2, 5, 6.
[120] Kachanov, L.M., Dokl. Akad. Nauk USSR 96 (1954) 2; Izv. Akad. Nauk USSR Mekh. 5 (1962).
[121] Kachanov, L.M., PMM 19 (1954) 3.
[122] Kachanov, L.M., Vest. Leningr. Univ. 19 (1959).
[123] Kachanov, L.M., PMM 23 (1959) 3.
[124] Koiter, W., Quart. Appl. Math. 11 (1953) 350.
[125] Klyushnikov, V.D., PMM 23 (1959) 4.
[126] Klyushnikov, V.D., Izv. Akad. Nauk USSR, Mekh. 6 (1964).
[127] Klyushnikov, V.D., MTT 4 (1966).
[128] Koopman, D. and Lance, R., JMPS 13 (1965) 77.
[129] Levy, M., Compt. Rend. Acad. Sci. Paris 70 (1870) 473.
[130] Levy, M., J. Math. Pures Appl. 16 (1871) 369.
[131] Lee, E.H. and Tupper, S.J., JAM 21 (1954) 63.
[132] Lippman, G., JMPS 10 (1962) 111.
[133] Lurye, A.I., Tr. Leningr. Politekn..Inst. 1 (1946).
[134] Markov, A.A., PMM 11 (1947) 3.
[135] Martin, D., JAM 32 (1965) 1.
[136] Von Mises, R., Gött. Nachricht, Math. – Phys. 582 (1913).
[137] Mosolov, P.P. and Myasnikov, V.P., PMM 29 (1965) 3.
[138] Nayar, E., Rykhlevskii, Ya. and Shapiro, G.S., Bull. Polsk. Akad. Nauk Ser. tekh. nauk. 14 (1966) 9.
[139] Novozhilov, V.V., PMM 15 (1951) 2.
[140] Novozhilov, V.V., PMM 27 (1963) 5.
[141] Novozhilov, V.V. and Kadashevich, Yu.I., PMM 2 (1958) 1.
[142] Novozhilov, V.V., PMM 15 (1951) 2.
[143] Pearson, C.J., Aero. Sci. 17 (1950) 417.
[144] Prandtl, L., Proc. 1st Intern. Congr. Appl. Math. (Delft) 1924, p. 43.
[145] Rabotnov, Yu. N., Inzh. Sbor. 11 (1952).
[146] Rakhamatulin, H.A., PMM 9 (1945) 1.
[147] Rzhanitsin, A.R., Izv. Akad. Nauk USSR, Mekh. 2 (1959).
[148] Rozenblum, V.I., Izv. Akad. Nauk USSR, otd. tekh. nauk. 6 (1958).
[149] Rozenblum, V.I., PMTF 5 (1965).

[150] Rozenblum, V.I., MTT 4 (1966).
[151] Rykhlevskii, Ya., Mekhanika 3 (1967).
[152] Symonds, P., JAM 20 (1953) 475.
[153] Symonds, P., Proc. 2nd U.S. Nat. Congr. Appl. Mech., 1954.
[154] Sedov, L.I., Vest. Akad. Nauk 7 (1960).
[155] Sedov, L.I., PMM 23 (1959) 2.
[156] Saint Venant, B. de, Compt. Rend 70 (1870) 473.
[157] Saint Venant, B. de, J. Math. Pures Appl. 16 (1871) 308.
[158] Sokolovskii, V.V., PMM 12 (1948) 3.
[159] Sokolovskii, V.V., Inzh. Zhurn. 1 (1961) 3.
[160] Tamuzh, V.P., PMM 26 (1962) 4.
[161] Feinberg, S.M., PMM 12 (1948) 1.
[162] Haar, A., von Karman, T., Gött. Nachricht. Math. Phys. 204 (1909).
[163] Hill, R., JMPS 4 (1956) 247; 5 (1956) 1; 5 (1957) 153.
[164] Hill, R. and Sewell, M.J., JMPS 8 (1960) 105.
[165] Hill, R., JMPS 5 (1956) 66.
[166] Hill, R., JMPS 10 (1962) 80.
[167] Khristianovich, S.A., Matem. Sbor. nov. ser. 1 (43) (1936) 4.
[168] Shapiro, G.S., PMM 10 (1946) 5–6.
[169] Shapiro, G.S., PMM 23 (1959) 1.
[170] Shapiro, G.S., Izv. Akad. Nauk USSR, Mekh. 4 (1963).
[171] Shanley, F., J. Aero Sci. 14 (1947) 261.
[172] Shield, R., Proc. Roy. Soc. A 233 (1955) 267.
[173] Cherepanov, G.P., PMM 26 (1962) 4.
[174] Cherepanov, G.P., PMM 27 (1963) 3.
[175] Yamanoto, I., Quart. Appl. Math. 10 (1952) 215.
[176] Allen, D. and Sopwith, D., Proc. Roy. Soc. A 205 (1951) 1080.
[177] Cox, A., Eason, H. and Hopkins, H., Phil. Trans. Roy. Soc. A 254 (1961) 1036.
[178] Edelman, F. and Drucker, D., J. Franklin Inst. 6 (1951).
[179] Voigt, W., Abh. Ges. Wiss. Göttingen, 1890.
[180] Ford, H. and Lianis, G., Z. Angew. Math. Phys. 8 (1957) 360.
[181] Greenberg, H., Quart. Appl. Math. 7 (1949) 85.
[182] Hill, R., JMPS 1 (1952) 1.
[183] Hodge, P.G., Prager Anniv. Vol., N.Y., 1963.
[184] Johnson, W. and Sowerby, R., Intern. J. Mech. Sci. 9 (1967) 7.
[185] Kobayashi, S. and Thomsen, E., Intern. J. Mech. Sci. 7 (1965) 2.
[186] Maxwell, J., Phil. Trans. Roy. Soc. A 157 (1867) 52.
[187] Meyerhof, A. and Chaplin, J., Brit. J. Appl. Phys. 4 (1953) 1.
[188] Neal, B., Phil. Trans. Roy. Soc. A 242 (1950) 846.
[189] Pflüger, A., Ing. Arch. 20 (1952) 5.
[190] Phillipps, A., Quart. Appl. Math. 7 (1949) 195.
[191] Prager, W., von Karman Mem. Volume, 1965.
[192] Taylor, G.I., J. Inst. Civil Eng. 2 (1946) 486.
[193] Tekinalp, B., JMPS 5 (1957) 2.
[194] Thomas, T., JMPS 1 (1953) 2.
[195] Ziegler, H., Ing. Arch. 20 (1952) 1.

## D. Updated references

[196] Body, D.B., Elastic-plastic plane-strain solution with separable stress field, J. Appl. Mech. 36 (1969) 528.

[197] Brooks, D.C., The elasto-plastic behaviour of a circular bar loaded by axial force and torque in the strain-hardening range, Intern. J. Mech. Sciences 11 (1969) 75.

[198] Brovman, M.Ya., Computation of plane plastic deformation for statically indeterminate problems, Mashinovedênie 5 (1967) 61.

[199] Bykovtsev, G.I., Ivlev, D.D. and Myasnyankin, Yu.M., Relations on surfaces of discontinuity of stress in three-dimensional perfectly rigid-plastic solids, Soviet Phys. Doklady 12 (1968) 1167.

[200] Chakrabarty, J. and Alexander, J.M., Plastic instability of thick-walled tubes with closed ends, Intern. J. Mech. Science 11 (1969) 175.

[201] Collins, I.F., An optimum loading criterion for rigid-plastic materials, J. Mech. Phys. Sol. 16 (1968) 73.

[202] Collins, I.F., Compression of a rigid perfectly plastic strip between parallel smooth rotating dies, Quart. J. Mech. Appl. Math. 23 (1970) 329.

[203] Devecpeck, M.L. and Weinstein, A.S., Experimental investigation of work hardening effects in wedge flattening with relation to nonhardening theory, J. Mech. Phys. Solids 18 (1970) 213.

[204] Dietrich, L. and Szczepinski, W., Plastic yielding of axially-symmetric bars with nonsymmetric V-notch, Acta Mechanica 4 (1967) 230.

[205] Dillon, O.W. and Kravotchil, J., A strain gradient theory of plasticity, Intern. J. Solids Structures 6 (1970) 1513.

[206] Drucker, D.C. and Chen, W.F., On the use of simple discontinuous fields to bound limit loads, in: J. Heymann and A. Leckie (eds.), Engineering Plasticity (Cambridge, 1968).

[207] Eisenberg, M.A. and Phillips, A., On nonlinear kinematic hardening, Acta Mechanica 5 (1968) 1.

[208] Ellyin F., The effect of yield surfaces on the limit pressure of intersecting shells, Intern. J. Solids Structures 5 (1969) 713.

[209] Ewing, D., Calculations on the bending of rigid/plastic notched bars, J. Mech. Phys. Solids 16 (1968) 205.

[210] Ivlev, D.D., On strain theories of plasticity for singular loading surfaces, J. Appl. Math. Mech. 31 (1967) 895.

[211] Khristianovich, S.A. and Shemyakin, E.I., Theory of ideal plasticity, Mekh. Tverd. Tela 4 (1967) 86.

[212] Lehmann, T., Seitz, H. and Thermann, K., Theory of Plasticity, V.D.I. Zeitschrift 109 (1967) 1580.

[213] Macherauch, E., The basic principles of plastic deformation, Z. für Metallkunde 61 (1970) 617.

[214] Maier, G., Complementary plastic-work theorems in piecewise-linear elastoplasticity, Intern. J. Solids Structures 5 (1969) 261.

[215] Maier, G., A minimum principle for incremental elastoplasticity with non-associated flow laws, J. Mech. Phys. Solids 18 (1970) 319.

[216] Palmer, A.C., Maier, G. and Drucker, D.C., Normality relations and convexity of yield surfaces for unstable materials or structural elements, J. Appl. Mech. 34 (1967) 464.

[217] Peronne, N., Impulsively-loaded strain hardened rate-sensitive rings and tubes, Intern. J. Solids Structures 6 (1970) 1119.

[218] Ramsey, H., On the stability of elastic-plastic and rigid-plastic plates of arbitrary thickness, Intern. J. Solids Structures 5 (1969) 921.

[219] Rice, J.R. and Rosengren, G.F., Plane strain deformation near a crack tip in a power-law hardening material, J. Mech. Phys. Solids 16 (1968) 1.

[220] Samanta, S.K., Slip-line field for extrusion through cosine-shaped dies, J. Mech. Phys. Solids 18 (1970) 311.

[221] Sawczuk, A., On yield criteria and collapse modes for plates, Intern. J. Nonlinear Mech. 2 (1967) 233.

[222] Sawczuk, A. and Hodge, P.G., Limit analysis and yield-line theory, J. Appl. Mech. 35 (1968) 357.

[223] Sayir, M. and Ziegler, H., Prandtl's indentation problem, Ing. Arch. 36 (1968) 294.

[224] Sukhikh, L.I., Elastic-plastic torsion of a bar with a longitudinal notch, Prikl. Mekh. 4 (1968) 123.

[225] Szczepinski, W. and Miastowski, J., Plastic straining of notched bars with intermediate thickness and small shoulder ratio, Intern. J. Nonlinear Mech. 3 (1968) 83.

[226] Thomas, T.Y., Collapse of thick hollow cylinders by external pressure, J. Math. Mech. 17 (1968) 987.

[227] Thomas, T.Y., Stress-strain relations for crystals containing plastic deformation, Proc. Nat. Acad. Sci. 60 (1968) 1102.

[228] Thomason, P.F., An analysis of necking in axisymmetric tension specimens, Intern. J. Mech. Sciences 11 (1969) 481.

[229] Vaisblat, M.B., Application of the variational principle to the solution of some problems in the theory of plasticity, Prikl. Mekh. 3 (1967) 125.

[230] Valanis, K.C., On the thermodynamic foundation of classical plasticity, Acta Mechanica 9 (1970) 278.

[231] Zadoyan, M.A., Plane and axisymmetric flow of plastic mass between rough moving cylinders, Izv. Akad. Nauk Armyansk. SSR. Mekh. 19 (1966) 22.

# Subject Index

A CATALOG OF SELECTED

# DOVER BOOKS

IN SCIENCE AND MATHEMATICS

# Astronomy

BURNHAM'S CELESTIAL HANDBOOK, Robert Burnham, Jr. Thorough guide to the stars beyond our solar system. Exhaustive treatment. Alphabetical by constellation: Andromeda to Cetus in Vol. 1; Chamaeleon to Orion in Vol. 2; and Pavo to Vulpecula in Vol. 3. Hundreds of illustrations. Index in Vol. 3. 2,000pp. 6⅛ x 9¼.
Vol. I: 23567-X
Vol. II: 23568-8
Vol. III: 23673-0

EXPLORING THE MOON THROUGH BINOCULARS AND SMALL TELESCOPES, Ernest H. Cherrington, Jr. Informative, profusely illustrated guide to locating and identifying craters, rills, seas, mountains, other lunar features. Newly revised and updated with special section of new photos. Over 100 photos and diagrams. 240pp. 8¼ x 11. 24491-1

THE EXTRATERRESTRIAL LIFE DEBATE, 1750–1900, Michael J. Crowe. First detailed, scholarly study in English of the many ideas that developed from 1750 to 1900 regarding the existence of intelligent extraterrestrial life. Examines ideas of Kant, Herschel, Voltaire, Percival Lowell, many other scientists and thinkers. 16 illustrations. 704pp. 5⅜ x 8½. 40675-X

THEORIES OF THE WORLD FROM ANTIQUITY TO THE COPERNICAN REVOLUTION, Michael J. Crowe. Newly revised edition of an accessible, enlightening book recreates the change from an earth-centered to a sun-centered conception of the solar system. 242pp. 5⅜ x 8½. 41444-2

A HISTORY OF ASTRONOMY, A. Pannekoek. Well-balanced, carefully reasoned study covers such topics as Ptolemaic theory, work of Copernicus, Kepler, Newton, Eddington's work on stars, much more. Illustrated. References. 521pp. 5⅜ x 8½.
65994-1

A COMPLETE MANUAL OF AMATEUR ASTRONOMY: Tools and Techniques for Astronomical Observations, P. Clay Sherrod with Thomas L. Koed. Concise, highly readable book discusses: selecting, setting up and maintaining a telescope; amateur studies of the sun; lunar topography and occultations; observations of Mars, Jupiter, Saturn, the minor planets and the stars; an introduction to photoelectric photometry; more. 1981 ed. 124 figures. 26 halftones. 37 tables. 335pp. 6½ x 9¼.
42820-6

AMATEUR ASTRONOMER'S HANDBOOK, J. B. Sidgwick. Timeless, comprehensive coverage of telescopes, mirrors, lenses, mountings, telescope drives, micrometers, spectroscopes, more. 189 illustrations. 576pp. 5⅝ x 8¼. (Available in U.S. only.)
24034-7

STARS AND RELATIVITY, Ya. B. Zel'dovich and I. D. Novikov. Vol. 1 of *Relativistic Astrophysics* by famed Russian scientists. General relativity, properties of matter under astrophysical conditions, stars, and stellar systems. Deep physical insights, clear presentation. 1971 edition. References. 544pp. 5⅝ x 8¼. 69424-0

# Chemistry

THE SCEPTICAL CHYMIST: The Classic 1661 Text, Robert Boyle. Boyle defines the term "element," asserting that all natural phenomena can be explained by the motion and organization of primary particles. 1911 ed. viii+232pp. 5⅜ x 8½. 42825-7

RADIOACTIVE SUBSTANCES, Marie Curie. Here is the celebrated scientist's doctoral thesis, the prelude to her receipt of the 1903 Nobel Prize. Curie discusses establishing atomic character of radioactivity found in compounds of uranium and thorium; extraction from pitchblende of polonium and radium; isolation of pure radium chloride; determination of atomic weight of radium; plus electric, photographic, luminous, heat, color effects of radioactivity. ii+94pp. 5⅜ x 8½. 42550-9

CHEMICAL MAGIC, Leonard A. Ford. Second Edition, Revised by E. Winston Grundmeier. Over 100 unusual stunts demonstrating cold fire, dust explosions, much more. Text explains scientific principles and stresses safety precautions. 128pp. 5⅜ x 8½. 67628-5

THE DEVELOPMENT OF MODERN CHEMISTRY, Aaron J. Ihde. Authoritative history of chemistry from ancient Greek theory to 20th-century innovation. Covers major chemists and their discoveries. 209 illustrations. 14 tables. Bibliographies. Indices. Appendices. 851pp. 5⅜ x 8½. 64235-6

CATALYSIS IN CHEMISTRY AND ENZYMOLOGY, William P. Jencks. Exceptionally clear coverage of mechanisms for catalysis, forces in aqueous solution, carbonyl- and acyl-group reactions, practical kinetics, more. 864pp. 5⅜ x 8½. 65460-5

ELEMENTS OF CHEMISTRY, Antoine Lavoisier. Monumental classic by founder of modern chemistry in remarkable reprint of rare 1790 Kerr translation. A must for every student of chemistry or the history of science. 539pp. 5⅜ x 8½. 64624-6

THE HISTORICAL BACKGROUND OF CHEMISTRY, Henry M. Leicester. Evolution of ideas, not individual biography. Concentrates on formulation of a coherent set of chemical laws. 260pp. 5⅜ x 8½. 61053-5

A SHORT HISTORY OF CHEMISTRY, J. R. Partington. Classic exposition explores origins of chemistry, alchemy, early medical chemistry, nature of atmosphere, theory of valency, laws and structure of atomic theory, much more. 428pp. 5⅜ x 8½. (Available in U.S. only.) 65977-1

GENERAL CHEMISTRY, Linus Pauling. Revised 3rd edition of classic first-year text by Nobel laureate. Atomic and molecular structure, quantum mechanics, statistical mechanics, thermodynamics correlated with descriptive chemistry. Problems. 992pp. 5⅜ x 8½. 65622-5

FROM ALCHEMY TO CHEMISTRY, John Read. Broad, humanistic treatment focuses on great figures of chemistry and ideas that revolutionized the science. 50 illustrations. 240pp. 5⅜ x 8½. 28690-8

# Engineering

DE RE METALLICA, Georgius Agricola. The famous Hoover translation of greatest treatise on technological chemistry, engineering, geology, mining of early modern times (1556). All 289 original woodcuts. 638pp. 6¾ x 11. 60006-8

FUNDAMENTALS OF ASTRODYNAMICS, Roger Bate et al. Modern approach developed by U.S. Air Force Academy. Designed as a first course. Problems, exercises. Numerous illustrations. 455pp. 5⅜ x 8½. 60061-0

DYNAMICS OF FLUIDS IN POROUS MEDIA, Jacob Bear. For advanced students of ground water hydrology, soil mechanics and physics, drainage and irrigation engineering, and more. 335 illustrations. Exercises, with answers. 784pp. 6⅛ x 9¼. 65675-6

THEORY OF VISCOELASTICITY (Second Edition), Richard M. Christensen. Complete, consistent description of the linear theory of the viscoelastic behavior of materials. Problem-solving techniques discussed. 1982 edition. 29 figures. xiv+364pp. 6⅛ x 9¼. 42880-X

MECHANICS, J. P. Den Hartog. A classic introductory text or refresher. Hundreds of applications and design problems illuminate fundamentals of trusses, loaded beams and cables, etc. 334 answered problems. 462pp. 5⅜ x 8½. 60754-2

MECHANICAL VIBRATIONS, J. P. Den Hartog. Classic textbook offers lucid explanations and illustrative models, applying theories of vibrations to a variety of practical industrial engineering problems. Numerous figures. 233 problems, solutions. Appendix. Index. Preface. 436pp. 5⅜ x 8½. 64785-4

STRENGTH OF MATERIALS, J. P. Den Hartog. Full, clear treatment of basic material (tension, torsion, bending, etc.) plus advanced material on engineering methods, applications. 350 answered problems. 323pp. 5⅜ x 8½. 60755-0

A HISTORY OF MECHANICS, René Dugas. Monumental study of mechanical principles from antiquity to quantum mechanics. Contributions of ancient Greeks, Galileo, Leonardo, Kepler, Lagrange, many others. 671pp. 5⅜ x 8½. 65632-2

STABILITY THEORY AND ITS APPLICATIONS TO STRUCTURAL MECHANICS, Clive L. Dym. Self-contained text focuses on Koiter postbuckling analyses, with mathematical notions of stability of motion. Basing minimum energy principles for static stability upon dynamic concepts of stability of motion, it develops asymptotic buckling and postbuckling analyses from potential energy considerations, with applications to columns, plates, and arches. 1974 ed. 208pp. 5⅜ x 8½. 42541-X

METAL FATIGUE, N. E. Frost, K. J. Marsh, and L. P. Pook. Definitive, clearly written, and well-illustrated volume addresses all aspects of the subject, from the historical development of understanding metal fatigue to vital concepts of the cyclic stress that causes a crack to grow. Includes 7 appendixes. 544pp. 5⅜ x 8½. 40927-9

ROCKETS, Robert Goddard. Two of the most significant publications in the history of rocketry and jet propulsion: "A Method of Reaching Extreme Altitudes" (1919) and "Liquid Propellant Rocket Development" (1936). 128pp. 5⅜ x 8½. 42537-1

STATISTICAL MECHANICS: Principles and Applications, Terrell L. Hill. Standard text covers fundamentals of statistical mechanics, applications to fluctuation theory, imperfect gases, distribution functions, more. 448pp. 5⅜ x 8½. 65390-0

ENGINEERING AND TECHNOLOGY 1650–1750: Illustrations and Texts from Original Sources, Martin Jensen. Highly readable text with more than 200 contemporary drawings and detailed engravings of engineering projects dealing with surveying, leveling, materials, hand tools, lifting equipment, transport and erection, piling, bailing, water supply, hydraulic engineering, and more. Among the specific projects outlined–transporting a 50-ton stone to the Louvre, erecting an obelisk, building timber locks, and dredging canals. 207pp. 8⅜ x 11¼. 42232-1

THE VARIATIONAL PRINCIPLES OF MECHANICS, Cornelius Lanczos. Graduate level coverage of calculus of variations, equations of motion, relativistic mechanics, more. First inexpensive paperbound edition of classic treatise. Index. Bibliography. 418pp. 5⅜ x 8½. 65067-7

PROTECTION OF ELECTRONIC CIRCUITS FROM OVERVOLTAGES, Ronald B. Standler. Five-part treatment presents practical rules and strategies for circuits designed to protect electronic systems from damage by transient overvoltages. 1989 ed. xxiv+434pp. 6⅛ x 9¼. 42552-5

ROTARY WING AERODYNAMICS, W. Z. Stepniewski. Clear, concise text covers aerodynamic phenomena of the rotor and offers guidelines for helicopter performance evaluation. Originally prepared for NASA. 537 figures. 640pp. 6⅛ x 9¼. 64647-5

INTRODUCTION TO SPACE DYNAMICS, William Tyrrell Thomson. Comprehensive, classic introduction to space-flight engineering for advanced undergraduate and graduate students. Includes vector algebra, kinematics, transformation of coordinates. Bibliography. Index. 352pp. 5⅜ x 8½. 65113-4

HISTORY OF STRENGTH OF MATERIALS, Stephen P. Timoshenko. Excellent historical survey of the strength of materials with many references to the theories of elasticity and structure. 245 figures. 452pp. 5⅜ x 8½. 61187-6

ANALYTICAL FRACTURE MECHANICS, David J. Unger. Self-contained text supplements standard fracture mechanics texts by focusing on analytical methods for determining crack-tip stress and strain fields. 336pp. 6⅛ x 9¼. 41737-9

STATISTICAL MECHANICS OF ELASTICITY, J. H. Weiner. Advanced, self-contained treatment illustrates general principles and elastic behavior of solids. Part 1, based on classical mechanics, studies thermoelastic behavior of crystalline and polymeric solids. Part 2, based on quantum mechanics, focuses on interatomic force laws, behavior of solids, and thermally activated processes. For students of physics and chemistry and for polymer physicists. 1983 ed. 96 figures. 496pp. 5⅜ x 8½. 42260-7

# Mathematics

FUNCTIONAL ANALYSIS (Second Corrected Edition), George Bachman and Lawrence Narici. Excellent treatment of subject geared toward students with background in linear algebra, advanced calculus, physics, and engineering. Text covers introduction to inner-product spaces, normed, metric spaces, and topological spaces; complete orthonormal sets, the Hahn-Banach Theorem and its consequences, and many other related subjects. 1966 ed. 544pp. 6⅛ x 9¼. 40251-7

ASYMPTOTIC EXPANSIONS OF INTEGRALS, Norman Bleistein & Richard A. Handelsman. Best introduction to important field with applications in a variety of scientific disciplines. New preface. Problems. Diagrams. Tables. Bibliography. Index. 448pp. 5⅜ x 8½. 65082-0

VECTOR AND TENSOR ANALYSIS WITH APPLICATIONS, A. I. Borisenko and I. E. Tarapov. Concise introduction. Worked-out problems, solutions, exercises. 257pp. 5⅜ x 8¼. 63833-2

THE ABSOLUTE DIFFERENTIAL CALCULUS (CALCULUS OF TENSORS), Tullio Levi-Civita. Great 20th-century mathematician's classic work on material necessary for mathematical grasp of theory of relativity. 452pp. 5⅜ x 8¼. 63401-9

AN INTRODUCTION TO ORDINARY DIFFERENTIAL EQUATIONS, Earl A. Coddington. A thorough and systematic first course in elementary differential equations for undergraduates in mathematics and science, with many exercises and problems (with answers). Index. 304pp. 5⅜ x 8½. 65942-9

FOURIER SERIES AND ORTHOGONAL FUNCTIONS, Harry F. Davis. An incisive text combining theory and practical example to introduce Fourier series, orthogonal functions and applications of the Fourier method to boundary-value problems. 570 exercises. Answers and notes. 416pp. 5⅜ x 8½. 65973-9

COMPUTABILITY AND UNSOLVABILITY, Martin Davis. Classic graduate-level introduction to theory of computability, usually referred to as theory of recurrent functions. New preface and appendix. 288pp. 5⅜ x 8½. 61471-9

ASYMPTOTIC METHODS IN ANALYSIS, N. G. de Bruijn. An inexpensive, comprehensive guide to asymptotic methods–the pioneering work that teaches by explaining worked examples in detail. Index. 224pp. 5⅜ x 8½ 64221-6

APPLIED COMPLEX VARIABLES, John W. Dettman. Step-by-step coverage of fundamentals of analytic function theory–plus lucid exposition of five important applications: Potential Theory; Ordinary Differential Equations; Fourier Transforms; Laplace Transforms; Asymptotic Expansions. 66 figures. Exercises at chapter ends. 512pp. 5⅜ x 8½. 64670-X

INTRODUCTION TO LINEAR ALGEBRA AND DIFFERENTIAL EQUATIONS, John W. Dettman. Excellent text covers complex numbers, determinants, orthonormal bases, Laplace transforms, much more. Exercises with solutions. Undergraduate level. 416pp. 5⅜ x 8½. 65191-6

INTRODUCTORY REAL ANALYSIS, A.N. Kolmogorov, S. V. Fomin. Translated by Richard A. Silverman. Self-contained, evenly paced introduction to real and functional analysis. Some 350 problems. 403pp. 5⅜ x 8½. 61226-0

APPLIED ANALYSIS, Cornelius Lanczos. Classic work on analysis and design of finite processes for approximating solution of analytical problems. Algebraic equations, matrices, harmonic analysis, quadrature methods, more. 559pp. 5⅜ x 8½. 65656-X

AN INTRODUCTION TO ALGEBRAIC STRUCTURES, Joseph Landin. Superb self-contained text covers "abstract algebra": sets and numbers, theory of groups, theory of rings, much more. Numerous well-chosen examples, exercises. 247pp. 5⅜ x 8½. 65940-2

QUALITATIVE THEORY OF DIFFERENTIAL EQUATIONS, V. V. Nemytskii and V.V. Stepanov. Classic graduate-level text by two prominent Soviet mathematicians covers classical differential equations as well as topological dynamics and ergodic theory. Bibliographies. 523pp. 5⅜ x 8½. 65954-2

THEORY OF MATRICES, Sam Perlis. Outstanding text covering rank, nonsingularity and inverses in connection with the development of canonical matrices under the relation of equivalence, and without the intervention of determinants. Includes exercises. 237pp. 5⅜ x 8½. 66810-X

INTRODUCTION TO ANALYSIS, Maxwell Rosenlicht. Unusually clear, accessible coverage of set theory, real number system, metric spaces, continuous functions, Riemann integration, multiple integrals, more. Wide range of problems. Undergraduate level. Bibliography. 254pp. 5⅜ x 8½. 65038-3

MODERN NONLINEAR EQUATIONS, Thomas L. Saaty. Emphasizes practical solution of problems; covers seven types of equations. ". . . a welcome contribution to the existing literature. . . . "–*Math Reviews*. 490pp. 5⅜ x 8½. 64232-1

MATRICES AND LINEAR ALGEBRA, Hans Schneider and George Phillip Barker. Basic textbook covers theory of matrices and its applications to systems of linear equations and related topics such as determinants, eigenvalues, and differential equations. Numerous exercises. 432pp. 5⅜ x 8½. 66014-1

MATHEMATICS APPLIED TO CONTINUUM MECHANICS, Lee A. Segel. Analyzes models of fluid flow and solid deformation. For upper-level math, science, and engineering students. 608pp. 5⅜ x 8½. 65369-2

ELEMENTS OF REAL ANALYSIS, David A. Sprecher. Classic text covers fundamental concepts, real number system, point sets, functions of a real variable, Fourier series, much more. Over 500 exercises. 352pp. 5⅜ x 8½. 65385-4

SET THEORY AND LOGIC, Robert R. Stoll. Lucid introduction to unified theory of mathematical concepts. Set theory and logic seen as tools for conceptual understanding of real number system. 496pp. 5⅜ x 8¼. 63829-4

# Math–Decision Theory, Statistics, Probability

ELEMENTARY DECISION THEORY, Herman Chernoff and Lincoln E. Moses. Clear introduction to statistics and statistical theory covers data processing, probability and random variables, testing hypotheses, much more. Exercises. 364pp. 5⅜ x 8½. 65218-1

STATISTICS MANUAL, Edwin L. Crow et al. Comprehensive, practical collection of classical and modern methods prepared by U.S. Naval Ordnance Test Station. Stress on use. Basics of statistics assumed. 288pp. 5⅜ x 8½. 60599-X

SOME THEORY OF SAMPLING, William Edwards Deming. Analysis of the problems, theory, and design of sampling techniques for social scientists, industrial managers, and others who find statistics important at work. 61 tables. 90 figures. xvii +602pp. 5⅜ x 8½. 64684-X

LINEAR PROGRAMMING AND ECONOMIC ANALYSIS, Robert Dorfman, Paul A. Samuelson and Robert M. Solow. First comprehensive treatment of linear programming in standard economic analysis. Game theory, modern welfare economics, Leontief input-output, more. 525pp. 5⅜ x 8½. 65491-5

PROBABILITY: An Introduction, Samuel Goldberg. Excellent basic text covers set theory, probability theory for finite sample spaces, binomial theorem, much more. 360 problems. Bibliographies. 322pp. 5⅜ x 8½. 65252-1

GAMES AND DECISIONS: Introduction and Critical Survey, R. Duncan Luce and Howard Raiffa. Superb nontechnical introduction to game theory, primarily applied to social sciences. Utility theory, zero-sum games, n-person games, decision-making, much more. Bibliography. 509pp. 5⅜ x 8½. 65943-7

INTRODUCTION TO THE THEORY OF GAMES, J. C. C. McKinsey. This comprehensive overview of the mathematical theory of games illustrates applications to situations involving conflicts of interest, including economic, social, political, and military contexts. Appropriate for advanced undergraduate and graduate courses; advanced calculus a prerequisite. 1952 ed. x+372pp. 5⅜ x 8½. 42811-7

FIFTY CHALLENGING PROBLEMS IN PROBABILITY WITH SOLUTIONS, Frederick Mosteller. Remarkable puzzlers, graded in difficulty, illustrate elementary and advanced aspects of probability. Detailed solutions. 88pp. 5⅜ x 8½. 65355-2

PROBABILITY THEORY: A Concise Course, Y. A. Rozanov. Highly readable, self-contained introduction covers combination of events, dependent events, Bernoulli trials, etc. 148pp. 5⅜ x 8¼. 63544-9

STATISTICAL METHOD FROM THE VIEWPOINT OF QUALITY CONTROL, Walter A. Shewhart. Important text explains regulation of variables, uses of statistical control to achieve quality control in industry, agriculture, other areas. 192pp. 5⅜ x 8½. 65232-7

# Math–Geometry and Topology

ELEMENTARY CONCEPTS OF TOPOLOGY, Paul Alexandroff. Elegant, intuitive approach to topology from set-theoretic topology to Betti groups; how concepts of topology are useful in math and physics. 25 figures. 57pp. 5⅜ x 8½. 60747-X

COMBINATORIAL TOPOLOGY, P. S. Alexandrov. Clearly written, well-organized, three-part text begins by dealing with certain classic problems without using the formal techniques of homology theory and advances to the central concept, the Betti groups. Numerous detailed examples. 654pp. 5⅜ x 8½. 40179-0

EXPERIMENTS IN TOPOLOGY, Stephen Barr. Classic, lively explanation of one of the byways of mathematics. Klein bottles, Moebius strips, projective planes, map coloring, problem of the Koenigsberg bridges, much more, described with clarity and wit. 43 figures. 210pp. 5⅜ x 8½. 25933-1

CONFORMAL MAPPING ON RIEMANN SURFACES, Harvey Cohn. Lucid, insightful book presents ideal coverage of subject. 334 exercises make book perfect for self-study. 55 figures. 352pp. 5⅝ x 8¼. 64025-6

THE GEOMETRY OF RENÉ DESCARTES, René Descartes. The great work founded analytical geometry. Original French text, Descartes's own diagrams, together with definitive Smith-Latham translation. 244pp. 5⅜ x 8½. 60068-8

PRACTICAL CONIC SECTIONS: The Geometric Properties of Ellipses, Parabolas and Hyperbolas, J. W. Downs. This text shows how to create ellipses, parabolas, and hyperbolas. It also presents historical background on their ancient origins and describes the reflective properties and roles of curves in design applications. 1993 ed. 98 figures. xii+100pp. 6½ x 9¼. 42876-1

THE THIRTEEN BOOKS OF EUCLID'S ELEMENTS, translated with introduction and commentary by Thomas L. Heath. Definitive edition. Textual and linguistic notes, mathematical analysis. 2,500 years of critical commentary. Unabridged. 1,414pp. 5⅜ x 8½. Three-vol. set. Vol. I: 60088-2 Vol. II: 60089-0 Vol. III: 60090-4

GEOMETRY OF COMPLEX NUMBERS, Hans Schwerdtfeger. Illuminating, widely praised book on analytic geometry of circles, the Moebius transformation, and two-dimensional non-Euclidean geometries. 200pp. 5⅝ x 8¼. 63830-8

DIFFERENTIAL GEOMETRY, Heinrich W. Guggenheimer. Local differential geometry as an application of advanced calculus and linear algebra. Curvature, transformation groups, surfaces, more. Exercises. 62 figures. 378pp. 5⅜ x 8½. 63433-7

CURVATURE AND HOMOLOGY: Enlarged Edition, Samuel I. Goldberg. Revised edition examines topology of differentiable manifolds; curvature, homology of Riemannian manifolds; compact Lie groups; complex manifolds; curvature, homology of Kaehler manifolds. New Preface. Four new appendixes. 416pp. 5⅜ x 8½. 40207-X

# History of Math

THE WORKS OF ARCHIMEDES, Archimedes (T. L. Heath, ed.). Topics include the famous problems of the ratio of the areas of a cylinder and an inscribed sphere; the measurement of a circle; the properties of conoids, spheroids, and spirals; and the quadrature of the parabola. Informative introduction. clxxxvi+326pp; supplement, 52pp. 5⅜ x 8½. 42084-1

A SHORT ACCOUNT OF THE HISTORY OF MATHEMATICS, W. W. Rouse Ball. One of clearest, most authoritative surveys from the Egyptians and Phoenicians through 19th-century figures such as Grassman, Galois, Riemann. Fourth edition. 522pp. 5⅜ x 8½. 20630-0

THE HISTORY OF THE CALCULUS AND ITS CONCEPTUAL DEVELOPMENT, Carl B. Boyer. Origins in antiquity, medieval contributions, work of Newton, Leibniz, rigorous formulation. Treatment is verbal. 346pp. 5⅜ x 8½. 60509-4

THE HISTORICAL ROOTS OF ELEMENTARY MATHEMATICS, Lucas N. H. Bunt, Phillip S. Jones, and Jack D. Bedient. Fundamental underpinnings of modern arithmetic, algebra, geometry, and number systems derived from ancient civilizations. 320pp. 5⅜ x 8½. 25563-8

A HISTORY OF MATHEMATICAL NOTATIONS, Florian Cajori. This classic study notes the first appearance of a mathematical symbol and its origin, the competition it encountered, its spread among writers in different countries, its rise to popularity, its eventual decline or ultimate survival. Original 1929 two-volume edition presented here in one volume. xxviii+820pp. 5⅜ x 8½. 67766-4

GAMES, GODS & GAMBLING: A History of Probability and Statistical Ideas, F. N. David. Episodes from the lives of Galileo, Fermat, Pascal, and others illustrate this fascinating account of the roots of mathematics. Features thought-provoking references to classics, archaeology, biography, poetry. 1962 edition. 304pp. 5⅜ x 8½. (Available in U.S. only.) 40023-9

OF MEN AND NUMBERS: The Story of the Great Mathematicians, Jane Muir. Fascinating accounts of the lives and accomplishments of history's greatest mathematical minds–Pythagoras, Descartes, Euler, Pascal, Cantor, many more. Anecdotal, illuminating. 30 diagrams. Bibliography. 256pp. 5⅜ x 8½. 28973-7

HISTORY OF MATHEMATICS, David E. Smith. Nontechnical survey from ancient Greece and Orient to late 19th century; evolution of arithmetic, geometry, trigonometry, calculating devices, algebra, the calculus. 362 illustrations. 1,355pp. 5⅜ x 8½. Two-vol. set. Vol. I: 20429-4 Vol. II: 20430-8

A CONCISE HISTORY OF MATHEMATICS, Dirk J. Struik. The best brief history of mathematics. Stresses origins and covers every major figure from ancient Near East to 19th century. 41 illustrations. 195pp. 5⅜ x 8½. 60255-9

# Physics

OPTICAL RESONANCE AND TWO-LEVEL ATOMS, L. Allen and J. H. Eberly. Clear, comprehensive introduction to basic principles behind all quantum optical resonance phenomena. 53 illustrations. Preface. Index. 256pp. 5⅜ x 8½. 65533-4

QUANTUM THEORY, David Bohm. This advanced undergraduate-level text presents the quantum theory in terms of qualitative and imaginative concepts, followed by specific applications worked out in mathematical detail. Preface. Index. 655pp. 5⅜ x 8½. 65969-0

ATOMIC PHYSICS: 8th edition, Max Born. Nobel laureate's lucid treatment of kinetic theory of gases, elementary particles, nuclear atom, wave-corpuscles, atomic structure and spectral lines, much more. Over 40 appendices, bibliography. 495pp. 5⅜ x 8½. 65984-4

A SOPHISTICATE'S PRIMER OF RELATIVITY, P. W. Bridgman. Geared toward readers already acquainted with special relativity, this book transcends the view of theory as a working tool to answer natural questions: What is a frame of reference? What is a "law of nature"? What is the role of the "observer"? Extensive treatment, written in terms accessible to those without a scientific background. 1983 ed. xlviii+172pp. 5⅜ x 8½. 42549-5

AN INTRODUCTION TO HAMILTONIAN OPTICS, H. A. Buchdahl. Detailed account of the Hamiltonian treatment of aberration theory in geometrical optics. Many classes of optical systems defined in terms of the symmetries they possess. Problems with detailed solutions. 1970 edition. xv+360pp. 5⅜ x 8½. 67597-1

PRIMER OF QUANTUM MECHANICS, Marvin Chester. Introductory text examines the classical quantum bead on a track: its state and representations; operator eigenvalues; harmonic oscillator and bound bead in a symmetric force field; and bead in a spherical shell. Other topics include spin, matrices, and the structure of quantum mechanics; the simplest atom; indistinguishable particles; and stationary-state perturbation theory. 1992 ed. xiv+314pp. 6⅛ x 9¼. 42878-8

LECTURES ON QUANTUM MECHANICS, Paul A. M. Dirac. Four concise, brilliant lectures on mathematical methods in quantum mechanics from Nobel Prize–winning quantum pioneer build on idea of visualizing quantum theory through the use of classical mechanics. 96pp. 5⅜ x 8½. 41713-1

THIRTY YEARS THAT SHOOK PHYSICS: The Story of Quantum Theory, George Gamow. Lucid, accessible introduction to influential theory of energy and matter. Careful explanations of Dirac's anti-particles, Bohr's model of the atom, much more. 12 plates. Numerous drawings. 240pp. 5⅜ x 8½. 24895-X

ELECTRONIC STRUCTURE AND THE PROPERTIES OF SOLIDS: The Physics of the Chemical Bond, Walter A. Harrison. Innovative text offers basic understanding of the electronic structure of covalent and ionic solids, simple metals, transition metals and their compounds. Problems. 1980 edition. 582pp. 6⅛ x 9¼. 66021-4

HYDRODYNAMIC AND HYDROMAGNETIC STABILITY, S. Chandrasekhar. Lucid examination of the Rayleigh-Benard problem; clear coverage of the theory of instabilities causing convection. 704pp. 5⅝ x 8¼. 64071-X

INVESTIGATIONS ON THE THEORY OF THE BROWNIAN MOVEMENT, Albert Einstein. Five papers (1905–8) investigating dynamics of Brownian motion and evolving elementary theory. Notes by R. Fürth. 122pp. 5⅜ x 8½. 60304-0

THE PHYSICS OF WAVES, William C. Elmore and Mark A. Heald. Unique overview of classical wave theory. Acoustics, optics, electromagnetic radiation, more. Ideal as classroom text or for self-study. Problems. 477pp. 5⅜ x 8½. 64926-1

PHYSICAL PRINCIPLES OF THE QUANTUM THEORY, Werner Heisenberg. Nobel Laureate discusses quantum theory, uncertainty, wave mechanics, work of Dirac, Schroedinger, Compton, Wilson, Einstein, etc. 184pp. 5⅜ x 8½. 60113-7

ATOMIC SPECTRA AND ATOMIC STRUCTURE, Gerhard Herzberg. One of best introductions; especially for specialist in other fields. Treatment is physical rather than mathematical. 80 illustrations. 257pp. 5⅜ x 8½. 60115-3

AN INTRODUCTION TO STATISTICAL THERMODYNAMICS, Terrell L. Hill. Excellent basic text offers wide-ranging coverage of quantum statistical mechanics, systems of interacting molecules, quantum statistics, more. 523pp. 5⅜ x 8½. 65242-4

THEORETICAL PHYSICS, Georg Joos, with Ira M. Freeman. Classic overview covers essential math, mechanics, electromagnetic theory, thermodynamics, quantum mechanics, nuclear physics, other topics. xxiii+885pp. 5⅜ x 8½. 65227-0

PROBLEMS AND SOLUTIONS IN QUANTUM CHEMISTRY AND PHYSICS, Charles S. Johnson, Jr. and Lee G. Pedersen. Unusually varied problems, detailed solutions in coverage of quantum mechanics, wave mechanics, angular momentum, molecular spectroscopy, more. 280 problems, 139 supplementary exercises. 430pp. 6½ x 9¼. 65236-X

THEORETICAL SOLID STATE PHYSICS, Vol. I: Perfect Lattices in Equilibrium; Vol. II: Non-Equilibrium and Disorder, William Jones and Norman H. March. Monumental reference work covers fundamental theory of equilibrium properties of perfect crystalline solids, non-equilibrium properties, defects and disordered systems. Total of 1,301pp. 5⅜ x 8½. Vol. I: 65015-4 Vol. II: 65016-2

WHAT IS RELATIVITY? L. D. Landau and G. B. Rumer. Written by a Nobel Prize physicist and his distinguished colleague, this compelling book explains the special theory of relativity to readers with no scientific background, using such familiar objects as trains, rulers, and clocks. 1960 ed. vi+72pp. 23 b/w illustrations. 5⅜ x 8½. 42806-0 $6.95

A TREATISE ON ELECTRICITY AND MAGNETISM, James Clerk Maxwell. Important foundation work of modern physics. Brings to final form Maxwell's theory of electromagnetism and rigorously derives his general equations of field theory. 1,084pp. 5⅜ x 8½. Two-vol. set. Vol. I: 60636-8 Vol. II: 60637-6

QUANTUM MECHANICS: Principles and Formalism, Roy McWeeny. Graduate student-oriented volume develops subject as fundamental discipline, opening with review of origins of Schrödinger's equations and vector spaces. Focusing on main principles of quantum mechanics and their immediate consequences, it concludes with final generalizations covering alternative "languages" or representations. 1972 ed. 15 figures. xi+155pp. 5⅜ x 8½. 42829-X

INTRODUCTION TO QUANTUM MECHANICS WITH APPLICATIONS TO CHEMISTRY, Linus Pauling & E. Bright Wilson, Jr. Classic undergraduate text by Nobel Prize winner applies quantum mechanics to chemical and physical problems. Numerous tables and figures enhance the text. Chapter bibliographies. Appendices. Index. 468pp. 5⅜ x 8½. 64871-0

METHODS OF THERMODYNAMICS, Howard Reiss. Outstanding text focuses on physical technique of thermodynamics, typical problem areas of understanding, and significance and use of thermodynamic potential. 1965 edition. 238pp. 5⅜ x 8½. 69445-3

TENSOR ANALYSIS FOR PHYSICISTS, J. A. Schouten. Concise exposition of the mathematical basis of tensor analysis, integrated with well-chosen physical examples of the theory. Exercises. Index. Bibliography. 289pp. 5⅜ x 8½. 65582-2

THE ELECTROMAGNETIC FIELD, Albert Shadowitz. Comprehensive undergraduate text covers basics of electric and magnetic fields, builds up to electromagnetic theory. Also related topics, including relativity. Over 900 problems. 768pp. 5⅝ x 8¼. 65660-8

GREAT EXPERIMENTS IN PHYSICS: Firsthand Accounts from Galileo to Einstein, Morris H. Shamos (ed.). 25 crucial discoveries: Newton's laws of motion, Chadwick's study of the neutron, Hertz on electromagnetic waves, more. Original accounts clearly annotated. 370pp. 5⅜ x 8½. 25346-5

RELATIVITY, THERMODYNAMICS AND COSMOLOGY, Richard C. Tolman. Landmark study extends thermodynamics to special, general relativity; also applications of relativistic mechanics, thermodynamics to cosmological models. 501pp. 5⅜ x 8½. 65383-8

STATISTICAL PHYSICS, Gregory H. Wannier. Classic text combines thermodynamics, statistical mechanics, and kinetic theory in one unified presentation of thermal physics. Problems with solutions. Bibliography. 532pp. 5⅜ x 8½. 65401-X